FOR THE
IB DIPLOMA

Mathematics
APPLICATIONS AND INTERPRETATION SL

Paul Fannon
Vesna Kadelburg
Ben Woolley
Stephen Ward

**HODDER
EDUCATION**
AN HACHETTE UK COMPANY

The Publishers would like to thank the following for permission to reproduce copyright material.

Photo credits

p.xiii © https://en.wikipedia.org/wiki/Principia_Mathematica#/media/File:Principia_Mathematica_54-43.png/https://creativecommons.org/licenses/by-sa/3.0/; **p.xxiv** © R MACKAY/stock.adobe.com; **pp.2–3** *left to right* © 3dmentat - Fotolia ; © GOLFCPHOTO/stock.adobe.com; © Janis Smits/stock.adobe.com; **pp.22–23** © Reineg/stock.adobe.com; **p.26** © The Granger Collection/Alamy Stock Photo; **p.30** *left to right* © glifeisgood - stock.adobe.com; The Metropolitan Museum of Art, New York/H. O. Havemeyer Collection, Bequest of Mrs. H. O. Havemeyer, 1929/https://creativecommons.org/publicdomain/zero/1.0/; © Iarygin Andrii - stock.adobe.com; © Nino Marcutti/Alamy Stock Photo; **pp.48–9** *left to right* © Julija Sapic - Fotolia ; © Getty Images/iStockphoto/Thinkstock; © Marilyn barbone/stock.adobe.com; © SolisImages/stock.adobe.com; **pp.74–5** *left to right* © Evenfh/stock.adobe.com; © Andrey Bandurenko/stock.adobe.com; © Arochau/stock.adobe.com; © Tony Baggett/stock.adobe.com; © Peter Hermes Furian/stock.adobe.com; **p.93** *left to right* © Shawn Hempel/stock.adobe.com; © Innovated Captures/stock.adobe.com; **p.133** © Bettmann/Getty Images; **pp.178–9** *left to right* © Lrafael/stock.adobe.com; © Comugnero Silvana/stock.adobe.com; © Talaj/stock.adobe.com © tawhy/123RF ; **pp.200–201** *left to right* © Daniel Deme/WENN.com/Alamy Stock Photo; © Maxisport/stock.adobe.com; © The Asahi Shimbun/Getty Images; **pp.222–3** *left to right* © Shutterstock/Pete Niesen; © Luckybusiness - stock.adobe.com; © Shutterstock/Muellek Josef; **p.225** © Caifas/stock.adobe.com; **p.230** © Dmitry Nikolaev/stock.adobe.com; **pp.256–7** *left to right* © Alena Ozerova/stock.adobe.com © Alena Ozerova/123RF; © Vesnafoto/stock.adobe.com; © Drobot Dean/stock.adobe.com; **pp.278–9** *left to right* © Kesinee/stock.adobe.com; © Yanlev/stock.adobe.com; © Dmitrimaruta/stock.adobe.com; © ArenaCreative/stock.adobe.com; **p.284** © GL Archive/Alamy Stock Photo; **pp.296–7** *left to right* © Hanohiki/stock.adobe.com; © Chlorophylle/stock.adobe.com; © Nowyn/stock.adobe.com; **pp.308–9** *left to right* © Aleksey Stemmer/stock.adobe.com; © Brocreative/stock.adobe.com; © Tan Kian Khoon/stock.adobe.com; **pp.342–3** © Vlorzor/stock.adobe.com; © Jane/stock.adobe.com; © Benjaminnolte/stock.adobe.com; © Hendrik Schwartz/stock.adobe.com; **pp.404–5** © Ipopba/stock.adobe.com; © Alekss/stock.adobe.com; © Inti St. Clair/stock.adobe.com; © Vegefox.com/stock.adobe.com.

Acknowledgements

The Publishers wish to thank Pedro Monsalve Correa for his valuable comments and positive review of the manuscript.

All IB Past Paper questions are used with permission of the IB © International Baccalaureate Organization

Every effort has been made to trace all copyright holders, but if any have been inadvertently overlooked, the Publishers will be pleased to make the necessary arrangements at the first opportunity.

Although every effort has been made to ensure that website addresses are correct at time of going to press, Hodder Education cannot be held responsible for the content of any website mentioned in this book. It is sometimes possible to find a relocated web page by typing in the address of the home page for a website in the URL window of your browser.

Hachette UK's policy is to use papers that are natural, renewable and recyclable products and made from wood grown in well-managed forests and other controlled sources. The logging and manufacturing processes are expected to conform to the environmental regulations of the country of origin.

Orders: please contact Bookpoint Ltd, 130 Park Drive, Milton Park, Abingdon, Oxon OX14 4SE. Telephone: +44 (0)1235 827827. Fax: +44 (0)1235 400401. Email education@bookpoint.co.uk Lines are open from 9 a.m. to 5 p.m., Monday to Saturday, with a 24-hour message answering service. You can also order through our website: www.hoddereducation.com

ISBN: 9781510462380

© Paul Fannon, Vesna Kadelburg, Ben Woolley, Stephen Ward 2019

First published in 2019 by

Hodder Education,
An Hachette UK Company
Carmelite House
50 Victoria Embankment
London EC4Y 0DZ

www.hoddereducation.com

Impression number   10 9 8 7 6 5 4 3 2 1

Year   2023 2022 2021 2020 2019

All rights reserved. Apart from any use permitted under UK copyright law, no part of this publication may be reproduced or transmitted in any form or by any means, electronic or mechanical, including photocopying and recording, or held within any information storage and retrieval system, without permission in writing from the publisher or under licence from the Copyright Licensing Agency Limited. Further details of such licences (for reprographic reproduction) may be obtained from the Copyright Licensing Agency Limited, www.cla.co.uk

Cover photo: Under the Wave off Kanagawa (Kanagawa oki nami ura), also known as The Great Wave by Katsushika Hokusai, The Metropolitan Museum of Art, New York/H. O. Havemeyer Collection, Bequest of Mrs. H. O. Havemeyer, 1929/https://creativecommons.org/publicdomain/zero/1.0/

Illustrations by Integra Software Services and also the authors

Typeset in Integra Software Services Pvt. Ltd., Pondicherry, India

Printed in Italy

A catalogue record for this title is available from the British Library.

Contents

Introduction vi

The 'toolkit' and the mathematical exploration ix

Core SL content

Chapter 1   Core: Exponents and logarithms 2
- 1A Laws of exponents 4
- 1B Operations with numbers in the form $a \times 10k$, where $1 \leqslant a < 10$ and k is an integer 11
- 1C Logarithms 14

Chapter 2   Core: Sequences 22
- 2A Arithmetic sequences and series 24
- 2B Geometric sequences and series 33
- 2C Financial applications of geometric sequences and series ... 39

Chapter 3   Core: Functions 48
- 3A Concept of a function 50
- 3B Sketching graphs 61

Chapter 4   Core: Coordinate geometry 74
- 4A Equations of straight lines in two dimensions 76
- 4B Three-dimensional coordinate geometry 86

Chapter 5   Core: Geometry and trigonometry 92
- 5A Volumes and surface areas of three-dimensional solids 94
- 5B Rules of trigonometry 101
- 5C Applications of trigonometry 113

Chapter 6   Core: Statistics 130
- 6A Sampling 132
- 6B Summarizing data 139
- 6C Presenting data 150
- 6D Correlation and regression 159

Chapter 7   Core: Probability 178
- 7A Introduction to probability 180
- 7B Probability techniques 184

Chapter 8 Core: Probability distributions ... 200
- 8A Discrete random variables ... 202
- 8B Binomial distribution ... 207
- 8C The normal distribution ... 212

Chapter 9 Core: Differentiation ... 222
- 9A Limits and derivatives ... 224
- 9B Graphical interpretation of derivatives ... 230
- 9C Finding an expression for the derivative ... 240
- 9D Tangents and normals at a given point and their equations ... 245

Chapter 10 Core: Integration ... 256
- 10A Anti-differentiation ... 258
- 10B Definite integration and the area under a curve ... 262

Core SL content: Review Exercise ... 272

Additional Applications and interpretation SL content

Chapter 11 Applications and interpretation: Number and finance ... 278
- 11A Approximation ... 280
- 11B Further financial mathematics ... 287

Chapter 12 Applications and interpretation: Solving equations with technology ... 296
- 12A Systems of linear equations ... 298
- 12B Polynomial equations ... 301

Chapter 13 Applications and interpretation: Mathematical models ... 308
- 13A Linear models ... 310
- 13B Quadratic models ... 314
- 13C Exponential models ... 318
- 13D Direct and inverse variation and cubic models ... 323
- 13E Sinusoidal models ... 328
- 13F Modelling skills ... 332

Chapter 14 Applications and interpretation: Geometry ... 342
- 14A Arcs and sectors ... 344
- 14B Voronoi diagrams ... 349

Contents

Chapter 15 Applications and interpretation: Hypothesis testing 370
- 15A Chi-squared tests 372
- 15B *t*-tests 385
- 15C Spearman's rank correlation 391

Chapter 16 Applications and interpretation: Calculus 404
- 16A Maximum and minimum points 406
- 16B Optimization 409
- 16C Trapezoidal rule 412

Applications and interpretation SL: Practice Paper 1 420

Applications and interpretation SL: Practice Paper 2 424

Answers 427

Glossary 492

Index 494

Introduction

Welcome to your coursebook for Mathematics for the IB Diploma: Applications and interpretation SL. The structure and content of this coursebook follow the structure and content of the 2019 IB Mathematics: Applications and interpretation guide, with headings that correspond directly with the content areas listed in the guide.

This is also the first book required by students taking the higher level course. Students should be familiar with the content of this book before moving on to Mathematics for the IB Diploma: Applications and interpretation HL.

Using this book

The book begins with an introductory chapter on the 'toolkit', a set of mathematical thinking skills that will help you to apply the content in the rest of the book to any type of mathematical problem. This chapter also contains advice on how to complete your mathematical exploration.

The remainder of the book is divided into two sections. Chapters 1 to 10 cover the core content that is common to both Mathematics: Analysis and approaches and Mathematics: Applications and interpretation. Chapters 11 to 16 cover the remaining SL content required for Mathematics: Applications and interpretation.

Special features of the chapters include:

ESSENTIAL UNDERSTANDINGS

Each chapter begins with a summary of the key ideas to be explored and a list of the knowledge and skills you will learn. These are revisited in a checklist at the end of each chapter.

CONCEPTS

The IB guide identifies 12 concepts central to the study of mathematics that will help you make connections between topics, as well as with the other subjects you are studying. These are highlighted and illustrated with examples at relevant points throughout the book.

KEY POINTS

Important mathematical rules and formulae are presented as Key Points, making them easy to locate and refer back to when necessary.

WORKED EXAMPLES

There are many Worked Examples in each chapter, demonstrating how the Key Points and mathematical content described can be put into practice. Each Worked Example comprises two columns:

On the left, how to **think** about the problem and what tools or methods will be needed at each step.

On the right, what to **write**, prompted by the left column, to produce a formal solution to the question.

Using this book

Exercises

Each section of each chapter concludes with a comprehensive exercise so that students can test their knowledge of the content described and practise the skills demonstrated in the Worked Examples. Each exercise contains the following types of questions:

- **Drill questions:** These are clearly linked to particular Worked Examples and gradually increase in difficulty. Each of them has two parts – **a** and **b** – desgined such that if students get **a** wrong, **b** is an opportunity to have another go at a very similar question. If students get **a** right, there is no need to do **b** as well.
- **Problem-solving questions:** These questions require students to apply the skills they have mastered in the drill questions to more complex, exam-style questions. They are colour-coded for difficulty.

  1 Green questions are closely related to standard techniques and require a small number of processes. They should be approachable for all candidates.

  2 Blue questions require students to make a small number of tactical decisions about how to apply the standard methods and they will often require multiple procedures. They should be achievable for SL students aiming for higher grades.

  3 Red questions often require a creative problem-solving approach and extended technical procedures. They will stretch even advanced SL students and be challenging for HL students aiming for the top grades.

  4 Black questions go beyond what is expected in IB examinations, but provide an enrichment opportunity for the most advanced students.

The questions in the Mixed Practice section at the end of each chapter are similarly colour-coded, and contain questions taken directly from past IB Diploma Mathematics exam papers. There is also a review exercise halfway through the book covering all of the core content, and two practice examination papers at the end of the book.

Answers to all exercises can be found at the back of the book.

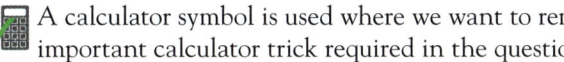 A calculator symbol is used where we want to remind you that there is a particularly important calculator trick required in the question.

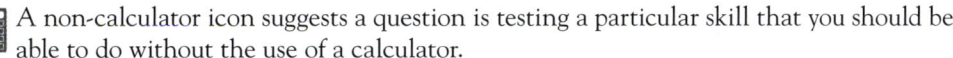 A non-calculator icon suggests a question is testing a particular skill that you should be able to do without the use of a calculator.

 The guide places great emphasis on the importance of technology in mathematics and expects you to have a high level of fluency with the use of your calculator and other relevant forms of hardware and software. Therefore, we have included plenty of screenshots and questions aimed at raising awareness and developing confidence in these skills, within the contexts in which they are likely to occur. This icon is used to indicate topics for which technology is particularly useful or necessary.

 **Making connections:** Mathematics is all about making links. You might be interested to see how something you have just learned will be used elsewhere in the course and in different topics, or you may need to go back and remind yourself of a previous topic.

Be the Examiner

These are activities that present you with three different worked solutions to a particular question or problem. Your task is to determine which one is correct and to work out where the other two went wrong.

Introduction

LEARNER PROFILE
Opportunities to think about how you are demonstrating the attributes of the IB Learner Profile are highlighted at the beginning of appropriate chapters.

 **TOOLKIT**
There are questions, investigations and activities interspersed throughout the chapters to help you develop mathematical thinking skills, building on the introductory toolkit chapter in relevant contexts. Although the ideas and skills presented will not be examined, these features are designed to give you a deeper insight into the topics that will be. Each toolkit box addresses one of the following three key topics: proof, modelling and problem solving.

Proof

Proofs are set out in a similar way to Worked Examples, helping you to gain a deeper understanding of the mathematical rules and statements you will be using and to develop the thought processes required to write your own proofs.

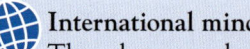 **International mindedness**
These boxes explore how the exchange of information and ideas across national boundaries has been essential to the progress of mathematics and to illustrate the international aspects of the subject.

You are the Researcher

This feature prompts you to carry out further research into subjects related to the syllabus content. You might like to use some of these ideas as starting points for your mathematical exploration or even an extended essay.

Tip
There are short hints and tips provided in the margins throughout the book.

TOK Links
Links to the interdisciplinary Theory of Knowledge element of the IB Diploma programme are made throughout the book.

Links to: Other subjects
Links to other IB Diploma subjects are made at relevant points, highlighting some of the real-life applications of the mathematical skills you will learn.

 Topics that have direct real-world applications are indicated by this icon.

There is a glossary at the back of the book. Glossary terms are **purple**.

About the authors

The authors are all University of Cambridge graduates and have a wide range of expertise in pure mathematics and in applications of mathematics, including economics, epidemiology, linguistics, philosophy and natural sciences.

Between them they have considerable experience of teaching IB Diploma Mathematics at Standard and Higher Level, and two of them currently teach at the University of Cambridge.

The toolkit and the mathematical exploration

Mathematics is about more than just arithmetic, geometry, algebra and statistics. It is a set of skills that are widely transferable. All of the IB Diploma Programme Mathematics courses allocate time to the development of these skills, collectively known as the 'toolkit', which will help you formulate an approach to any mathematical problem and form a deeper understanding of the real-life applications of mathematics.

In this chapter, we will look at four of these skills:

- problem solving
- proof
- modelling
- technology.

For each, we have provided some background information and activities to help you develop these skills. There will be many more activities interspersed at appropriate places throughout the book. We will then also look at how these skills can be demonstrated in your exploration.

For some students, part of the additional 'toolkit' time might be usefully spent practising their basic algebra skills, so we have also provided an exercise to assist with this.

This chapter has been designed as a useful resource of information and advice that will give you a good grounding in the required skills at the start of the course, but that you can also refer back to at appropriate times throughout your studies to improve your skills.

Problem Solving

Some people think that good mathematicians see the answers to problems straight away. This could not be further from the truth.

>
> **TOOLKIT: Problem Solving**
>
> Answer the following problem as fast as you can using your mental maths skills.
> A bottle and a cork cost $1.10. The bottle costs $1 more than the cork. How much is the cork?

The human brain is highly evolved to do things such as spot predators and food. There has not been a lot of evolutionary pressure to do mental maths, so most people get this type of problem wrong when they try to do it quickly (the most common answer is usually $0.10; the correct answer is actually $0.05). Good mathematicians do not just try to spot the correct answer, but work through a clear, analytical process to get to the right answer.

One famous mathematician, George Polya, tried to break down the process mathematicians go through when solving problems. He suggested four steps:

1. Understand the problem.
2. Make a plan.
3. Carry out the plan.
4. Look back on your work and reflect.

Most people put all their efforts into step 3, but often the others are vital in developing your problem-solving skills.

Polya also came up with a list of problem-solving strategies, called heuristics, which he used to guide him when he was stuck. For example, when dealing with a difficult problem; answer a simpler one.

TOOLKIT: Problem Solving

A strange mathematical prison has 100 cells each containing 1 prisoner. Initially they are all locked.
- On day 1 the guard turns the key in every cell door, therefore all the doors are unlocked.
- On day 2 the guard turns the key on the door of every cell that is a multiple of 2. This locks all the even numbered cells.
- On day 3 the guard turns the key on the door of every cell that is a multiple of 3.
- On day 4 the guard turns the key on the door of every cell that is a multiple of 4.

This continues in this way until the 100th day, on which the guard turns the key on every cell that is a multiple of 100 (i.e. just the 100th cell.) All prisoners whose cell doors are then opened are released. Which prisoners get released?

You might find that the cycle of mathematical inquiry below is a useful guide to the process you have to go through when tackling the problem above, along with many of the other problems in this section.

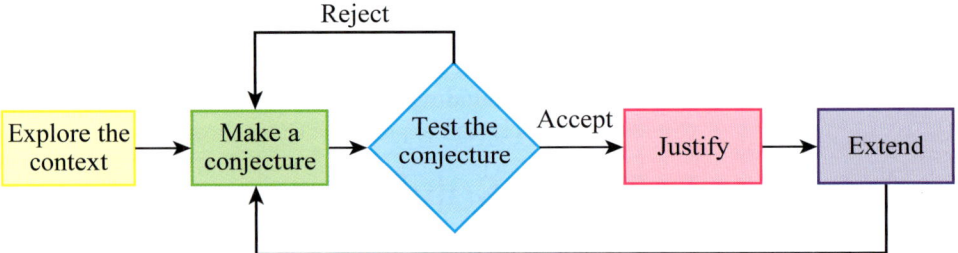

TOOLKIT: Problem Solving

What is the formula for the sum of the angles inside an n-sided polygon?

a Explore the context:

   What is a polygon? Is a circle a polygon? Does it have to be regular? Can it have obtuse angles? What about reflex angles? What is the smallest and largest possible value of n?

b Make a conjecture:

   Using your prior learning, or just by drawing and measuring, fill in the following table:

Shape	n	Sum of angles, S
Triangle		
Quadrilateral		
Pentagon		

   Based on the data, suggest a rule connecting n and S.

c Test your conjecture:

   Now, by drawing and measuring, or through research, see if your conjecture works for the next polygon. If your conjecture works, move on to **d**, otherwise, make a new conjecture and repeat **c**.

d Justify your conjecture:

If you connect one corner of your polygon to all the other corners, how many triangles does this form? Can you use this to justify your conjecture?

e Extend your conjecture:

Does the justification still work if the polygon has reflex angles? If the polygon is regular, what is each internal angle? What whole-number values can the internal angle of a regular polygon be? What is the external angle of a regular n-sided polygon? Which regular polygons can tesselate together?

One way of exploring the problem to help you form a conjecture is to just try putting some numbers in.

TOOLKIT: Problem Solving

Simplify $\cos^{-1}(\sin(x))$ for $0° < x < 90°$.

With a lot of problem solving, you will go through periods of not being sure whether you are on the right track and not knowing what to do next. Persistence itself can be a useful tool.

TOOLKIT: Problem Solving

Each term in the look-and-say sequence describes the digits of the previous term:

1, 11, 21, 1211, 111221, 312211, …

For example, the fourth term should be read as '1 two, 1 one', describing the digits in the third term. What is the 2000th term in this sequence? How would your answer change if the first element in the sequence was 2?

One of the key challenges when dealing with a difficult problem is knowing where to start. It is often useful to look for the most constrained part.

TOOLKIT: Problem Solving

The problem below is called a KenKen or Calcudoku. The numbers 1 to 5 are found exactly once in each row and column. The numbers in the cells connected by outlines can be used, along with the operation given, to form the stated result. For example, the box labelled 2÷ gives you the desired result and the operation used to get it. The two numbers must be able to be divided (in some order) to make 2, so it could be 1 and 2 or 2 and 4.

60×	2÷		1−	
		4+		2
4−	12×		2÷	4+
	10+			
2÷			4−	

Sometimes your immediate response to a hard problem is total panic, as you do not know where to begin. Always remember that the problems you will face will have a solution; it just might require some patience and careful thinking to get there.

TOOLKIT: Problem Solving

In the multiple-choice quiz below, each question refers to the quiz as a whole:

1   How many answers are A?
    A   0
    B   1
    C   2
    D   3
    E   4

2   The first question whose answer is A is question
    A   1
    B   2
    C   3
    D   4
    E   There are no 'A's

3   The previous answer is
    A   C
    B   D
    C   E
    D   A
    E   B

4   The only repeated answer is
    A   C
    B   B
    C   A
    D   E
    E   D

Proof

■ What makes a good proof?

Although in examinations the Applications and interpretation course does not have much focus on formal proof, for mathematicians, proof is their way of 'explaining'. Therefore, at appropriate places we will still present proofs which you might find provide a useful insight into how the theory is developed. Although you will not be expected to learn and reproduce these proofs, it is useful for your general problem solving skills to see some of the arguments and techniques used in mathematics.

Proof

In many ways, proof is the defining feature of mathematics. It is steps of logical reasoning beginning at a clear starting point (called an axiom). This requires precise, unambiguous communication so mathematical proofs shared between mathematicians often look scarily formal. For example, overleaf is a short section of Bertrand Russell's 1910 *Principia Mathematica* in which he used fundamental ideas about sets to show that 1 + 1 = 2.

*54·43. $\vdash :. \alpha, \beta \in 1 . \supset : \alpha \cap \beta = \Lambda . \equiv . \alpha \cup \beta \in 2$

Dem.

$\vdash . *54·26 . \supset \vdash :. \alpha = \iota`x . \beta = \iota`y . \supset : \alpha \cup \beta \in 2 . \equiv . x \neq y .$
[*51·231] $\equiv . \iota`x \cap \iota`y = \Lambda .$
[*13·12] $\equiv . \alpha \cap \beta = \Lambda$ (1)

$\vdash . (1) . *11·11·35 . \supset$
$\vdash :. (\exists x, y) . \alpha = \iota`x . \beta = \iota`y . \supset : \alpha \cup \beta \in 2 . \equiv . \alpha \cap \beta = \Lambda$ (2)

$\vdash . (2) . *11·54 . *52·1 . \supset \vdash .$ Prop

From this proposition it will follow, when arithmetical addition has been defined, that $1 + 1 = 2$.

However, you should not get too hung up on making proofs look too formal. A good proof is all about convincing a sceptical peer. A famous study by Celia Hoyles and Lulu Healy in 1999 showed that students often thought totally incorrect algebraic proofs were better than well-reasoned proofs that used words or diagrams. Good proof is about clearly communicating ideas, not masking them behind complicated notation.

TOOLKIT: Proof

Does the following diagram prove Pythagoras' Theorem?

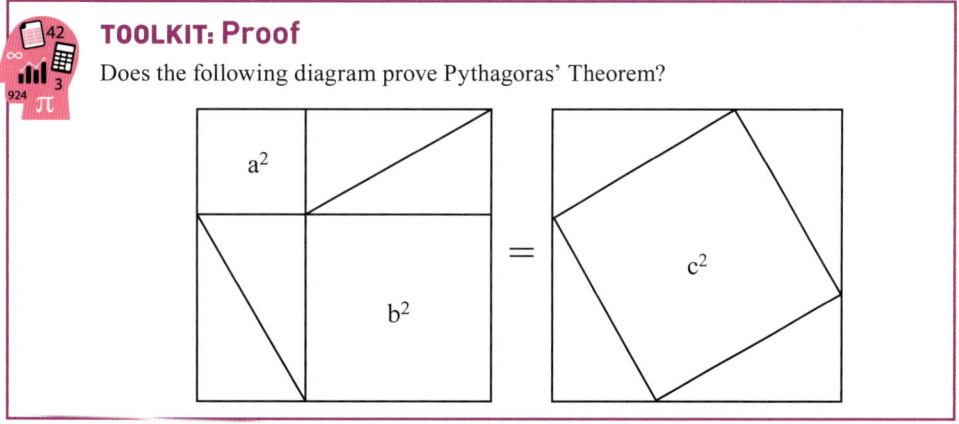

One of the problems you might encounter in developing proofs is not knowing what axioms you can start from. There is a certain amount of 'community knowledge' in knowing what acceptable starting points are. For example, if you were asked to prove that the area of a circle is πr^2, it would not be acceptable to say this is a well-known fact found in the formula book. However, if you were asked to prove that a circle is the shape with the largest area for a given perimeter then you could probably use $A = \pi r^2$ without proof.

The other issue is that too often in maths you are asked to prove obvious facts that you already know, for example proving that the sum of two odd numbers is even. However, the methods used to do that are helping with the development of precise reasoning which can be used to prove facts that are much less obvious.

TOOLKIT: Proof

Are there more positive even numbers or positive whole numbers?

Imagine creating a list of all the positive whole numbers alongside a list of all the positive even numbers:

$$1 \to 2$$
$$2 \to 4$$
$$3 \to 6$$
$$\vdots$$
$$n \to 2n$$

For every positive whole number, n, we can associate with it exactly one even number, $2n$. Therefore, there are the *same* number of whole numbers and even numbers.

Most people find this fact counter-intuitive. You might want to see if you can extend this method to ask:

a Are there more positive whole numbers or prime numbers?

b Are there more positive whole numbers or fractions between 0 and 1?

c Are there more positive whole number or decimals between 0 and 1?

You are the Researcher

Reasoning with infinity is famously problematic. There is a famous proof that the sum of all positive whole numbers is $-\frac{1}{12}$. Although massively counter-intuitive and usually considered invalid, this result has found applications in quantum theory. You might want to research this result, along with other astounding results of the great Indian mathematician Ramanujan.

One way of developing your own skills in proof is to critically appraise other attempted proofs. Things can look very plausible at first but have subtle errors.

TOOLKIT: Proof

Find the flaw in the following proof that $7 = 3$.

Start from the true statement:
$$-21 = -21$$
Rewrite this as:
$$49 - 70 = 9 - 30$$
This is equivalent to:
$$7^2 - 10 \times 7 = 3^2 - 10 \times 3$$
Add on 25 to both sides:
$$7^2 - 10 \times 7 + 25 = 3^2 - 10 \times 3 + 25$$
This can be expressed as:
$$7^2 - 2 \times 5 \times 7 + 5^2 = 3^2 - 2 \times 5 \times 3 + 5^2$$
Both sides are perfect squares and can be factorized:
$$(7-5)^2 = (3-5)^2$$
Square rooting both sides:
$$7 - 5 = 3 - 5$$
Therefore $7 = 3$.

You are the Researcher

There is a famous proof that all triangles are equilateral. See if you can find this proof and determine the flaw in it.

Proof

TOOLKIT: Proof

Simplify each of the following expressions:

$\mathbb{N} \cup \mathbb{Z}^+$

$\mathbb{N} \cap \mathbb{Z}^+$

$\mathbb{Q} \cap \overline{\mathbb{Q}}$

$\mathbb{Q} \cup \overline{\mathbb{Q}}$

■ Sets, logic and the language of proof

Unfortunately, the precision of the notation used in mathematics can sometimes be intimidating. For example, the different notation for number sets:

$\mathbb{Z}$: Integers: $\{\ldots -1, 0, 1, 2, 3 \ldots\}$

$\mathbb{Z}^+$: Positive integers: $\{1, 2, 3 \ldots\}$

$\mathbb{N}$: Natural numbers: $\{0, 1, 2 \ldots\}$

$\mathbb{Q}$: Rationals: $\{-\frac{1}{3}, 0, 0.25, 4 \ldots\}$

$\overline{\mathbb{Q}}$: Irrationals: $\{-\pi, \sqrt{2}, 1 + \sqrt{3} \ldots\}$

$\mathbb{R}$: Reals: $\{-3, 0, \frac{1}{5}, \sqrt{2}, 8 \ldots\}$

> In common English usage, zero is *neither* positive nor negative. However, in some countries, for example, some parts of France, zero is considered *both* positive and negative. How much of maths varies between countries?

Thinking about sets (possibly visualizing them using Venn diagrams) is a very powerful way of describing many mathematical situations.

One of the main purposes for working on sets and proof is to develop logic.

TOOLKIT: Proof

A barber in a town shaves all those people who do not shave themselves, and only those people. Does the barber shave himself?

This is an example of a paradox often attributed to the mathematician and philosopher Bertrand Russell. Can you see why it causes a paradox?

TOOLKIT: Proof

A set of cards has numbers on one side and letters on the other. James claims that if there is a vowel on one side there must be an even number on the other side. Which of the following cards must be turned over to test James's claim?

| A | B | 1 | 2 |

This question, called the Watson selection test, formed part of a psychological study into deductive reasoning and very few people gave the correct answer. However, when they were asked the following question, which is logically identical, nearly everybody gave the correct answer.

Four people are drinking in a bar. If someone is drinking beer, they must be aged over 21. Which of the following people need to be investigated further to check that everybody is complying with this rule?

| Beer drinker | Coke drinker | 16-year-old person | 28-year-old person |

One of the conclusions from this study was that people are generally weak at abstract reasoning. The important message for you is, where possible, take abstract ideas and try to put them into a context.

(The answer in both cases is the yellow and pink cards.)

The previous example demonstrates that many people do not have an intuitive approach to formal logic. To help develop this, it is useful to have some terminology to work with.

A logical statement, A, is something which is either true or false, e.g. A = 'The capital of Italy is Rome'.

The negation of a logical statement, $\overline{A}$, is something which is true when A is false and vice versa, e.g. $\overline{A}$ = 'The capital of Italy is not Rome'. Sometimes you have to be careful here – if the statement is 'the cup is full', the negation is 'the cup is not full', rather than 'the cup is empty'.

An implication is the basic unit of logic. It is of the form 'if A then B'. People are often not precise enough with this logic and confuse this implication with other related statements:

The converse: 'if B then A'

The inverse: 'if $\overline{A}$ then $\overline{B}$'

The contrapositive: 'if $\overline{B}$ then $\overline{A}$'.

TOOLKIT: Proof

For each of the following statements, decide if the converse, inverse or contrapositive are always true:
- **If** a shape is a rectangle with equal sides, **then** it is a square.
- **If** x is even, **then** x^2 is even.
- **If** two shapes are similar, **then** all corresponding angles are equal.
- **If** two numbers are both positive, **then** their product is positive.
- **If** a shape is a square, **then** it is a parallelogram.
- **If** a number is less than 5, **then** it is less than 10.

Logic gets even harder when it is taken out of abstract mathematical thought and put into real-life contexts. This is often implicitly presented in the form of a syllogism:

Premise 1: All IB students are clever.

Premise 2: You are an IB student.

Conclusion: Therefore, you are clever.

The above argument is logically consistent. This does not necessarily mean that the conclusion is true – you could dispute either of the premises, but if you agree with the premises then you will be forced to agree with the conclusion.

TOOLKIT: Proof

Decide if each of the following arguments is logically consistent.
- If Darren cheats he will get good grades. He has got good grades; therefore, he must have cheated.
- If you work hard you will get a good job. Therefore, if you do not work hard you will not get a good job.
- All HL Maths students are clever. No SL Maths students are HL Maths students; therefore, no SL maths students are clever.
- All foolish people bought bitcoins. Since Jamila bought bitcoins, she is foolish.
- All of Paul's jokes are either funny or clever but not both. All of Paul's maths jokes are clever, therefore none of them are funny.

TOK Links

Is it more important for an argument to be logically consistent or the conclusion to be true?

Modelling

■ Creating models

Mathematics is an idealized system where we use strict rules to manipulate expressions and solve equations, yet it is also extremely good at describing real-world situations. The real world is very complicated, so often we need to ignore unnecessary details and just extract the key aspects of the situation we are interested in. This is called creating a model. The list of things we are assuming (or ignoring) is called the modelling assumptions. Common modelling assumptions include:

- Treating objects as just being a single point, called a particle. This is reasonable if they are covering a space much bigger than their size, for example, modelling a bird migrating.
- Assuming that there is no air resistance. This is reasonable if the object is moving relatively slowly, for example, a person walking.
- Treating individuals in a population as all being identical. This is reasonable if the population is sufficiently large that differences between individuals average out.

TOOLKIT: Modelling

None of the following statements are always true, but they are all sometimes useful. When do you think these would be good modelling assumptions? When would they be a bad idea to use?
- The Earth is a sphere.
- The Earth is flat.
- Parallel lines never meet.
- The sum of the angles in a triangle add up to 180 degrees.
- Each possible birthday of an individual is equally likely.
- The average height of people is 1.7 m.
- The Earth takes 365 days to orbit the Sun.
- The Earth's orbit is a circle.
- The more individuals there are in a population, the faster the population will grow.

One of the most difficult things to do when modelling real-world situations is deciding which variables you want to consider.

TOOLKIT: Modelling

What variables might you consider when modelling:
- the population of rabbits on an island
- the height reached by a basketball thrown towards the net
- the average income of a teacher in the UK
- the optimal strategy when playing poker
- the volume of water passing over the Niagara Falls each year
- the profit a company will make in 3 years' time
- the result of the next national election
- the thickness of cables required on a suspension bridge.

Can you put the variables in order from most important to least important?

Once you have decided which variables are important, the next step is to suggest how the variables are linked. This can be done in two different ways:

The toolkit and the mathematical exploration

TOK Links

Is theory-led or data-led modelling more valid?

- **Theory-led:** This is where some already accepted theory – such as Newton's laws in physics, supply–demand theory in economics or population dynamics in geography – can be used to predict the form that the relationship should take.
- **Data-led:** If there is no relevant theory, then it might be better to look at some experimental data and use your knowledge of different functions to suggest an appropriate function to fit the data.

Validating models

Once you have a model you need to decide how useful it is. One way of doing this is to see how well it fits the data. Unfortunately, there is a balance to be struck between the complexity of the model and how well it fits – the more parameters there are in a model, the better the fit is likely to be. For example, a cubic model of the form $y = ax^3 + bx^2 + cx + d$ has four parameters (a, b, c and d) and will fit any data set at least as well as a linear function $y = ax + b$ with two parameters (a and b).

TOK Links

There is a very important idea in philosophy known as Occam's razor. This basically says that, all else being equal, the simpler explanation is more likely to be the correct one.

TOOLKIT: Modelling

Consider the following set of data, showing the speed of a ball (in metres per second) against the time since the ball was dropped.

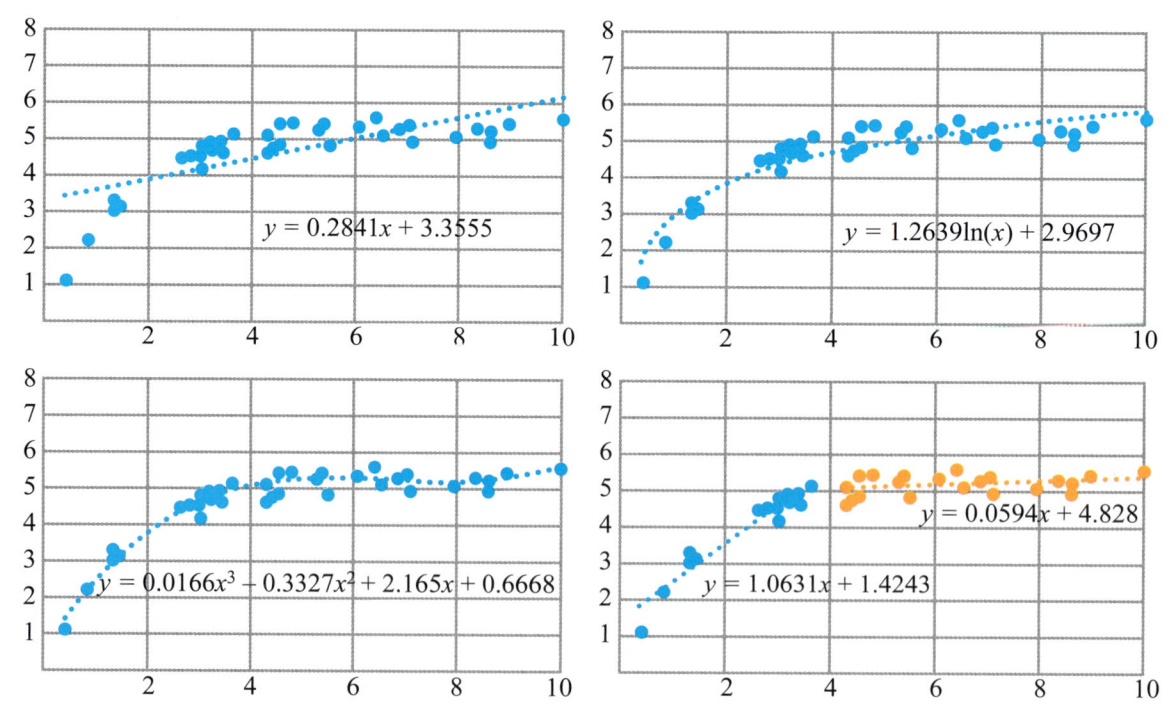

What criteria might you use to decide between the different models used?

Based on the data, Sam says that when $x = 6$, she thinks that $y = 5.1$.

Olesya says that when $x = 6$, she thinks that y is between 4.5 and 5.5.

What are the benefits or drawbacks of each of their ways of reporting their findings?

The modelling cycle

In real-life applications you are unlikely to get a suitable model at the first attempt. Instead, there is a cycle of creating, testing and improving models.

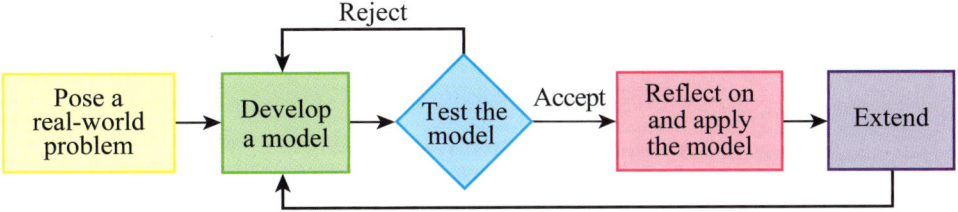

 **TOOLKIT: Modelling**

Carlos is creating a business plan for his bicycle repair business. Based on a survey, he believes that he will have 10 customers each week if he charges $30 for a repair and 20 customers each week if he charges $20.

a Pose a real-world problem:

   Can you predict how many customers Carlos will have each week (N) if he charges d for a repair?

b Develop a model:

   Use a formula of the form $N = ad + b$ to model the number of customers Carlos will have.

c Test the model:

   In his first week, Carlos charges $25 and gets 14 customers. Evaluate your model. Should it be improved? If so suggest a better model.

d Apply the model:

   Carlos believes he can cope with up to 30 customers each week. What is the least amount of money he should charge?

e Reflect on the model:

   Does it matter that your original model predicts a negative number of customers if Carlos charges $100 for a repair? How accurate does the model have to be? Is it more important for a new business to make a lot of money or to have a lot of customers? Is a single price for any repair a sensible business proposition?

f Extend:

   Use your model to predict the total income Carlos will get if he charges d. What price would you advise Carlos to charge?

 # Using technology

The main technology that you need to master is your Graphical Display Calculator (GDC), which you should use as much as possible early in the course to get accustomed to all its features. However, outside of examinations, there are many more technological tools which you might find helpful at various points, especially when writing your exploration. The following questions provide starting points to develop your skills with different types of software.

■ Computer algebra systems (CAS)

Most professional mathematicians make extensive use of CAS to avoid the more tedious elements of algebraic simplification. Various websites, such as Wolfram Alpha, and some calculators will do a lot of manipulation for you. This reflects the fact that, in modern mathematics, the skill is not algebraic manipulation, but more deciding which equation needs to be solved in the first place.

a Use CAS to simplify $\sin(\cos^{-1} x)$.

b Simplify $(\sin x)^4 + (\cos x)^4 + \frac{1}{2}(\sin(2x))^2$.

■ Spreadsheets

a Spreadsheets can be used to automate calculations.

A teacher uses the following boundaries to assign grades to internal tests.

Percentage, p	Grade
$p < 11$	Fail
$11 \leq p < 22$	1
$22 \leq p < 33$	2
$33 \leq p < 44$	3
$44 \leq p < 55$	4
$55 \leq p < 66$	5
$66 \leq p < 77$	6
$p \geq 77$	7

Design a spreadsheet that will allow a teacher to input raw marks on Paper 1 (out of 40) and Paper 2 (out of 50) and automatically calculate an overall percentage (reported to the nearest whole number) and grade.

b Spreadsheets are also useful for answering questions about modelling.

The change in speed of a ball dropped off a cliff with air resistance is modelled using the following rules.

The speed (in metres per second) at the end of each second is given by $0.5 \times$ previous speed $+ 10$. The distance travelled in each second is given by the average of the speed at the beginning of the second and the speed at the end. The initial speed is zero.

How far has the ball travelled after 20 seconds? What speed is achieved?

c Spreadsheets are very useful for investigating sequences.

i The first term of a sequence is 2. All subsequent terms are found by doing $1 - \dfrac{1}{\text{previous term}}$. Find the 100th term. What do you notice about this sequence?

ii The first term of a sequence is n. The subsequent terms are found using the following rules:

■ If the previous term is odd, the next term is 1 plus 3 times the previous term.

■ If the previous term is even, the next term is half the previous term.

Investigate the long-term behaviour of this sequence for various different values of n.

> **TOK Links**
>
> The sequence in **c ii** leads to a famous mathematical idea called the Collatz conjecture. It has been tested for all numbers up to about 87×2^{60} and no counterexamples have been found, but this is still not considered adequate for a proof in mathematics. What are the differences between 'mathematical proof' and 'scientific proof'?

The exploration

■ Dynamic geometry packages

Dynamic geometry packages are useful for exploring different geometric configurations by drawing a diagram using the given constraints, then seeing what stays the same while the unconstrained quantities are changed. They often allow you to make interesting observations that can then be proved; however, it is a skill in itself to turn a problem in words into a diagram in a dynamic geometry package.

a A triangle ABC has a right angle at C. The midpoint of AB is called M. How is the length of MC related to the length of AB? How could you prove your conjecture?

b Two touching circles share a common tangent which meets the first circle at A. The opposite end of the diameter of the first circle at A is called B. The tangent from B meets the other circle tangentially at C. Find the ratio $AB:BC$.

■ Programming

Programming is a great way to provide clear, formal instructions, that requires planning and accuracy. All of these are great mathematical skills. There are many appropriate languages to work in, such as Java, Python or Visual Basic. Alternatively, many spreadsheets have a programming capability.

a Write a program to find the first 1000 prime numbers.

b Write a program that will write any positive number as the sum of four square numbers (possibly including zero).

c Write a program that will decode the following message:

```
Jrypbzr gb Zngurzngvpf sbe gur Vagreangvbany
Onppnynherngr. Jr ubcr gung lbh rawbl guvf obbx
naq yrnea ybgf bs sha znguf!
```

> **Tip**
> You might want to do a frequency analysis on the message and compare it to the standard frequencies in the English language.

The exploration

The purpose of this maths course is much more than just to prepare for an examination. We are hoping you will develop an appreciation for the beauty of mathematics as well as seeing a wide range of applications. 20% of the final mark is awarded for an exploration – a piece of independent writing, approximately 12 to 20 pages long, that explores a mathematical topic. There are many different types of mathematical exploration, but some of the more common types include:

- creating a mathematical model for a situation
- using statistics to answer a question
- exploring the applications of a mathematical method
- solving a pure mathematics puzzle
- exploring the historical development of a mathematical idea.

Your exploration will be marked according to five criteria:

- presentation
- mathematical communication
- personal engagement
- reflection
- use of mathematics.

> **Tip**
> There are many successful explorations in each of these categories, but you should be aware that the first three of these types listed, involving models, statistics and applications of a method, are often felt by students to be easier to get higher marks, particularly in relation to the personal engagement and reflection assessment criteria.

Presentation

This is about having a well-structured project which is easy to follow. If someone else reads your presentation and can summarize what you have written in a few sentences, it is probably well-structured.

It needs to be:

- **Organized:** There should be an introduction, conclusion and other relevant sections which are clearly titled.
- **Coherent:** It should be clear how each section relates to each other section, and why they are in the order you have chosen. The aim of the project should be made clear in the introduction, and referred to again in the conclusion.
- **Concise:** Every graph, calculation and description needs to be there for a reason. You should not just be repeating the same method multiple times without a good reason.

Mathematical communication

> **TOK Links**
>
> Is mathematics a different language? What are the criteria you use to judge this?

As well as the general communication of your ideas, it is expected that by the end of this course you will be able to communicate in the technical language of mathematics.

Your mathematical communication needs to be:

- **Relevant:** You should use a combination of formulae, diagrams, tables and graphs. All key terms (which are not part of the IB curriculum) and variables should be defined.
- **Appropriate:** The notation used needs to be at the standard expected of IB mathematicians. This does not mean that you need to use formal set theory and logical implications throughout, but it does means that all graphs need to labelled properly and computer notation cannot be used (unless it is part of a screenshot from a computer output).

Common errors which cost marks include:

- ☐ 2^x instead of 2^x
- ☐ x*y instead of $x \times y$
- ☐ 2E12 instead of 2×10^{12}
- ☐ writing $\frac{1}{3} = 0.33$ instead of $\frac{1}{3} = 0.33$ (2 d.p.) or $\frac{1}{3} \approx 0.33$.

> **Tip**
>
> One common error from people using calculus is to call every derivative $\frac{dy}{dx}$. If you are looking at the rate of change of C over time, the appropriate notation is $\frac{dC}{dt}$.

- **Consistent:** Your use of appropriate notation needs to be across all of the exploration, not just in the mathematical proof parts. You also need to make sure that your notation is the same across all of the exploration. If you define the volume of a cone to be x in one part of the exploration it should not suddenly become X or V_c later (unless you have explained your good reason for doing so!).

One other part of consistency is to do with accuracy. If you have measured something to one significant figure as part of your exploration, it would not then be consistent to give your final answer to five significant figures.

Personal engagement

> **TOK Links**
>
> Although personal engagement is difficult to describe, it is often easy to see. Can you think of other situations in which there is wide consensus about something despite no clear criteria being applied?

Your exploration does not need to be a totally new, ground-breaking piece of mathematics, but it does need to have a spark of originality in it to be likely to score well in personal engagement. This might be using novel examples to illustrate an idea, applying a tool to something from your own experience or collecting your own data to test a model.

The exploration

Tip

Beware overcomplicated sources, such as many internet discussion boards. Personal engagement marks are often lost by people who clearly have not understood their source material.

Also, try to avoid the common topics of the golden ratio and the prisoner's dilemma. It is quite tricky to show personal engagement with these, although it is not impossible to do so.

The key to writing an exploration with great personal engagement is to write about a topic which really interests you. We recommend that from the very first day of your course you keep a journal with mathematical ideas you have met that you have found interesting. Some of these will hopefully develop into an exploration. Importantly, do not just think about this during your maths lessons – we want you to see maths everywhere! If you are struggling for inspiration, the following sources have been fruitful for our students in the past:

- podcasts, such as More or Less: Behind the Stats
- YouTube channels, such as Numberphile
- websites, such as Underground Mathematics, NRICH or Khan Academy
- magazines, such as *Scientific American* or *New Scientist*.

One of our top tips is to find an overlap between mathematics and your future studies or career. As well as looking good on applications, it will give you an insight that few people have. Every future career can make use of mathematics, sometimes in surprising ways!

Links to: Other subjects

Some topics we have seen that show great applications of mathematics relevant to other subjects include:
- **Psychology:** Do humans have an intuitive understanding of Bayesian probability?
- **Art:** Did Kandinsky use the golden ratio in his abstract art?
- **Modern languages:** Can you measure the distance between two languages? A phylogenetic analysis of European languages.
- **English literature:** Was Shakespeare sexist? A statistical analysis of the length of male and female character speeches.
- **Medicine:** How do doctors understand uncertainty in clinical tests?
- **Law:** What is the prosecutor's fallacy?
- **Economics:** Does the market model apply to sales in the school cafeteria?
- **Physics:** How accurate are the predictions made by Newton's laws for a paper aeroplane?
- **Chemistry:** Using logarithms to find orders of reactions.
- **Biology:** Did Mendel really cheat when he discovered genetics? A computer simulation.
- **History:** Was it worth it? A statistical analysis of whether greater mortality rates led to more land gained in WW1.
- **Politics:** How many people need to be included in a poll to predict the outcome of an election?
- **Geography:** Creating a model for the population of a city.

■ Reflection

Reflection is the point where you evaluate your results. It should be:

- **Meaningful:** For example, it should link to the aim of the exploration. It could also include commentary on what you have learned and acknowledge any limitations of your results.
- **Critical:** This might include looking at the implications of your results in other, related contexts; a consideration of how the assumptions made affect the ways in which the results can be interpreted; comparing the strengths and weaknesses of different mathematical methods (not all of which need to have been demonstrated in the exploration); and considering whether it is possible to interpret the results in any other way.
- Substantial: The reflection should be related to all parts of the exploration. The best way to do this is to include subsections on reflection in every appropriate section.

■ Use of mathematics

This is the area where your actual mathematical skills are assessed – things such as the technical accuracy of your algebra and the clarity of your reasoning. It is the only criterion that has a slight difference between the Standard Level and Higher Level.

Your work has to be:
- **Commensurate with the level of the course:** This means that it cannot just be on material from the prior knowledge if you want to score well.
- **Correct:** Although a few small slips which are not fundamental to the progress of the exploration are acceptable, you must make sure that you check your work thoroughly.
- **Thorough:** This means that all mathematical arguments are well understood and clearly communicated. Just obtaining the 'correct' answer is not sufficient to show understanding.

In addition, if you are studying Higher Level Mathematics then to get the top marks the work should be:
- **Sophisticated:** This means that the topic you choose must be sufficiently challenging that routes which would not be immediately obvious to most of your peers would be required.
- **Rigorous:** You should be able to explain why you are doing each process, and you should have researched the conditions under which this process holds and be able to demonstrate that they hold. Where feasible, you should prove any results that are central to your argument.

If you pick the right topic, explorations are a really enjoyable opportunity to get to know a mathematical topic in depth. Make sure that you stick to all the internal deadlines set by your teacher so that you give yourself enough time to complete all parts of the exploration and meet all of the assessment criteria.

Tip
Although it might sound impressive to write an exploration on the Riemann Hypothesis or General Relativity, it is just as dangerous to write an exploration on too ambitious a topic as one which is too simple. You will only get credit for maths which you can actually use.

Algebra and confidence

Mathematics, more than many other subjects, splits people into those who love it and those who hate it. We believe that one reason why some people dislike mathematics is because it is a subject where you can often be told you are wrong. However, often the reason someone has got the wrong answer is not because they misunderstand the topic currently being taught, but because they have made a simple slip in algebra or arithmetic which undermines the rest of the work. Therefore, for some mathematicians, a good use of toolkit time might be to revise some of the prior learning so that they have the best possible foundation for the rest of the course. The exercise below provides some possible questions.

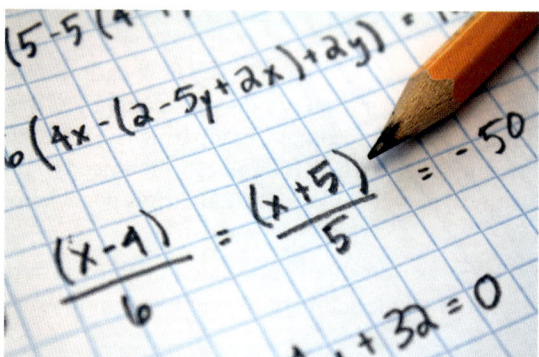

Algebra and confidence **XXV**

> **TOK Links**
> What are the rules of algebra? Can you know how to do something without having the technical language to describe it?

Algebra Practice

1. Simplify the following expressions.
   - **a** $3x + 5y + 8x + 10y$
   - **b** $3x^2 + 5xy + 7x^2 + 2xy$
   - **c** $5 + 2(x - 1)$
   - **d** $(2x + 1) - (x - 3)$
   - **e** $(xy) \times (yz)$
   - **f** $7xy \times 9x$
   - **g** $(2x)^3$
   - **h** $\dfrac{x}{2} - \dfrac{x}{3}$
   - **i** $\dfrac{3 + 6x}{3y}$
   - **j** $\dfrac{7xy}{21xz}$
   - **k** $\dfrac{x^2 + 3x}{2x + 6}$
   - **l** $\dfrac{x - y}{y - x}$
   - **m** $\dfrac{x}{3} \times \dfrac{x}{5}$
   - **n** $\dfrac{x + 1}{x + 2} \times \dfrac{2x + 4}{5}$
   - **o** $\dfrac{x}{3} \div \dfrac{y}{6}$

2. Expand the following brackets.
   - **a** $2x(x - 3)$
   - **b** $(x + 3)^2$
   - **c** $(x - 4)(x + 5)$
   - **d** $x(x + 1)(x + 2)$
   - **e** $(x + y + 1)(x + y - 1)$
   - **f** $(x + 1)(x + 3)(x + 5)$

3. Factorize the following expressions.
   - **a** $12 - 8y$
   - **b** $3x - 6y$
   - **c** $7x^2 - 14x$
   - **d** $5yx^2 + 10xy^2$
   - **e** $2z(x + y) + 5(x + y)$
   - **f** $3x(x - 2) + 2 - x$

4. Solve the following linear equations and inequalities.
   - **a** $5x + 9 = 2 - 3x$
   - **b** $3(x + 2) = 18$
   - **c** $4 - (3 - x) = 9 + 2x$
   - **d** $\dfrac{x - 1}{2} + 3 = x$
   - **e** $4 - 2x > 17$
   - **f** $10 - 2x \leqslant 5 + x$

5. Solve the following simultaneous equations.
   - **a** $x + y = 10$
     $x - y = 2$
   - **b** $2x + y = 7$
     $x - y = 2$
   - **c** $2x + 5y = 8$
     $3x + 2y = 1$

6. Evaluate the following expressions.
   - **a** $3x + 4$ when $x = -2$
   - **b** $(3 - x)(5 + 2x)$ when $x = 2$
   - **c** 3×2^x when $x = 2$
   - **d** $x^2 - x$ when $x = -1$
   - **e** $\dfrac{2}{x} + 3x$ when $x = \dfrac{1}{6}$
   - **f** $\sqrt{x^2 + 9}$ when $x = 4$

7. Rearrange the following formulae to make x the subject.
   - **a** $y = 2x + 4$
   - **b** $y(3 + x) = 2x$
   - **c** $y = \dfrac{x + 1}{x + 2}$
   - **d** $y = \dfrac{ax}{x - b}$
   - **e** $a = \sqrt{x^2 - 4}$
   - **f** $y = \dfrac{1}{x} - \dfrac{3}{2x}$

8. Simplify the following surd expressions.
   - **a** $(1 + 2\sqrt{2}) - (-1 + 3\sqrt{2})$
   - **b** $2\sqrt{3}\sqrt{5}$
   - **c** $\sqrt{2} \times \sqrt{8}$
   - **d** $\sqrt{3} + \sqrt{12}$
   - **e** $(1 - \sqrt{2})^2$
   - **f** $(2 + \sqrt{5})(2 - \sqrt{5})$

9. **HL only:** Rationalize the following denominators.
   - **a** $\dfrac{2}{\sqrt{2}}$
   - **b** $\dfrac{1}{\sqrt{2} - 1}$
   - **c** $\dfrac{1 + \sqrt{3}}{3 + \sqrt{3}}$

10. **HL only:** Solve the following quadratic equations.
    - **a** $x^2 + 5x + 6 = 0$
    - **b** $4x^2 - 9 = 0$
    - **c** $x^2 + 2x - 5 = 0$

11. **HL only:** Write the following as a single algebraic fraction.
    - **a** $\dfrac{1}{a} - \dfrac{1}{b}$
    - **b** $\dfrac{3}{x} + \dfrac{5}{x^2}$
    - **c** $x + \dfrac{13}{x - 1}$
    - **d** $\dfrac{1}{1 - x} + \dfrac{1}{1 + x}$
    - **e** $\dfrac{5 + a}{a} - \dfrac{a}{5 - a}$
    - **f** $\dfrac{x + 1}{x - 1} + \dfrac{x + 2}{x - 2}$

1 Core: Exponents and logarithms

ESSENTIAL UNDERSTANDINGS

- Number and algebra allow us to represent patterns, show equivalences and make generalizations which enable us to model real-world situations.
- Algebra is an abstraction of numerical concepts and employs variables to solve mathematical problems.

In this chapter you will learn…

- how to use the laws of exponents with integer exponents
- how to perform operations with numbers in the form $a \times 10^k$, where $1 \leq a < 10$ and k is an integer
- about the number e
- about logarithms
- how to solve simple exponential equations.

CONCEPTS

The following key concepts will be addressed in this chapter:
- Different **representations** of numbers enable quantities to be compared and used for computational purposes with ease and accuracy.
- Numbers and formulae can appear in different, but equivalent forms, or **representations**, which can help us establish identities.

Tip
The notation $k \in \mathbb{Z}$ means that k is an integer.

PRIOR KNOWLEDGE

Before starting this chapter, you should already be able to complete the following:

1 Evaluate the following:
   a 3^4
   b 5×2^3

2 Write the following values in the form $a \times 10^k$ where $1 \leq a < 10$ and $k \in \mathbb{Z}$.
   a 342.71
   b 0.00856

3 Express 64 in the form 2^k, where $k \in \mathbb{Z}$.

■ Figure 1.1 How do we compare very large numbers?

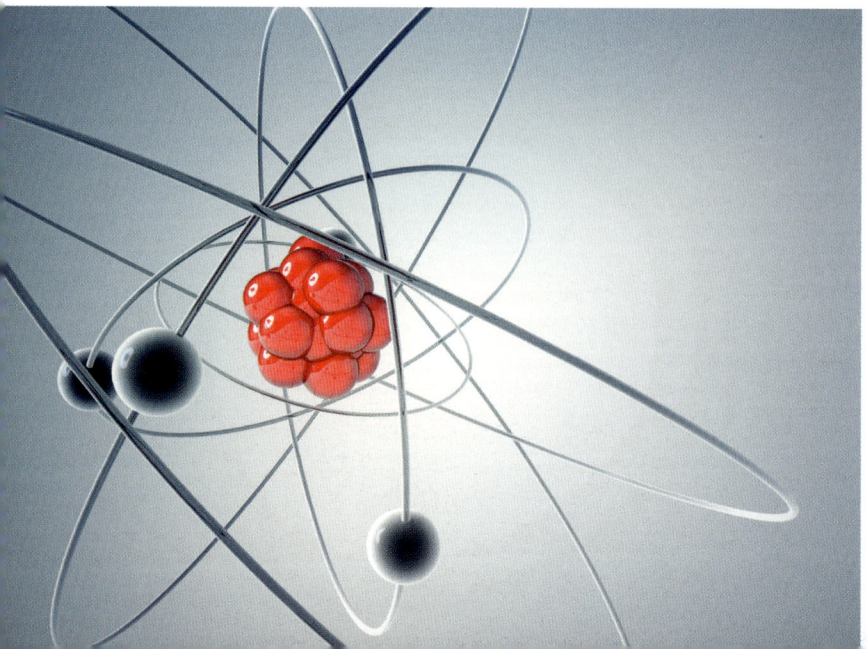

Core: Exponents and logarithms

When you first study maths, it is easy to think it is a subject that is all about numbers. But actually, the study of numbers is just one area in which we apply the logic of maths. Once you have experience of numbers, you will be looking to make links: for example, $3^2 + 4^2 = 5^2$, but is this true in general for consecutive numbers? Of course the answer is 'no', but $3^2 \times 4^2 = 12^2$ gives you a rule that works when applied to multiplications.

In this chapter we will be generalizing some patterns you are probably already aware of. You will then be applying these rules to numbers represented in different ways. You might wonder why we need a different form for writing numbers, but try writing down the number of atoms in the sun. When numbers are very large or very small, it is useful to have a more convenient form to express them in.

Finally, you will see how to solve equations where the unknown is in the exponent, such as $10^x = 7$. In so doing you will meet a number that has wide-ranging applications in fields including science, economics and engineering.

Starter Activity

Look at the pictures in Figure 1.1. In small groups discuss whether there are more atoms in a jug of water than there are jugs of water in the Atlantic Ocean.

Now look at this problem:

A model suggests that the level of carbon dioxide (CO_2) in the atmosphere in parts per million is given by 400×1.05^n, where n is the number of years after 2018.

Use a spreadsheet to estimate when this model predicts the level of CO_2 in the atmosphere will reach 1000 parts per million.

LEARNER PROFILE – Inquirers

Is mathematics invented or discovered? Is mathematics designed to mirror reality?

1A Laws of exponents

> **TOK Links**
>
> In expressions like a^n, a is called the **base** and n is called the **exponent**, although you may see it referred to as power, index or indice. Does having a label help you understand it? Would it be better if everybody used the same label?

In your previous work you have may have noticed that $2^3 \times 2^2 = (2 \times 2 \times 2) \times (2 \times 2) = 2^5$. Based on specific examples such as this you can generalize to a formula.

> **Tip**
>
> Be aware that you will often have to use these rules 'backwards'. For example, if you see 2^{12} you can rewrite it as $(2^3)^4$ or as $(2^4)^3$.

> **KEY POINT 1.1**
>
> - $a^m \times a^n = a^{m+n}$
> - $\dfrac{a^m}{a^n} = a^{m-n}$
> - $(a^m)^n = a^{mn}$

If we set m and n equal in the second law in Key Point 1.1 then it follows that $a^0 = 1$ for any (non-zero) value of a.

>  **TOOLKIT: Problem Solving**
>
> Anything raised to the power 0 is 1, but 0 to any power is 0. So, what is the value of 0^0?

> **WORKED EXAMPLE 1.1**
>
> Simplify $x^6 \times x^3$.
>
> Use $a^m \times a^n = a^{m+n}$ .... $x^6 \times x^3 = x^{6+3}$
>
> $= x^9$

> **WORKED EXAMPLE 1.2**
>
> Simplify $\dfrac{y^8}{y^2}$.
>
> Use $\dfrac{a^m}{a^n} = a^{m-n}$ .... $\dfrac{y^8}{y^2} = y^{8-2}$
>
> $= y^6$

> **WORKED EXAMPLE 1.3**
>
> Simplify $(7^3)^5$. You do not need to evaluate your result.
>
> Use $(a^m)^n = a^{mn}$ …… $(7^3)^5 = 7^{3 \times 5}$
> $= 7^{15}$

If the expression has numbers as well as algebraic values, just multiply or divide the numbers separately. This follows because we can do multiplication in any convenient order.

> **WORKED EXAMPLE 1.4**
>
> Simplify $2b^3 \times 7b^4$.
>
> You can reorder the multiplication as $2 \times 7 \times b^3 \times b^4$ …… $2b^3 \times 7b^4 = 14b^3 b^4$
>
> Then $b^3 \times b^4 = b^{3+4}$ …… $= 14b^7$

> **WORKED EXAMPLE 1.5**
>
> Simplify $\dfrac{6c^9}{2c^4}$.
>
> You can use your knowledge of fractions to write the expression as a numeric fraction multiplied by an algebraic fraction …… $\dfrac{6c^9}{2c^4} = \dfrac{6}{2} \times \dfrac{c^9}{c^4}$
>
> Use $\dfrac{c^9}{c^4} = c^{9-4}$ …… $= 3c^5$

The laws in Key Point 1.1 work for negative as well as positive integers. But what is the meaning of a negative exponent? In the second law in Key Point 1.1 we can set $m = 0$ and use the fact that $a^0 = 1$ to deduce a very important rule.

> **KEY POINT 1.2**
>
> - $a^{-n} = \dfrac{1}{a^n}$

> **WORKED EXAMPLE 1.6**
>
> Without a calculator write 2^{-4} as a fraction in its simplest terms.
>
> Use $a^{-n} = \dfrac{1}{a^n}$ …… $2^{-4} = \dfrac{1}{2^4}$
> $= \dfrac{1}{16}$

WORKED EXAMPLE 1.7

Write $4x^{-3}$ as a fraction.

Apply the power -3 to x (but not to 4) .... $4x^{-3} = 4 \times \dfrac{1}{x^3}$

$$= \dfrac{4}{x^3}$$

You will often need to apply an exponent to a product or fraction. We can use the fact that multiplication can be reordered to help suggest a rule. For example,

$$(ab)^3 = (ab)(ab)(ab) = (a \times a \times a) \times (b \times b \times b) = a^3 b^3$$

This suggests the following generalization.

KEY POINT 1.3

- $(ab)^n = a^n \times b^n$
- $\left(\dfrac{a}{b}\right)^n = \dfrac{a^n}{b^n}$

WORKED EXAMPLE 1.8

Simplify $(2x^2 y^{-4})^5$.

Apply the power 5 to each term in the product .... $(2x^2 y^{-4})^5 = 2^5 (x^2)^5 (y^{-4})^5$

$$= 32 x^{10} y^{-20}$$

WORKED EXAMPLE 1.9

Simplify $\left(\dfrac{2u}{3v^2}\right)^{-3}$.

Apply the power -3 to each part of the fraction .... $\left(\dfrac{2u}{3v^2}\right)^{-3} = \dfrac{2^{-3} u^{-3}}{3^{-3} (v^2)^{-3}}$

$$= \dfrac{\dfrac{1}{8} \times \dfrac{1}{u^3}}{\dfrac{1}{27} \times \dfrac{1}{v^6}}$$

$$= \dfrac{\left(\dfrac{1}{8u^3}\right)}{\left(\dfrac{1}{27v^6}\right)}$$

Four level fractions are easiest dealt with by dividing two normal fractions. Flip the second fraction and multiply

$$= \dfrac{1}{8u^3} \div \dfrac{1}{27v^6}$$

$$= \dfrac{1}{8u^3} \times \dfrac{27v^6}{1}$$

$$= \dfrac{27v^6}{8u^3}$$

Tip

Perhaps an easier way to do Worked Example 1.9 is to use the last rule from Key Point 1.1 to write the expression as $\left(\left(\dfrac{2u}{3v^2}\right)^3\right)^{-1}$, but it is good to practise working with four level fractions too!

Be the Examiner 1.1

Simplify $(2x^2 y^3)^4$.

Which is the correct solution? Identify the errors made in the incorrect solutions.

Solution 1	Solution 2	Solution 3
$(2x^2 y^3)^4 = 2^4 (x^2)^4 (y^3)^4$ $= 16 x^6 y^7$	$(2x^2 y^3)^4 = 2^4 (x^2)^4 (y^3)^4$ $= 8 x^8 y^{12}$	$(2x^2 y^3)^4 = 2^4 (x^2)^4 (y^3)^4$ $= 16 x^8 y^{12}$

More complicated expressions may have to be simplified before the laws of exponents are applied.

WORKED EXAMPLE 1.10

Simplify $\dfrac{16a^5 b^2 - 12ab^4}{4ab^2}$.

You can split a fraction up if the top is a sum or a difference
$$\dfrac{16a^5 b^2 - 12ab^4}{4ab^2} = \dfrac{16a^5 b^2}{4ab^2} - \dfrac{12ab^4}{4ab^2}$$

Simplify numbers, as and bs separately
$$= 4a^4 - 3b^2$$

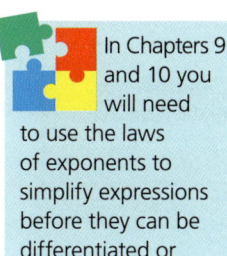

In Chapters 9 and 10 you will need to use the laws of exponents to simplify expressions before they can be differentiated or integrated.

You can only apply the laws of exponents when the bases are the same. Sometimes you can rewrite one of the bases to achieve this.

WORKED EXAMPLE 1.11

Express $3^4 \times 9^5$ in the form 3^k, for some integer k.

Write 9 as 3^2
$$3^4 \times 9^5 = 3^4 \times (3^2)^5$$
$$= 3^4 \times 3^{10}$$
$$= 3^{14}$$
$$\therefore k = 14$$

This technique can be used to solve some equations.

WORKED EXAMPLE 1.12

Solve $2^{x+6} = 8^x$.

Write 8 as 2^3
$$2^{x+6} = 8^x$$
$$2^{x+6} = (2^3)^x$$
$$2^{x+6} = 2^{3x}$$

Since the bases are the same, the exponents must be equal
$$\therefore x + 6 = 3x$$
$$2x = 6$$
$$x = 3$$

An equation like this with the unknown (x) in the power is called an **exponential equation**. In Section 1C, you will see how to solve more complicated examples using logarithms.

The laws of exponents and your prior knowledge of algebra can be applied to modelling real-life situations.

> **WORKED EXAMPLE 1.13**
>
> The length of a baby fish is modelled by $L = 2t^2$ where t is the age in days and L is the length in cm. Its mass in grams is modelled by $M = 4L^3$.
>
> a Find and simplify an expression for M in terms of t.
> b Find the age of the fish when the model predicts a mass of 1000 g.
> c Explain why the model is unlikely to still hold after 100 days.
>
> Substitute $L = 2t^2$ into the expression for M .... **a** $M = 4L^3$
>
> $\qquad\qquad = 4(2t^2)^3$
>
> Apply the exponent 3 to 2 and to t^2 .... $= 4\left(2^3 \left(t^2\right)^3\right)$
>
> $\qquad\qquad = 4(8t^6)$
>
> $\qquad\qquad = 32t^6$
>
> You need to find the value of t when $M = 1000$ .... **b** When $M = 1000$,
>
> $\qquad\qquad 1000 = 32t^6$
>
> $\qquad\qquad t^6 = \dfrac{1000}{32}$
>
> Take the sixth root of both sides to find t .... $t = \sqrt[6]{\dfrac{1000}{32}} = 1.77$ days
>
> **c** The model predicts that the fish will continue growing, whereas in reality it is likely that after 100 days the fish will be growing far more slowly, if at all.

Exercise 1A

For questions 1 to 4, use the method demonstrated in Worked Example 1.1 to simplify the expressions. If numerical, you do not need to evaluate your result.

1 a $x^7 \times x^4$
   b $x^5 \times x^7$

2 a $y^3 \times y^5$
   b $z^5 \times z^5$

3 a $a^6 \times a$
   b $a^{10} \times a$

4 a $5^7 \times 5^{10}$
   b $2^{12} \times 2^{12}$

For questions 5 to 8, use the method demonstrated in Worked Example 1.2 to simplify the expressions. If numerical, you do not need to evaluate your result.

5 a $\dfrac{x^4}{x^3}$
   b $\dfrac{x^8}{x^5}$

6 a $y^8 \div y^4$
   b $z^9 \times z^3$

7 a $\dfrac{b^7}{b}$
   b $\dfrac{b^9}{b}$

8 a $11^{12} \div 11^4$
   b $7^{10} \div 7^5$

For questions 9 to 12, use the method demonstrated in Worked Example 1.3 to simplify the expressions. If numerical, you do not need to evaluate your result.

9 a $\left(x^3\right)^5$
   b $\left(x^4\right)^8$

10 a $\left(y^4\right)^4$
   b $\left(z^5\right)^5$

11 a $\left(c^7\right)^2$
   b $\left(c^2\right)^7$

12 a $\left(3^5\right)^{10}$
   b $\left(13^7\right)^4$

1A Laws of exponents

For questions 13 to 15, use the method demonstrated in Worked Example 1.4 to simplify the expressions.

13   a $12x^2 \times 4x^5$ 14   a $a \times 3a^2$ 15   a $5x^2yz \times 4x^3y^2$

    b $3x^4 \times 5x^3$ b $b^2 \times 5b^2$ b $6x^7yz^2 \times 2xz^3$

For questions 16 to 19, use the method demonstrated in Worked Example 1.5 to simplify the expressions.

16   a $\dfrac{10x^{10}}{5x^5}$ 17   a $\dfrac{8x}{16x^4}$ 18   a $15x^2 \div 9x^4$ 19   a $14x^3y^5 \div 2xy^2$

    b $\dfrac{9x^9}{3x^3}$ b $\dfrac{5x^2}{20x^3}$ b $21x^5 \div 28x^3$ b $6x^5yz^2 \div 3x^2y^2z$

For questions 20 to 23, use the method demonstrated in Worked Example 1.6 to write the expression as a fraction in its simplest terms.

20   a 10^{-1} 21   a 3^{-3} 22   a $\left(\dfrac{3}{4}\right)^{-1}$ 23   a $\left(\dfrac{2}{3}\right)^{-2}$

    b 7^{-1} b 5^{-2} b $\left(\dfrac{5}{7}\right)^{-1}$ b $\left(\dfrac{2}{5}\right)^{-3}$

For questions 24 to 26, use the method demonstrated in Worked Example 1.7 to write the expression as a fraction in its simplest terms.

24   a 7×3^{-2} 25   a $6x^{-1}$ 26   a $3^{-2} \div 2^{-3}$

    b 5×2^{-4} b $10x^{-4}$ b $4^{-3} \div 3^{-4}$

For questions 27 to 29, use the method demonstrated in Worked Example 1.8 to simplify each expression.

27   a $(3u^{-2})^3$ 28   a $(2a^{-5})^{-2}$ 29   a $(5x^2y^3)^2$

    b $(2v^{-3})^5$ b $(3b^{-7})^{-3}$ b $(3a^2b^{-2})^4$

For questions 30 to 32, use the method demonstrated in Worked Example 1.9 to simplify each expression.

30   a $\left(\dfrac{x}{3}\right)^3$ 31   a $\left(\dfrac{3x^2}{2y^3}\right)^3$ 32   a $\left(\dfrac{3u}{4v^2}\right)^{-2}$

    b $\left(\dfrac{5}{x^2}\right)^2$ b $\left(\dfrac{5uv^3}{7b^4}\right)^2$ b $\left(\dfrac{2a^3}{3b^2}\right)^{-3}$

For questions 33 to 35, use the method demonstrated in Worked Example 1.10 to simplify each expression.

33   a $\dfrac{6x^2 - 21x^7}{3x}$ 34   a $\dfrac{15u^3v + 18u^2v^3}{3uv}$ 35   a $\dfrac{10p^3q - 6pq^3}{2p^2q}$

    b $\dfrac{15y^4 + 25y^6}{5y^3}$ b $\dfrac{20a^5b^7c^2 - 16a^4b^2c^3}{4a^3b^2c}$ b $\dfrac{14s^4t^3 + 21s^5t^7}{7s^2t^5}$

For questions 36 to 39, use the method demonstrated in Worked Example 1.11 to express each value in the form a^b for the given prime number base a.

36   a   Express 9^4 as a power of 3.   b   Express 27^8 as a power of 3.

37   a   Express $2^5 \times 4^2$ as a power of 2.   b   Express $5^4 \times 125^2$ as a power of 5.

38   a   Express $4^7 \div 8^3$ as a power of 2.   b   Express $27^5 \div 9^2$ as a power of 3.

39   a   Express $8^3 \times 2^7 + 4^8$ as a power of 2.   b   Express $16^2 - 4^8 \div 8^3$ as a power of 2.

For questions 40 to 43, use the method demonstrated in Worked Example 1.12 to solve each equation to find the unknown value.

40   a   Solve $3^x = 81$.   41   a   Solve $2^{x+4} = 8$.   42   a   Solve $7^{3x-5} = 49$.   43   a   Solve $3^{2x+5} = \dfrac{1}{27}$.

    b   Solve $5^x = 125$.   b   Solve $3^{x-3} = 27$.   b   Solve $4^{2x-7} = 16$.   b   Solve $2^{3x+5} = \dfrac{1}{16}$.

44 Simplify $\dfrac{4x^2 + 8x^3}{2x}$. **45** Simplify $\dfrac{(2x^2y)^3}{8xy}$. **46** Write $(2ab^{-2})^{-3}$ without brackets or negative indices.

47 The number of people suffering from a disease 'D' in a country is modelled by $D = 1\,000\,000n^{-2}$ where n is the amount spent on prevention (in millions of dollars).

    a Rearrange the equation to find n in terms of D.

    b According to the model, how many people will have the disease if $2 million dollars is spent on prevention?

    c How much must be spent to reduce the number of people with the disease to 10 000?

48 A computer scientist analyses two different methods for finding the prime factorization of a number. They both take a time T microseconds that depends on the number of digits (n).

Method A: $T_A = k_A n^3$

Method B: $T_B = k_B n^2$

Both methods take 1000 microseconds to factorize a five digit number.

    a Find the values of k_A and k_B.

    b Find and simplify an expression for $\dfrac{T_A}{T_B}$.

    c Which method would be quicker at factorizing a 10 digit number? Justify your answer.

49 Solve $10 + 2 \times 2^x = 18$. **50** Solve $9^x = 3^{x+5}$. **51** Solve $5^{x+1} = 25 \times 5^{2x}$.

52 Solve $8^x = 2 \times 4^{2x}$. **53** Solve $25^{2x+4} = 125 \times 5^{x-1}$.

54 The pressure (P) in a gas is equal to $0.8T$ where T is the temperature measured in kelvin. The air resistance, R, in newtons, of an aeroplane is modelled by $R = 5P^2$.

    a Find an expression for R in terms of T.

    b If the air resistance has a magnitude of 200 000 newtons, find the temperature in kelvin.

55 A boat travels 3 km at a speed of v km per hour.

    a Find an expression for the time taken.

The boat uses up $0.5v^2$ litres of petrol per hour.

    b Find an expression for the amount of petrol used in the 3 km journey.

The boat has 60 litres of fuel.

    c Find the maximum speed the boat can travel at if it is to complete the 3 km journey.

56 Solve the simultaneous equations:

$8^x 2^y = 1$ and $\dfrac{4^x}{2^y} = 32$

57 Solve $6^x = 81 \times 2^x$. **58** Solve $32 + 2^{x-1} = 2^x$.

59 Find all solutions to $(x-2)^{x+5} = 1$.

60 Determine, with justification, which is larger out of 2^{7000} and 5^{3000}.

61 What is the last digit of $316^{316} + 631^{631}$?

1B Operations with numbers in the form $a \times 10^k$, where $1 \leq a < 10$ and k is an integer

> **Tip**
>
> The notation $k \in \mathbb{Z}$ means that k is an integer.

You already know that it can be useful to write very large or very small numbers in the form $a \times 10^k$ where $1 \leq a < 10$ and $k \in \mathbb{Z}$. This is often referred to as **standard index form**, standard form or scientific notation.

> **You are the Researcher**
>
> Extremely large numbers require other ways of representing them. You might want to research tetration and the types of number – such as Graham's Number – which require this notation for them to be written down.

You now need to be able to add, subtract, multiply and divide numbers in this form.

WORKED EXAMPLE 1.14

Without a calculator, write $(3 \times 10^7) \times (4 \times 10^{-3})$ in the form $a \times 10^k$ where $1 \leq a < 10$ and $k \in \mathbb{Z}$.

Reorder so that the respective parts of each number are together.
$$(3 \times 10^7) \times (4 \times 10^{-3}) = (3 \times 4) \times (10^7 \times 10^{-3})$$

$10^7 \times 10^{-3} = 10^{7+(-3)}$
$$= 12 \times 10^4$$

$$12 \times 10^4 = (1.2 \times 10) \times 10^4 = 1.2 \times 10^5$$

WORKED EXAMPLE 1.15

Show that if $x = 3 \times 10^7$ and $y = 4 \times 10^{-2}$, then $\frac{x}{y} = 7.5 \times 10^8$.

Split off the powers of 10
$$\frac{x}{y} = \frac{3 \times 10^7}{4 \times 10^{-2}}$$
$$= \frac{3}{4} \times \frac{10^7}{10^{-2}}$$
$$= 0.75 \times 10^{7-(-2)}$$
$$= 0.75 \times 10^9$$

Change the number into standard form
$$= (7.5 \times 10^{-1}) \times 10^9$$
$$= 7.5 \times 10^{-1+9}$$
$$= 7.5 \times 10^8$$

TOK Links

In Worked Example 1.15, we made explicit the rules of indices used, but just assumed that $\frac{3}{4} = 0.75$ was obvious. When you were 11 years old you might have had to explain this bit too, but part of mathematics is knowing the mathematical level and culture of your audience. If you continue with mathematics, then in a few years' time you would not be expected to explain the rules of indices anymore as everybody reading your explanations will probably know them. Is this unique to mathematics or do explanations in every area of knowledge depend on the audience?

With a 'show that' question, like that in Worked Example 1.15, you need to be able to explain each step in the calculation; most of the time, however, you will be able to do this type of arithmetic on a calculator, as shown below.

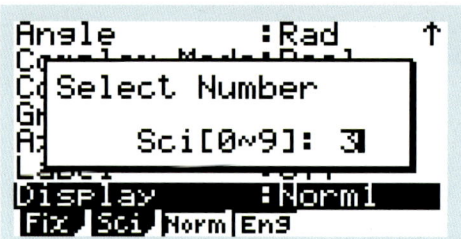
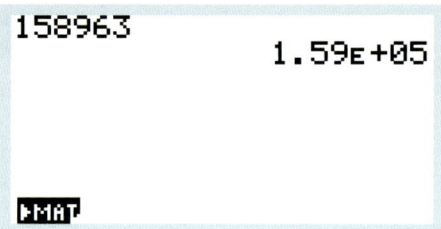

Note that in this mode, the output 1.59E+05 shown on the right means 1.59×10^5.

You can choose the number of significant figures to which the number is rounded (here 3 was chosen, as shown on the left).

WORKED EXAMPLE 1.16

Show that $3.2 \times 10^{19} + 4.5 \times 10^{20} = 4.82 \times 10^{20}$.

Write 3.2×10^{19} as 0.32×10^{20} so that there is clearly a factor of 10^{20} .... $3.2 \times 10^{19} + 4.5 \times 10^{20} = 0.32 \times 10^{20} + 4.5 \times 10^{20}$

$$= (0.32 + 4.5) \times 10^{20}$$

Change the number into standard form .... $= 4.82 \times 10^{20}$

In the 'show that' questions you need to be able to explain all the steps without referring to a calculator. Most of the time, however, you will be able to do this type of arithmetic on a calculator.

CONCEPTS – REPRESENTATION

5^{20} and $9.5367... \times 10^{13}$ are both **representations** of the same number. However, they have different uses. The second probably gives you a better sense of the scale of the number, but the former might be more useful in solving $5^{2x} = 5^{20}$ or comparing it to 3^{18}. One major skill in mathematics is deciding which representation to use in which problem.

Exercise 1B

Questions 1 to 4 are designed to remind you of your prior learning.

1. **a** Write 3.2×10^4 as a normal number.
   b Write 6.92×10^6 as a normal number.
2. **a** Write 4.8×10^{-2} as a decimal.
   b Write 9.85×10^{-4} as a decimal.
3. **a** Write the value 612.07 in standard index form.
   b Write the value 3076.91 in standard index form.
4. **a** Write the value 0.003 061 7 in standard index form.
   b Write the value 0.022 19 in standard index form.

For questions 5 to 7, use the method demonstrated in Worked Example 1.14 to write the given expressions in standard index form. Do not use a calculator.

5. **a** $(2 \times 10^4) \times (3.4 \times 10^3)$
   b $(3.2 \times 10^5) \times (3 \times 10^6)$
6. **a** $(5 \times 10^4) \times (2 \times 10^{-5})$
   b $(3 \times 10^{-2}) \times (4 \times 10^{-4})$
7. **a** $(5 \times 10^{10})^2$
   b $(6 \times 10^6)^2$

For questions 8 to 10, use the method demonstrated in Worked Example 1.15 to write the given expressions in standard index form. Do not use a calculator.

8. **a** $\dfrac{6 \times 10^2}{3 \times 10^4}$
   b $\dfrac{8 \times 10^6}{2 \times 10^{10}}$
9. **a** $\dfrac{1 \times 10^5}{2 \times 10^4}$
   b $\dfrac{2 \times 10^5}{8 \times 10^2}$
10. **a** $(8.4 \times 10^{14}) \div (4 \times 10^3)$
    b $(9.3 \times 10^{15}) \div (3 \times 10^6)$

For questions 11 to 13, use the method demonstrated in Worked Example 1.16 to write the given expressions in standard index form. Do not use a calculator.

11. **a** $1 \times 10^4 + 2 \times 10^5$
    b $3 \times 10^8 + 2 \times 10^6$
12. **a** $8 \times 10^5 - 4 \times 10^4$
    b $9 \times 10^{14} - 9 \times 10^{12}$
13. **a** $(2.1 \times 10^3) + (3.8 \times 10^4)$
    b $(5.7 \times 10^{13}) + (4.3 \times 10^{12})$

14. Show that if $a = 4 \times 10^6$, $b = 5 \times 10^{-3}$, then $a \times b = 2 \times 10^4$.
15. Show that if $c = 1.4 \times 10^3$, $d = 5 \times 10^8$, then $c \times d = 7 \times 10^{11}$.
16. Show that if $a = 4 \times 10^6$, $b = 5 \times 10^{-3}$, then $\dfrac{a}{b} = 8 \times 10^8$.
17. Show that if $c = 1.4 \times 10^3$, $d = 2 \times 10^8$, then $\dfrac{c}{d} = 7 \times 10^{-6}$.
18. Show that if $a = 4.7 \times 10^6$, $b = 7.1 \times 10^5$, then $a - b = 3.99 \times 10^6$.
19. Show that if $c = 3.98 \times 10^{13}$, $d = 4.2 \times 10^{14}$, then $d - c = 3.802 \times 10^{14}$.
20. Let $p = 12.2 \times 10^7$ and $q = 3.05 \times 10^5$.
    a Write p in the form $a \times 10^k$ where $1 \leq a < 10$ and $k \in \mathbb{Z}$.
    b Evaluate $\dfrac{p}{q}$.
    c Write your answer to part **b** in the form $a \times 10^k$ where $1 \leq a < 10$ and $k \in \mathbb{Z}$.
21. The number of atoms in a balloon is approximately 6×10^{23}. Theoretical physics predicts that there are approximately 10^{80} atoms in the known universe. What proportion of atoms in the known universe are found in the balloon?
22. 12 grams of carbon contains 6.02×10^{23} atoms. What is the mass (in grams) of one atom of carbon? Give your answer in the form $a \times 10^k$ where $1 \leq a < 10$ and $k \in \mathbb{Z}$.
23. The diameter of a uranium nucleus is approximately 15 fm where 1 fm is 10^{-15} m.
    a Write the diameter (in metres) in the form $a \times 10^k$ where $1 \leq a < 10$ and $k \in \mathbb{Z}$.

    If the nucleus is modelled as a sphere, then the volume is given by $V = \dfrac{1}{6}\pi d^3$ where d is the diameter.

    b Estimate the volume of a uranium nucleus in metres cubed.

24 The area of Africa is approximately $3.04 \times 10^{13}\, m^2$. The area of Europe is approximately $1.02 \times 10^{13}\, m^2$. The population of Africa is approximately 1.2 billion and the population of Europe is 741 million.
   a How many times bigger is Africa than Europe?
   b Write the population of Europe in the form $a \times 10^k$ where $1 \leq a < 10$ and $k \in \mathbb{Z}$.
   c Does Africa or Europe have more people per metre squared? Justify your answer.

25 You are given that
$$(3 \times 10^a) \times (5 \times 10^b) = c \times 10^d$$
where $1 \leq c < 10$ and $d \in \mathbb{Z}$.
   a Find the value of c.
   b Find an expression for d in terms of a and b.

26 You are given that
$$(2 \times 10^a) \div (5 \times 10^b) = c \times 10^d$$
where $1 \leq c < 10$ and $d \in \mathbb{Z}$.
   a Find the value of c.
   b Find an expression for d in terms of a and b.

27 $x = a \times 10^p$ and $y = b \times 10^q$ where $4 < a < b < 9$. When written in standard form, $xy = c \times 10^r$. Express r in terms of p and q.

1C Logarithms

■ Introduction to logarithms

If you want to find the positive value of x for which $x^2 = 5$ you can use the square root function: $x = \sqrt{5} \approx 2.236$.

There is also a function that will let you find the value of x such that, say, $10^x = 5$. That function is logarithm with base 10: $x = \log_{10} 5 \approx 0.699$.

Although base 10 logarithms are common, any positive base other than 1 can be used.

> **Tip**
> Usually $\log_{10} x$ will just be written as $\log x$.

> **KEY POINT 1.4**
> $a = b^x$ is equivalent to $\log_b a = x$.

> **TOK Links**
> The study of logarithms is usually attributed to the Scottish mathematician John Napier. Do you think he discovered something which already existed or invented something new?

As 10^x is always positive, note that there is no answer to a question such as '10 raised to what exponent gives –2?' Therefore, you can only take logarithms of positive numbers.

WORKED EXAMPLE 1.17

Without a calculator, calculate the value of log 1000.

Let the value of log 1000 be x ·········· $x = \log 1000$

Use $a = b^x$ is equivalent to $\log_b a = x$ ·········· $\therefore 10^x = 1000$

You can see (or experiment to find) that $x = 3$ ·········· $\log 1000 = 3$

WORKED EXAMPLE 1.18

Find the exact value of y if $\log_{10}(y + 1) = 3$.

Use $a = b^x$ is equivalent to $\log_b a = x$ ·········· $\log_2 (y + 1) = 3$

$$y + 1 = 10^3 = 1000$$
$$y = 999$$

Another very common base for logarithms is the number e ≈ 2.718 28, which is an irrational number a bit like π.

The logarithm with base e is called the natural logarithm and is written as ln x.

WORKED EXAMPLE 1.19

Make x the subject of $\ln(3x - 2) = y$.

Use $a = b^x$ is equivalent to $\log_b a = x$, with $b = e$ ·········· $\ln(3x - 2) = y$

$$3x - 2 = e^y$$
$$x = \frac{e^y + 2}{3}$$

Logarithms can be treated algebraically with all the usual rules applying.

WORKED EXAMPLE 1.20

Simplify $\dfrac{3\ln x - \ln x}{\ln x}$.

Since $3y - y = 2y$ we can simplify the numerator ·········· $\dfrac{3\ln x - \ln x}{\ln x} = \dfrac{2\ln x}{\ln x}$

Divide top and bottom by $\ln x$ ·········· $= 2$

Since the process of taking a logarithm with base b reverses the process of raising b to an exponent, you have the following important results.

> **KEY POINT 1.5**
>
> For base 10:
> - $\log 10^x = x$
> - $10^{\log x} = x$
>
> For base e:
> - $\ln e^x = x$
> - $e^{\ln x} = x$

WORKED EXAMPLE 1.21

Without using a calculator, find the exact value of $e^{3\ln 2}$.

Use the law of exponents $a^{mn} = (a^m)^n$ .... $e^{3\ln 2} = (e^{\ln 2})^3$

$e^{\ln 2} = 2$ .... $= 2^3$

$= 8$

■ Numerical evaluation of logarithms using technology

You need to be able to use your graphical display calculator (GDC) to evaluate logarithms.

WORKED EXAMPLE 1.22

Using a calculator, find the value of $\ln(345678)$, giving your answer in standard index form to three significant figures.

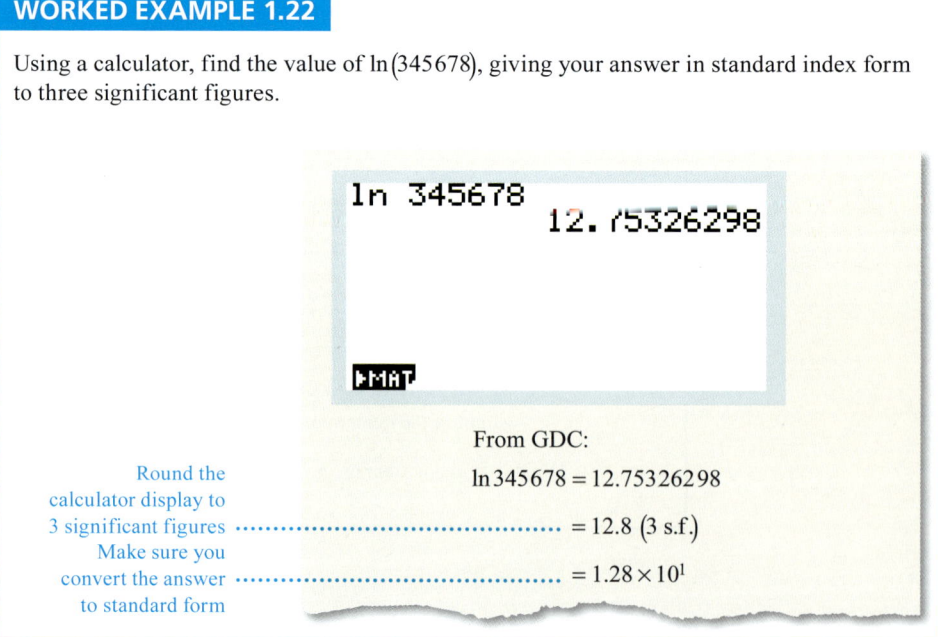

From GDC:

$\ln 345678 = 12.75326298$

Round the calculator display to 3 significant figures .... $= 12.8$ (3 s.f.)

Make sure you convert the answer to standard form .... $= 1.28 \times 10^1$

1C Logarithms

■ Solving exponential equations

You can use technology to find approximate solutions to exponential equations.

> **WORKED EXAMPLE 1.23**
>
> Solve the equation $e^x = 4$, giving your answer to three significant figures.
>
> Use $a = b^x$ is equivalent to $\log_b a = x$, with $b = e$ .... $x = \ln 4$
>
> Evaluate using your GDC .... $= 1.39$ (3 s.f.)

Sometimes you need to find exact solutions.

> **WORKED EXAMPLE 1.24**
>
> Solve $10^{x-2} = 21$, giving your answer in an exact form.
>
> Use $a = b^x$ is equivalent to $\log_b a = x$ .... $x - 2 = \log 21$
>
> Rearrange .... $x = 2 + \log 21$

You might have to rearrange an equation first.

> **WORKED EXAMPLE 1.25**
>
> Solve $e^{3x+2} = 4e$, giving your answer correct to three significant figures.
>
> This equation is not of the form $a = b^x$ so you cannot just introduce logarithms immediately. Instead, aim to express everything as a power of e .... $e^{3x+2} = 4e^x$
>
> $e^{3x+2} = e^{\ln 4} e^x$
>
> Use the $a^m a^n = a^{m+n}$ on the RHS .... $e^{3x+2} = e^{\ln 4 + x}$
>
> Since the bases are the same, the exponents must be equal as well .... $3x + 2 = \ln 4 + x$
>
> $2x = \ln 4 - 2$
>
> $x = \dfrac{\ln 4 - 2}{2}$
>
> $= -0.307$ (3 s.f.)

Tip

There are many possible approaches to answering the question in Worked Example 1.25. For example, you could also have divided both sides by e^x and then taken natural logs of both sides.

Exercise 1C

For questions 1 to 4, without using a calculator, find the exact value of each of the given logarithms, using the method demonstrated in Worked Example 1.17.

1. a $\log 10$
   b $\log 100$
2. a $\log 100\,000$
   b $\log 1\,000\,000$
3. a $\log 1$
   b $\log 0.1$
4. a $\log 0.01$
   b $\log 0.0001$

For questions 5 to 8, without using a calculator, find the exact value of each of the given logarithms, using the method demonstrated in Worked Example 1.17. Note that these logs are to a base other than 10, but practising these will enhance your understanding.

5. a $\log_2 2$
   b $\log_2 4$
6. a $\log_3 27$
   b $\log_3 81$
7. a $\log_5 1$
   b $\log_5 0.2$
8. a $\log_2 0.25$
   b $\log_2 0.125$

For questions 9 to 11, without using a calculator, find the exact value of x in each of the given logarithms, using the method demonstrated in Worked Example 1.18.

9. a $\log x = 2$
   b $\log x = 3$
10. a $\log(2x - 4) = 5$
    b $\log(3x + 4) = 3$
11. a $\log(2x - 3) = -1$
    b $\log(x + 2) = -2$

For questions 12 to 14, rearrange each equation to find an expression for x, using the method demonstrated in Worked Example 1.19.

12. a $\ln x = 2$
    b $\ln x = 5$
13. a $\ln x = y + 1$
    b $\ln x = y^2$
14. a $\ln(2x + 4) = y - 3$
    b $\ln\left(\frac{1}{2}x - 6\right) = 2y + 1$

For questions 15 to 17, simplify each expression using the method demonstrated in Worked Example 1.20.

15. a $\log x + 4\log x$
    b $10\log x - 5\log x$
16. a $\dfrac{2(\log 3x)^2}{\log 3x}$
    b $\dfrac{\log 2x}{(\log 2x)^2}$
17. a $\dfrac{\ln x + (\ln x)^2}{\ln x} - \ln x$
    b $\dfrac{\ln x - (\ln x)^3}{(\ln x)^2} + \ln x$

For questions 18 to 25, simplify each expression using the method demonstrated in Worked Example 1.21 and Key Point 1.5.

18. a $\log 10^{-15}$
    b $\log 10^{17}$
19. a $\ln e^{4.5}$
    b $\ln e^{-1.5}$
20. a $10^{\log 13}$
    b $10^{\log 7}$
21. a $e^{\ln 3}$
    b $e^{\ln 7}$
22. a $10^{3\log 2}$
    b $10^{2\log 3}$
23. a $e^{2\ln 5}$
    b $e^{4\ln 3}$
24. a $10^{-\log 3}$
    b $10^{-\log 6}$
25. a $e^{-5\ln 2}$
    b $e^{-3\ln 4}$

For questions 26 to 28, use the method demonstrated in Worked Example 1.22 (that is, use technology) to evaluate the following to three significant figures.

26. a $\log 124.7$
    b $\log 1399.8$
27. a $\ln 245.3$
    b $\ln 17.9$
28. a $\log 0.5$
    b $\log 0.04$

For questions 29 to 31, use the method demonstrated in Worked Example 1.23 to solve the given equations, giving an exact answer.

29. a $10^x = 5$
    b $10^x = 7$
30. a $10^x = 0.2$
    b $10^x = 0.06$
31. a $e^x = 3$
    b $e^x = 7$

For questions 32 to 34, use the method demonstrated in Worked Example 1.24 to solve the given equations to find x, giving your answer in an exact form.

32. a $10^{x-2} = 7$
    b $10^{x+4} = 13$
33. a $10^{x-2} = 70$
    b $10^{x+4} = 1300$
34. a $e^{x+2} = k - 2$
    b $e^{x+2} = 2k + 1$

1C Logarithms

For questions 35 and 36, use the method demonstrated in Worked Example 1.25 to solve the given equations, giving your answers correct to three significant figures.

35 a $10^x = 5 \times 10^{-x}$
  b $10^{2x+1} = 4 \times 10^x$

36 a $e^{2x} = 6e^{2-x}$
  b $e^{3x-1} = 4e^{1-x}$

37 Find the exact solution of $1 + 2\log x = 9$.

38 Find the exact solution of $\log(3x+4) = 3$.

39 Simplify $\ln(e^a e^b)$.

40 Use technology to solve $10^x = 5$.

41 Use technology to solve $3 \times 10^x = 20$.

42 Use technology to solve $2 \times 10^x + 6 = 20$.

43 Find the exact solution of $5 \times 20^x = 8 \times 2^x$.

44 One formula for the pH of a solution is given by $pH = -\log[H^+]$ where $[H^+]$ is the concentration of H^+ ions in moles per litre.
  a A solution contains 2.5×10^{-8} moles per litre of H^+ ions. Find the pH of this solution.
  b A solution of hydrochloric acid has a pH of 1.9. Find the concentration of H^+ ions.

45 The radioactivity (R) of a substance after a time t days is modelled by $R = 10 \times e^{-0.1t}$.
  a Find the initial ($t = 0$) radioactivity.
  b Find the time taken for the radioactivity to fall to half of its original value.

46 The population of bacteria (B) at time t hours after being added to an agar dish is modelled by $B = 1000 \times e^{0.1t}$
  a Find the number of bacteria
    i initially
    ii after 2 hours.
  b Find an expression for the time it takes to reach 3000. Use technology to evaluate this expression.

47 The population of penguins (P) after t years on two islands is modelled by:
First island: $P = 200 \times e^{0.1t}$
Second island: $P = 100 \times e^{0.25t}$.
How many years are required before the population of penguins on both islands is equal?

48 The decibel scale measures the loudness of sound. It has the formula $L = 10\log(10^{12} I)$ where L is the noise level in decibels and I is the sound intensity in watts per metre squared.
  a The sound intensity inside a car is 5×10^{-7} watts per metre squared. Find the noise level in the car.
  b The sound intensity in a factory is 5×10^{-6} watts per metre squared. Find the noise level in the factory.
  c What is the effect on the noise level of multiplying the sound intensity by 10?
  d Any noise level above 90 decibels is considered dangerous to a human ear. What sound intensity does this correspond to?

49 a Write 20 in the form e^k.
  b If $20^x = 7$ find an exact expression for x in terms of natural logarithms.

50 Solve the simultaneous equations
$\log(xy) = 3$ and $\log\left(\dfrac{x}{y}\right) = 1$

51 Evaluate $\log 10x - \log x$ where x is a positive number.

Checklist

- You should know how to use the laws of exponents with integer exponents:
  - ☐ $a^m \times a^n = a^{m+n}$
  - ☐ $(a^m)^n = a^{mn}$
  - ☐ $(ab)^n = a^n \times b^n$
  - ☐ $\dfrac{a^m}{a^n} = a^{m-n}$
  - ☐ $a^{-n} = \dfrac{1}{a^n}$
  - ☐ $\left(\dfrac{a}{b}\right)^n = \dfrac{a^n}{b^n}$

- You should be able to apply these rules of exponents to numbers in standard form.

- You should know the definition of a logarithm and be able to work with logarithms including those with base e:
  - ☐ $a = b^x$ is equivalent to $\log_b a = x$

- For base 10:
  - ☐ $\log 10^x = x$
  - ☐ $10^{\log x} = x$

- For base e:
  - ☐ $\ln e^x = x$
  - ☐ $e^{\ln x} = x$

- You should be able to use logarithms to solve simple exponential equations.

■ Mixed Practice

1 A rectangle is 2680 cm long and 1970 cm wide.
  a Find the perimeter of the rectangle, giving your answer in the form $a \times 10^k$, where $1 \leq a < 10$ and $k \in \mathbb{Z}$.
  b Find the area of the rectangle, giving your answer correct to the nearest thousand square centimetres.

Mathematical Studies SL May 2009 TZ2 Paper 1 Q1

2 Simplify $\dfrac{(3xy^2)^2}{(xy)^3}$.

3 Write $(3x^2 y^{-3})^{-2}$ without brackets or negative indices.

4 Zipf's law in geography is a model for how the population of a city (P) in a country depends on the rank of that city (R), that is, whether it is the largest ($R = 1$), second largest ($R = 2$), and so on. The suggested formula is $P = kR^{-1}$.
In a particular country the second largest city has a population of 2 000 000.
  a Find the value of k.
  b What does the model predict to be the size of the fourth largest city?
  c What does the model predict is the rank of the city with population 250 000?

5 Find the exact solution of $8^x = 2^{x+6}$.

6 Show that if $a = 3 \times 10^8$ and $b = 4 \times 10^4$ then $ab = 1.2 \times 10^{13}$.

7 Show that if $a = 1 \times 10^9$ and $b = 5 \times 10^{-4}$ then $\dfrac{a}{b} = 2 \times 10^{12}$.

8 Show that if $a = 3 \times 10^4$ and $b = 5 \times 10^5$ then $b - a = 4.7 \times 10^5$.

9 The speed of light is approximately $3 \times 10^8 \,\text{m s}^{-1}$. The distance from the Sun to the Earth is 1.5×10^{11} m. Find the time taken for light from the Sun to reach the Earth.

10 Find the exact solution of $\log(x + 1) = 2$.

11 Find the exact solution to $\ln(2x) = 3$.

12 Use technology to solve $e^x = 2$.

13 Use technology to solve $5 \times 10^x = 17$.

Mixed Practice 21

14 Rearrange to make x the subject of $5e^x - 1 = y$.

15 a Given that $2^m = 8$ and $2^n = 16$, write down the value of m and of n.
b Hence or otherwise solve $8^{2x+1} = 16^{2x-3}$.

Mathematics SL May 2015 TZ1 Paper 1 Q3

16 You are given that $(7 \times 10^a) \times (4 \times 10^b) = c \times 10^d$
where $1 \leq c < 10$ and $d \in \mathbb{Z}$.
a Find the value of c.
b Find an expression for d in terms of a and b.

17 You are given that $(6 \times 10^a) \div (5 \times 10^b) = c \times 10^d$
where $1 \leq c < 10$ and $d \in \mathbb{Z}$.
a Find the value of c.
b Find an expression for d in terms of a and b.

18 The Henderson–Hasselbach equation predicts that the pH of blood is given by:

$$\text{pH} = 6.1 + \log\left(\frac{[\text{HCO}_3^-]}{[\text{H}_2\text{CO}_3]}\right)$$

where $[\text{HCO}_3^-]$ is the concentration of bicarbonate ions and $[\text{H}_2\text{CO}_3]$ is the concentration of carbonic acid (created by dissolved carbon dioxide). Given that the bicarbonate ion concentration is maintained at 0.579 moles per litre, find the range of concentrations of carbonic acid that will maintain blood pH at normal levels (which are between 7.35 and 7.45).

19 In attempting to set a new record, skydiver Felix Baumgartner jumped from close to the edge of the Earth's atmosphere. His predicted speed (v) in metres per second at a time t seconds after he jumped was modelled by: $v = 1350(1 - e^{-0.007t})$.
a Find the predicted speed after one second.
b Baumgartner's aim was to break the speed of sound (300 ms^{-1}). Given that he was in free fall for 600 seconds, did he reach the speed of sound? Justify your answer.

20 The Richter scale measures the strength of earthquakes. The strength (S) is given by $S = \log A$, where A is the amplitude of the wave measured on a seismograph in micrometres.
a If the amplitude of the wave is 1000 micrometres, find the strength of the earthquake.
b If the amplitude on the seismograph multiplies by 10, what is the effect on the strength of the earthquake?
c The 1960 earthquake in Chile had a magnitude of 9.5 on the Richter scale. Find the amplitude of the seismograph reading for this earthquake.

21 Solve $3 \times 20^x = 2^{x+1}$.

22 Solve the simultaneous equations:
$$9^x \times 3^y = 1$$
$$\frac{4^x}{2^y} = 16$$

23 Solve the simultaneous equations:
$$\log(xy) = 0$$
$$\log\left(\frac{x^2}{y}\right) = 3$$

2 Core: Sequences

ESSENTIAL UNDERSTANDINGS

- Number and algebra allow us to represent patterns, show equivalences and make generalizations which enable us to model real-world situations.
- Algebra is an abstraction of numerical concepts and employs variables to solve mathematical problems.

In this chapter you will learn…
- how to use the formula for the nth term of an arithmetic sequence
- how to use the formula for the sum of the first n terms of an arithmetic sequence
- how to use the formula for the nth term of a geometric sequence
- how to use the formula for the sum of the first n terms of a geometric sequence
- how to work with sigma notation for sums of arithmetic and geometric sequences
- how to apply arithmetic and geometric sequences to real-world problems.

CONCEPTS

The following key concepts will be addressed in this chapter:
- **Modelling** real-life situations with the structure of arithmetic and geometric sequences and series allows for prediction, analysis and interpretation.
- Formulae are a **generalization** made on the basis of specific examples, which can then be extended to new examples.
- Mathematical financial models such as compounded growth allow computation, evaluation and interpretation of debt and investment both **approximately** and accurately.

■ Figure 2.1 The early development of a fertilized egg.

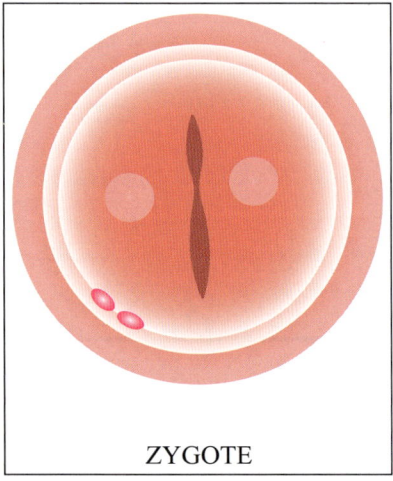
ZYGOTE

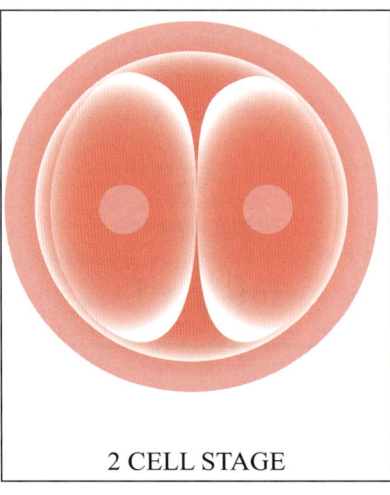
2 CELL STAGE

4 CELL STAGE

Core: Sequences

> **PRIOR KNOWLEDGE**
>
> Before starting this chapter, you should already be able to complete the following:
>
> 1. Solve the simultaneous equations
> $2x + 5y = 3$
> $3x - y = 13$
> 2. Evaluate the following:
>   a 5×2^4 b $4 \times (-3)^2$
> 3. Find all solutions to $x^4 = 16$

One area of overlap between mathematics, science and art is patterns. While these might seem purely aesthetic in art, some artists have made use of mathematics in their drawings and sculptures, from the ancient Greeks to Leonardo da Vinci, to in modern times the graphic artist Escher.

In science, patterns appearing in experimental data have led to advances in theoretical understanding.

In mathematics, an understanding of simple relationships can help in modelling more complex patterns in the future.

Starter Activity

Look at Figure 2.1. In small groups discuss whether these images represent doubling or halving.

Now look at this problem:

In early 2018, a Canadian woman won the jackpot on the lottery after buying a scratch card for the first time on her 18th birthday. She was given the choice of taking a C$ 1 million lump sum or receiving C$ 50 000 a year for life – both options would be tax free.

Which option would you choose in this situation?

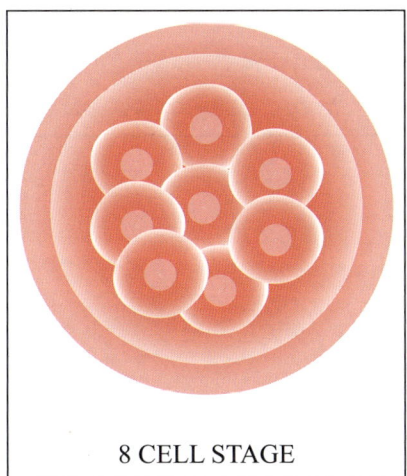

8 CELL STAGE

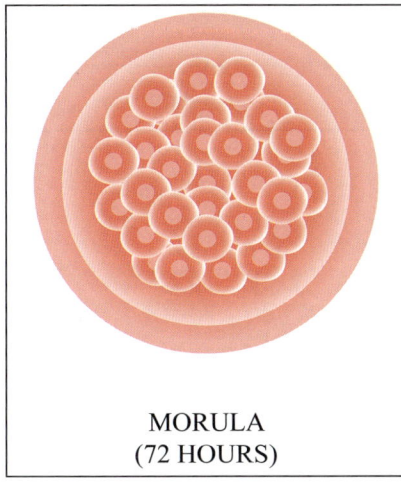

MORULA
(72 HOURS)

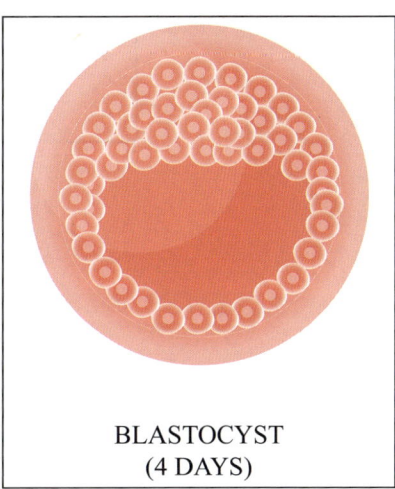

BLASTOCYST
(4 DAYS)

2A Arithmetic sequences and series

■ Use of the formula for the nth term of an arithmetic sequence

An **arithmetic sequence** is formed by adding or subtracting the same number to get to the next term. For example:

1, 4, 7, 10, 13, ...   has a common difference of + 3

19, 15, 11, 7, 3, ...   has a common difference of −4

In general, if the common difference is d (which can be negative), then

1st term	2nd term	3rd term	4th term	5th term
u_1	$u_1 + d$	$u_1 + 2d$	$u_1 + 3d$	$u_1 + 4d$

From this, we can suggest a formula for the nth term.

Tip

The standard notation for sequences is to use u_1 for the first term, u_2 for the second term and so on. The nth term is therefore u_n.

> **KEY POINT 2.1**
>
> For an arithmetic sequence with common difference d,
> $$u_n = u_1 + (n-1)d.$$

> **CONCEPTS – GENERALIZATION**
>
> This formula has not been formally proved – that would require a method called mathematical induction. However, do you think it is a convincing **generalization** of the pattern? Sometimes systematic listing in the correct way allows easy generalization. Would it have been more or less obvious if it had been written as $u_n = u_1 + nd - d$?

> **WORKED EXAMPLE 2.1**
>
> An arithmetic sequence has first term 7 and common difference 8.
>
> Find the 10th term.
>
> Use $u_n = u_1 + (n-1)d$ .... $u_n = 7 + 8(n-1)$
>   with $u = 7$ and $d = 8$
>   So,
> Let $n = 10$ .... $u_{10} = 7 + 8(10 - 1) = 79$

2A Arithmetic sequences and series

> **WORKED EXAMPLE 2.2**
>
> The third term of an arithmetic sequence is 10 and the 15th term is 46.
>
> Find the first term and the common difference.
>
> Use $u_n = u_1 + (n-1)d$ to write the first bit of information as an equation
>
> $u_3 = 10$
> Therefore, $u_1 + 2d = 10$
>
> And likewise with the second bit of information
>
> $u_{15} = 46$
> Therefore, $u_1 + 14d = 46$
>
> Now solve simultaneously with your GDC
>
> From the GDC,
> $u_1 = 4, d = 3$
>
>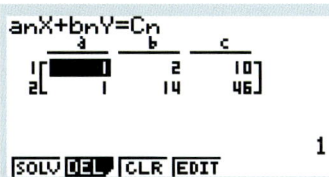

> **WORKED EXAMPLE 2.3**
>
> Find the number of terms in the arithmetic sequence 1, 4, 7,... , 100.
>
> This is an arithmetic sequence with first term 1 and common difference 3
> $u_1 = 1, d = 3$
>
> Find the formula for the nth term
> So, $u_n = 1 + 3(n-1)$
>
> Set this equal to 100 and solve for n
> $1 + 3(n-1) = 100$
> $3(n-1) = 99$
> $n - 1 = 33$
> $n = 34$

■ Use of the formula for the sum of the first n terms of an arithmetic sequence

It is also useful to have a formula for the sum of the first n terms of an arithmetic sequence. This sum is sometimes called an **arithmetic series**. There are two versions of this.

> **Tip**
>
> Just as u_1 is the first term, u_2 the second term and so on, S_2 is the sum of the first two terms, S_3 the sum the first three terms and so on. The sum of the first n terms is therefore S_n.

> **KEY POINT 2.2**
>
> For an arithmetic sequence with common difference d,
> - $S_n = \dfrac{n}{2}[2u_1 + (n-1)d]$
>
> or
> - $S_n = \dfrac{n}{2}(u_1 + u_n)$

The second formula in Key Point 2.2 follows directly from the first:

$$S_n = \frac{n}{2}[2u_1 + (n-1)d]$$
$$= \frac{n}{2}[u_1 + u_1 + (n-1)d]$$
$$= \frac{n}{2}(u_1 + u_n)$$

Proof 2.1

Prove that for an arithmetic sequence with common difference d, $S_n = \frac{n}{2}[2u_1 + (n-1)d]$.

Write out the first couple of terms and the last couple
$$S_n = a + [a+d] + \ldots + [a+(n-2)d] + [a+(n-1)d]$$

Write down the sum again but in the opposite order
$$S_n = [a+(n-1)d] + [a+(n-2)d] + \ldots + [a+d] + a$$

Now adding the two expressions gives n identical terms
$$2S_n = [2a+(n-1)d] + [2a+(n-1)d] + \ldots + [2a+(n-1)d] + [2a+(n-1)d]$$

$$2S_n = n[2a+(n-1)d]$$
$$S_n = \frac{n}{2}[2a+(n-1)d]$$

Carl Friedrich Gauss (1777–1855) was one of the most renowned mathematicians of the nineteenth century. There is a famous story of how, when he was about nine years old, his teacher wanted to keep the class quiet and so asked them to add up all the integers from 1 to 100. Within seconds Gauss had replied with the correct answer, much to the annoyance of his teacher! It is thought he had used a very similar method to the above proof to arrive at the answer.

WORKED EXAMPLE 2.4

An arithmetic sequence has first term 5 and common difference −2.

Find the sum of the first 10 terms.

$$u_1 = 5, d = -2$$

Use $S_n = \frac{n}{2}[2u_1 + (n-1)d]$
$$S_{10} = \frac{10}{2}[2 \times 5 + (10-1) \times (-2)]$$
$$= -40$$

2A Arithmetic sequences and series

WORKED EXAMPLE 2.5

An arithmetic sequence has first term 10 and last term 1000. If there are 30 terms, find the sum of all the terms.

$$u_1 = 10, \ u_n = 1000$$

Use $S_n = \dfrac{n}{2}[u_1 + u_n]$

$$S_{30} = \dfrac{30}{2}[10 + 1000]$$
$$= 15\,150$$

■ Use of sigma notation for sums of arithmetic sequences

Instead of writing out the terms of a sequence in a sum, you will often see this expressed in a shorthand form using sigma notation.

KEY POINT 2.3

● $\displaystyle\sum_{r=1}^{r=n} u_r = u_1 + u_2 + u_3 + \ldots + u_n$

The value of r at the bottom of the sigma (here $r = 1$) shows where the counting starts.

The value of r at the top of the sigma (here $r = n$) shows where the counting stops.

Tip

There is nothing special about the letter r here – any letter could be used.

WORKED EXAMPLE 2.6

Evaluate

$$\sum_{r=3}^{10}(5r + 2).$$

Substitute the first few values of r into the formula: $r = 3, r = 4, r = 5$

$$\sum_{r=3}^{10}(5r + 2) = (5 \times 3 + 2) + (5 \times 4 + 2) + (5 \times 5 + 2) + \ldots$$

This is the sum of an arithmetic sequence with $u_1 = 17$, $d = 5$ and $n = 8$

$$= 17 + 22 + 27 + \ldots$$

Use $S_n = \dfrac{n}{2}[2u_1 + (n-1)d]$

$$= \dfrac{8}{2}[2 \times 17 + (8-1)5]$$
$$= 276$$

Tip

Be careful! In Worked Example 2.6 it is easy to think that there are seven terms in the sequence but there are actually eight ($u_3, u_4, u_5, u_6, u_7, u_8, u_9$ and u_{10}).

Although you will be expected to identify the kind of sum in Worked Example 2.6 as being arithmetic and then show how you have found the value using the formula for S_n, it is possible to do this directly on your calculator.

```
Σ(5R+2,R,3,10)
                    276



FMin FMax Σ( logab       ▷
```

If you know the sum of the terms of an arithmetic sequence you can find the number of terms using technology.

WORKED EXAMPLE 2.7

The sum of the first n terms of the sequence 3, 7, 11, 15, ... is 465. Find n.

This is an arithmetic sequence with first term 3 and common difference 4 .... $u_1 = 3, d = 4$

So, $S_n = \frac{n}{2}[2 \times 3 + (n-1)4]$

Find and (if you want to) simplify the formula for the sum of the first n terms .... $= \frac{n}{2}[6 + 4n - 4]$

$= \frac{n}{2}[2 + 4n]$

$= n + 2n^2$

The sum of the first n terms is 465 .... $S_n = 465$

Therefore, $2n^2 + n = 465$

Use the table function on your GDC to search for the smallest positive value of n that gives 465 .... From GDC, $n = 15$

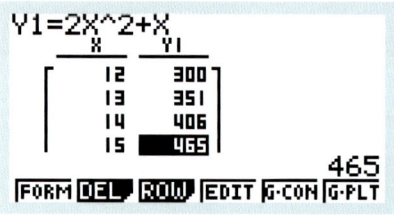

Applications of arithmetic sequences

Arithmetic sequences can be used to model any process where something tends to change by a constant amount. In particular, you should know that **simple interest** means that the same amount is added into an account each year.

WORKED EXAMPLE 2.8

A savings account pays 10% simple interest each year. If Rashid invests €3000 in this account, how much will he have after 5 years?

10% of 3000 = 300, so 300 will be added on each year. It is a good idea to write out the first couple of terms to see what is happening

The amount in the account after
1 year: 3000 + 300 = 3300
2 years: 3300 + 300 = 3600

This is an arithmetic sequence with $u_1 = 3300$ and $d = 300$.

Use $u_n = u_1 + (n-1)d$ .... $u_5 = 3300 + 4 \times 300$

$= 4500$

2A Arithmetic sequences and series

Be the Examiner 2.1

Ben starts a new job on 1 January 2020 earning £20 000. His salary will go up by £750 on 1 January each year. What will he earn in 2030?

Which is the correct solution? Identify the errors made in the incorrect solutions.

Solution 1	Solution 2	Solution 3
$u_{10} = 20000 + (10-1)750$ $= 26750$ He will earn £26 750.	$S_{10} = \frac{10}{2}[2 \times 20000 + (10-1)750]$ $= 233750$ He will earn £233 750.	$u_{11} = 20000 + (11-1)750$ $= 27500$ He will earn £27 500.

■ Analysis, interpretation and prediction where a model is not perfectly arithmetic in real life

If the difference between data terms isn't exactly the same, an arithmetic sequence can still be used to model the process.

WORKED EXAMPLE 2.9

The data below show the number of schools taking the IB diploma in different years:

2013	2014	2015	2016	2017
2155	2211	2310	2487	2667

Modelling the data as an arithmetic sequence, predict the number of schools taking the IB in 2023.

The difference between years is not constant, so all we can do is take the average difference between the first and last given terms ⋯⋯⋯⋯ $d = \frac{2667 - 2155}{4} = 128$

Use $u_n = u_1 + (n-1)d$, with $u_1 = 2155$ and $d = 128$. Note that 2023 is the 11th term of the sequence ⋯⋯⋯⋯ $u_{11} = 2155 - (11-1)128$
$= 3435$

In this example we assumed that the arithmetic sequence had to be fitted to the first (and last) terms in the given data. You will see in Chapter 6 that linear regression provides a better way of modelling the arithmetic sequence that does not require this assumption.

CONCEPTS – MODELLING

What are the underlying assumptions in applying an arithmetic **model** to this data? How valid do you think these assumptions are? Do you think this model is more useful for predicting future behaviour or analysing past behaviour to identify anomalous years?

TOK Links

Is all knowledge concerned with the identification and use of patterns? One sequence that you may be interested in researching further is the Fibonacci sequence, in which each term is the sum of the two preceding terms: 1, 1, 2, 3, 5, 8, 13, 21… This sequence of numbers, and the related golden ratio, can be seen occurring naturally in biological contexts, as well as in human endeavours such as art and architecture. You might like to investigate how the photographs on the covers of all four Hodder Education books for IB Diploma Mathematics exhibit this sequence of numbers.

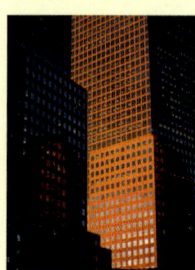

Exercise 2A

For questions 1 and 2, use the method demonstrated in Worked Example 2.1. to find the required term of the arithmetic sequence.

1. a $u_1 = 5$, $d = 4$. Find u_{12}.
   b $u_1 = 11$, $d = 3$. Find u_{20}.
2. a $u_1 = 13$, $d = -7$. Find u_6.
   b $u_1 = 11$, $d = -3$. Find u_{18}.

For questions 3 to 5, use the method demonstrated in Worked Example 2.2 to find the required feature of the arithmetic sequence.

3. a $u_1 = 7$, $u_5 = 15$. Find d.
   b $u_1 = 19$, $u_9 = 75$. Find d.
4. a $u_1 = 10$, $u_{21} = 0$. Find d.
   b $u_1 = 100$, $u_9 = 80$. Find d.
5. a $u_8 = 12$, $u_{11} = 33$. Find u_1 and d.
   b $u_7 = 11$, $u_{12} = 56$. Find u_1 and d.

For questions 6 to 8, use the method demonstrated in Worked Example 2.1 and Worked Example 2.2 to find a formula for the nth term of the arithmetic sequence described.

6. a $u_1 = 8$, $d = -3$
   b $u_1 = 10$, $d = 4$
7. a $u_1 = 3$, $u_5 = 11$
   b $u_1 = 10$, $u_{11} = 50$
8. a $u_2 = 6$, $u_5 = 21$
   b $u_3 = 14$, $u_6 = -15$

For questions 9 to 10, use the method demonstrated in Worked Example 2.3 to find the number of terms in each of the given arithmetic sequences.

9. a 2, 8, 14,…, 200
   b 5, 9, 13, …, 97
10. a 13, 7, 1, …, –59
    b 95, 88, 81, …, –101

For questions 11 to 13, use the method demonstrated in Worked Example 2.4 (and Worked Example 2.2) to find the given sum.

11. a $u_1 = 9$, $d = 4$. Find S_{12}.
    b $u_1 = 17$, $d = 2$. Find S_{20}.
12. a $u_1 = 34$, $d = -7$. Find S_{11}.
    b $u_1 = 9$, $d = -11$. Find S_8.
13. a $u_1 = 7$, $u_6 = 32$. Find S_{11}.
    b $u_1 = 9$, $u_5 = 41$. Find S_{12}.

For questions 14 and 15, use the method demonstrated in Worked Example 2.5 to find the given sum.

14. a $u_1 = 9$, $u_8 = 37$ Find S_8.
    b $u_1 = 4$, $u_{12} = 61$ Find S_{12}.
15. a $u_1 = 27$, $u_{10} = -36$ Find S_{10}.
    b $u_1 = 50$, $u_{200} = -50$ Find S_{200}.

For questions 16 to 18, use the method demonstrated in Worked Example 2.6 to evaluate the given sum.

16. a $\displaystyle\sum_{r=1}^{8} 4$
    b $\displaystyle\sum_{r=1}^{11} 9$
17. a $\displaystyle\sum_{r=1}^{7} (2r + 7)$
    b $\displaystyle\sum_{r=1}^{10} (3r + 4)$
18. a $\displaystyle\sum_{r=5}^{17} (6r - 4)$
    b $\displaystyle\sum_{r=3}^{8} (12r - 7)$

For questions 19 and 20, use the method demonstrated in Worked Example 2.7 to find the number of terms in the arithmetic sequence.

19   a   4, 9, 14... $S_n = 442$
     b   8, 9, 10... $S_n = 125$

20   a $u_1 = 20, d = -1, S_n = 200$
     b $u_1 = 100, d = -5, S_n = 1000$

21   An arithmetic sequence has first term 7 and common difference 11.
     a   Find the 20th term of the sequence.
     b   Find the sum of the first 20 terms.

22   The first term of an arithmetic sequence is 3 and the second term is 7.
     a   Write down the common difference.
     b   Find the eighth term of the sequence.
     c   Find the sum of the first 15 terms.

23   In an arithmetic sequence, the second term is 13 and the common difference is 5.
     a   Write down the first term.
     b   Find the sum of the first 10 terms.

24   The first term of an arithmetic sequence is -8 and the 16th term is 67.
     a   Find the common difference.
     b   Find the 25th term.

25   Sam invests £300 at 4% simple interest. She does not withdraw any money.
     a   How much does she have in her account at the end of the first year?
     b   How much does she have in her account at the end of the 10th year?

26   On Daniel's first birthday, his grandparents started a saving account with £100. On each of his subsequent birthdays they put £10 in the account. What is the total amount they have saved for Daniel after they have deposited money on his 21st birthday? You may ignore any interest accrued.

27   The height of each step in a stairway follows an arithmetic sequence. The first step is 10 cm off the ground and each subsequence step is 20 cm higher. If the staircase is 270 cm high, how many steps does it have?

28   An arithmetic sequence has first term 11 and last term 75.
     a   Given the common difference is 8, find the sum of the terms.
     b   Given the common difference is 4, find the sum of the terms.

29   In an arithmetic sequence, the 10th term is 26 and the 30th term is 83. Find the 50th term.

30   In an arithmetic sequence, the first term is 8, the common difference is 3 and the last term is 68.
     a   How many terms are there in this sequence?
     b   Find the sum of all the terms of the sequence.

31   The 7th term of an arithmetic sequence is 35 and the 18th term is 112.
     a   Find the common difference and the first term of the sequence.
     b   Find the sum of the first 18 terms.

32   An arithmetic sequence has $u_{10} = 16$ and $u_{30} = 156$.
     a   Find the value of u_{50}.
     b   Find $\sum_{r=1}^{20} u_r$.

33   Find $\sum_{r=1}^{16} (5r - 3)$.

34   Three consecutive terms of an arithmetic sequence are $3x + 1, 2x + 1, 4x - 5$. Find the value of x.

35   The fifth term of an arithmetic sequence is double the second term, and the seventh term of the sequence is 28. Find the 11th term.

36   The sum of the first three terms of an arithmetic sequence equal the tenth term, and the seventh term of the sequence is 27. Find the 12th term.

37 A sequence is defined by $u_{n+1} = u_n + 4$. If $u_1 = 10$ find $\sum_{r=1}^{20} u_r$.

38 **a** The arithmetic sequence U has first term 7, common difference 12 and last term 139. Find the sum of terms in U.

   b The arithmetic sequence V has the same first and last terms as sequence U but has common difference 6. Find the sum of terms in V.

   c The arithmetic sequence W has the same first term and common difference as V, and the same number of terms as U. Find the sum of terms in W.

39 Joe plays a game five times and scores 53, 94, 126, 170 and finally 211.

   Modelling the data as an arithmetic sequence, predict the score he might get on his 10th attempt, if his pattern continues consistently.

40 A survey of the number of ducks seen on a lake on the first four days of the inward migration period records 3, 12, 23 and then 33 ducks on the lake.

   Modelling the data as an arithmetic sequence, predict the number of ducks seen on the sixth day, if the pattern continues consistently.

41 The first three terms of an arithmetic sequence are 1, a, $3a + 5$. Evaluate the fourth term of the sequence.

42 In an effort to reduce his screen time, Stewart reduced his consumption by 5 minutes each day. On day one of the program he looks at a screen for 200 minutes.

   a Which is the first day on which his screen time is reduced to zero?

   b For how many minutes in total has Stewart looked at a screen during this program?

43 As part of a health program Theo tries to increase the number of steps he takes each day by 500 steps.

   On the first day he walks 1000 steps. Assume that he sticks to this plan.

   a How long will it take him to get to 10 000 steps in a day?

   b How long will it take him to complete a total of 540 000 steps.

44 The sum of the first n terms of an arithmetic sequence is denoted by S_n. Given that $S_n = 2n^2 + n$,

   a find the first two terms of the sequence

   b hence find the 50th term.

45 The sum of the first n terms of a sequence is given by $S_n = n^2 + 4n$. Find an expression for the nth term of the sequence.

46 The ninth term of an arithmetic sequence is four times as large than the third term. Find the value of the first term.

47 What is the sum of all three-digit multiples of 7?

48 What is the sum of all whole numbers from 0 up to (and including) 100 which are multiples of 5 but not multiples of 3.

49 A rope of length 4 m is cut into 10 sections whose lengths form an arithmetic sequence. The largest resulting part is 3 times larger than the smallest part. Find the size of the smallest part.

50 The first four terms of an arithmetic sequence are 4, a, b, $a - b$. Find the sixth term.

51 **a** A sequence has formula $u_n = a + nd$. Prove that there is a constant different between consecutive terms of the sequence.

   b A sequence has formula $u_n = an^2 + bn$. Prove that the differences between consecutive terms forms an arithmetic sequence.

52 Alessia writes down a list of the first n positive integers in order.

   a Show that by the time she has written down the number 19 she has written down 29 digits.

   b Her final list contains a total of 342 digits. What was the largest number in her list?

2B Geometric sequences and series

■ Use of the formula for the nth term of a geometric sequence

A **geometric sequence** has a common ratio between each term. For example,

3, 6, 12, 24, 48, ...   has a common ratio of 2

80, 40, 20, 10, 5, ...   has a common ratio of $\frac{1}{2}$

2, –6, 18, –54, 162, ...   has a common ratio of –3

In general, if the common ratio is r (which can be negative), then

1st term	2nd term	3rd term	4th term	5th term
u_1	$u_1 r$	$u_1 r^2$	$u_1 r^3$	$u_1 r^4$

From this, we can suggest a formula for the nth term.

> **KEY POINT 2.4**
>
> For a geometric sequence with common ratio r, $u_n = u_1 r^{n-1}$.

WORKED EXAMPLE 2.10

A geometric sequence has first term 3 and common ratio –2. Find the 10th term.

Use $u_n = u_1 r^{n-1}$ with $u_1 = 3$ and $r = -2$ .... $u_n = 3(-2)^{n-1}$

So, $u_{10} = 3(-2)^9 = -1536$

WORKED EXAMPLE 2.11

The second term of a geometric sequence is 6 and the fourth term is 96. Find the possible values of the first term and the common ratio.

Use $u_n = u_1 r^{n-1}$ to write the first bit of information as an equation .... $u_2 = 6$
Therefore, $u_1 r^1 = 6$

Do likewise with the second bit of information .... $u_4 = 96$
Therefore, $u_1 r^3 = 96$

These are non-linear simultaneous equations. We can solve them by dividing them .... Therefore $\dfrac{u_1 r^3}{u_1 r} = \dfrac{96}{6}$

So $r^2 = 16$

There are two numbers which square to make 16 .... $r = \pm 4$

Substitute each of these values into $u_1 r = 6$ .... If $r = 4$ then $u_1 = 1.5$
If $r = -4$ then $u_1 = -1.5$

You will see in Chapter 3 that these equations can also be solved directly on your GDC.

WORKED EXAMPLE 2.12

Find the number of terms in the geometric sequence 1, 2, 4, 8, ..., 512.

This is a geometric sequence with first term 1 and common ratio 2 .... $u_1 = 1, r = 2$

Find the formula for the nth term .... So, $u_n = 1 \times 2^{n-1}$

$u_n = 512$

The nth term is 512 .... Therefore, $2^{n-1} = 512$

Use the table function on your GDC to search for the value of n that gives 512 .... From GDC, $n = 10$

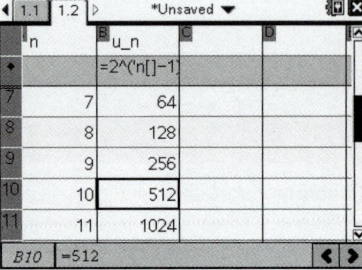

 Note that you could also solve the equation $2^{n-1} = 512$ using logs with the methods of Section 1C.

■ Use of the formula for the sum of the first n terms of a geometric sequence

Just as for arithmetic sequences, there is a formula for finding the sum of the first n terms of a geometric sequence. This sum is sometimes called a **geometric series**.

KEY POINT 2.5

For a geometric sequence with common ratio r:

- $S_n = \dfrac{u_1(1 - r^n)}{1 - r}, \quad r \neq 1$

or

- $S_n = \dfrac{u_1(r^n - 1)}{r - 1}, \quad r \neq 1$

Tip

The second formula in Key Point 2.5 follows from the first on multiplication of the numerator and denominator by −1.

2B Geometric sequences and series

> **Proof 2.2**
>
> Prove that for a geometric sequence with common ratio r,
>
> $$S_n = \frac{u_1(1-r^n)}{1-r}.$$
>
> Write out the first few terms and the last few $\quad S_n = u_1 + u_1 r + u_1 r^2 + \ldots + u_1 r^{n-2} + u_1 r^{n-1}$
>
> Multiply through by r $\quad rS_n = u_1 r + u_1 r^2 + \ldots + u_1 r^{n-2} + u_1 r^{n-1} + u_1 r^n$
>
> Subtract to remove all the terms in common $\quad S_n - rS_n = u_1 - u_1 r^n$
>
> Factorize both sides $\quad S_n(1-r) = u_1(1-r^n)$
>
> $$S_n = \frac{u_1(1-r^n)}{1-r}$$

At what point in the proof above do you use the fact that $r \neq 1$? Can you find a formula for the sum of the first n terms when $r = 1$?

WORKED EXAMPLE 2.13

A geometric sequence has first term 16 and common ratio -0.5.

Find the sum of the first eight terms to three significant figures.

$$u_1 = 16, r = -0.5$$

Use $S_n = \frac{u_1(1-r^n)}{1-r}$

$$S_8 = \frac{16(1-(-0.5)^8)}{1-(-0.5)}$$

$$= 10.6 \text{ (3 s.f.)}$$

■ Use of sigma notation for sums of geometric sequences

WORKED EXAMPLE 2.14

Evaluate

$$\sum_{r=2}^{8} 2 \times 4^r.$$

Substitute the first few values of r into the formula: $r = 2, r = 3, r = 4$

$$\sum_{r=2}^{8} 2 \times 4^r = (2 \times 4^2) + (2 \times 4^3) + (2 \times 4^4) + \ldots$$

This is the sum of a geometric sequence with $u_1 = 32, r = 4$ and $n = 7$

$$= 32 + 128 + 512 + \ldots$$

Use $S_n = \frac{u_1(1-r^n)}{1-r}$

$$= \frac{32(1-4^7)}{1-4}$$

$$= 174752$$

As with sums of arithmetic sequences in Section 2A, you can check the calculation in Worked Example 2.14 on your GDC:

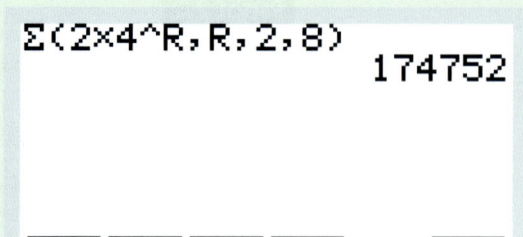

Applications of geometric sequences

Geometric sequences can be used to model any process where something tends to change by a constant factor.

Often this will be expressed in terms of a percentage change.

> **KEY POINT 2.6**
>
> An increase of $r\%$ is equivalent to multiplying by the factor $1 + \dfrac{r}{100}$.

EXAMPLE 2.15

A model predicts that the number of students taking the IB in a region will increase by 5% each year. If the number of IB students in the region is currently 12 000, predict the number taking the IB in 5 years' time.

A 5% increase corresponds to multiplication by 1.05. It is a good idea to write out the first couple of terms to see what is happening

Number of IB students in
1 year's time: $12\,000 \times 1.05$
2 years' time: $12\,000 \times 1.05^2$

This is a geometric sequence with
$u_1 = 12\,000 \times 1.05$ and $r = 1.05$

Use $u_n = u_1 r^{n-1}$

$u_5 = u_1 r^4$
$= (12\,000 \times 1.05) \times 1.05^4$
$\approx 15\,315$

You are the Researcher

A Farey sequence of order n is a list of reduced fractions in order of size between 0 and 1 which have a denominator of at most n. They have many algebraic and geometric properties and applications which you might like to research.

Exercise 2B

For questions 1 to 3, use the method demonstrated in Worked Example 2.10 to find the required term of the geometric sequence.

1. a $u_1 = 5, r = 4$. Find u_{12}.
   b $u_1 = 2, r = 3$. Find u_7.

2. a $u_1 = 5, r = -5$. Find u_6.
   b $u_1 = 11, r = -2$. Find u_{18}.

3. a $u_1 = 32, r = -\frac{1}{2}$. Find u_{11}.
   b $u_1 = 54, r = -\frac{1}{3}$. Find u_{10}.

For questions 4 to 6, use the method demonstrated in Worked Example 2.11 to find all possible values of the first term and common ratio for the following geometric sequences.

4. a $u_3 = 28, u_8 = 896$
   b $u_5 = 108, u_8 = 2916$

5. a $u_3 = 12, u_9 = 768$
   b $u_5 = 45, u_7 = 405$

6. a $u_2 = 6, u_6 = -96$
   b $u_8 = 56, u_{12} = 3.5$

For questions 7 and 8, use the method demonstrated in Worked Example 2.12 to find the number of terms in each of the geometric sequences described.

7. a $u_1 = 3, r = 5, u_n = 9375$
   b $u_1 = 17, r = 2, u_n = 2176$

8. a $3, 6, 12, \ldots, 12\,288$
   b $5, 15, 45, \ldots, 10\,935$

For questions 9 to 11, use the method demonstrated in Worked Example 2.13 to find the required sum of the geometric sequence described.

9. a $u_1 = 5, r = 3$. Find S_7.
   b $u_1 = 3, r = 4$. Find S_6.

10. a $u_1 = 96, r = 0.5$. Find S_7.
    b $u_1 = 162, r = \frac{1}{3}$. Find S_5.

11. a $u_1 = 192, r = -\frac{1}{4}$. Find S_5.
    b $u_1 = 216, r = -\frac{1}{3}$. Find S_6.

For questions 12 to 15, use the formula for the sum of terms from a geometric sequence and the method demonstrated in Worked Example 2.14 to evaluate the following expressions.

12. a $\sum_{r=1}^{5} 3^r$
    b $\sum_{r=1}^{4} 7^r$

13. a $\sum_{r=1}^{8} 2 \times 5^r$
    b $\sum_{r=1}^{9} 11 \times 3^r$

14. a $\sum_{r=3}^{8} 7 \times 3^r$
    b $\sum_{r=4}^{11} 8 \times 5^r$

15. a $\sum_{r=2}^{7} 72 \times \left(-\frac{1}{2}\right)^r$
    b $\sum_{r=6}^{10} 24\,057 \times \left(-\frac{1}{3}\right)^r$

16. A geometric sequence has first term 128 and common ratio 0.5.
    a Find the eighth term of the sequence.
    b Find the sum of the first eight terms.

17. The first term of a geometric sequence is 3 and the second term is 6.
    a Write down the common ratio.
    b Find the sixth term.
    c Find the sum of the first 10 terms.

18. In a geometric sequence the second term is 24 and the fifth term is 81.
    a Find the common ratio.
    b Find the value of the seventh term.

19. A student studies the amount of algae in a pond over time. She estimates the area of algae on the pond on day 1 of her study is $15\,\text{cm}^2$, and she predicts that the area will double every eight days.
    According to her projection, what area will the algae cover at the start of the ninth week?

20 A yeast culture is grown in a sugar solution. The concentration of sugar in the solution at the start of the culture is 1.2 mg ml⁻¹, and halves every 2 days. What is the concentration of the sugar solution after 12 days?

21 The first term of a geometric sequence is 8 and the sum of the first two terms is 12. Find the sum of the first five terms.

22 The time taken for a computer learning program to identify a face decreases by 20% every time it sees that face. The first time the face is observed it takes 5 seconds to identify it. How long does it take to identify a face on the 10th attempt?

23 During a drought, the volume of water in a reservoir decreases by 8% every day. At the beginning of day one of the drought it contains 5000 m³:

  a Find the volume at the start of the fifth day.

  b How many days does it take to use up 3000 m³ of water?

24 The fifth term of a geometric sequence is eight times larger than the second term. Find the ratio $\frac{S_8}{u_1}$.

25 According to a common legend, the game of chess was invented by Sissa ben Dahir of the court of King Shiram in India. The King was so impressed he asked Sissa to name his reward. Sissa asked for one grain of wheat on the first square of the chessboard, 2 on the second, 4 on the third and so on. The King thought that Sissa was being very foolish to ask for so little.

  a How many grains would be on the 64th square of the chessboard?

  b How many grains would be on the chessboard in total?

  c The annual worldwide production of wheat is approximately 7.5×10^{14} g. If one grain of wheat has a mass of 0.1 g, how many years would it take to produce enough wheat to satisfy Sissa's reward? Give your answer to the nearest 50 years.

TOK Links

How useful is your intuition as a way of knowing for dealing with very large numbers?
Does relating new knowledge to established knowledge to put big numbers in some more meaningful context (such as in part **c** of this question) improve your understanding?

26 Find an expression for the *n*th term of the geometric sequence $xy^2, y^3, x^{-1}y^4 \ldots$

27 If $u_1 = 3$ and $u_{n+1} = 2u_n$ find $\sum_{r=1}^{10} u_r$.

28 The first three terms of a geometric sequence are $1, x, 2x^2 + x$.

Find the value of the 10th term.

29 Evaluate $\sum_{r=1}^{10} \frac{6^r}{2^r}$.

30 A basketball is dropped vertically. The first bounce reaches a height of 0.6 m and subsequent bounces are 80% of the height of the previous bounce.

  a Find the height of the fifth bounce.

  b Find the total distance travelled by the ball from top of the first bounce to the top of the fifth bounce.

  c Suggest why this model is no longer accurate by the 20th bounce.

2C Financial applications of geometric sequences and series

■ Compound interest

One common application of geometric sequences is **compound interest.** This is calculated by applying a percentage increase to an initial sum, then applying the same percentage increase to the new sum and so on.

Interest rates are often quoted annually but then compounded over different periods – for example, monthly, quarterly or half yearly. The quoted annual interest rate is split equally amongst each of these periods.

Tip

Be warned – this means that the actual annual interest rate is not always the same as the quoted one when it is compounded over shorter time intervals.

> **WORKED EXAMPLE 2.16**
>
> A savings account has an annual interest rate of 3% compounded monthly.
>
> Find the amount in the account 18 months after $5000 is invested.
>
> The monthly interest rate will be $\frac{3}{12} = 0.25\%$
>
> 0.25% is equivalent to multiplying by $1 + \frac{0.25}{100} = 1.0025$
>
> After 18 months the amount in the account will be $5000 \times 1.0025^{18} = \$5229.85$

 This calculation can also be done on your GDC using the TVM (Time Value of Money) package.

The variables you need to enter are:

- n – the **number** of time periods (here 18 months)
- I% – the annual interest rate
- PV – the present value. Conventionally this is negative for an investment.
- FV is the future value, which is the answer in Worked Example 2.16. Additionally, if you enter this quantity, the calculator can be used to find one of the other values.
- P/Y – the number of payments per year (here 12 as each period is 1 month). Some calculators might allow you to specify separately P/Y as the number of payments made each year and C/Y as the number of compounding periods per year. You should always make both of these equal for all problems in the IB.

> **Tip**
>
> In the TVM package, depreciation corresponds to a negative interest rate.

■ Annual depreciation and inflation

An asset, such as a car, will tend to lose value or **depreciate**. The method of calculating the value of an asset after depreciating is exactly the same as for compound interest, except that the percentage change will be negative. For example, an annual depreciation rate of 15% would correspond to multiplication by $1 - \frac{15}{100} = 0.85$.

Over time, prices tend to rise. The average percentage increase in prices over a year is known as the **inflation rate**.

The effect of inflation is that the value of money decreases or depreciates over time. The value of money after being adjusted for inflation is known as the **value in real terms**. If nothing else is said then you should assume that this is relative to the initial time.

For example, if the inflation rate is 2% in a given year, then £100 at the beginning of the year will be worth, in real terms, £98 at the end of the year.

> ### KEY POINT 2.7
>
> The percentage change in real terms, $r\%$, is given by $r = c - i$, where $c\%$ is the given percentage change and $i\%$ is the inflation rate.

> ### CONCEPTS – APPROXIMATION
>
> In fact, the method in Key Point 2.7 for finding the value in real terms is just an **approximation** – which is very good for small percentage changes. The true value would be found by multiplication by $\dfrac{1 + \frac{c}{100}}{1 + \frac{i}{100}}$.

> ### WORKED EXAMPLE 2.17
>
> An investment account offers 2% annual interest. Jane invests £1000 in this account for 5 years. During those 5 years there is inflation of 2.5% per annum.
>
> What is the value in real terms of Jane's investment after those 5 years.
>
> | | |
> |---|---|
> | Use $r = c - i$ to find the annual real terms percentage change | The annual real terms percentage change is $2 - 2.5 = -0.5\%$ |
> | A decrease of 0.5% means multiplication by $1 - \frac{0.5}{100} = 0.995$ | So, after 5 years the value in real terms is $1000 \times 0.995^5 \approx £975$ |

> **Tip**
>
> This can also be done using the TVM package with an annual interest rate of −0.5%.

> ### CONCEPTS – APPROXIMATION
>
> In Worked Example 2.17, which quantities are exact and which are uncertain? Would it be appropriate to give the answer to two decimal places, as is often done when doing financial arithmetic? Compare the confidence you have in your model in Worked Examples 2.16 and 2.17.

Exercise 2C

For questions 1 to 3, use the TVM package on your calculator or the method demonstrated in Worked Example 2.16 to find the final value of the investment.

1. **a** $2000 invested at 6% compounded annually for 3 years.
   b $3000 invested at 4% compounded annually for 18 years.
2. **a** $500 invested at 2.5% compounded quarterly for 6 years.
   b $100 invested at 3.5% compounded twice-yearly for 10 years.
3. **a** $5000 invested at 5% compounded monthly for 60 months.
   b $800 invested at 4% compounded monthly for 100 months.

For questions 4 and 5, use the TVM package on your calculator to find how long is required to achieve the given final value.

4. **a** $100 invested at 5% compounded annually. £1000 required.
   b $500 invested at 4% compounded annually. £600 required.
5. **a** $300 invested at 2% compounded monthly. £400 required.
   b $1000 invested at 1.5% compounded monthly. £1100 required.

For questions 6 and 7, use the TVM package on your calculator to find the annual interest rate required to achieve the given final value. Assume that interest is compounded annually.

6. **a** €500 invested for 10 years. €1000 required.
   b €500 invested for 10 years. €800 required.
7. **a** €100 invested for 5 years. €200 required.
   b €100 invested for 10 years. €200 required.

For questions 8 and 9, use the TVM package on your calculator or the method demonstrated in Worked Example 2.16 to find the final value of an asset after the given depreciation. Assume this depreciation is compounded annually.

8. **a** Initial value $1200, annual depreciation 10% for 10 years.
   b Initial value $1200, annual depreciation 20% for 10 years.
9. **a** Initial value £30 000, annual depreciation 15% for 5 years.
   b Initial value £30 000, annual depreciation 25% for 5 years.

For questions 10 and 11, use the TVM package on your calculator or the method demonstrated in Worked Example 2.17 to find the real value of cash (that is, no interest paid) subject to the inflation described, relative to the initial value.

10. **a** $100 after 1 year of inflation at 3%.
    b $300 after 1 year of inflation at 2%.
11. **a** £1000 after 10 years of inflation at 3%.
    b £5000 after 10 years of inflation at 5%.

For questions 12 and 15, use the TVM package on your calculator or the method demonstrated in Worked Example 2.17 to find the real value of an investment subject to the inflation described. All interest is compounded annually.

12. **a** $100 after 1 year of interest at 5% and inflation at 2%.
    b $5000 after 1 year of interest at 4% and inflation at 1.4%.
13. **a** £100 after 1 year of interest at 4% and inflation at 10%.
    b £2500 after 1 year of interest at 3% and inflation at 5%.
14. **a** €1000 after 10 years of interest at 3.5% and inflation at 3%.
    b €500 after 5 years of interest at 5% and inflation at 2%.
15. **a** $1000 after 10 years of interest at 5% and inflation at 10%.
    b $5000 after 8 years of interest at 3% and inflation at 5%.

16. A savings account pays interest at 3% per year (compounded annually). If £800 is deposited in the account, what will the balance show at the end of 4 years?

17 A bank charges 5% annual interest on loans. If a seven-year loan of £10 000 is taken, with no repayments until the seven years end, how much will need to be paid at the end of the loan term?

18 A savings account has an annual interest rate of 4% compounded monthly. Find the account balance 30 months after $8000 is invested.

19 A bank loan has a published annual interest rate of 5.8%, compounded monthly. What is the balance 18 months after a loan of €15 000 is taken out, if no payments are made?

20 A car is bought for £20 000. Each year there is 15% depreciation. Find the expected value of the car after 5 years.

21 £1000 is invested at an annual interest rate of 6%.
   a Is the return better if the interest is compounded annually or monthly?
   b Over 10 years, what is the difference in the return between these two compounding methods?

22 A car is bought for £15 000. It depreciates by 20% in the first year and 10% in subsequent years. There is an average of 2.5% inflation each year. What is the real terms value of the car after 5 years? Give your answer to the nearest £10.

23 A company buys a manufacturing machine for $20 000, with an expected lifetime of 8 years and scrap value $1500. The depreciation rate is set at 30%. Complete the table of asset values for the 8 years. Give all values to the nearest dollar.

Year	Start-year value ($)	Depreciation expense ($)	End-year value ($)
1	20 000		
2			
3			
4			
5			
6			
7			
8			

24 £1000 is invested for 5 years in an account paying 6.2% annual interest. Over the same period, inflation is judged to be 3.2% annually. What is the real terms percentage increase in value of the investment at the end of the 5 years?

25 A bank advertises a loan charging 12% annual interest compounded monthly. What is the equivalent annual interest rate if compounded annually?

26 A government pledges to increase spending on education by 5% each year.
   a If the inflation rate is predicted to 2.5% each year, what is the real terms increase per year?
   b If this is pledged for a 5-year period, what is the overall real terms percentage increase in spending on education?

27  In 1923 Germany suffered from hyperinflation. The annual inflation rate was approximately 10^{11}%.

Use the **exact** formula for inflation given in the Key Concept box on page 40 to answer the following:
   a At the beginning of 1923 a loaf of bread costs approximately 250 marks. How much did it cost at the end of 1923?
   b Siegfried had 2 000 000 marks in savings at the beginning of 1923. He invested it in a savings account paying 20% interest. How much was it worth in real terms at the end of 1923?
   c Anna had a mortgage of 25 million marks at the beginning of 1923. Interest of 15% was added to this mortgage during 1923. Anna made no payments. What is the debt in real terms by the end of 1923?

Checklist

TOOLKIT: Problem Solving

Pick any natural number. If it is even, halve it. If it is odd, multiply by three, then add one, then continually repeat this process with the result. What number will you end up with? Use a spreadsheet to come up with a conjecture. For further practice using spreadsheets to investigate sequences, refer back to the introductory toolkit chapter.

This result has been suspected for some time, but it is still unproven. You might like to see if you can prove it for any special cases, or investigate how long it takes the sequence to first reach its endpoint for different starting values.

Checklist

- You should know the formula for the nth term of an arithmetic sequence:
  - □ $u_n = u_1 + (n-1)d$
- You should know the two versions of the formula for the sum of the first n terms of an arithmetic sequence:
  - □ $S_n = \frac{n}{2}[2u_1 + (n-1)d]$ or
  - □ $S_n = \frac{n}{2}(u_1 + u_n)$
- You should know the formula for the nth term of a geometric sequence:
  - □ $u_n = u_1 r^{n-1}$
- You should know the two versions of the formula for the sum of the first n terms of a geometric sequence:
  - □ $S_n = \frac{u_1(1-r^n)}{1-r}, \quad r \neq 1$ or
  - □ $S_n = \frac{u_1(r^n - 1)}{r-1}, \quad r \neq 1$
- You should be able to work with arithmetic and geometric sequences given in sigma notation:
  - □ $\sum_{r=1}^{r=n} u_r = u_1 + u_2 + u_3 + \ldots + u_n$
    - — The value of r at the bottom of the sigma (here $r = 1$) shows where the counting starts.
    - — The value of r at the top of the sigma (here $r = n$) shows where the counting stops.
- You should be able to apply arithmetic and geometric sequences to real-world problems. This will often involve percentage change:
  - □ A change of $r\%$ is equivalent to multiplying by the factor $1 + \frac{r}{100}$
  - □ An annual inflation rate of $i\%$ is equivalent to multiplying by a factor $\dfrac{1}{1 + \frac{i}{100}}$ each year to find the value in real terms.

Mixed Practice

1 Pierre invests 5000 euros in a fixed deposit that pays a nominal annual interest rate of 4.5%, compounded *monthly*, for 7 years.
   a Calculate the value of Pierre's investment at the end of this time. Give your answer correct to two decimal places.

Carla has 7000 dollars to invest in a fixed deposit which is compounded *annually*. She aims to double her money after 10 years.
   b Calculate the minimum annual interest rate needed for Carla to achieve her aim.

Mathematical Studies SL May 2015 TZ1 Paper 1 Q10

2 Only one of the following four sequences is arithmetic and only one of them is geometric.

$$a_n = 1, 2, 3, 5, \ldots$$
$$b_n = 1, \frac{3}{2}, \frac{9}{4}, \frac{27}{8}, \ldots$$
$$c_n = 1, \frac{1}{2}, \frac{1}{3}, \frac{1}{4}, \ldots$$
$$d_n = 1, 0.95, 0.90, 0.85, \ldots$$

   a State which sequence is
      i arithmetic,
      ii geometric.
   b For **another** geometric sequence $e_n = -6, -3, -\frac{3}{2}, -\frac{3}{4}, \ldots$
      i write down the common ratio;
      ii find the **exact** value of the 10th term. Give your answer as a fraction.

Mathematical Studies SL May 2015 TZ2 Paper 1 Q9

3 In an arithmetic sequence $u_8 = 10$, $u_9 = 12$.
   a Write down the value of the common difference.
   b Find the first term.
   c Find the sum of the first 20 terms.

4 In a geometric sequence the first term is 2 and the second term is 8.
   a Find the common ratio.
   b Find the fifth term.
   c Find the sum of the first eight terms.

5 A company projects a loss of $100 000 in its first year of trading, but each year it will make $15 000 more than the previous year. In which year does it first expect to make a profit?

6 A car has initial value $25 000. It falls in value by $1500 each year. How many years does it take for the value to reach $10 000?

7 One week after being planted, a sunflower is 20 cm tall and it subsequently grows by 25% each week.
   a How tall is it 5 weeks after being planted?
   b In which week will it first exceed 100 cm?

8 Kunal deposits 500 euros in a bank account. The bank pays a nominal annual interest rate of 3% compounded quarterly.
   a Find the amount in Kunal's account after 4 years, assuming no further money is deposited. Give your answer to two decimal places.
   b How long will it take until there has been a total of 100 euros paid in interest?

9 At the end of 2018 the world's population was 7.7 billion. The annual growth rate is 1.1%. If this growth rate continues
  a estimate the world's population at the end of 2022
  b what is the first year in which the population is predicted to exceed 9 billion?

10 The second term of an arithmetic sequence is 30. The fifth term is 90.
  a Calculate
   i the common difference of the sequence;
   ii the first term of the sequence.

The first, second and fifth terms of this arithmetic sequence are the first three terms of a geometric sequence.
  b Calculate the seventh term of the **geometric** sequence.

Mathematical Studies SL May 2015 TZ1 Paper 1 Q7

11 The sum of the first n terms of an arithmetic sequence is given by $S_n = 6n + n^2$.
  a Write down the value of
   i S_1;
   ii S_2.

The nth term of the arithmetic sequence is given by u_n.
  b Show that $u_2 = 9$.
  c Find the common difference of the sequence.
  d Find u_{10}.
  e Find the lowest value of n for which u_n is greater than 1000.
  f There is a value of n for which $u_1 + u_2 + \ldots + u_n = 1512$.
      Find the value of n.

Mathematical Studies SL May 2015 TZ2 Paper 2 Q3

12 The first three terms of an arithmetic sequence are x, $2x + 4$, $5x$. Find the value of x.

13 The audience members at the first five showings of a new play are: 24, 34, 46, 55, 64.
  a Justify that the sequence is approximately arithmetic.
  b Assuming that the arithmetic sequence model still holds, predict the number of audience members at the sixth showing.

14 Evaluate $\sum_{r=1}^{12} \dfrac{6^r}{3^r}$.

15 According to a business plan, a company thinks it will sell 100 widgets in its first month trading, and 20 more widgets each month than the previous month. How long will it take to sell a total of 400 widgets?

16 Find an expression for the nth term of the geometric sequence a^2b^2, a^4b, a^6.

17 When digging a rail tunnel there is a cost of $10 000 for the first metre. Each additional metre costs $500 dollars more per metre (so the second metre costs $10 500). How much does it cost to dig a tunnel 200 m long?

18 When Elsa was born, her grandparents deposited $100 in her savings account and then on subsequent birthdays they deposit $150 then $200 then $250 and so on in an arithmetic progression. How much have they deposited in total just after her 18th birthday?

19 £5000 is invested for 3 years in an account paying 5.8% annual interest. Over the same period, inflation is judged to be 2.92% annually. What is the real terms percentage increase in value of the investment at the end of the 3 years?

20 A company purchases a computer for $2000. It assumes that it will depreciate in value at a rate of 10% annually. If inflation is predicted to be 2% annually, what is the real terms value of the computer after 4 years?

21 Cameron invests $1000. He has a choice of two schemes:

Scheme A offers $25 every year.

Scheme B offers 2% interest compounded annually.

Over what periods of investment (in whole years) is scheme A better than scheme B?

22 In a game, n small pumpkins are placed 1 metre apart in a straight line. Players start 3 metres before the first pumpkin.

Each player **collects** a single pumpkin by picking it up and bringing it back to the start. The nearest pumpkin is collected first. The player then collects the next nearest pumpkin and the game continues in this way until the signal is given for the end.

Sirma runs to get each pumpkin and brings it back to the start.
a Write down the distance, a_1, in metres that she has to run in order to **collect** the first pumpkin.
b The distances she runs to **collect** each pumpkin form a sequence $a_1, a_2, a_3, \ldots$.
  i Find a_2.   **ii** Find a_3.
c Write down the common difference, d, of the sequence.

The final pumpkin Sirma **collected** was 24 metres from the start.
d **i** Find the total number of pumpkins that Sirma **collected**.
   ii Find the total distance that Sirma ran to **collect** these pumpkins.

Peter also plays the game. When the signal is given for the end of the game he has run 940 metres.
e Calculate the total number of pumpkins that Peter **collected**.
f Calculate Peter's distance from the start when the signal is given.

Mathematical Studies SL November 2014 Paper 2 Q5

23 The seventh, third and first terms of an arithmetic sequence form the first three terms of a geometric sequence.

The arithmetic sequence has first term a and non-zero common difference d.
a Show that $d = \frac{a}{2}$.

The seventh term of the arithmetic sequence is 3. The sum of the first n terms in the arithmetic sequence exceeds the sum of the first n terms in the geometric sequence by at least 200.
b Find the least value of n for which this occurs.

Mathematics HL November 2014 Paper 2 Q7

Mixed Practice

24 An athlete is training for a marathon. She considers two different programs. In both programs on day 1 she runs 10 km.

In program A she runs an additional 2 km each day compared to the previous day.

In program B she runs an additional 15% each day compared to the previous day.
  a In which program will she first reach 42 km in a day. On what day of the program does this occur?
  b In which program will she first reach a total of 90 km run. On what day of the program does this occur?

25 A teacher starts on a salary of £25 000. Each year the teacher gets a pay rise of £1500. The teacher is employed for 30 years.
  a Find their salary in their final year.
  b Find the total they have earned during their teaching career.
  c If the inflation rate is on average 1.5% each year, find the real value of their final salary at the end of their final year in terms of the value at the beginning of their career, giving your answer to the nearest £100.

26 The 10th term of an arithmetic sequence is two times larger than the fourth term.
Find the ratio: $\frac{u_1}{d}$.

27 If a, b, c, d are four consecutive terms of an arithmetic sequence, prove that $2(b-c)^2 = bc - ad$.

3 Core: Functions

ESSENTIAL UNDERSTANDINGS

- Models are depictions of real-life events using expressions, equations or graphs while a function is defined as a relation or expression involving one or more variables.
- Creating different representations of functions to model the relationships between variables, visually and symbolically as graphs, equations and/or tables represents different ways to communicate mathematical ideas.

In this chapter you will learn…
- about function notation
- how to determine the domain and range of a function
- about the use of functions in modelling real-life events
- about the idea of an inverse function
- how to draw graphs of inverse functions
- how to sketch a graph from given information
- how to find key features of graphs using a GDC
- how to find intersections of graphs using a GDC
- how to solve equations using a GDC.

CONCEPTS

The following key concepts will be addressed in this chapter:
- Different **representations** of functions – symbolically and visually as graphs, equations and tables – provide different ways to communicate mathematical **relationships**.
- Moving between different forms to **represent** functions allows for deeper understanding and provides different approaches to problem solving.

■ **Figure 3.1** Can we quantify the relationships between inputs and outputs in the examples shown?

Core: Functions

> **PRIOR KNOWLEDGE**
>
> Before starting this chapter, you should already be able to complete the following:
>
> 1 Use technology to sketch the graph of $y = x^2 + x - 12$.
> 2 Find the values of x for which $2x - 3 \geqslant 0$.
> 3 Evaluate on your calculator $\ln 1$.
> 4 Solve $0.1 = e^x$.
> 5 Sketch $y = |x - 1|$ using technology.

In mathematics, a process that takes inputs and generates outputs is called a function. These functions can be represented in many different forms, such as a graph, algebraically or in a table. Sometimes it will be necessary to use one of these for a particular reason, but often you can choose the one that is most appropriate to work with for the task at hand.

Functions have a wide range of uses in mathematics. They are studied as objects in their own right, but knowing the properties of a wide range of functions also helps you to model many real-world situations.

Starter Activity

Look at Figure 3.1. In small groups identify any inputs and outputs you can think of that are related to each of the images.

Now look at this problem:

A particular mathematical rule is defined to be the sum of a number and all its digits.

If the answer is 19, what was the number?

3A Concept of a function

■ Function notation

A function is a rule that maps each input value, x, to a single output value, $f(x)$.

For example, the rule 'square the input value and then add 3' would be written in function notation as:

- $f(x) = x^2 + 3$

or alternatively as:

- $f : x \mapsto x^2 + 3$

A function can be

- one-to-one: when each output value comes from a single input value, for example, $y = x^3$

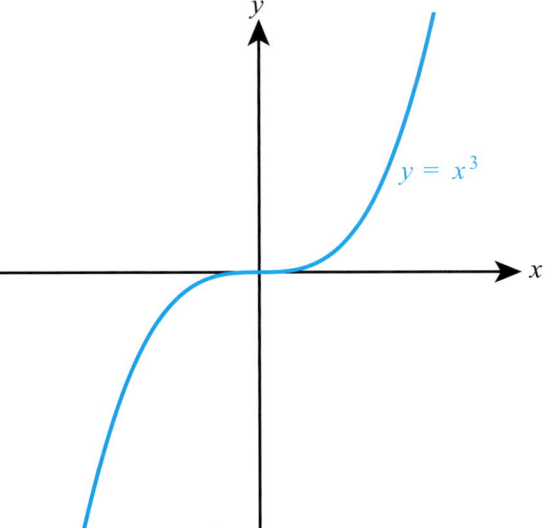

- many-to-one: when an output value can come from more than one different input value, for example, $y = x^2$.

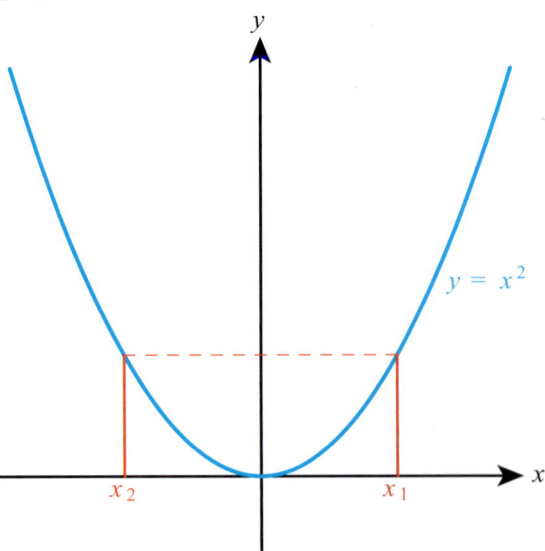

3A Concept of a function

> **CONCEPTS – REPRESENTATION**
>
> Does the graphical **representation** or the algebraic representation of the function make it more obvious whether the function is one-to-one? Which one would be more useful in finding y when $x = 13$? One key skill is deciding which representation of a function is most useful in solving different types of problems.

The notation f(x) was first used by the Swiss mathematician Leonhard Euler in 1750.

WORKED EXAMPLE 3.1

If f(x) = $2x + 7$ find f(3).

Substitute $x = 3$ into $2x + 7$ ·············· f(3) = 2(3) + 7
= 13

WORKED EXAMPLE 3.2

Determine, with reasons, if either of the following graphs shows a function of the form $y = $ f(x).

a
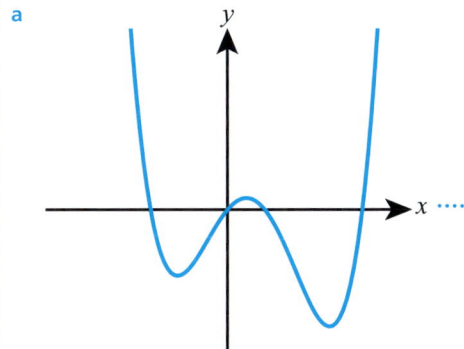

a There are no values of x that map to more than one output, so this is a function.

b
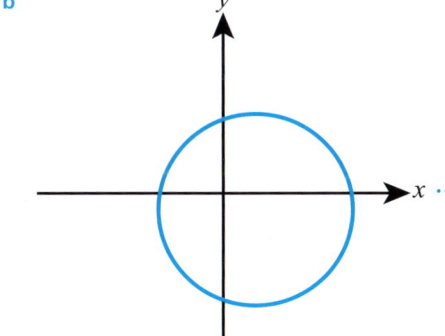

b There are values of x that map to two different outputs, so this is not a function.

■ Domain and range

As well as knowing what the function does to the input value, you also need to know what inputs are allowed.

> **KEY POINT 3.1**
>
> The domain of a function is the set of all allowed input values.

Tip

If no domain is specified, you can assume that the domain is over all real numbers, sometimes written as $x \in \mathbb{R}$.

There are three main things which you should consider when determining the largest possible domain of a function.

- You can never square root a negative number.
- You can never divide by zero.
- You can never take the logarithm of a negative number or zero.

WORKED EXAMPLE 3.3

Find the largest possible domain of $f(x) = \sqrt{2-x}$.

The expression under the square root must be greater than or equal to zero

Largest possible domain is given by
$$2 - x \geq 0$$
$$x \leq 2$$

Be the Examiner 3.1

Find the largest possible domain of $f(x) = \dfrac{1}{x-2}$.

Which is the correct solution? Identify the errors made in the incorrect solutions.

Solution 1	Solution 2	Solution 3
Largest possible domain is given by: $x - 2 > 0$ $x > 2$	Largest possible domain is given by: $x - 2 \neq 0$ $x \neq 2$	Largest possible domain is given by: $x \neq -2$

Once you know the domain of a function, you can find the possible output values.

> **KEY POINT 3.2**
>
> The range of a function is the set of all possible outputs.

To find the range it is best to start with a graph of the function.

> **WORKED EXAMPLE 3.4**
>
> Find the range of $f(x) = \sqrt{2-x}$, $x \leq 0$.
>
>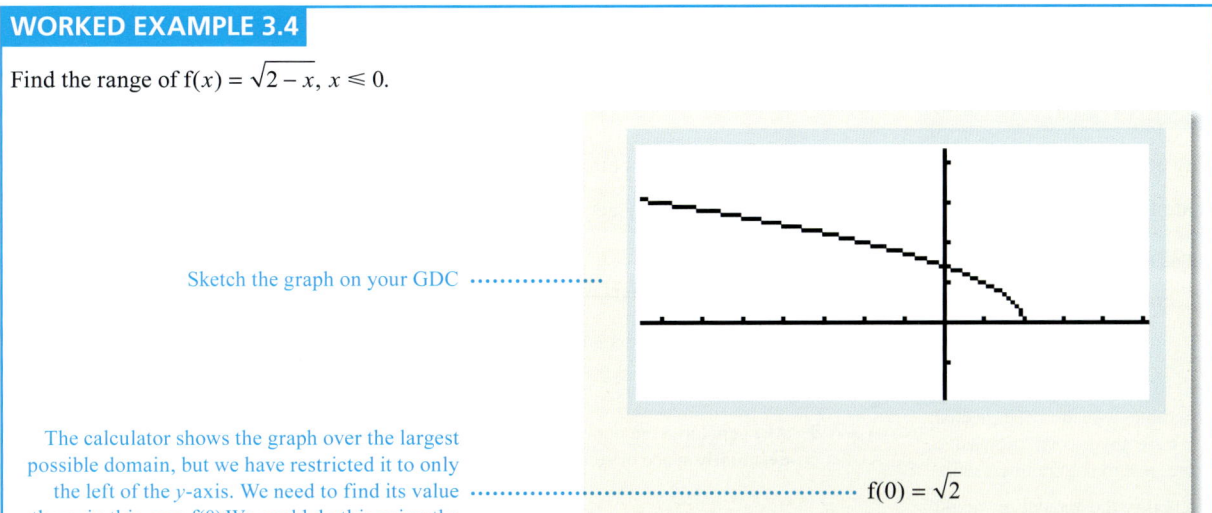
>
> Sketch the graph on your GDC
>
> The calculator shows the graph over the largest possible domain, but we have restricted it to only the left of the y-axis. We need to find its value there, in this case $f(0)$. We could do this using the calculator, or substitute numbers into the formula
>
> $f(0) = \sqrt{2}$
>
> From the graph we can see that the restricted function is always greater than its value at $x = 0$
>
> Range is $f(x) \geq \sqrt{2}$

■ The concept of a function as a mathematical model

Functions can often be created and used to predict the results observed in real life from a particular process – this is known as producing a mathematical model of a process. By understanding the underlying theory of the process involved, the particular function needed can be derived.

For example, Zipf's Law relates the population of a city to the rank of that city's population (largest, second largest, etc).

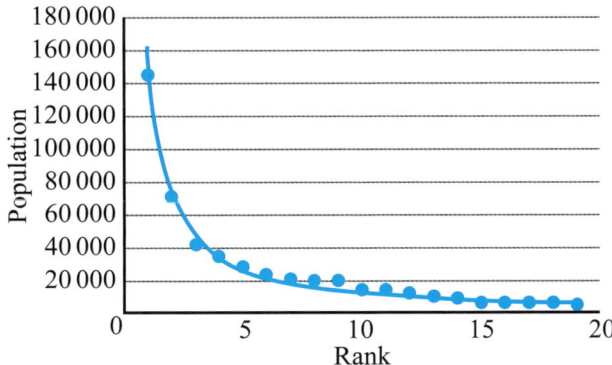

This model does not give exactly the results observed in real life, as you can see from the observed data points and the predictions made by the graph of the function. Just because a model is not perfect does not mean that it is useless. You need to understand what the limitations of any given model might be. Things to consider:

- Check at large and small values that you get a sensible result.
- Check whether the range is sensible in the given context.
- Think about how accurate the model needs to be, for example, the dosage of drugs for a patient compared to the volume of water in a swimming pool.

- Think about whether there is some theoretical underpinning for the model, such as, does the function used follow from a mathematical derivation, or is the function just chosen in order to fit the observed data.

WORKED EXAMPLE 3.5

The population of pigeons, P(t) on an island at a time t years after being introduced is modelled by $P(t) = 100t^2$.

State four reasons why this model is unlikely to be perfectly accurate.

See if there are any particular values of t that do not give sensible values for the number of pigeons

1 $P(0) = 0$, which says that there are no pigeons at the start of the time period. This is not consistent with pigeons being introduced to the island.

2 $P\left(\dfrac{1}{3}\right) = \dfrac{100}{9}$, so the model predicts a non-integer number of pigeons after 4 months.

Think about the biological situation – you are not expected to have any special knowledge here, just to use common sense

3 As t gets large the population grows without limit. This is not realistic as eventually space or resources such as food will run out.

In reality the population would increase more quickly in the spring/summer, for example

4 The model does not allow for any randomness in the growth rate of the population, or any differences due to seasonality.

TOK Links

Even though the model is not perfect, this does not mean it is not useful. It might give reasonable estimates of the approximate population over the first 10 years. In more advanced work, models do not predict single numbers, but a range of possible values the output can take. Which is better – a precise answer unlikely to be perfectly correct or an imprecise answer which is likely to include the correct answer?

■ Informal concept of an inverse function

Tip

Note that $f^{-1}(x)$ *does not* mean $\dfrac{1}{f(x)}$, as you might be tempted to think from the laws of exponents for numbers.

The inverse function, f^{-1}, of a function f reverses the effect of f. For example, if $f(2) = 4$, then $f^{-1}(4) = 2$.

It is important to note that only one-to-one functions have inverses. For example, the many-to-one function $f(x) = x^2$ does not have an inverse because if you know the output is 9, it is impossible to determine whether the input was -3 or 3.

WORKED EXAMPLE 3.6

If $f(x) = 3x + 5$ find $f^{-1}(11)$.

To find $f^{-1}(11)$, solve $f(x) = 11$

$$3x + 5 = 11$$
$$3x = 6$$
$$x = 2$$
So, $f^{-1}(11) = 2$

3A Concept of a function

■ Graphical interpretation of an inverse function

When you find the inverse of a function, you are making the input of f the output of f^{-1} and the output of f the input of f^{-1}.

Graphically you are swapping the x and y coordinates, which is achieved by reflecting the graph of $y = f(x)$ in the line $y = x$.

KEY POINT 3.3

The graph of $y = f^{-1}(x)$ is a reflection of the graph $y = f(x)$ in the line $y = x$.

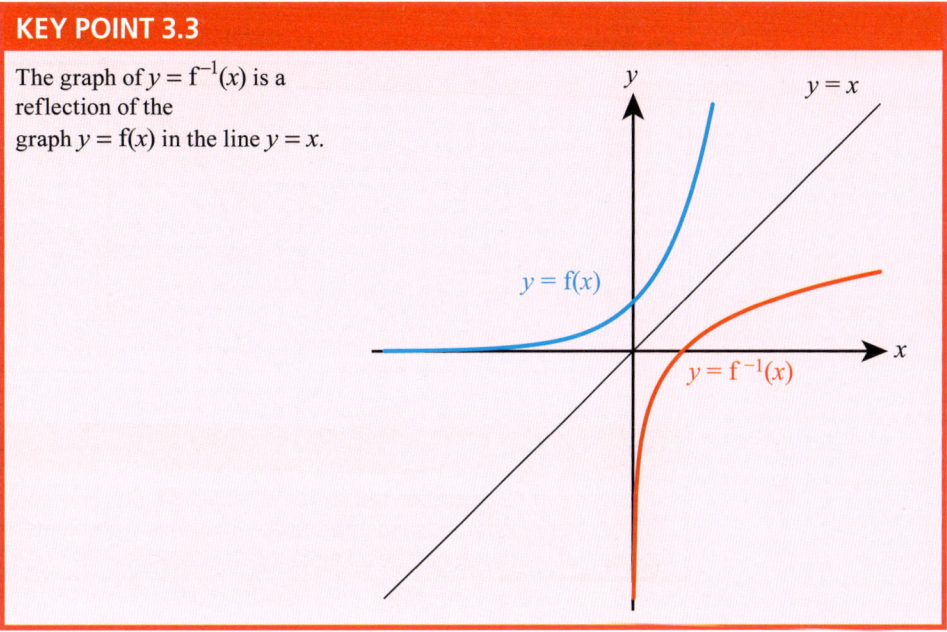

WORKED EXAMPLE 3.7

Sketch the inverse function of the following graph.

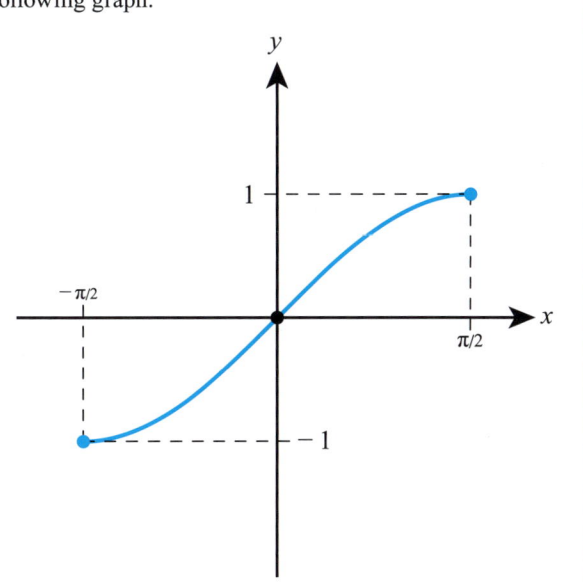

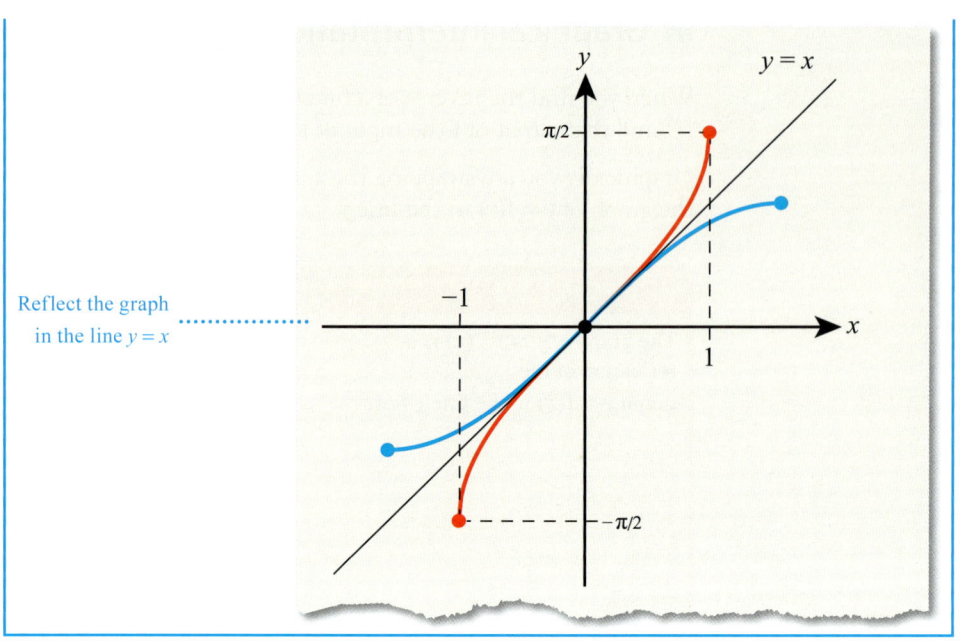

Reflect the graph in the line $y = x$

> **CONCEPTS – REPRESENTATION**
>
> Functions can be **represented** by rules, tables, graphs or mapping diagrams. Different representations are appropriate in different problems and choosing which representation to use can be a key tactical decision.

Exercise 3A

For questions 1 to 4, use the method shown in Worked Example 3.1.

1. If $f(x) = x + 5$, find
   a f(4)
   b f(9)

2. If $f(x) = 5x - 8$, find
   a f(3)
   b f(11)

3. If $f(x) = 7 - 3x$, find
   a f(8)
   b f(-4)

4. If $f(x) = x^3 - 2x^2 + 1$, find
   a f(-2)
   b f(-3)

For questions 5 to 7, use the method shown in Worked Example 3.2 to determine whether the following graphs represent functions.

5 a 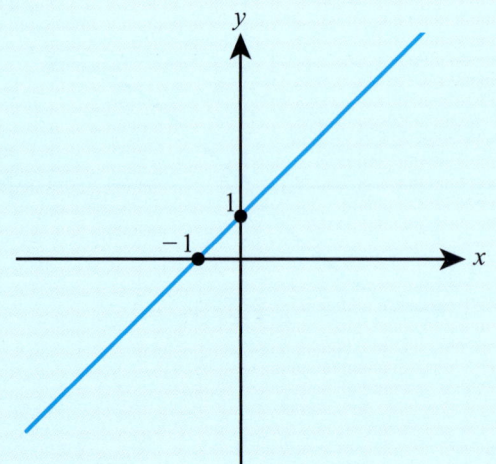   b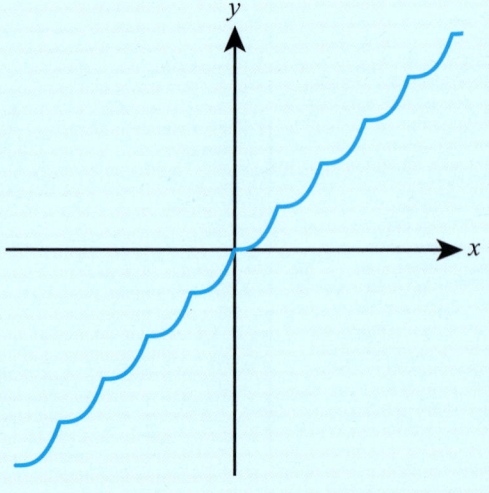

6 a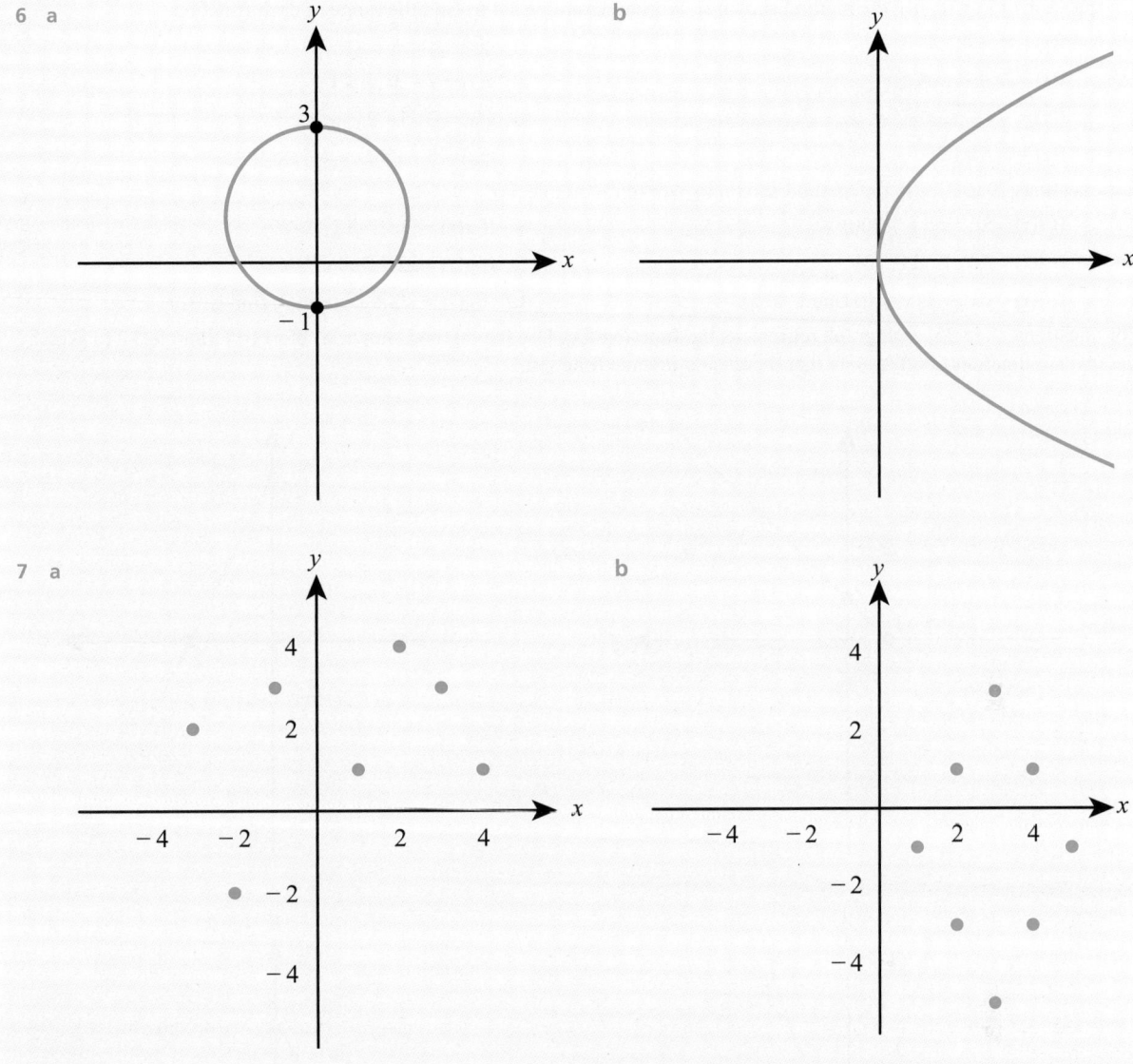

For questions 8 and 9, use technology to sketch the graph and then the method shown in Worked Example 3.2 to determine whether the following equations for y describe a function of x.

8 a $y = \pm\sqrt{x}$, $x \geq 0$

  b $y = \sqrt{x^2}$, $x \in \mathbb{R}$

9 a $y = x^{\frac{1}{3}}$, $x \in \mathbb{R}$

  b $y = \dfrac{2}{x^2}$, $x \neq 0$

For questions 10 to 14, use the method shown in Worked Example 3.3 to find the largest possible domain of the given real functions.

10 a $f(x) = 3x^2 - 2$

   b $g(x) = 7x^3 - 2x$

11 a $f(x) = \dfrac{1}{x}$

   b $f(x) = -\dfrac{3}{x}$

12 a $f(x) = 2\sqrt{x+5}$

   b $g(x) = 3\sqrt{2x-7}$

13 a $f(x) = \ln(5x+3)$

   b $g(x) = \ln(2x-8)$

14 a $f(x) = \dfrac{3}{5-2x}$

   b $g(x) = \dfrac{2x}{3+x}$

For questions 15 to 18, use the method shown in Worked Example 3.4 to find the range of each function.

15 a $f(x) = x^2 - 2, x \in \mathbb{R}$ b $g(x) = 3x^2 + 7, x \in \mathbb{R}$
16 a $f(x) = 2x + 8, x \leq 5$ b $g(x) = 7 - 3x, x > 1$
17 a $f(x) = \sqrt{4 + x}, x \geq -4$ b $g(x) = 3\sqrt{5 - 2x}, x < 2.5$
18 a $f(x) = 2 + \ln(x - 7), x \geq 8$ b $g(x) = 3 - \ln(2x + 1), x > 0$

For questions 19 to 21, use the method shown in Worked Example 3.6.

19 a If $f(x) = 5x$, find $f^{-1}(200)$. b If $g(x) = 7x$, find $g^{-1}(-49)$.
20 a If $f(x) = 3x^5$, find $f^{-1}(3)$. b If $g(x) = 2x^3$, find $g^{-1}(16)$.
21 a If $f(x) = 6x^2 + 1, x \leq 0$, find $f^{-1}(55)$. b If $g(x) = \sqrt{3 + x^2}, x \leq 0$, find $g^{-1}(2)$.

For questions 22 to 25, each graph represents the function f(x). Use the method shown in Worked Example 3.7 to sketch the inverse function $f^{-1}(x)$ or state that there is no inverse function.

22 a 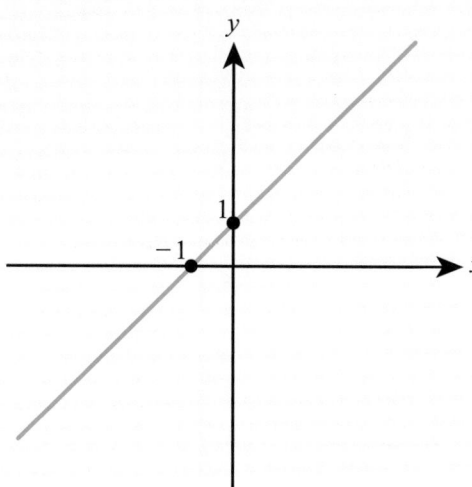 b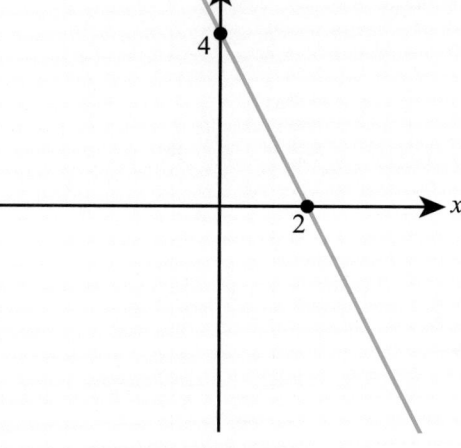

23 a 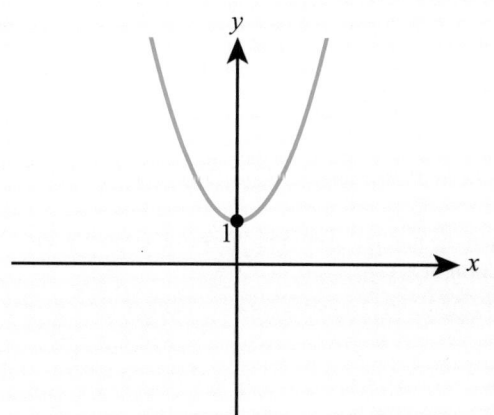 b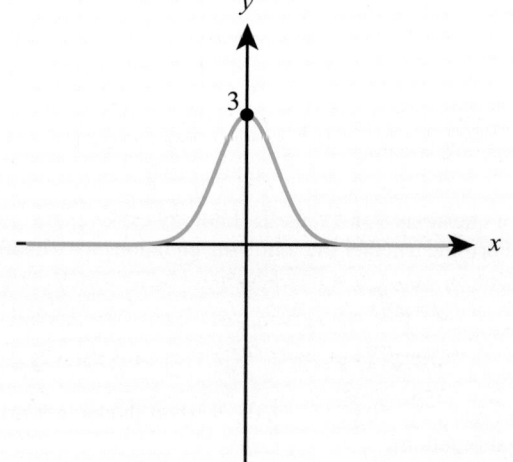

24 a 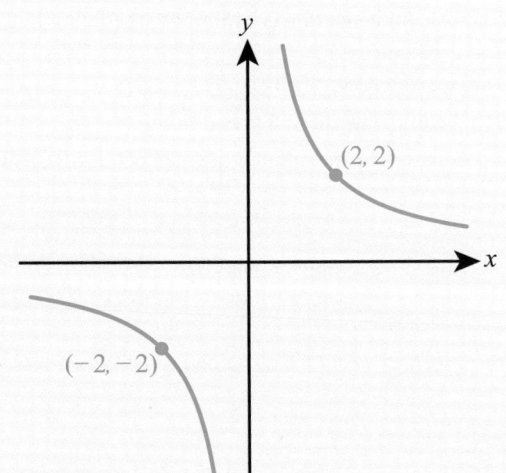 b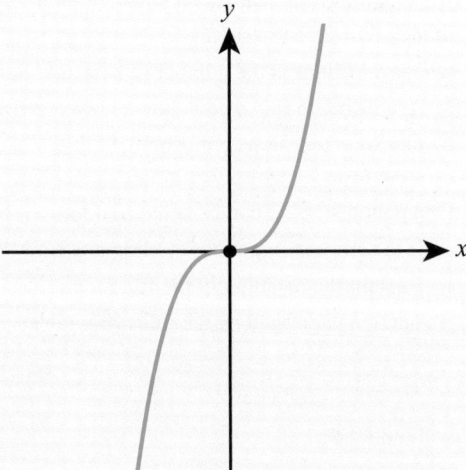

25 a 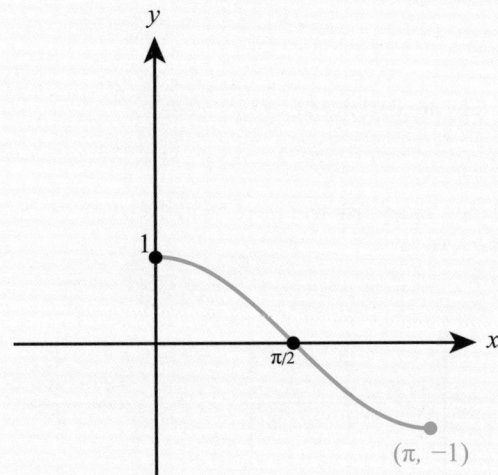 b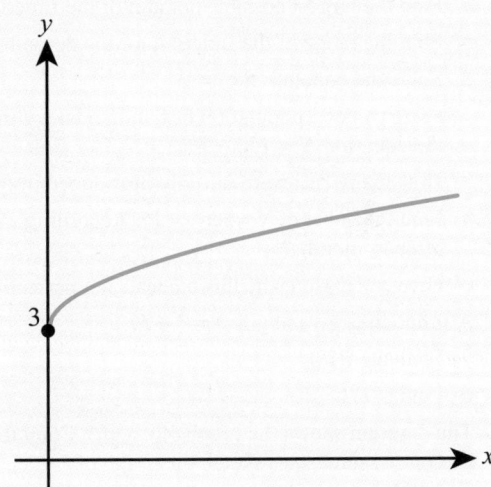

26 Given a function $g(x) = 4x - 5$,
 a evaluate $g(-2)$
 b solve the equation $g(x) = 7$.

27 Given the function $h(x) = \dfrac{x-5}{3}$, solve the equation $h(x) = 12$.

28 The speed, $v\,\text{m s}^{-1}$, of a car at the time t seconds is modelled by the function $v(t) = 3.8t$.
 a Find the speed of the car after 1.5 seconds.
 b Comment on whether this model should be used to predict the speed of the car after 30 seconds.

29 a Write down the largest possible domain of the function $f(x) = \dfrac{3}{(x-5)^2}$.
 b Evaluate $f(2)$.

30 A function is defined for all real numbers by $q(x) = 3x^2 - 2$.
 a Evaluate $q\left(\dfrac{1}{2}\right)$, giving your answer as an exact fraction.
 b Find the range of the function.
 c Solve the equation $q(x) = 46$.

31 A technology company wants to model the number of smartphones in the world. They propose the following model: $N = 2.3e^{0.098t} + 1.2$

where N billions is the number of smartphones at the time t years from now.

a According to this model, how many smartphones will there be in the world in 5 years' time?

b Give one reason why this model might not give a correct prediction for the number of smartphones 50 years from now.

32 One pound sterling can be converted into $1.30.

a If the amount in pounds sterling is x find the function f(x) which tells you the amount this is worth in dollars.

b Explain the meaning of the function $f^{-1}(x)$.

33 A function is defined by the following table:

x	0	1	2	3	4	5	6
f(x)	3	6	4	2	0	1	5

a Find $f^{-1}(6)$.

b Solve f(x) = x + 2.

34 a Find the largest possible domain of the function $f(x) = \sqrt{2x - 5}$.

b Find the range of the function for the domain from part **a**.

c Solve the equation $f(x) = 3$.

35 The size (N) of a population of fish in a lake, t months after they are first introduced, is modelled by the equation $N = 150 - 90e^{-0.1t}$.

a How many fish were initially introduced into the lake?

b Find the value of N when t = 15. According to this model, how many fish will there be in the lake after 15 months?

c Give one reason why this model is not perfectly accurate.

36 A function is given by $f(x) = \frac{x}{2} + 5$.

a Evaluate f(18).

b Find $f^{-1}(7)$.

37 The diagram shows the graph of y = g(t). Copy the graph and on the same axes sketch the graph of $y = g^{-1}(t)$.

38 A function is defined by $g(x) = \log_3 x$.

a State the largest possible domain of g.

b Evaluate g(81).

c Find $g^{-1}(-2)$.

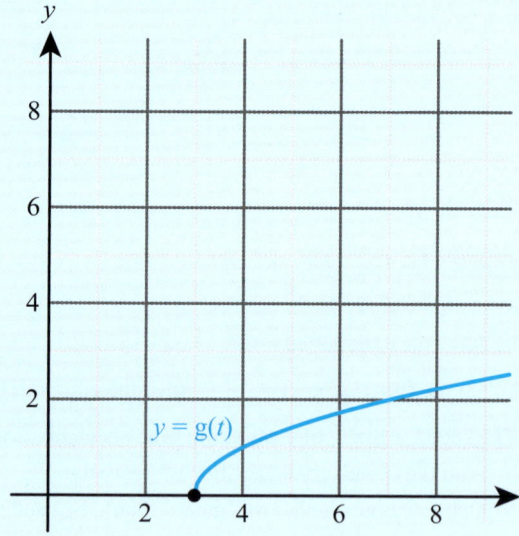

3B Sketching graphs

39 Find the largest possible domain of the function n(x) = log$_5$(3x − 15).

40 a Find the largest possible domain of the function h(x) = log(7 − 3x).
   b Find h^{-1}(2).

41 For a function f(x) = 10 − 3x defined on the domain $x \leq 2$:
   a Evaluate f(−3).
   b Find the range of the function.
   c Explain why the equation f(x) = 1 has no solutions.

42 The diagram shows the graph of y = f(x).
   a Find f(4).
   b Find f^{-1}(4).
   c Copy the graph and on the same axes sketch the graph of y = f^{-1}(x).

43 A function is defined by N(t) = 3e$^{-0.4t}$.
   a Evaluate N(7).
   b Find N^{-1}(2.1).

44 a Find the largest domain of the function
   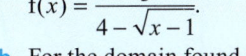
   b For the domain found in part a find the range of the function.

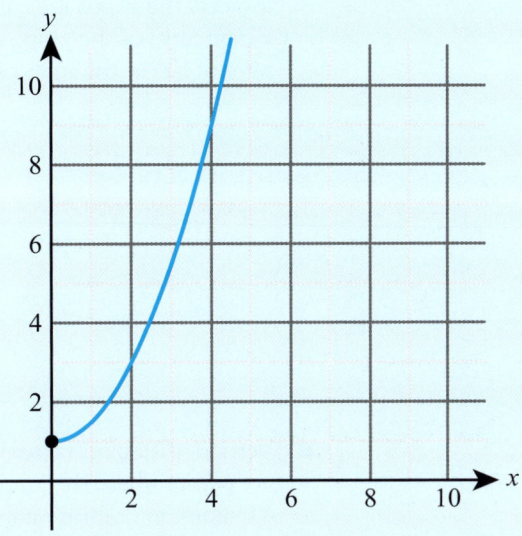

45 A function is defined by f(x) = x^3 + x − 8.
   a Use technology to sketch y = f(x).
   b Sketch on the same axis y = f^{-1}(x).
   c Solve f(x) = f^{-1}(x).

3B Sketching graphs

■ Key features of graphs

There are several important features of the graph y = f(x) that you need to be able to identify.

- **Intercepts** – these are the points where the graph meets the axes. Where it meets the *y*-axis is called the *y*-intercept. The points where it meets the *x*-axis are called the *x*-intercepts of the graph or the zeros of f(*x*) or the roots of f(*x*) = 0.

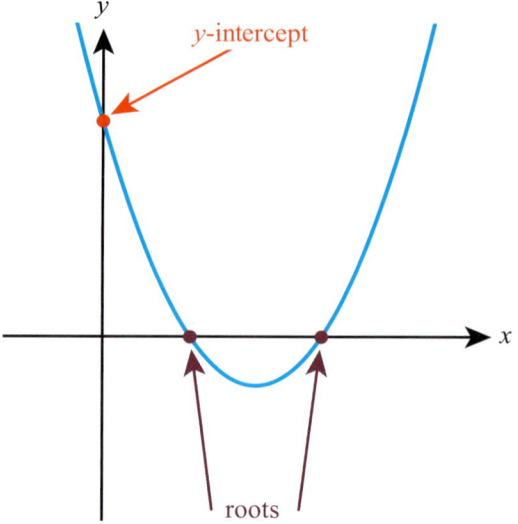

- Vertices (singular: vertex) – these are points where the graph reaches a maximum or minimum point and changes direction. For example, points *A* and *B* are both **vertices of the graph**:

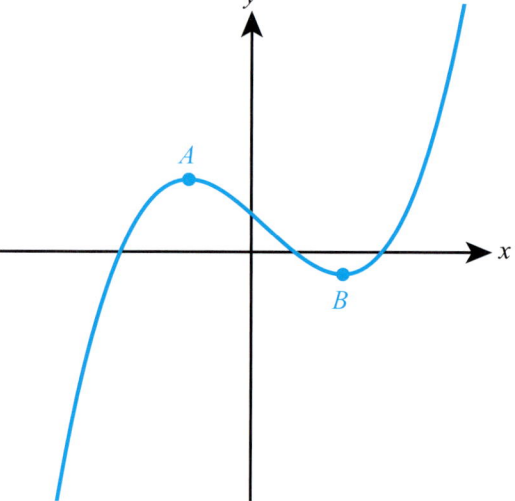

- **Asymptotes** – these are lines to which a graph tends but never reaches. For example, the two dotted lines are both asymptotes. The red one is a vertical asymptote and the green one a horizontal asymptote:

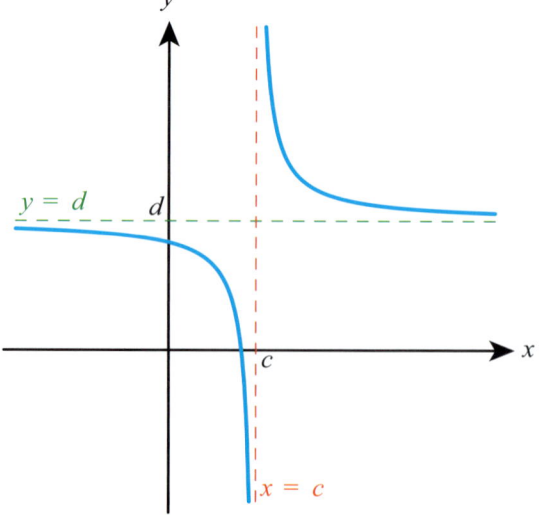

3B Sketching graphs

Tip

For most functions you will meet, any lines of symmetry will pass through either the vertices or the vertical asymptotes of the graph. This is the easiest way to find them.

- Symmetries – in this course you might be asked to find vertical lines of symmetry.

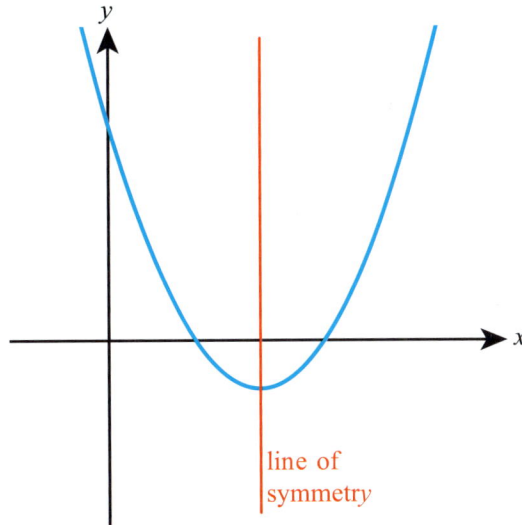

line of symmetry

You are the Researcher

There are many other types of symmetry in graphs, such as rotational symmetry and translational symmetry. You might want to think about how these can be described using function notation. A particularly important type of symmetry is enlargement symmetry, when zooming in on a graph produces a similar graph. This creates a type of picture called a fractal, that as well as being beautiful has applications in fields from compressing computer files to analysing animals' pigmentation.

 You can use your GDC to help you draw graphs and identify these key features.

WORKED EXAMPLE 3.8

Use technology to sketch $y = \dfrac{x-3}{x+4}$, labelling all **axis intercepts** and asymptotes.

Draw the graph and find the y-intercept:

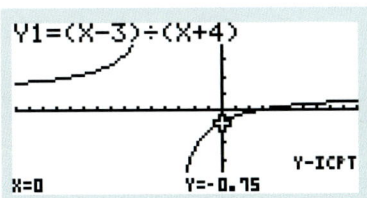

y-intercept is $(0, -0.75)$

The x-intercept can be found using the appropriate function on your calculator, such as, 'Root' or 'G-solve'

x-intercept is $(3, 0)$

Zoom out to get a better view of the shape of the graph. There looks to be one vertical asymptote and one horizontal asymptote. Use 'Trace' or the equivalent function on your GDC to find these ($x = -4$, $y =$ ERROR means the vertical asymptote is at $x = -4$)

Vertical asymptote at $x = -4$
Horizontal asymptote at $y = 1$

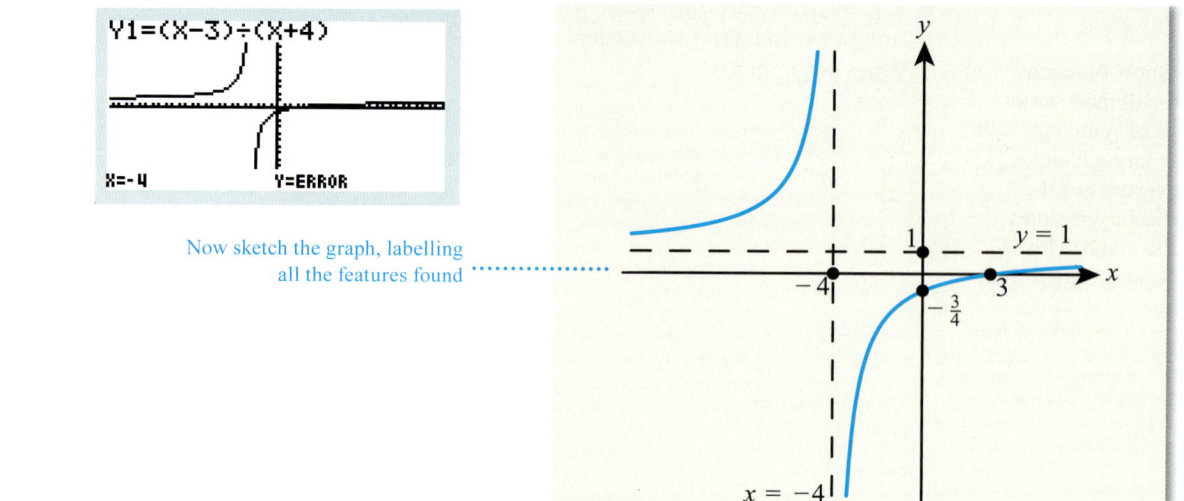

Now sketch the graph, labelling all the features found

WORKED EXAMPLE 3.9

Find the minimum value of $f(x) = xe^x$. Hence find the range of $f(x)$.

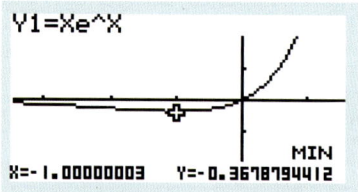

The minimum value of $f(x)$ is -0.368

Hence the range is $f(x) \geqslant -0.368$

WORKED EXAMPLE 3.10

a Find the coordinates of the vertices on the graph of $f(x) = x^4 - 4x^3 + 2x^2 + 4x - 3$.

b Given that the curve $y = f(x)$ has a line of symmetry, find its equation.

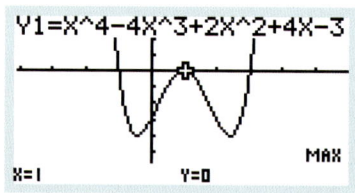

The vertices are $(1, 0)$, $(-0.414, -4)$, $(2.414, -4)$

From the graph you can see that there is a line of symmetry passing vertically through the maximum point

The line of symmetry is $x = 1$

3B Sketching graphs

■ Creating a sketch from information given

WORKED EXAMPLE 3.11

Sketch a graph with the following properties:
- The range is $f(x) \leq 0$ or $f(x) > 1$
- There are vertical asymptotes at $x = \pm 2$
- The only axis intercept is the origin, which is also a vertex.

Step 1: Range $f(x) \leq 0$ or $f(x) > 1$ gives a horizontal asymptote at $y = 1$

There are also two vertical asymptotes at $x = \pm 2$

Draw the asymptotes and identify the regions corresponding to the range

Step 2: Vertex at O (together with lower part of range) means the curve must lie in the lower part of the graph for $-2 < x < 2$

Fill in the graph for $-2 < x < 2$ as it must tend to $-\infty$ as x approaches either vertical asymptote

A possible graph is:

Step 3: One or both of the parts of the graph given by the intervals $x < -2$ and $x > 2$ must fill in the upper part of the range

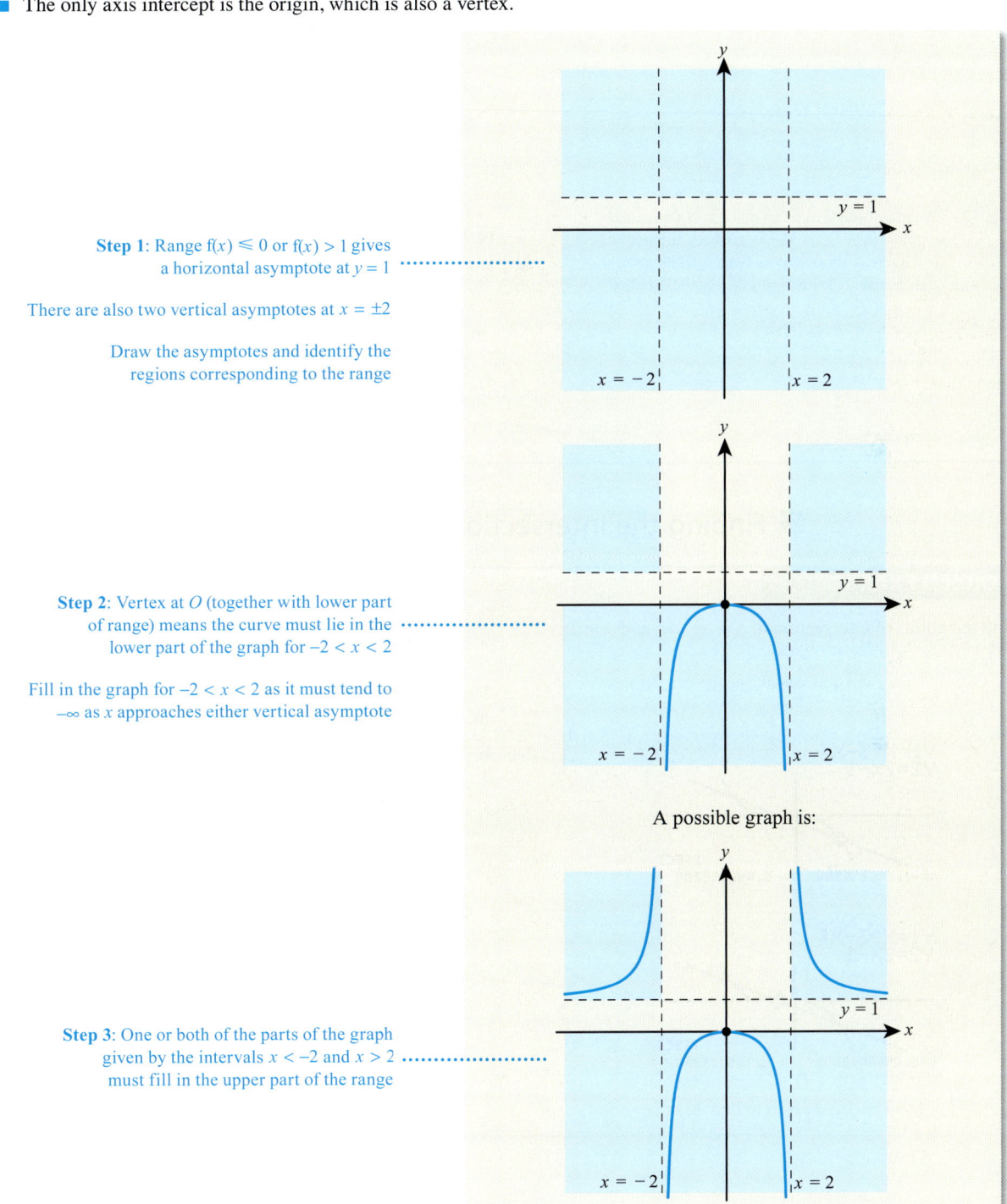

Creating a sketch from context

WORKED EXAMPLE 3.12

The function $N(d)$ gives the total number of seeds distributed within a distance d of the centre of the tree. The tree is modelled as a cylinder of radius 0.2 m. The tree releases a total of 300 seeds. There are only 50 seeds within 1 m of the tree, and half are distributed more than 3 m from the tree.

The graph starts at 0.2 m and has a horizontal asymptote at $N(d) = 300$. In context, it must be increasing as d is increasing. The points $(1, 50)$ and $(3, 150)$ should also be labelled

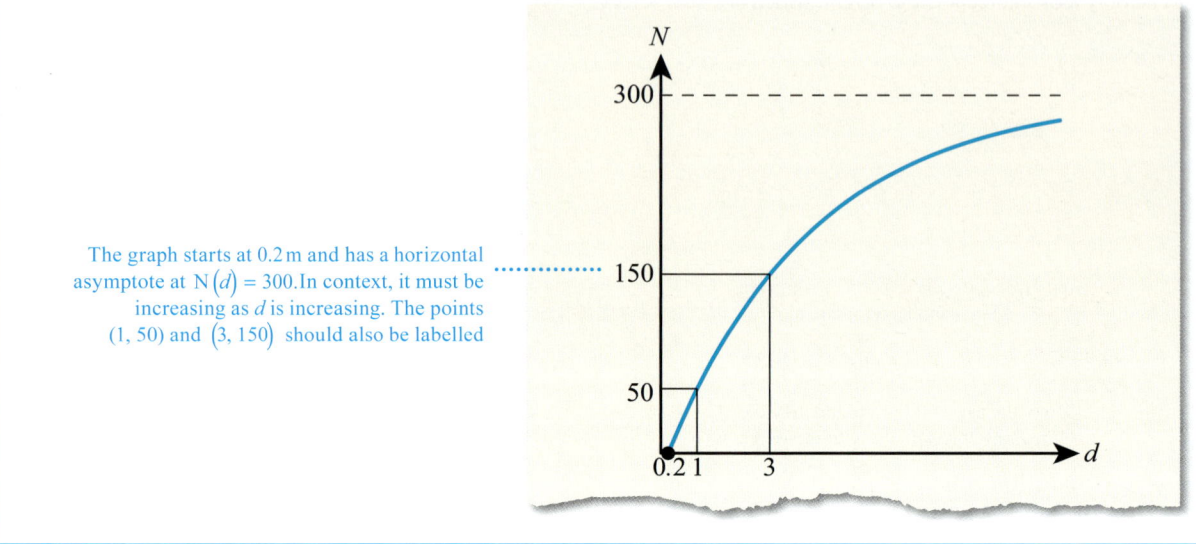

Finding the intersection of graphs using technology

WORKED EXAMPLE 3.13

Find the points of intersection of $y = x^3 - x^2$ and $y = 2x - 1$.

Draw both graphs and find the coordinates of their points of intersection

The points of intersection are $(-1.25, -3.49)$, $(0.445, -0.110)$ and $(1.80, 2.60)$.

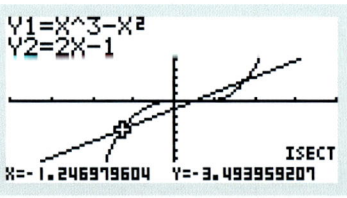

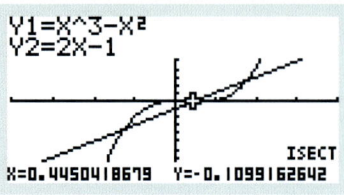

Move the cursor right to find the other two

3B Sketching graphs

■ Solving equations using graphs

Instead of just being asked to solve an equation, you might instead be asked to find:

- The **roots of an equation** – these are exactly the same thing as the solutions of the equation.
- The **zeros of a function** – these are the values of x for which $f(x) = 0$.

WORKED EXAMPLE 3.14

Find the zeros of the function $f(x) = \dfrac{1}{x} + x - \sqrt{x} - 2$.

Graph the function and find any roots .... The zeros are $x = 0.450$ and $x = 3.63$.

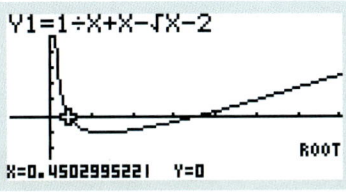

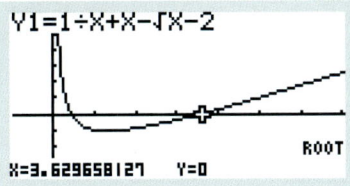

Move the cursor right to find the second root.
You should zoom out enough to make sure that
you have not missed any other roots

> 🌐 A report by Charles Godfrey from the Fourth International Congress of Mathematicians lamented that the reliance of students on modern technology to solve equations meant that 'there was a danger that the boys would become helpless in dealing with the most straight-forward algebraic expressions'. You might not find this surprising until you find out that this was a conference from 1908 and the modern technology that was causing this concern was graph paper! This new method of solving equations had only recently become feasible with a massive reduction in the price of graph paper and there were serious concerns that there were not enough teachers who were able to use this revolutionary and innovative method.

WORKED EXAMPLE 3.15

Find the root(s) of the equation $x^2 = e^x$.

Rearrange as a single function equal to zero

$$x^2 = e^x$$
$$x^2 - e^x = 0$$

Now graph the function $f(x) = x^2 - e^x$ and find any roots .... The root is $x = -0.703$

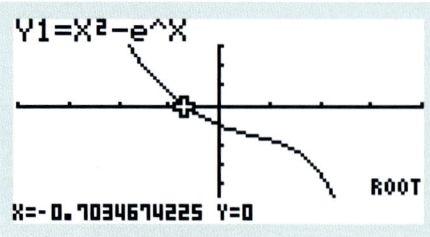

CONCEPTS – REPRESENTATION AND RELATIONSHIPS

An alternative to the approach in Worked Example 3.15 would be to draw the graphs $y = x^2$ and $y = e^x$ and find their point of intersection. You need to use the fact that the solutions of $x^2 - e^x = 0$ and $x^2 = e^x$ are equivalent and both have a **relationship** to their graphical **representation**.

All equations can be solved graphically, but not all equations can be solved algebraically. For example, in Worked Example 3.15 there is no way of exactly solving this equation and representing the answer using functions from school mathematics. However, such equations do exist in real-world situations, so choosing an appropriate representation of both the problem and the solution is vital.

Exercise 3B

For questions 1 to 3, show the graph of the given function on your calculator and then use the method shown in Worked Example 3.8 to sketch the curve in the interval $-5 \leq x \leq 5$ and $-5 \leq y \leq 5$, labelling all axis intercepts and asymptotes.

1. a $f(x) = \dfrac{1}{x+2}$
   b $g(x) = \dfrac{x-1}{2x}$

2. a $f(x) = \dfrac{x+3}{3x-3}$
   b $g(x) = \dfrac{x-1}{x+2}$

3. a $f(x) = \dfrac{3x^2 - 12}{x^2}$
   b $g(x) = 5xe^{-x}$

For questions 4 to 6, use the method demonstrated in Worked Example 3.9 to find the range of each function, using technology to determine maximum and minimum values where needed.

4. a $f(x) = 3x^2 - 4x + 5$
   b $g(x) = 7 - 5x - 2x^2$

5. a $f(x) = \ln(x^2 + 3x + 3)$
   b $g(x) = \ln(2x^2 + 2x + 5)$

6. a $f(x) = xe^{1-x^2}$
   b $g(x) = e^{1-x^2} \cos 4x$

For questions 7 and 8, use technology and the method demonstrated in Worked Example 3.10 to find the coordinates of any vertices on the graph of each curve. Where there is a vertical line of symmetry, state its equation.

7. a $f(x) = 5x^2 - 7x + 3$
   b $g(x) = 5 - 4x - 7x^2$

8. a $f(x) = x^4 + 4x^3 + 5x^2 + 2x + 2$
   b $g(x) = 4x^4 - 8x^3 + 4x^2 - 1$

3B Sketching graphs

For questions 9 to 11, use the method demonstrated in Worked Example 3.11 to sketch a graph with the given properties.

9   a The graph of f(x) has a zero at 2 and no vertices. It has a single vertical asymptote at $x = 3$ and a horizontal asymptote at $y = 1$. The range of f(x) is f(x) ≠ 1.

   b The graph of g(x) has a zero at −3 and no vertices. It has a single vertical asymptote at $x = 1$ and a horizontal asymptote at $y = -2$. The range of g(x) is g(x) ≠ −2.

10 a The range of f(x) is 0 < f(x) < 3. The graph has no vertices and has y-axis intercept 2. The function is always increasing, so f(a) < f(b) whenever $a < b$.

   b The graph of g(x) has zeros at 1 and −2 and no vertices. It has a single vertical asymptote at $x = 0$ and for large positive and negative values of x, g(x) ≈ x + 1 The range of g(x) is $\mathbb{R}$.

11 a The graph of f(x) has zeros at 3 and at 1 and no vertices. It has a single vertical asymptote at $x = 2$ and a horizontal asymptote at $y = 2$. The range of f(x) is f(x) < 2.

   b The graph of g(x) has zeros at 1 and 3 and a vertex at (2, −2). It has vertical asymptotes at $x = 0$ and $x = 4$ and range g(x) ⩾ −2 or g(x) < −8.

For questions 12 to 14, use technology to find the points of intersection of the following pairs of equations; refer to the method shown in Worked Example 3.13. Give your answers to three significant figures.

12 a $y = x^2 - 7$ and $y = x^3 + x$

   b $y = 2x^3 + 1$ and $y = x^2 - 3x - 2$

13 a $y = x^3 + 3x^2 + 3$ and $y = x^2 + x + 5$

   b $y = x^4 - 4x^3 + 11x - 3$ and $y = x^2 - 5x + 9$

14 a $y = x^4 + x^3 - 2x^2 - 2$ and $y = x^3 - x^2$

   b $y = x^3 - 3x^2 + 2x + 7$ and $y = x^3 + x^2 - 6x + 3$

For questions 15 and 16, use your calculator to find the zeros of the following functions, drawing on the method shown in Worked Example 3.14. Give your answers correct to three significant figures.

15 a $f(x) = x^3 - 5\sqrt{x} + 1$

   b $g(x) = 5x\sqrt{x} - 7x + 1$

16 a $f(x) = 3\ln x - x + 1$

   b $g(x) = e^x - 2x - 7$

For questions 17 and 18, use your calculator to find the roots of the following equations, drawing on the method shown in Worked Example 3.15. Give your answers correct to three significant figures.

17 a $x^3 = 4\sqrt{x} - 1$

   b $7x\sqrt{x} = x^2 + 1$

18 a $x^2 + e^{3x} = xe^x + 5$

   b $\ln(2x^2 + 1) = 3 - x - x^2$

19 Sketch the graph of $y = \ln(x^2 + 3x + 5)$ and write down the coordinates of its vertex.

20 Sketch the graph of $y = 7 - 3e^{-0.5x}$. Label any axis intercepts and find the equation of the horizontal asymptote.

21 a Find the largest possible domain of the function $f(x) = \dfrac{3x - 1}{x + 2}$.

   b Sketch the graph of $y = f(x)$, stating the equations of any asymptotes.

22 Sketch a graph with a y-intercept 3, no vertical asymptotes and horizontal asymptote $y = 1$.

23 Find the coordinates of the point of intersection of the graph $y = 5 - x$ and $y = \dfrac{1}{2}e^x$.

24 Find the coordinates of the maximum point on the graph of $y = \dfrac{10}{x^2 + 2x + 5}$.

25 A manufacturing firm uses the following model for their monthly profit: $P = 9.4q - 0.02q^2 - 420$ where P is the profit and q is the number of items produced that month.

   How many items a month should the firm produce in order to maximize the profit?

26 Find the equation of the line of symmetry of the graph of $y = \dfrac{1}{x^2 + 2x + 3}$.

27 Sketch the graph of $y = \dfrac{1}{x^2 - 4}$. State the equations of all the asymptotes.

28 Sketch a graph with the following properties:
   - It has a vertical asymptote $x = 1$ and a horizontal asymptote $y = 3$.
   - Its only axis intercepts are (−1, 0) and (0, −3).

29 A graph is used to show the number of students, N, who scored fewer than m marks in a test. 80 students took the test, which was marked out of 50 marks. The lowest mark was 15 and the highest mark was 45. Exactly half the students scored fewer than 30 marks. Sketch one possible graph showing this information.

30 A car is moving at a speed of 26 m s^{-1} when it starts to brake. After 0.5 seconds its speed has reduced to 15 m s^{-1} and after another 0.5 seconds it has reduced to 10 m s^{-1}. Sketch a possible graph showing how the speed of the car changes with time.

31 A small business wants to model how the money spent on advertising affects their monthly profit. After collecting some data, they found that:
- with no money spent on advertising in a month, they made a monthly loss of $200
- when they spent $120 or $400 on advertising in a month, their monthly profit was $350
- they made the largest profit, $600, when they spent $180 on advertising in a month.

Sketch one possible graph showing how the monthly profit varies with money spent on advertising.

32 Solve the equation $x^2 = \dfrac{1}{x+1}$.

33 Find all the roots of the equation $4 - x^2 = \dfrac{1}{x+1}$.

34 Find the zeros of the function $f(x) = x^4 - 2x^3 - 5x^2 + 12x - 4$.

35 Find all the roots of the equation $|4 - x^2| = x^2$.

36 Find the maximum value of the function $f(x) = xe^{-0.4x}$.

37 Sketch the graph of $y = \dfrac{\ln(x-2)}{x}$.

38 Solve the equation $\ln|x-1| = |\ln(x-1)|$.

TOOLKIT: Problem Solving

Are Cartesian graphs the best way of representing functions? The website https://imaginary.org/ challenges this concept by using 3D images, animations and even music to convey information about functions and equations. There is often a link between simple equations and beautiful representations. See if you can explore the Surfer program to model mathematical sculptures of architecture, the natural world or abstract ideas.

Checklist

- You should understand function notation. A function can be written as, for example,
  - $f(x) = 3x - 2$ or
  - $f : x \mapsto 3x - 2$

- You should know how to find the domain and range of a function.
  - The domain of a function is the set of all allowed input values.
  - The range of a function is the set of all possible outputs.

- You should understand that functions can be used in modelling real-life events, and that there are often limitations of such models.

- You should understand that an inverse function reverses the effect of the original function.
  - The graph of $y = f^{-1}(x)$ is a reflection of the graph $y = f(x)$ in the line $y = x$.

- You should be able to use your GDC to:
  - sketch the shape of graphs
  - find key features of graphs such as axis intercepts, vertices, asymptotes
  - find intersections of graphs
  - solve equations.

Mixed Practice

1 Let $G(x) = 95e^{(-0.02x)} + 40$, for $20 \leq x \leq 200$.
  a Sketch the graph of G.
  b Robin and Pat are planning a wedding banquet. The cost per guest, G dollars, is modelled by the function $G(n) = 95e^{(-0.02n)} + 40$, for $20 \leq n \leq 200$, where n is the number of guests.

  Calculate the **total** cost for 45 guests.

Mathematics SL May 2015 TZ1 Paper 2 Q5

2 Find the zeros of the function $f(x) = x - 3\ln x$.

3 **a** Find the largest possible domain of the function $g(x) = \sqrt{x+5}$.
  b Solve the equation $g(x) = \dfrac{1}{x+3}$.

4 Find the coordinates of the points of intersection of the graph $y = 3 - x^2$ and $y = 2e^x$.

5 Sketch the graph of $y = |x^2 - 4x - 5|$, showing the coordinates of all axis intercepts and any maximum and minimum points.

6 **a** Sketch the graph of $y = \dfrac{3x-1}{x-3}$.
  a Hence find the domain and range of the function $f(x) = \dfrac{3x-1}{x-3}$.

7 A speed of a car, $v\,\text{m s}^{-1}$, at time t seconds, is modelled by the equation $v = 18te^{-0.2t}$.
  a Find the speed of the car after 1.5 seconds.
  b Find two times when the speed of the car is $10\,\text{m s}^{-1}$.
  c How long does it take for the car to reach the maximum speed?

8 The speeds of two runners are modelled by the equations: $v_1 = 8 - 6e^{-0.5}t$, $v_2 = 2 + 3x^2 - x^3$

  where v is in m s^{-1} and t is the time in seconds, with $0 \leq t \leq 2$.
  a Show that both runners start at the same speed.
  b After how long will they be running at the same speed again?

9 Water has a lower boiling point at higher altitudes. The relationship between the boiling point of water (T) and the height above sea level (h) can be described by the model $T = -0.0034h + 100$ where T is measured in degrees Celsius (°C) and h is measured in **metres** from sea level.
  a Write down the boiling point of water at sea level.
  b Use the model to calculate the boiling point of water at a height of 1.37 km above sea level.

  Water boils at the top of Mt. Everest at 70°C.
  c Use the model to calculate the height above sea level of Mt. Everest.

Mathematical Studies SL May 2012 TZ2 Paper 1 Q6

10 To convert temperature in degrees Celsius, x to degrees Fahrenheit, $f(x)$ a website says to multiply by 1.8 then add 32.
  a Find an expression for $f(x)$ in terms of x.
  b What is the interpretation of $f^{-1}(x)$ in this context?

11 **a** $f : x \to 3x - 5$ is a mapping from the set S to the set T as shown on the right. Find the values of p and q.
  b A function g is such that $g(x) = \dfrac{3}{(x-2)^2}$.

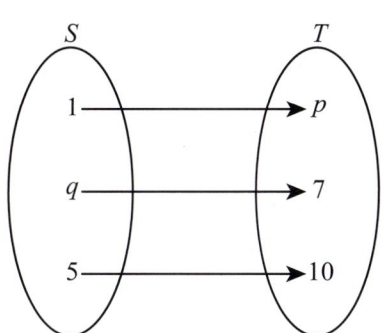

i State the domain of the function g(x).
  ii State the range of the function g(x).
  iii Write down the equation of the vertical asymptote.

  Mathematical Studies SL November 2007 Paper 1 Q12

12 A function is defined on the domain $x \geq 7$ by the equation $f(x) = 3x - 1$.
  a Find the range of f.
  b Find $f^{-1}(35)$.

13 Solve the equation $\dfrac{2x-1}{x+4} = 4 - x$.

14 Find all the roots of the equation $|x - 2| = \dfrac{1}{x}$.

15 Find the domain and range of the the function $h(x) = \dfrac{18}{x^2 - 9}$.

16 Find the range of the function $f(x) = x^2 - 7x + 3$ defined on the domain $0 < x < 6$.

17 A function is defined by $f(x) = 5e^x - 4x$.
  a Find the smallest possible value of f(x).
  b Solve the equation $f^{-1}(x) = 2$.

18 The function $g(x) = 8 - 3x^2$ is defined on the domain $-3 < x < 2$.
  a Find the range of the function.
  b Solve the equation $g(x) = 5$.
  c Explain why the equation $g(x) = -20$ has no solutions.

19 Sketch a possible graph of a function which is defined for all real values of x, has a horizontal asymptote $y = 5$, and crosses the coordinate axes at $(0, 3)$ and $(-1, 0)$.

20 A cup of tea initially has temperature $90°\,C$ and is left to cool in a room of temperature $20°\,C$. Sketch the graph showing how the temperature of the tea changes with time.

21 In a biology experiment, there are initially 600 bacteria. The population of bacteria increases but, due to space constraints, it can never exceed 2000. Sketch a possible graph showing how the number of bacteria changes with time.

22 A child pushes a toy car in the garden, starting from rest, and then lets it go. The speed of the car, $v\,\text{m}\,\text{s}^{-1}$, after t seconds is modelled by the equation $v = 3xe^{-x}$.
  a Find the maximum speed of the car.
  b Suggest why this is not a good model for the speed of the car after 20 seconds.

23 Solve the equation $\left| e^{2x} - \dfrac{1}{x+2} \right| = 2$.

Mathematics HL May 2005 TZ2 Paper 2 Q2

24 a Sketch the graph of $y = \dfrac{x - 12}{\sqrt{x^2 - 4}}$.
  b Write down
    i the x-intercept;
    ii the equations of all asymptotes.

Mathematics HL May 2005 TZ1 Paper 1 Q15

25 Solve the equation $e^x = x^3 - 2$.

26 A function is defined by $f(x) = \dfrac{x^2 - 9x}{5x + 1}$ for $x > 1$. Find the minimum value of this function.

Mixed Practice

27 A function is defined by $f(x) = \dfrac{10x^2 + 7}{x^2 - 4x}$.
  a Write down the largest possible domain of the function.
  b Find the range of the function for the domain from part **a**.

28 a Find the largest domain of the function $g(x) = \dfrac{2x}{3 + \ln x}$.
  b For the domain found in part **a** find the range of the function.

29 A function is defined by $g(x) = 2x + \ln(x - 2)$.
  a State the domain and range of g.
  b Sketch the graph of $y = g(x)$ and $y = g^{-1}(x)$ on the same set of axes.
  c Solve the equation $g(x) = g^{-1}(x)$.

4 Core: Coordinate geometry

ESSENTIAL UNDERSTANDINGS

- Geometry allows us to quantify the physical world, enhancing our spatial awareness in two and three dimensions.

In this chapter you will learn…
- how to find the gradient and intercepts of straight lines
- how to find the equation of a straight line in different forms
- about the gradients of parallel and perpendicular straight lines
- how to find the point of intersection of two straight lines
- how to find the distance between two points in three dimensions
- the midpoint of two points in three dimensions.

CONCEPTS

The following key concepts will be addressed in this chapter:
- The properties of shapes are highly dependent on the dimension they occupy in **space**.
- The **relationships** between the length of the sides and the size of the angles in a triangle can be used to solve many problems involving position, distance, angles and area.

PRIOR KNOWLEDGE

Before starting this chapter, you should already be able to complete the following:

1 Find the length AB in the following triangle:

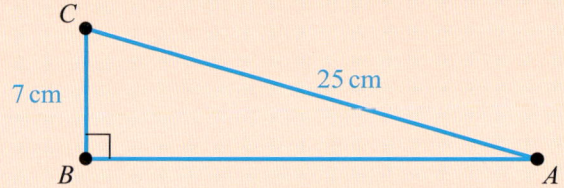

2 The points P and Q have coordinates $(2, -3)$ and $(-4, 5)$ respectively. Find
   a the distance between P and Q
   b the midpoint of PQ.

■ Figure 4.1 Can we use mathematics to model the gradients of these slopes?

Core: Coordinate geometry

The trajectory of a plane taking off or the spokes on a wheel are approximately straight lines, and it can be useful to describe these in an accurate way. For example, an air traffic controller would need to know whether the straight-line trajectory of any particular plane misses all the other planes that are nearby. But a straight-line model won't be perfectly accurate – does this mean the model isn't useful?

As well as physical situations, there are many other relationships that can be modelled by straight lines, for example, the distance travelled and fare paid for a taxi journey.

> ## Starter Activity
> Look at the images in Figure 4.1 and discuss the following questions: What is the best measure of steepness? Can we always tell whether a slope is going up or down?
>
> **Now look at this problem:**
>
> Consider the following data:
>
> | x | 0 | 1 | 2 | 3 | 4 |
> |---|---|---|---|---|---|
> | y | 18 | 22 | 26 | 30 | 34 |
>
> Can you predict the value of y when $x = 1.5$ if
>
> a y represents the temperature of water in a kettle in °C and x the time in seconds after turning it on?
>
> b y represents the number of bees in a hive in thousands and x represents time in years after the hive is created?

> ## LEARNER PROFILE – Open-minded
> Is mathematics a universal language? Why is mathematical notation so similar in different countries when other aspects of language differ so widely? Why are there still differences? What does 1,245 mean in your country? Does it mean the same everywhere? Is one convention better than another?

4A Equations of straight lines in two dimensions

■ Gradient and intercepts

The **gradient** of a straight line is a measure of how steep the line is – the larger the size of the number, the steeper the line.

The gradient can be positive, negative or zero:

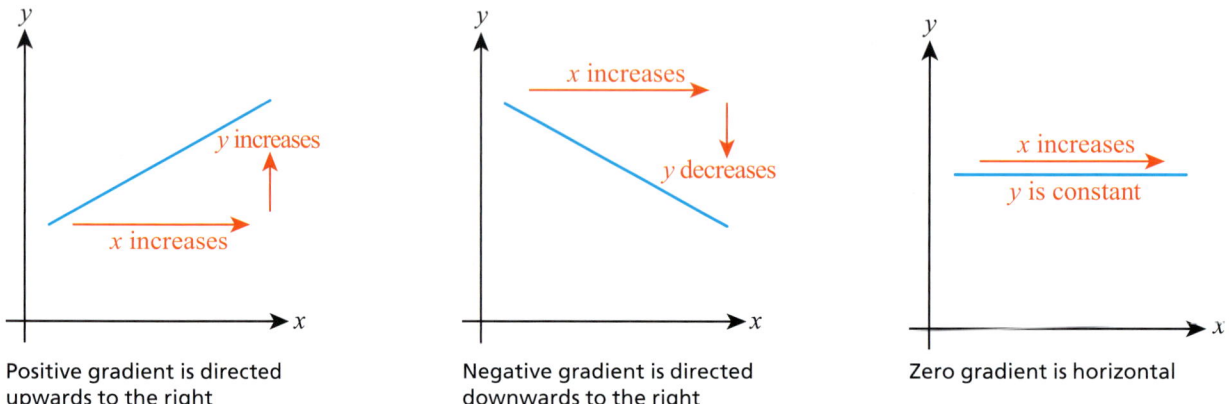

Positive gradient is directed upwards to the right

Negative gradient is directed downwards to the right

Zero gradient is horizontal

More precisely, the gradient is the change in the *y*-coordinate for every 1 unit of increase in the *x*-coordinate, or equivalently, the change in *y* divided by the change in *x*.

Given two points on the line, (x_1, y_1) and (x_2, y_2), the change in *y* is $(y_2 - y_1)$ and the change in *x* is $(x_2 - x_1)$.

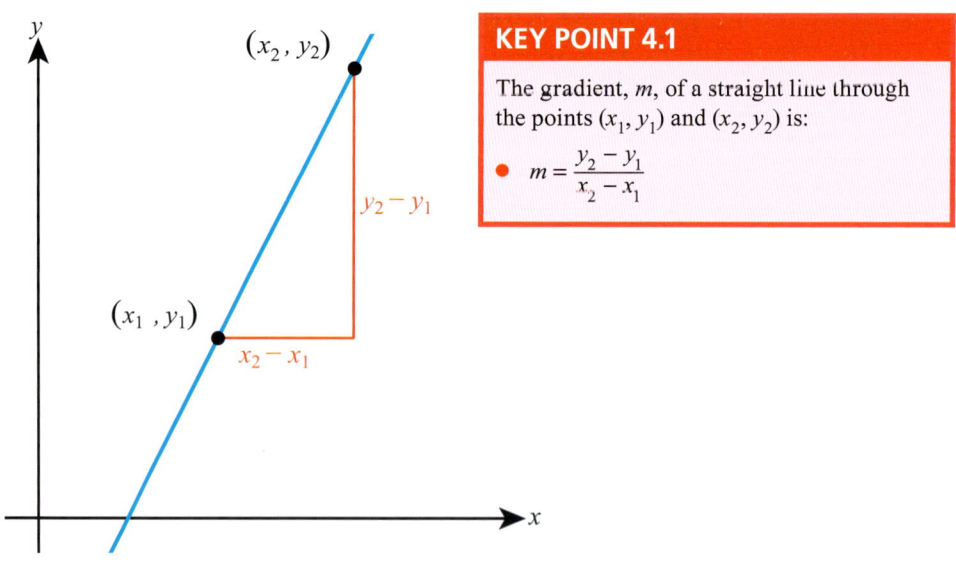

KEY POINT 4.1

The gradient, *m*, of a straight line through the points (x_1, y_1) and (x_2, y_2) is:

- $m = \dfrac{y_2 - y_1}{x_2 - x_1}$

4A Equations of straight lines in two dimensions

WORKED EXAMPLE 4.1

Find the gradient of the line connecting (3, 2) and (5, −4).

Use $m = \dfrac{y_2 - y_1}{x_2 - x_1}$ with $(x_1, y_1) = (3, 2)$ and $(x_2, y_2) = (5, -4)$ .... $m = \dfrac{-4 - 2}{5 - 3}$
$= -3$

Be the Examiner 4.1

Find the gradient of the line through the points (3, −2) and (−4, 7).

Which is the correct solution? Identify the errors made in the incorrect solutions.

Solution 1	Solution 2	Solution 3
$m = \dfrac{-2 - 7}{3 - (-4)}$	$m = \dfrac{7 - 2}{-4 - 3}$	$m = \dfrac{-2 - 3}{7 - (-4)}$
$= -\dfrac{9}{7}$	$= -\dfrac{5}{7}$	$= -\dfrac{5}{11}$

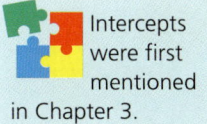

Intercepts were first mentioned in Chapter 3.

WORKED EXAMPLE 4.2

Find the x and y intercepts of the line $2x + 3y - 6 = 0$.

The x-intercept will occur where $y = 0$ ....
When $y = 0$:
$2x - 6 = 0$
$x = 3$
So x-intercept is (3, 0)

The y-intercept will occur where $x = 0$ ....
When $x = 0$:
$3y - 6 = 0$
$y = 2$
So y-intercept is (0, 2)

4 Core: Coordinate geometry

■ Different forms of the equation of a straight line

There are various ways of expressing the equation of a straight line.

> **Tip**
> This is sometimes referred to as 'gradient–intercept form'.

> **KEY POINT 4.2**
> The straight line with gradient m and y-intercept c has equation $y = mx + c$.

Is maths really an 'international language'? Different countries historically have slightly different preferences for the letters they use to refer to the gradient and intercept of a straight line:

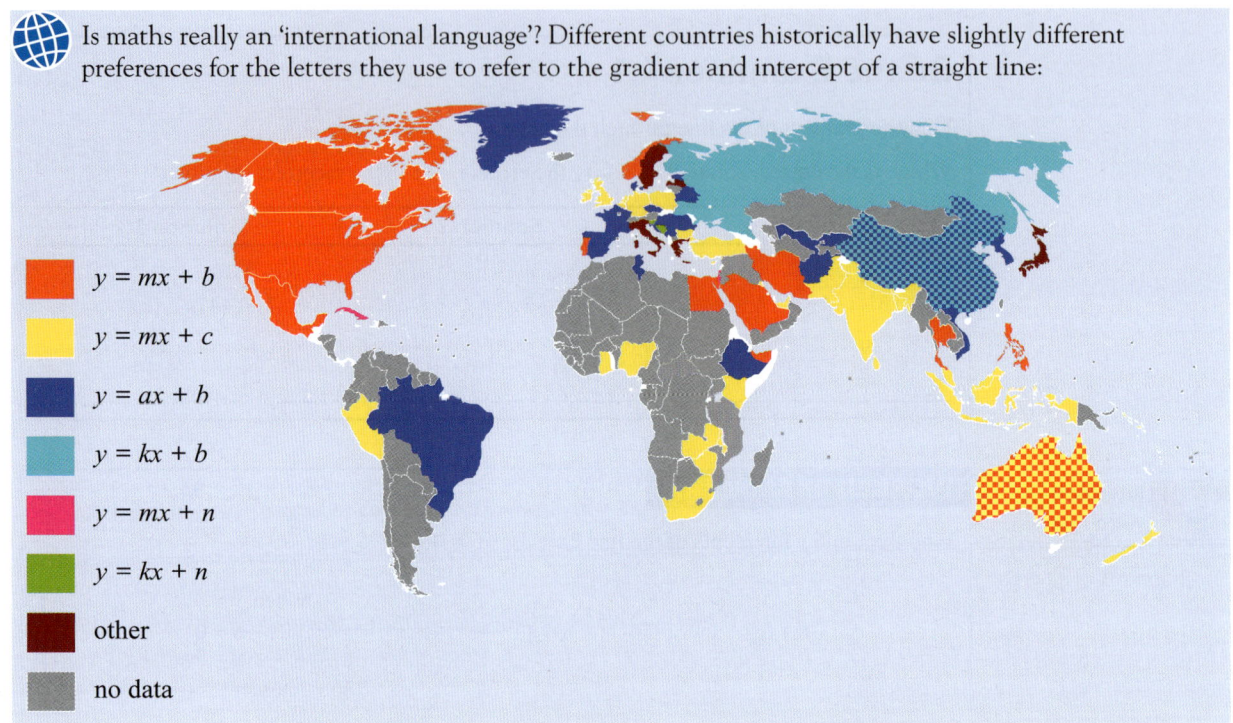

- $y = mx + b$
- $y = mx + c$
- $y = ax + b$
- $y = kx + b$
- $y = mx + n$
- $y = kx + n$
- other
- no data

> **CONCEPTS – SPACE**
> Key Point 4.2 actually only gives the equation of a line in two-dimensional **space**. In 3D it is much harder to describe a line using Cartesian coordinates. The number of dimensions the shape that we are trying to describe has, changes the rules we can use.

> **WORKED EXAMPLE 4.3**
>
> Find the equation of the line with gradient 4 and y-intercept -3 in the form $ax + by + d = 0$.
>
> Since we have information about gradient and y-intercept, we should use $y = mx + c$ .... $m = 4$ and $c = -3$
> Therefore, equation is $y = 4x - 3$
>
> Rearrange into the required form .... $-4x + y + 3 = 0$

4A Equations of straight lines in two dimensions

Tip

The form $ax + by + c = 0$ is sometimes referred to as the 'general form'.

> **WORKED EXAMPLE 4.4**
>
> a Find the equation of the line $3x + 4y + 6 = 0$ in the form $y = mx + c$.
>
> b Hence write down the gradient and y-intercept of the line.
>
> Rearrange to make y the subject
>
> a $3x + 4y + 6 = 0$
> $4y = -3x - 6$
> $y = -\frac{3}{4}x - \frac{3}{2}$
>
> Once in the form $y = mx + c$, the gradient is m and the y-intercept is c
>
> b Therefore,
> gradient $= -\frac{3}{4}$
> y-intercept $= -\frac{3}{2}$

If you know a point on the line and the gradient, there is another, more direct form of the equation of the line to use instead of $y = mx + c$.

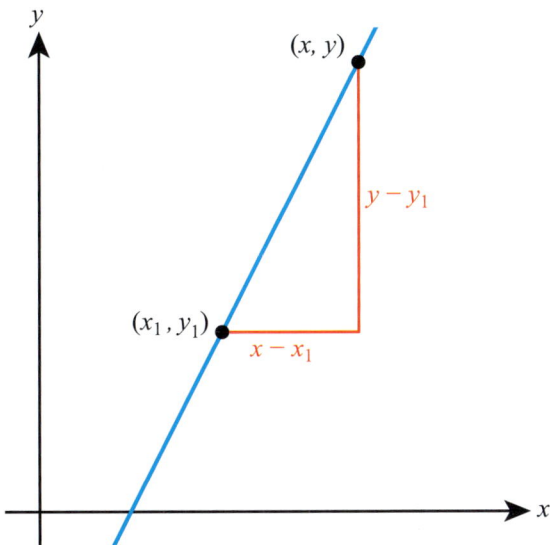

The point (x, y) is any general point on the line. Using this and the known point (x_1, y_1) you can write the gradient as

$$m = \frac{y - y_1}{x - x_1}$$

Rearranging this gives an alternative form of the equation of a straight line.

Tip

This is sometimes referred to as 'point–gradient form'.

> **KEY POINT 4.3**
>
> The straight line with gradient m, passing through the point (x_1, y_1), has equation
> $y - y_1 = m(x - x_1)$.

WORKED EXAMPLE 4.5

Find the equation of the line with gradient 2 through the point (3, 5).

The question does not ask for the answer in any particular form, so it can be left like this

$$y - y_1 = m(x - x_1)$$
$$y - 5 = 2(x - 3)$$

WORKED EXAMPLE 4.6

Find the equation of the line connecting (−1, 1) and (3, −7).

First find the gradient using $m = \dfrac{y_2 - y_1}{x_2 - x_1}$

$$m = \dfrac{-7 - 1}{3 - (-1)}$$
$$= -2$$

Then find the equation of the line using $y - y_1 = m(x - x_1)$. You can choose either of the points to be (x_1, y_1)

$$y - y_1 = m(x - x_1)$$
$$y - 1 = -2(x - (-1))$$
$$y - 1 = -2(x + 1)$$

■ Parallel lines

KEY POINT 4.4

Parallel lines have the same gradient.

WORKED EXAMPLE 4.7

Find the equation of the line parallel to $2x + y + 6 = 0$ through the point (2, 5).

First rearrange into the form $y = mx + c$

$$3x + y + 7 = 0$$
$$y = -3x - 7$$

Any line parallel to $y = -3x - 7$ will have gradient −3

Equation of line parallel to this and passing through (2, 5) is

$$y - 5 = -3(x - 2)$$

■ Perpendicular lines

The gradients of perpendicular lines are also related.

Tip

Key Point 4.5 is often used in the form $m_1 = -\dfrac{1}{m_2}$.

KEY POINT 4.5

If two lines with gradients m_1 and m_2 are perpendicular, then $m_1 m_2 = -1$.

Proof 4.1

Prove that if a line l_1 has gradient m, then a perpendicular line l_2 has gradient $-\frac{1}{m}$.

If line l_1 has gradient m, then a right-angled triangle can be drawn as shown with horizontal side of length 1 and vertical side of length m ····· The line l_1 has gradient $m = \frac{m}{1}$:

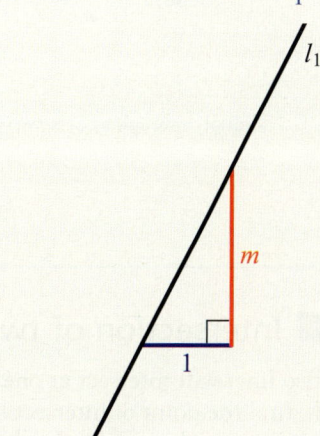

The right-angled triangle is also rotated by 90° making the horizontal side length m and the vertical side length -1 (since it is going down) ····· Rotating l_1 through 90° gives a perpendicular line l_2:

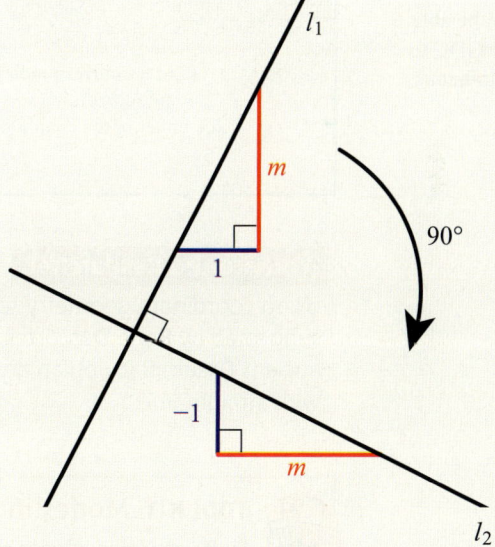

The gradient is $\frac{\text{change in } y}{\text{change in } x}$ ····· l_2 has gradient $\frac{-1}{m}$

WORKED EXAMPLE 4.8

Find the equation of the line perpendicular to $y = 5x + 2$ through the point $(1, 6)$. Give your answer in the form $ax + by + c = 0$, where $a, b, c \in \mathbb{Z}$.

Use $m_1 = -\dfrac{1}{m_2}$ .... Gradient of any line perpendicular to $y = 5x + 2$ is $-\dfrac{1}{5}$.
Equation of line through $(1, 6)$ is

$$y - 6 = -\dfrac{1}{5}(x - 1)$$

Multiply through by 5 .... $5y - 30 = -(x - 1)$
$5y - 30 = -x + 1$
$x + 5y - 31 = 0$

■ Intersection of two lines

Two lines will intersect at one point as long as the lines are not parallel or identical. To find the point of intersection, you need to solve the equations of the lines simultaneously.

Tip

Although you should be able to do this algebraically, as demonstrated in Worked Example 4.9, you will most often just need to be able to use your GDC to solve simultaneous equations.

WORKED EXAMPLE 4.9

Find the point of intersection of $y = 3x + 2$ and $y = x + 4$.

Where the lines intersect, the y-coordinates must be equal, so replace y in the second equation with the expression for y from the first .... $3x + 2 = x + 4$

Solve for x .... $2x = 2$
$x = 1$

Find the corresponding y value by substituting $x = 1$ into either of the equations .... When $x = 1$, $y = 3(1) + 2 = 5$
So the coordinates of the intersection point are $(1, 5)$

You are the Researcher

Using coordinate geometry to represent straight lines is a relatively recent innovation, suggested by René Descartes (1596–1650). You might like to research how the ancient Greek, Babylonian and Chinese mathematicians all described and worked with straight lines.

TOOLKIT: Modelling

Find a roadmap of your country and model the major roads using straight lines. Use this model to calculate the shortest routes between two points. What assumptions are you making in your model? How could you improve the model?

4A Equations of straight lines in two dimensions

Exercise 4A

For questions 1 to 3, use the method demonstrated in Worked Example 4.1 to find the gradient of the line connecting points A and B.

1 a $A(1, 5)$ and $B(3, 7)$
   b $A(2, 2)$ and $B(5, 8)$

2 a $A(4, 1)$ and $B(2, 7)$
   b $A(7, 3)$ and $B(5, 5)$

3 a $A(3, 3)$ and $B(7, 5)$
   b $A(-1, 8)$ and $B(5, 5)$

For questions 4 to 6, use the method demonstrated in Worked Example 4.2 to find the coordinates of the points where the line crosses the x- and y-axes.

4 a $y = 3x - 6$
   b $y = 8 - 2x$

5 a $2y = 5 - x$
   b $3y = 2x + 6$

6 a $4x + 3y - 12 = 0$
   b $2x - 3y - 9 = 0$

For questions 7 to 9, use the method demonstrated in Worked Example 4.3 to find the equation of the straight line with given gradient and y-intercept. Give your answer in the form $ax + by + d = 0$.

7 a Gradient 2, y-intercept 3
   b Gradient 5, y-intercept 1

8 a Gradient -2, y-intercept 4
   b Gradient -3, y-intercept 7

9 a Gradient $\frac{1}{2}$, y-intercept $\frac{7}{2}$
   b Gradient $-\frac{1}{3}$, y-intercept 3

For questions 10 to 12, use the method demonstrated in Worked Example 4.4 to rewrite the line equation in the form $y = mx + c$ and state the gradient and y-intercept of the line.

10 a $2x - y + 5 = 0$
    b $3x - y + 4 = 0$

11 a $2x + 3y + 6 = 0$
    b $5x + 2y - 10 = 0$

12 a $3x + 5y - 7 = 0$
    b $11x + 2y + 5 = 0$

For questions 13 to 15, use the method demonstrated in Worked Example 4.5 to give the equation of the line.

13 a Gradient 2 through $(1, 4)$
    b Gradient 3 through $(5, 2)$

14 a Gradient -5 through $(-1, 3)$
    b Gradient -2 through $(2, -1)$

15 a Gradient $\frac{2}{3}$ through $(1, -1)$
    b Gradient $-\frac{3}{4}$ through $(3, 1)$

For questions 16 to 19, use the method demonstrated in Worked Example 4.6 to find the equation of the line connecting points A and B in the form $ax + by + c = 0$.

16 a $A(3, 1), B(5, 5)$
    b $A(-2, 3), B(1, 9)$

17 a $A(3, 7), B(5, 5)$
    b $A(-2, 4), B(-1, 1)$

18 a $A(5, 1), B(1, 3)$
    b $A(-2, 3), B(6, -1)$

19 a $A(3, 1), B(7, -2)$
    b $A(-2, 7), B(3, -1)$

For questions 20 to 24, use the method demonstrated in Worked Example 4.7 to find the equations of the required lines.

20 a Parallel to $y = 3x + 2$ through $(0, 4)$
    b Parallel to $y = -x + 2$ through $(0, 9)$

21 a Parallel to $y = 1.5x + 4$ through $(1, 5)$
    b Parallel to $y = 0.5x - 3$ through $(-1, -3)$

22 a Parallel to $x - y - 3 = 0$ through $(2, -4)$
    b Parallel to $2x - y - 5 = 0$ through $(3, -1)$

23 a Parallel to $x + 3y - 1 = 0$ through $(1, 1)$
    b Parallel to $x + 5y + 4 = 0$ through $(-3, -1)$

24 a Parallel to $2x - 5y - 3 = 0$ through $(5, 2)$
    b Parallel to $3x + 4y - 7 = 0$ through $(-2, 1)$

For questions 25 to 29, use the method demonstrated in Worked Example 4.8 to find the equations of the required lines.

25 a Perpendicular to $y = x + 2$ through $(0, 5)$
    b Perpendicular to $y = 3x + 2$ through $(0, 8)$

26 a Perpendicular to $y = \frac{1}{4}x + 2$ through $(1, 5)$
    b Perpendicular to $y = -\frac{1}{3}x$ through $(2, 4)$

27 a Perpendicular to $x - 5y - 3 = 0$ through $(-1, 2)$
    b Perpendicular to $x - 2y - 3 = 0$ through $(2, 3)$

28 a Perpendicular to $2x + y - 7 = 0$ through $(1, 3)$
    b Perpendicular to $3x + y + 1 = 0$ through $(-2, -4)$

29 a Perpendicular to $5x + 2y = 0$ through $(-1, 7)$
    b Perpendicular to $3x - 2y = 0$ through $(3, -3)$

For questions 30 to 33, use the method demonstrated in Worked Example 4.9 or technology to find the point of intersection of the given lines.

30 a $x + 2y = 5$ and $x - 2y = -3$
    b $2x + 3y = 13$ and $x - y = 4$
32 a $y = 2x + 1$ and $y = x + 3$
    b $y = 3x + 5$ and $y = x - 1$

31 a $y = x + 2$ and $y = 4 - x$
    b $y = x + 3$ and $y = 7 - x$
33 a $y = 3 - 2x$ and $y = 3x + 1$
    b $y = 7 - 3x$ and $y = 2x - 4$

34 a Calculate the gradient of the line passing through the points $A(1, -7)$ and $B(5, 0)$.

b Find the equation of the line through the point $C(8, 3)$ which is perpendicular to the line AB. Give your answer in the form $y = mx + c$.

35

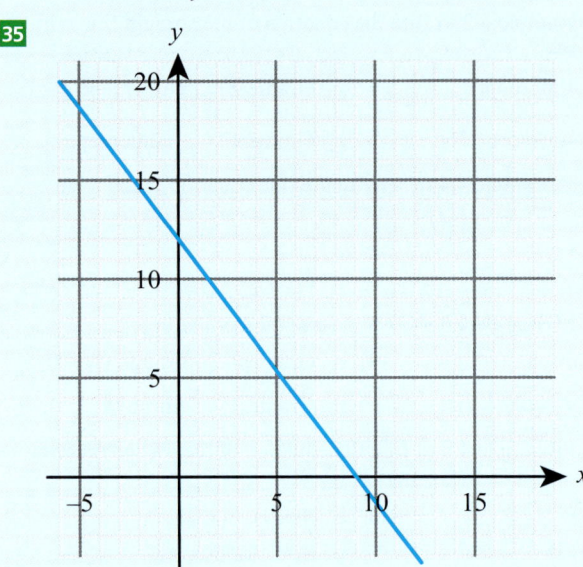

a Calculate the gradient of the line shown in the diagram.

b Find the equation of the line in the form $ax + by = c$, where a, b and c are positive integers.

36 A line has equation $5x + 7y = 17$.

a Find the gradient of the line.

b Find the x-intercept of the line.

37 A straight line connects points $A(-3, 1)$ and $B(7, 3)$.

a Find the gradient of the line.

b Find the equation of the line in the form $ax + by = c$, where a, b and c are integers.

38 A straight line l_1 has equation $7x + 2y = 42$.

a Find the gradient of the line l_1.

b Determine whether the point $P(8, -5)$ lies on the line l_1.

c Find the equation of the line l_2 which passes through P and is perpendicular to l_1. Give your answer in the form $y = mx + c$.

39 A mountain is modelled as a cone of height 8800 m and base width 24 000 m. Based on this model, what is the gradient of the mountain?

40 Point P has coordinates $(0, 11)$. Point Q is 8 units right of P. Point R has coordinates $(k, 3)$ and is directly below Q.
  a Write down the coordinates of Q.
  b Write down the value of k.
  c Write down the equation of the line QR.
  d Find the area of the triangle PQR.

41 Line l_1 has gradient $-\frac{3}{2}$ and crosses the y-axis at the point $(0, -6)$.
Line l_2 passes through the points $(-5, 1)$ and $(7, -2)$.
  a Write down the equation of l_1.
  b Find the equation of l_2, giving your answer in the form $ax + by = c$.
  c Find the coordinates of the point of intersection of l_1 and l_2.

42 A drawbridge consists of a straight plank pivoted at one end.
When the other end of the bridge is 3.5 m above the horizontal, the plank is at a gradient of 0.7. Find the length of the plank.

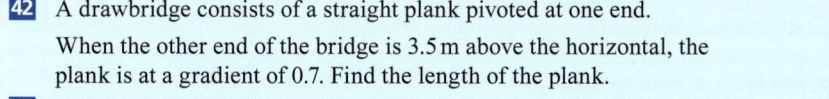

43 A balloon initially has a volume of 0.1 litres. It is being inflated at a rate of 0.5 litres per second.
  a Find an expression for the volume of the balloon after t seconds.
  b The balloon pops when its volume reaches 5 litres. How long would this take?

44 Hooke's law states that the force (F Newtons) required to stretch a spring x metres is given by $F = kx$.
  a State the units of k.
  b If the numerical value of k is 0.3, find the force required to extend the spring by 0.06 m.
  c Would a stiffer spring have a larger, smaller or the same value of k?
  d The spring breaks when a force of 0.14 N is applied. How much has it extended at the point of breaking?

45 A mobile phone contract has a fixed charge of \$5 per month plus \$1 per 100 minutes spent talking.
The monthly cost is \$$C$ when Joanna talks for m minutes.
  a Find a relationship of the form $C = am + b$
  b Find the cost if Joanna talks for a total of 3 hours in a month.
  c An alternative contract has no fixed charge, but charges \$1 per 50 minutes spent talking. How many minutes should Joanna expect to talk each month for the first contract to be better value?

46 A company receives £10 per item it sells. It has fixed costs of £2000 per month.
  a Find an expression for the profit P if the company sells n items.
  b How many items must the company sell to make a profit of £1500?
  c The company refines the model it is using. It believes that its fixed costs are actually £1200 but it has a variable cost of £2 per item sold. Find a new expression for the profit in terms of n.
  d At what value of n do the two models predict the same profit? Based on the new model, will the company have to sell more or fewer items to make a profit of £1500?

47 Consider a quadrilateral with vertices $A(-4, 3)$, $B(3, 8)$, $C(5, -1)$ and $D(-9, -11)$.
  a Show that the sides AB and CD are parallel.
  b Show that $ABCD$ is not a parallelogram.

48 A car can drive up a road with maximum incline 0.3. Find the minimum length of road which is required for a car to go up a mountain road from sea level to a point 400 m above sea level.

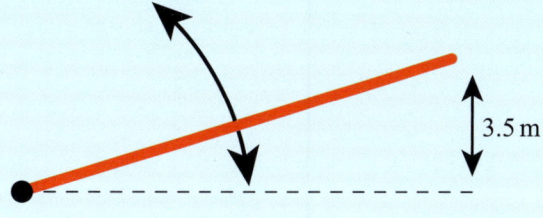

4B Three-dimensional coordinate geometry

■ The distance between two points

The idea of using Pythagoras' theorem to find the distance between two points in two dimensions can be extended to three dimensions.

> **KEY POINT 4.6**
>
> The distance d between the points (x_1, y_1, z_1), and (x_2, y_2, z_2) is:
>
> - $d = \sqrt{(x_2 - x_1)^2 + (y_2 - y_1)^2 + (z_2 - z_1)^2}$

Proof 4.2

Prove that the distance, d, between (x_1, y_1, z_1) and (x_2, y_2, z_2) is

$d = \sqrt{(x_2 - x_1)^2 + (y_2 - y_1)^2 + (z_2 - z_1)^2}$.

Consider the two points (x_1, y_1, z_1) and (x_2, y_2, z_2) as being at opposite ends of a diagonal of a cuboid:

Use Pythagoras in 2D for a right-angled triangle lying in the base of the cuboid to find the length of the diagonal

The length of the diagonal of the base of the cuboid, l, is related to the sides of the base by

$l^2 = (x_2 - x_1)^2 + (y_2 - y_1)^2$

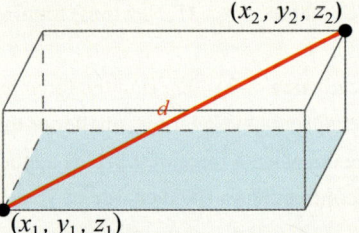

Then use Pythagoras in 2D again for the right-angled triangle with base l and hypotenuse d

Then, $d^2 = l^2 + (z_2 - z_1)^2$
$= (x_2 - x_1)^2 + (y_2 - y_1)^2 + (z_2 - z_1)^2$

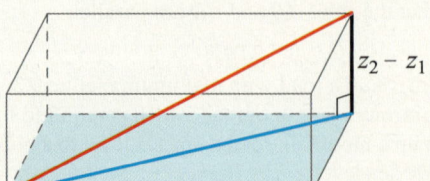

Therefore, $d = \sqrt{(x_2 - x_1)^2 + (y_2 - y_1)^2 + (z_2 - z_1)^2}$

4B Three-dimensional coordinate geometry

> **WORKED EXAMPLE 4.10**
>
> Find the distance between $(1, 2, 3)$ and $(3, -2, -4)$.
>
> Use $d = \sqrt{(x_2 - x_1)^2 + (y_2 - y_1)^2 + (z_2 - z_1)^2}$ .... $d = \sqrt{(3-1)^2 + (-2-2)^2 + (-4-3)^2}$
>
> $ = \sqrt{4 + 16 + 49}$
> $ = \sqrt{69}$
> $ = 8.31$ (3 s.f.)

■ The midpoint

The midpoint of the line connecting two points can be thought of as the average of the coordinates of the two points.

> **KEY POINT 4.7**
>
> The midpoint, M, of the points (x_1, y_1, z_1) and (x_2, y_2, z_2) is:
>
> - $M = \left(\dfrac{x_1 + x_2}{2}, \dfrac{y_1 + y_2}{2}, \dfrac{z_1 + z_2}{2} \right)$

> **WORKED EXAMPLE 4.11**
>
> Find the midpoint of $(3, 4, 8)$ and $(5, 5, -5)$.
>
> Use $M = \left(\dfrac{x_1 + x_2}{2}, \dfrac{y_1 + y_2}{2}, \dfrac{z_1 + z_2}{2} \right)$ .... $M = \left(\dfrac{3+5}{2}, \dfrac{4+5}{2}, \dfrac{8+(-5)}{2} \right)$
>
> $ = (4, 4.5, 1.5)$

Exercise 4B

For questions 1 to 3, use techniques from your prior learning to find the exact distance between the two points given.

1. **a** $(0, 0)$ and $(4, 3)$
   b $(0, 0)$ and $(5, 12)$
2. **a** $(7, 5)$ and $(2, -7)$
   b $(4, -1)$ and $(-2, 7)$
3. **a** $(3, 5)$ and $(-2, 3)$
   b $(-2, 5)$ and $(4, -2)$

For questions 4 to 6, use the method demonstrated in Worked Example 4.10 to find the exact distance between the two points given.

4. **a** $(0, 0, 0)$ and $(1, 2, 2)$
   b $(0, 0, 0)$ and $(2, 6, 9)$
5. **a** $(4, 1, 3)$ and $(6, 5, 7)$
   b $(-3, 2, -1)$ and $(3, -1, 5)$
6. **a** $(1, 5, -3)$ and $(4, -3, 2)$
   b $(-4, -3, 7)$ and $(2, 2, 9)$

For questions 7 to 9, use the method demonstrated in Worked Example 4.11 to find the midpoint of line segment AB.

7. **a** $A(1, 1)$ and $B(7, -3)$
   b $A(5, 2)$ and $B(3, 10)$
8. **a** $A(1, -3, 7)$ and $B(5, -5, -1)$
   b $A(2, 4, -8)$ and $B(-4, 0, 6)$
9. **a** $A(4, 3, 2)$ and $B(7, 1, -8)$
   b $A(-2, -2, 11)$ and $B(-9, 9, 3)$

10. **a** Find the coordinates of the midpoint of the line segment connecting points $A(-4, 1, 9)$ and $B(7, 0, 2)$.
    b Find the length of the segment AB.
11. Point A has coordinates $(4, -1, 2)$. The midpoint of AB has coordinates $(5, 1, -3)$. Find the coordinates of B.
12. The midpoint of the line segment connecting the points $P(-4, a, 1)$ and $Q(b, 1, 8)$ is $M(8, 2, c)$. Find the values of a, b and c.

13 Points A and B have coordinates $(3, -18, 8)$ and $(2, -2, 11)$. Find the distance of the midpoint of AB from the origin.

14 A birdhouse of height 1.6 m is located at the origin. A bird flies in a straight line from the tree house to the top of a 8.3 m tall tree located at the point with coordinates $(14, 3)$. The flight takes 3.5 seconds. Find the average speed of the bird.

15 The distance of the point $(k, 2k, 5k)$ from the origin is 30. Find the positive value of k.

16 The distance between points $(k, k, 0)$ and $(1, -1, 3k)$ is $\sqrt{46}$. Find the two possible values of k.

17 The point $(2a, a, 5a)$ is twice as far from the origin as the point $(-4, 1, 7)$. Find the positive value of a.

18 The midpoint of the line segment connecting points $P(3a + 1, 2a)$ and $Q(5 - b, b + 3)$ has coordinates $(4, -5)$.

   a Find the values of a and b.
   b Find the equation of the line connecting P and Q.

19 Line l_1 has equation $4x - 7y = 35$.

   a Find the gradient of the line l_1.
   b Line l_2 passes through the point $N(-4, 2)$ and is perpendicular to l_1. Find the equation of l_2 in the form $ax + by = c$, where a, b and c are integers.
   c Find the coordinates of P, the point of intersection of the lines l_1 and l_2.
   d Hence find the shortest distance from point N to the line l_1.

20 A tent has vertices $A(0, 0, 0)$, $B(3, 0, 0)$, $C(3, 2, 0)$, $D(0, 2, 0)$, $E(0.2, 1, 1.8)$ and $F(2.8, 1, 1.8)$. All lengths are in metres.

   a State the height of the tent.
   b An extra support needs to be added from the midpoint of BC to the corner E. Find the length of this support.

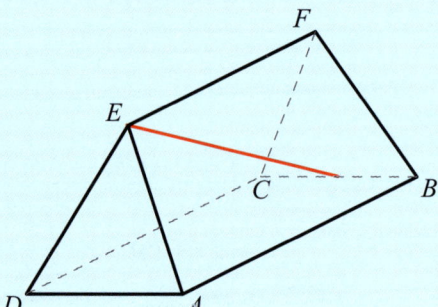

21 $A(1, 4)$ and $C(5, 10)$ are opposite vertices of a square $ABCD$.

   a Find the midpoint of AC.
   b Find the coordinates of B and D.

22 The points $A(1, 1)$ and $C(8, 8)$ form the opposite ends of the diagonal of rhombus $ABCD$.

   a Using the fact that each diagonal of a rhombus perpendicularly bisects the other, find the equation of the line BD.
   b If the gradient of line segment AB is $\frac{4}{3}$, find the coordinates of B and D.
   c Find the length of each side of the rhombus.

23 An empty room is 3 m by 4 m by 5 m. A spider and a fly start in one corner of the room.

   a The fly travels to the furthest corner of the room by flying and stops there. What is the shortest distance it can travel to get there?
   b The spider then chases after the fly by crawling along any wall. What is the shortest distance it can travel to get to the fly's final position?

Mixed Practice 89

Checklist

- You should be able to find the gradient of a straight line given two points on the line: $m = \dfrac{y_2 - y_1}{x_2 - x_1}$
- You should be able to work with the equation of a straight line in several different forms:
  - ☐ Gradient–intercept form: $y = mx + c$
  - ☐ General form: $ax + by + c = 0$
  - ☐ Point–gradient form: $y - y_1 = m(x - x_1)$
- You should know that parallel lines have the same gradient.
- You should know that if two lines with gradients m_1 and m_2 are perpendicular, then $m_1 m_2 = -1$.
- You should be able to find the intersection of two straight lines by solving simultaneous equations.
- You should be able to find the distance between two points in three dimensions:
  - ☐ $d = \sqrt{(x_2 - x_1)^2 + (y_2 - y_1)^2 + (z_2 - z_1)^2}$
- You should be able to find the midpoint of two points in three dimensions:
  - ☐ $M = \left(\dfrac{x_1 + x_2}{2}, \dfrac{y_1 + y_2}{2}, \dfrac{z_1 + z_2}{2} \right)$

■ Mixed Practice

1 $P(4, 1)$ and $Q(0, -5)$ are points on the coordinate plane.
  a Determine the
    i coordinates of M, the midpoint of P and Q
    ii gradient of the line drawn through P and Q
    iii gradient of the line drawn through M, perpendicular to PQ.

  The perpendicular line drawn through M meets the y-axis at $R(0, k)$.
  b Find k.

  Mathematical Studies SL May 2007 Paper 1 Q10

2 The midpoint, M, of the line joining $A(s, 8)$ to $B(-2, t)$ has coordinates $M(2, 3)$.
  a Calculate the values of s and t.
  b Find the equation of the straight line perpendicular to AB, passing through the point M.

  Mathematical Studies SL November 2007 Paper 1 Q13

3 The straight line, L_1, has equation $2y - 3x = 11$. The point A has coordinates $(6, 0)$.
  a Give a reason why L_1 **does not** pass through A.
  b Find the gradient of L_1.
  L_2 is a line perpendicular to L_1. The equation of L_2 is $y = mx + c$.
  c Write down the value of m.
  L_2 **does** pass through A.
  d Find the value of c.

  Mathematical Studies SL May 2013 Paper 1 TZ1 Q10

4 Two points have coordinates $A(3, 6)$ and $B(-1, 10)$.
   a Find the gradient of the line AB.

   Line l_1 passes through A and is perpendicular to AB.
   b Find the equation of l_1.
   c Line l_1 crosses the coordinate axes at the points P and Q. Calculate the area of the triangle OPQ, where O is the origin.

5 Two points have coordinates $P(-1, 2)$ and $Q(6, -4)$.
   a Find the coordinates of the midpoint, M, of PQ.
   b Calculate the length of PQ.
   c Find the equation of a straight line which is perpendicular to PQ and passes through M.

6 Two points have coordinates $P(-1, 2, 5)$ and $A(6, -4, 3)$.
   a Find the coordinates of the midpoint, M, of PQ.
   b Calculate the length of PQ.

7 The line l_1 with equation $x + 2y = 6$ crosses the y-axis at P and the x-axis at Q.
   a Find the coordinates of P and Q.
   b Find the exact distance between P and Q.
   c Find the point where the line $y = x$ meets l_1.

8 The line connecting points $M(3, -5)$ and $N(-1, k)$ has equation $4y + 7x = d$.
   a Find the gradient of the line.
   b Find the value of k.
   c Find the value of d.

9 The vertices of a quadrilateral have coordinates $A(-3, 8)$, $B(2, 5)$, $C(1, 6)$ and $D(-4, 9)$.
   a Show that $ABCD$ is a parallelogram.
   b Show that $ABCD$ is not a rectangle.

10 Show that the triangle with vertices $(-2, 5)$, $(1, 3)$ and $(5, 9)$ is right angled.

11 The distance of the point $(-4, a, 3a)$ from the origin is $\sqrt{4160}$. Find two possible values of a.

12 The midpoint of the line joining points $A(2, p, 8)$ and $B(-6, 5, q)$ has coordinates $(-2, 3, -5)$.
   a Find the values of p and q.
   b Calculate the length of the line segment AB.

13 The cross section of a roof is modelled as an isosceles triangle. The width of the house is 8 m and the height of the roof is 6 m. Find the gradient of the side of the roof.

14 The lines with equations $y = \frac{1}{2}x - 3$ and $y = 2 - \frac{2}{3}x$ intersect at the point P. Find the distance of P from the origin.

15 Triangle ABC has vertices $A(-4, 3)$, $B(5, 0)$ and $C(4, 7)$.
   a Show that the line l_1 with equation $y = 3x$ is perpendicular to AB and passes through its midpoint.
   b Find the equation of the line l_2 which is perpendicular to AC and passes through its midpoint.
   c Let S be the intersection of the lines l_1 and l_2. Find the coordinates of S and show that it is the same distance from all three vertices.

16 Health and safety rules require that ramps for disabled access have a maximum gradient of 0.2. A straight ramp is required for accessing a platform 2 m above ground level. What is the closest distance from the platform the ramp can start?

Mixed Practice

17 The point $C(2, 0, 2)$ is plotted on the diagram on the right.
  a Copy the diagram and plot the points $A(5, 2, 0)$ and $B(0, 3, 4)$.
  b Calculate the coordinates of M, the midpoint of AB.
  c Calculate the length of AB.

  Mathematical Studies SL November 2005 Paper 1 Q15

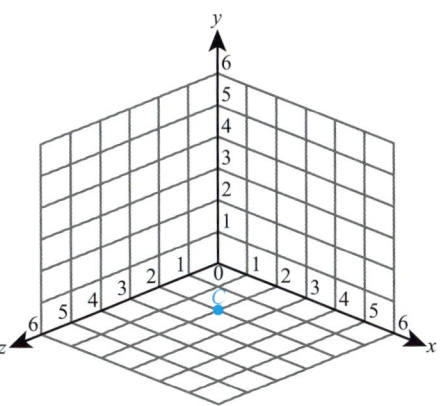

18 Two points have coordinates $A(-7, 2)$ and $B(5, 8)$.
  a Calculate the gradient of AB.
  b Find the coordinates of M, the midpoint of AB.
  c Find the equation of the line l_1 which passes through M and is perpendicular to AB. Give your answer in the form $ax + by = c$, where a, b and c are integers.
  d Show that the point $N(1, 1)$ lies on l_1.
  e Hence find the perpendicular distance from N to the line AB.

19 A straight line has equation $3x - 7y = 42$ and intersects the coordinate axes at points A and B.
  a Find the area of the triangle AOB, where O is the origin.
  b Find the length of AB.
  c Hence find the perpendicular distance of the line from the origin.

20 Line l_1 has equation $y = 5 - \frac{1}{2}x$ and crosses the x-axis at the point P. Line l_2 has equation $2x - 3y = 9$ and intersects the x-axis at the point Q. Let R be the intersection of the lines l_1 and l_2. Find the area of the triangle PQR.

21 $A(8, 1)$ and $C(2, 3)$ are opposite vertices of a square $ABCD$.
  a Find the equation of the diagonal BD.
  b Find the coordinates of B and D.

22 Quadrilateral $ABCD$ has vertices with coordinates $(-3, 2)$, $(4, 3)$, $(9, -2)$ and $(2, -3)$. Prove that $ABCD$ is a rhombus but not a square.

23 A car drives up a straight road with gradient 0.15. How far has the car travelled when it has climbed a vertical distance of 20 m?

24 A bridge consists of two straight sections, each pivoted at one end. When both sections of the bridge are raised 6 m above the horizontal, each section has gradient 0.75. Find the distance, d, between the two closest end points of the two sections.

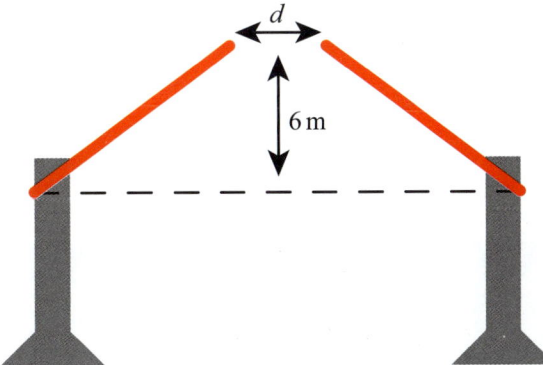

5 Core: Geometry and trigonometry

ESSENTIAL UNDERSTANDINGS

- Geometry and trigonometry allow us to quantify the physical world, enhancing our spatial awareness in two and three dimensions. This topic provides us with the tools for analysis, measurement and transformation of quantities, movements and relationships.

In this chapter you will learn…
- how to find the volume and surface area of three-dimensional solids
- how to work with trigonometric ratios and inverse trigonometric ratios
- how to find the angle between two intersecting lines in two dimensions
- how to use the sine rule to find sides and angles in non-right-angled triangles
- how to use the cosine rule to find sides and angles in non-right-angled triangles
- how to find the area of a triangle when you do not know the perpendicular height
- how to find the angle between two intersecting lines in three-dimensional shapes
- how to find the angle between a line and a plane in three-dimensional shapes
- how to construct diagrams from given information
- how to use trigonometry in questions involving bearings
- how to use trigonometry in questions involving angles of elevation and depression.

CONCEPTS

The following key concepts will be addressed in this chapter:
- Volume and surface area of shapes are determined by formulae, or general mathematical **relationships** or rules expressed using symbols or variables.
- The **relationships** between the length of the sides and the size of the angles in a triangle can be used to solve many problems involving position, distance, angles and area.

■ Figure 5.1 How else are triangles used in the real world?

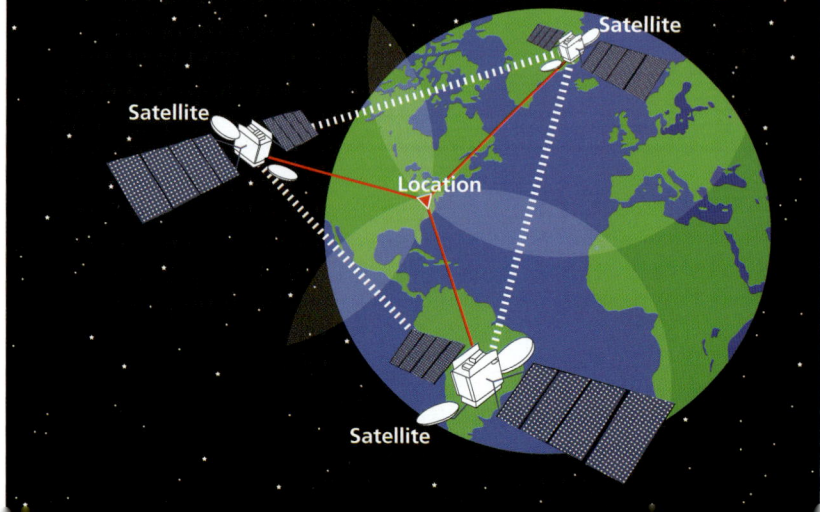

Core: Geometry and trigonometry

> **PRIOR KNOWLEDGE**
>
> Before starting this chapter, you should already be able to complete the following:
>
> 1 Find the angle θ:
>
>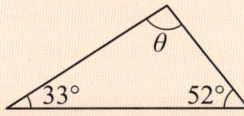
>
> 2 Find the length AC:
>
>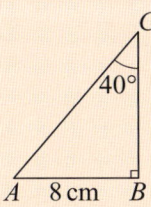
>
> 3 Find the area of the following triangle:
>
>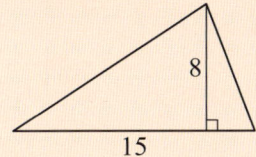
>
> 4 Find the angles α and β:
>
>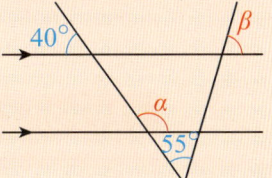
>
> 5 Point A is on a bearing of 230° from point B. Find the bearing of B from A.
>
> 6 Find the volume and surface area of the solid cylinder shown:
>
>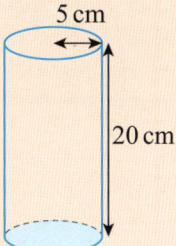

Geometry was initially all about measuring the real world, with applications from architecture to astronomy. But now mathematicians consider geometry to be the study of what stays the same when other things change. There are many different branches of geometry that seemingly have little to do with the real world that classical geometry models.

Starter Activity

Look at Figure 5.1. In small groups, discuss why triangles are of such fundamental importance in the real world.

Now look at this problem:

Draw any right-angled triangle containing an angle of 50°.

Label a as the shortest side of your triangle and b as the longest (the hypotenuse).

With a ruler, measure the lengths of the sides a and b and then work out:

a $\quad a + b$ b $\quad a - b$ c $\quad ab$ d $\quad \dfrac{a}{b}$

Compare your answers with other people's answers. What do you notice?

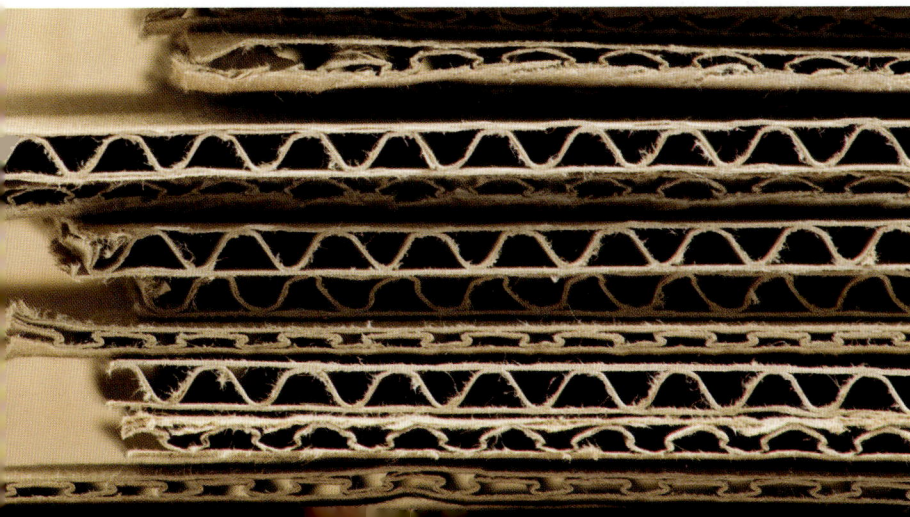

5A Volumes and surface areas of three-dimensional solids

In addition to the volume and surface area of cuboids, cylinders and prisms that you already know, you also need the formulae for the volume and surface area of other common three-dimensional shapes such as spheres, cones and pyramids.

KEY POINT 5.1

Shape	Volume	Surface area
Sphere of radius r 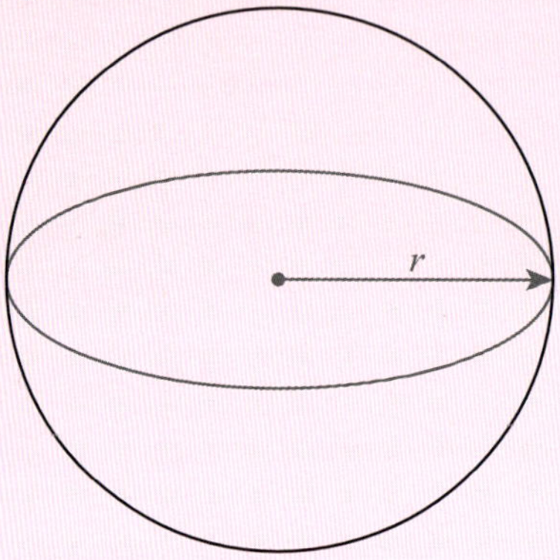	$\frac{4}{3}\pi r^3$	$4\pi r^2$
Cone of base radius r, height h and slope length l 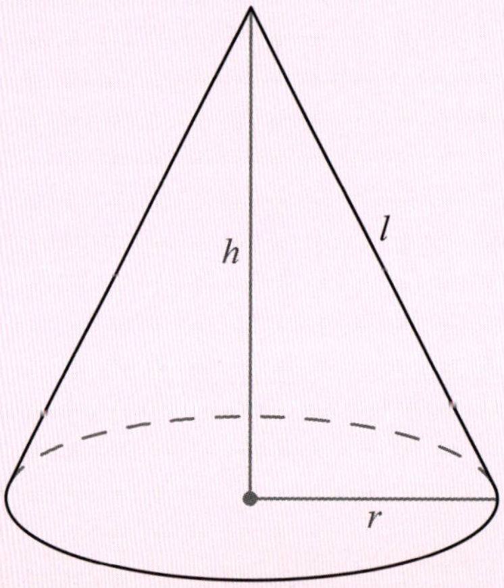	$\frac{1}{3}\pi r^2 h$	$\pi rl + \pi r^2$
Pyramid of base area B and height h 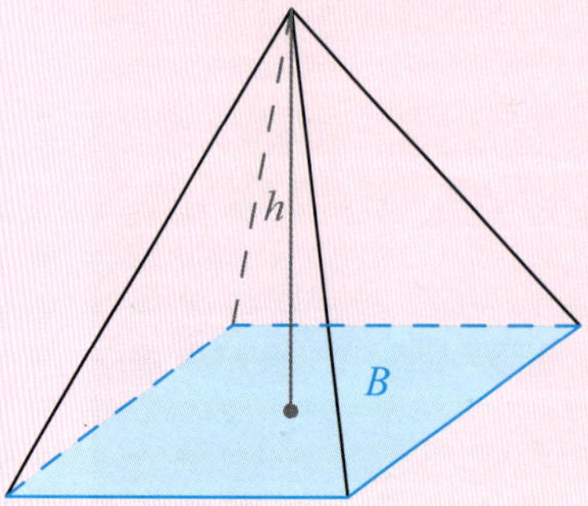	$\frac{1}{3}Bh$	Area of triangular sides + B

Tip

The surface area of a cone is in two parts. The curved surface area is πrl and the base area is πr^2.

5A Volumes and surface areas of three-dimensional solids

WORKED EXAMPLE 5.1

Find the surface area of the hemisphere shown.

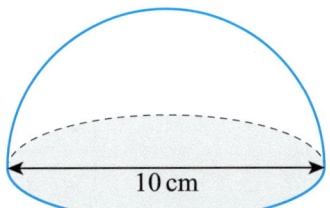

10 cm

The surface area is made up of half a sphere and a circle

$$\text{Surface area} = \frac{4\pi r^2}{2} + \pi r^2$$

The diameter is 10 cm so the radius is 5 cm

$$= \frac{4\pi(5)^2}{2} + \pi(5)^2$$

$$= 75\pi$$

$$\approx 236 \text{ cm}^2$$

WORKED EXAMPLE 5.2

The following shape models an ice cream as a hemisphere attached to a cone.

Find the volume of this shape.

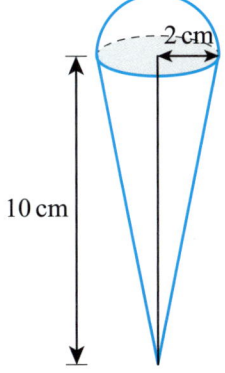

2 cm

10 cm

Add together the volume of the cone and the volume of the hemisphere

$$\text{Volume} = \frac{1}{3}\pi r^2 h + \frac{2}{3}\pi r^3$$

The radius of the cone and the hemisphere are both 2

$$= \frac{1}{3}\pi(2)^2(10) + \frac{2}{3}\pi(2)^3$$

$$= \frac{40}{3}\pi + \frac{16}{3}\pi$$

$$= \frac{56}{3}\pi$$

$$\approx 58.6 \text{ cm}^3$$

You might have to first use your knowledge of Pythagoras' theorem to find unknown lengths.

> **WORKED EXAMPLE 5.3**
>
> Find the surface area and volume of this solid pyramid.
>
>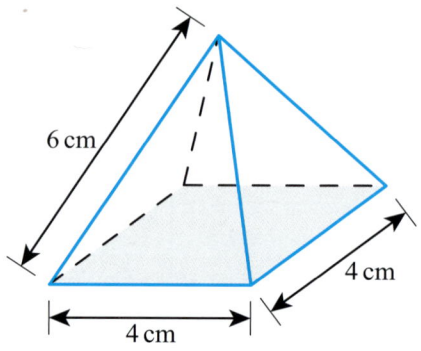
>
>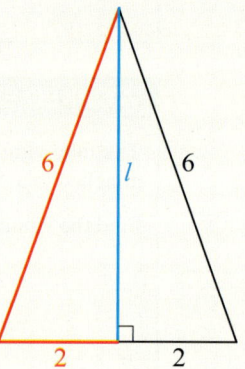
>
> To find the area of each triangular side, we need their perpendicular height. For this, use Pythagoras
>
> $l^2 = 6^2 - 2^2$
>
> So, $l = \sqrt{32} = 4\sqrt{2}$
>
> Use Area = $\frac{1}{2}bh$ to find the area of each triangle
>
> Area of triangular side = $\frac{1}{2} \times 4 \times 4\sqrt{2}$
>
> $= 8\sqrt{2}$
>
> There are four triangular sides and a square base
>
> So area of pyramid = $4 \times 8\sqrt{2} + 4^2$
>
> $= 32\sqrt{2} + 16$
>
> $\approx 61.3 \text{cm}^2$
>
>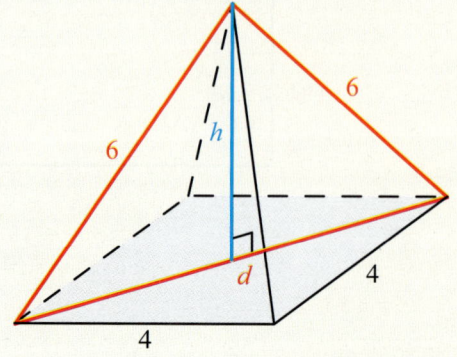
>
> For the volume, we need the perpendicular height of the pyramid. First find the length of the diagonal of the base
>
> $d^2 = 4^2 + 4^2$
>
> So, $d = \sqrt{32} = 4\sqrt{2}$

> **Tip**
> A 'right pyramid' is a special type of pyramid where the apex of the pyramid is directly above the centre of the base.

The right-angled triangle including the perpendicular height will have a base that is half of the diagonal, that is, $2\sqrt{2}$

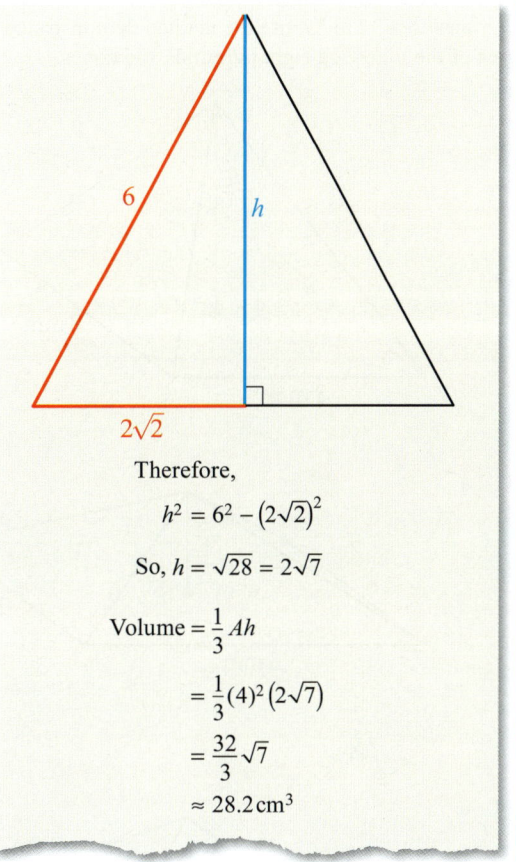

Therefore,
$h^2 = 6^2 - (2\sqrt{2})^2$

So, $h = \sqrt{28} = 2\sqrt{7}$

Volume $= \frac{1}{3}Ah$

$= \frac{1}{3}(4)^2(2\sqrt{7})$

$= \frac{32}{3}\sqrt{7}$

$\approx 28.2\,\text{cm}^3$

Exercise 5A

For questions 1 to 4, use Key Point 5.1 and the method demonstrated in Worked Example 5.1 to find the curved surface area and volume of the following solids.

1. a A sphere with radius 1.8 cm
    b A sphere with radius 11.2 m
2. a A sphere with diameter 1.6 m
    b A sphere with diameter 0.2 m
3. a A hemisphere with radius 5.2 m
    b A hemisphere with radius 4 km
4. a A cone with base radius 3 cm and height 4.1 cm
    b A cone with base radius 0.2 mm and height 7 mm

For questions 5 to 8, use Key Point 5.1 and the method demonstrated in Worked Example 5.1 to find the exact total surface area and volume of the following solids.

5. a A sphere with radius 2 cm
    b A sphere with radius 3 m
6. a A sphere with diameter 8 m
    b A sphere with diameter 12 m
7. a A hemisphere with radius $\frac{1}{2}$ m
    b A hemisphere with radius $\frac{2}{3}$ mm
8. a A cone with base radius 5 cm and height 12 cm
    b A cone with base radius 4 cm, height 3 cm

For questions 9 to 11, use Key Point 5.1 to calculate the volume of the following shapes.

9. a A triangular-based pyramid with base area 6 cm² and height 3 cm
    b A triangular-based pyramid with base area 8 cm² and height 6 cm
10. a A square-based pyramid with base area 5 cm² and height 12 cm
    b A square-based pyramid with base area 9 cm² and height 6 cm
11. a A cone with base area 8 cm² and height 10 cm
    b A cone with base area 2 cm² and height 4 cm

For questions 12 to 15, use the method demonstrated in Worked Example 5.3 to calculate the volume and total surface area of the following right pyramids and cones.

12 a

b

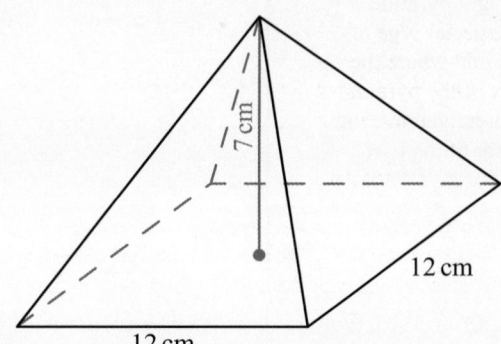

13 a

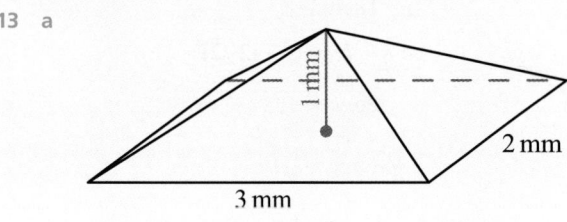

b

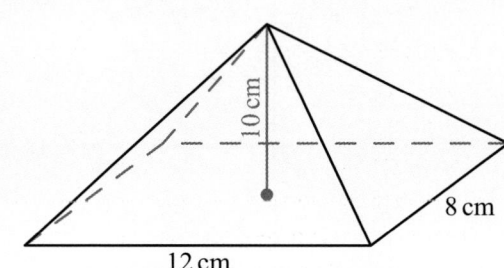

14 a

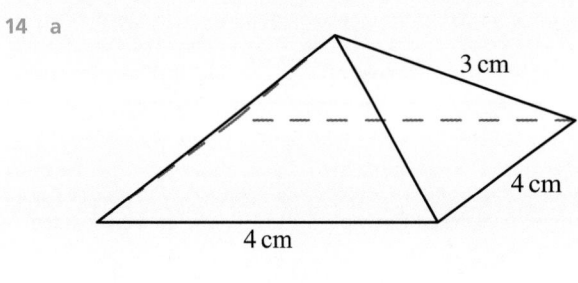

b

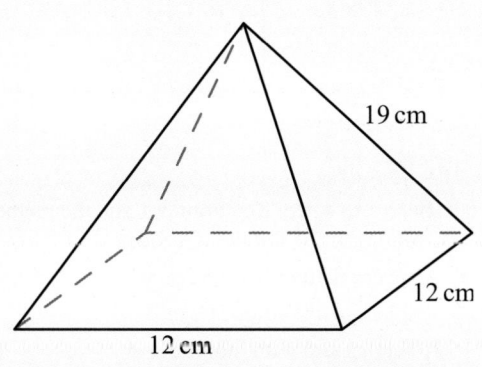

15 a

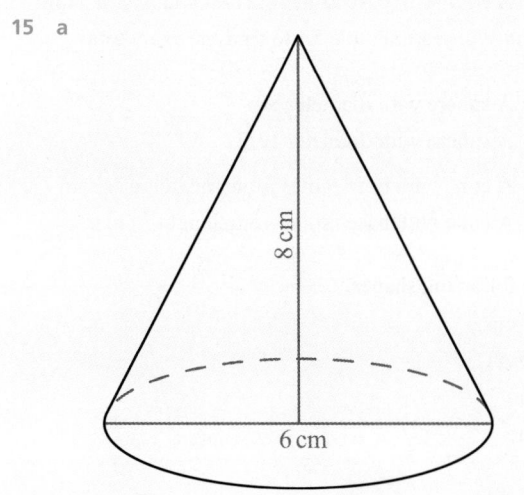

b

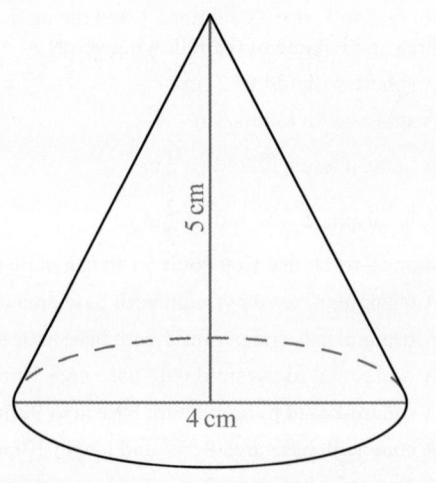

16 Find the surface area of a sphere with diameter 15 cm.
17 A solid hemisphere has radius 3.2 cm. Find its total surface area.
18 The base of a conical sculpture is a circle of diameter 3.2 m. The height of the sculpture is 3.1 m. Find the volume of the sculpture.
19 Find the volume of a solid hemisphere of radius 4 cm.
20 Find the volume of a sphere with diameter 18.3 cm.
21 A cylinder with length 1 m and radius 5 cm has a cone of the same radius and length 10 cm attached to one end. Find the surface area and volume of this compound shape.
22 A fence post can be modelled as a cylinder with a hemisphere on top. The cylinder and the hemisphere have the same diameter of 28 cm. The total height of the post is 73 cm. Find the volume of the whole post, giving your answer in standard form.
23 A cone has vertical height of 12 cm and slant height of 17 cm.
  a Find the radius of the base of the cone.
  b Find the total surface area of the cone.
24 A cylinder has height 12.3 cm and volume 503.7 cm^3.
  a Find the radius of the base of the cylinder.
  b Find the height of a cone which has the same radius and same volume as the cylinder.
25 A metal ornament is made in the shape of a cone with radius 4.7 cm and height 8.3 cm.
  a Find the volume of the ornament.
  The ornament is melted down and made into a ball.
  b Find the radius of the ball.
26 A toy rocket can be modelled as a cone on top of a cylinder. The cone and the cylinder both have the diameter of 18 cm. The height of the cylinder is 23 cm and the height of the whole toy is 35 cm.
  a Find the slant height of the cone.
  b Find the total surface area of the toy rocket.
27 The base of a pyramid is a square of side 12 cm. Each side face is an isosceles triangle with sides 12 cm, 15 cm and 15 cm.
  a Find the area of each side face.
  b Hence find the total surface area of the pyramid.
  c Find the volume of the pyramid.
28 A cone with height 6 cm and radius 2 cm has a hemispherical hole with radius 1 cm bored into the centre of its base. Find the surface area and volume of this compound shape.

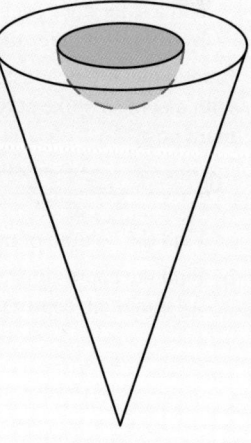

29 Two hemispheres, with A the centre of each, are joined at their flat surfaces. One has radius 8 mm and the other radius 10 mm. Find the total surface area and volume of this compound shape.

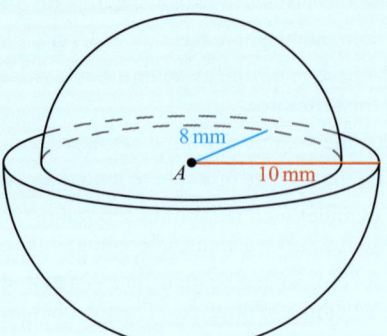

30 A conic frustum is formed by slicing the cone tip from a larger cone, with the slice parallel to the cone base. Find the volume and total surface area of a frustum formed by taking a right cone with radius 8 mm and length 30 mm and removing the cone tip with length 12 mm.

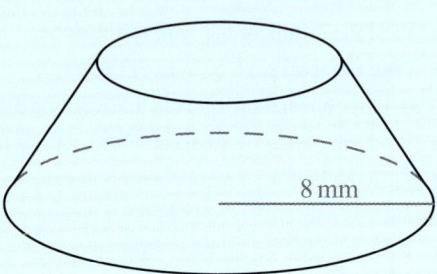

31 A square-based pyramid has height 24 cm and volume 1352 m³
  a Find the length of one side of the base.
  b Find the total surface area of the pyramid.

32 Find the surface area of the sphere whose volume is 354 m³.

33 A pharmaceutical company intends to make a new tablet, where each tablet takes the form of a cylinder capped at each end with a hemisphere. The initial tablet design has radius 2 mm and cylinder length 8 mm.
  a Find the surface area and volume of this shape.
  b After trialing the tablet, it is found that the radius of the tablet must be decreased by 10% for ease of swallowing, and a safer dose occurs when the volume is decreased by 10%. If it takes the same general shape as before, find the total length of the new tablet.

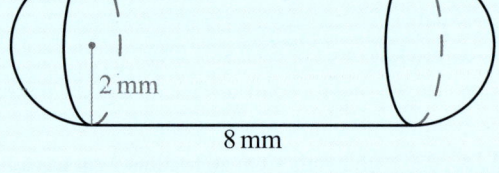

34 A solid metal fencepost has the shape shown below, consisting of a cylinder with radius 4 cm and length 1.8 m, with a conical spike at one end (length 0.1 m) and a hemisphere at the other. Both have the same radius as the main post.

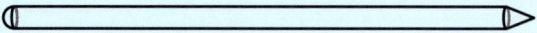

  a Find the volume of metal needed to manufacture a batch of 1000 such posts.
  b Find the approximate volume of paint needed for the batch, allowing that the paint is applied to produce a 0.4 mm thick coating. Justify any assumptions you make.

5B Rules of trigonometry

■ Revision of right-angled trigonometry

You are already familiar with using sine, cosine and tangent in right-angled triangles to find lengths and angles.

KEY POINT 5.2

- $\sin\theta = \dfrac{o}{h}$
- $\cos\theta = \dfrac{a}{h}$
- $\tan\theta = \dfrac{o}{a}$

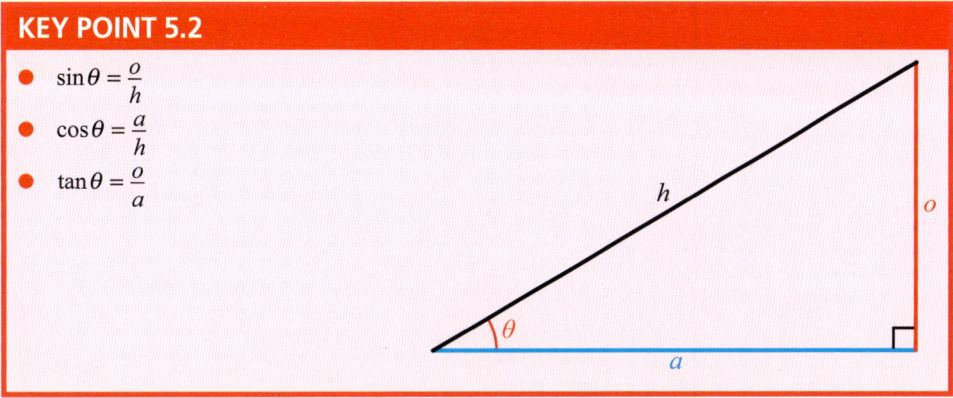

CONCEPTS – RELATIONSHIPS

One of the key ideas of geometry is that it is the study of what stays the same when things change. The **relationships** here reflect the fact that, for similar shapes, ratios of side lengths remain constant.

Tip

Remember that you need to use $\sin^{-1}$, $\cos^{-1}$, $\tan^{-1}$ when finding angles.

WORKED EXAMPLE 5.4

For the following triangle, find the size of angle A.

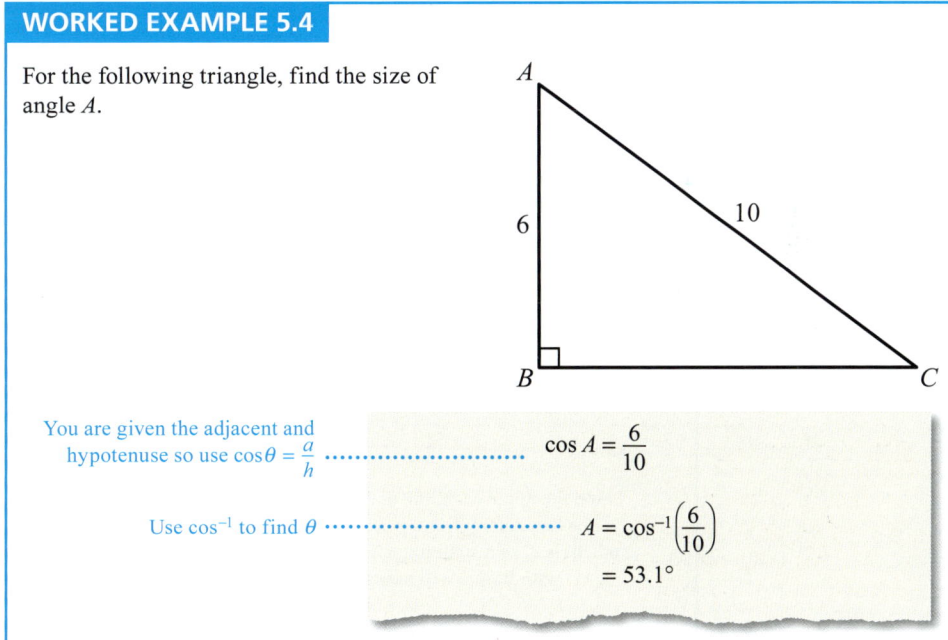

You are given the adjacent and hypotenuse so use $\cos\theta = \dfrac{a}{h}$

$\cos A = \dfrac{6}{10}$

Use $\cos^{-1}$ to find θ

$A = \cos^{-1}\left(\dfrac{6}{10}\right)$

$= 53.1°$

■ The size of an angle between intersecting lines

Given a straight line, you can always draw a right-angled triangle by taking a vertical from the line to the x-axis, and then use trigonometry to find the angle the line makes with the x-axis.

WORKED EXAMPLE 5.5

Find the angle between the line $y = 2x$ and the positive x-axis.

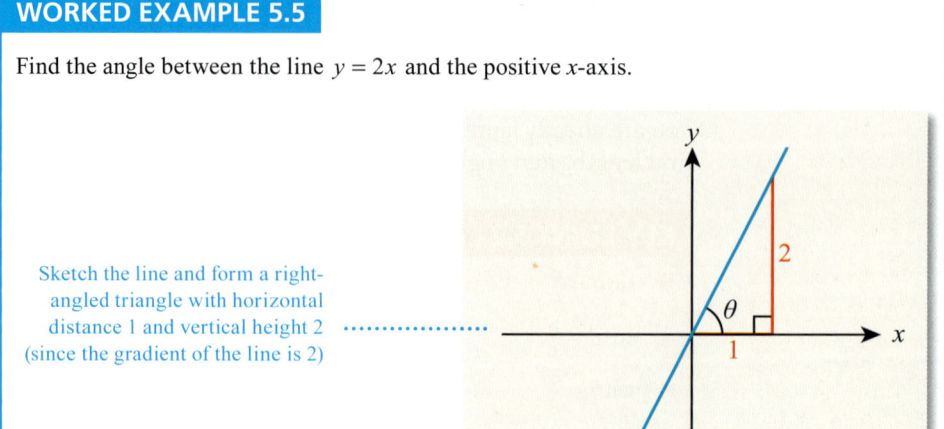

Sketch the line and form a right-angled triangle with horizontal distance 1 and vertical height 2 (since the gradient of the line is 2)

You know the opposite and adjacent sides to the angle A, so use $\tan \theta = \frac{o}{a}$

$$\tan \theta = \frac{2}{1}$$
$$\theta = \tan^{-1} 2 = 63.4°$$

Once you know how to find the angle between a line and the x-axis, you can use it to find the angle between two intersecting lines.

WORKED EXAMPLE 5.6

Find the acute angle between $y = 3x + 1$ and $y = 2 - x$.

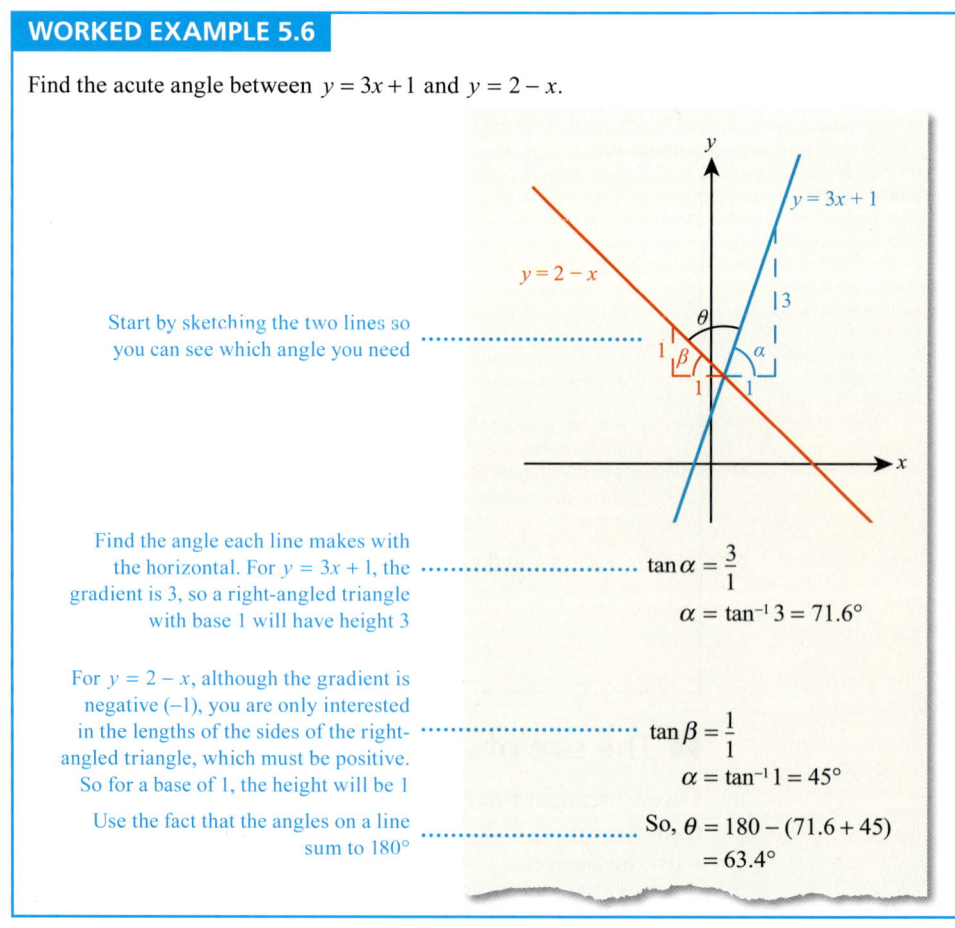

Start by sketching the two lines so you can see which angle you need

Find the angle each line makes with the horizontal. For $y = 3x + 1$, the gradient is 3, so a right-angled triangle with base 1 will have height 3

$$\tan \alpha = \frac{3}{1}$$
$$\alpha = \tan^{-1} 3 = 71.6°$$

For $y = 2 - x$, although the gradient is negative (−1), you are only interested in the lengths of the sides of the right-angled triangle, which must be positive. So for a base of 1, the height will be 1

$$\tan \beta = \frac{1}{1}$$
$$\alpha = \tan^{-1} 1 = 45°$$

Use the fact that the angles on a line sum to 180°

So, $\theta = 180 - (71.6 + 45)$
$$= 63.4°$$

5B Rules of trigonometry

■ The sine rule

If a triangle does not contain a right angle you cannot directly use your previous knowledge of trigonometry to find lengths or angles. However, there are some new rules which can be applied. The first of these is called the sine rule. The sine rule is useful if you know the length of a side and the size of the angle opposite that side (as well as either one other side or angle).

> **Tip**
>
> The first version of the sine rule is easier to work with when you want to find a missing side; the second version (which is just a rearrangement of the first) is easier to work with when you want to find a missing angle.

KEY POINT 5.3

The sine rule:

- $\dfrac{a}{\sin A} = \dfrac{b}{\sin B} = \dfrac{c}{\sin C}$

or, equivalently,

- $\dfrac{\sin A}{a} = \dfrac{\sin B}{b} = \dfrac{\sin C}{c}$

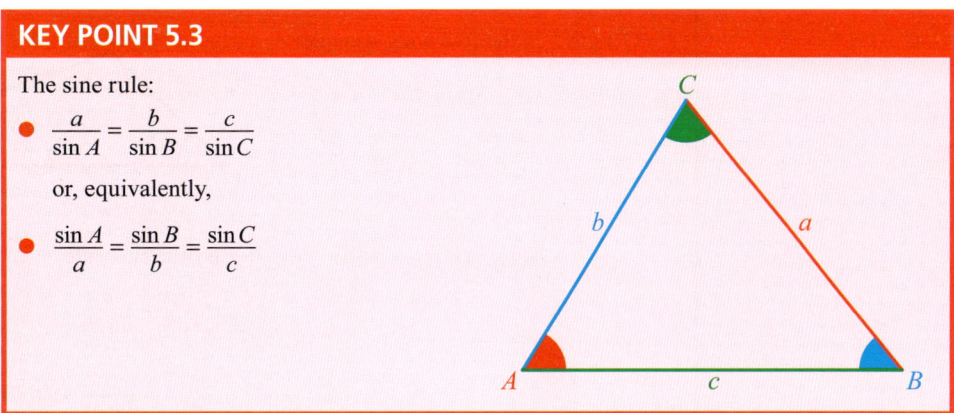

The diagram in Key Point 5.3 shows the convention that angles are labelled using capital letters and the opposite side is labelled using the equivalent lowercase letter.

> **Tip**
>
> Although the sine rule is quoted for all three sides and their corresponding angles, you will only ever need to use two of these at any one time.

Proof 5.1

Prove that in any triangle $\dfrac{a}{\sin A} = \dfrac{b}{\sin B}$.

Creating right-angled triangles allows you to use right-angled trigonometry ········ Divide the triangle ABC into two right-angled triangles:

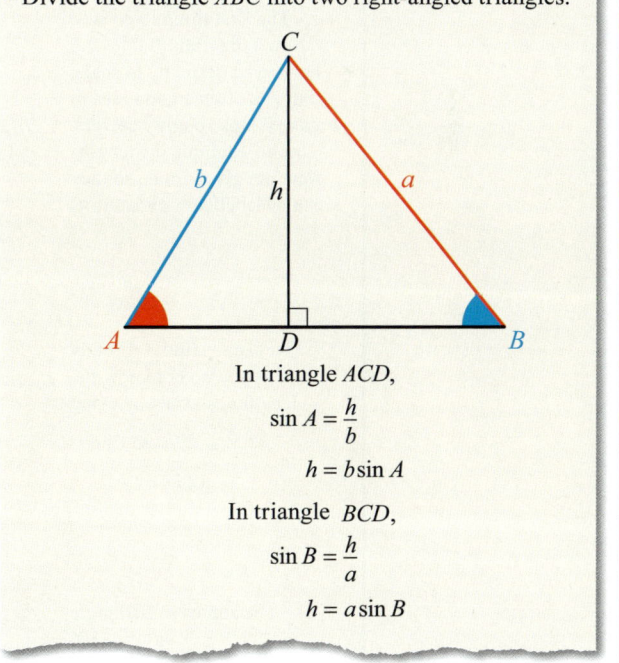

In triangle ACD,

$$\sin A = \frac{h}{b}$$

$$h = b \sin A$$

In triangle BCD,

$$\sin B = \frac{h}{a}$$

$$h = a \sin B$$

But h is common to both triangles so the two expressions can be equated .... So, $a\sin B = b\sin A$

$$\frac{a}{\sin A} = \frac{b}{\sin B}$$

WORKED EXAMPLE 5.7

Find length b in the following triangle.

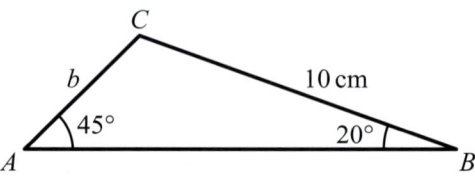

Use $\dfrac{a}{\sin A} = \dfrac{b}{\sin B}$ .... By the sine rule,

$$\frac{10}{\sin 45} = \frac{b}{\sin 20}$$

$$b = \frac{10}{\sin 45} \times \sin 20 = 4.84 \text{ cm}$$

WORKED EXAMPLE 5.8

In triangle ABC, angle $B = 70°$, $c = 2.6$ and $b = 3.2$. Find angle A.

The first thing to do is to draw a diagram. It does not have to be perfectly to scale, but it is often a good idea to make it look roughly correct. In this example the lengths were not given units, so they can be labelled without units

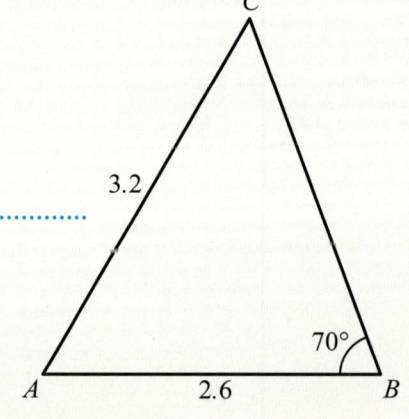

You cannot find A directly as you do not know the length a, so start by using $\dfrac{\sin B}{b} = \dfrac{\sin C}{c}$ to find C

By the sine rule,

$$\frac{\sin 70}{3.2} = \frac{\sin C}{2.6}$$

$$\sin C = \frac{\sin 70}{3.2} \times 2.6$$

$$C = \sin^{-1}\left(\frac{\sin 70}{3.2} \times 2.6\right) = 49.8°$$

The angles in a triangle sum to 180° .... So,

$A = 180 - 70 - 49.8$

$= 60.2°$

The cosine rule

The cosine rule is needed if you are given two sides and the angle between them, or all three sides but no angle. In these cases, you cannot use the sine rule.

> **KEY POINT 5.4**
>
> The cosine rule:
> - $a^2 = b^2 + c^2 - 2bc \cos A$
>   or, equivalently,
> - $\cos A = \dfrac{b^2 + c^2 - a^2}{2bc}$

Tip

The second version is just a rearrangement of the first to make it easier to find a missing angle.

Tip

You can use the cosine rule with any letter as the subject on the left-hand side of the formula, as long as you make sure the angle matches this. For example, it could be written as
$b^2 = a^2 + c^2 - 2ac \cos B$
or
$c^2 = a^2 + b^2 - 2ab \cos C$.

> **Proof 5.2**
>
> Prove that $a^2 = b^2 + c^2 - 2bc \cos A$.
>
> Divide the triangle ABC into two right-angled triangles:
>
>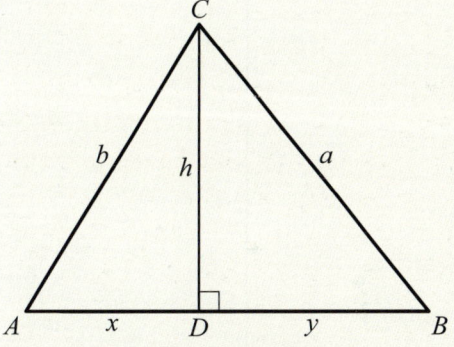
>
> *Again, start by creating two right-angled triangles*
>
> *Start by writing down an expression for a^2* — In triangle BCD,
> $$a^2 = h^2 + (c-x)^2$$
>
> *There are two variables we do not want (h and x) so we need two equations from the other triangle to eliminate these variables* — In triangle ACD,
> $$h^2 = b^2 - x^2$$
> and
> $$\cos A = \frac{x}{b}$$
> $$x = b \cos A$$
>
> So,
> *Expand the brackets in the expression from BCD*
> $$a^2 = h^2 + (c-x)^2$$
> $$= h^2 + c^2 - 2cx + x^2$$
> *Substitute in the two expressions from the second triangle*
> $$= b^2 - x^2 + c^2 - 2cb \cos A + x^2$$
> $$= b^2 + c^2 - 2bc \cos A$$

WORKED EXAMPLE 5.9

Find length c in the following diagram.

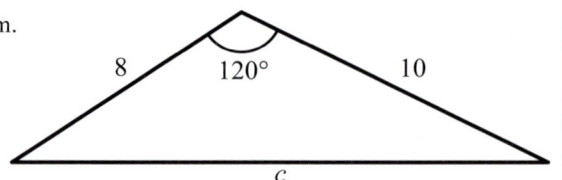

Use $c^2 = a^2 + b^2 - 2ab \cos C$ (remember the angle must match the side you choose as the subject on the LHS)

By the cosine rule,
$$c^2 = 10^2 + 8^2 - 2(10)(8)\cos 120$$
$$= 244$$

Remember to square root to find c

So, $c = 15.6$

WORKED EXAMPLE 5.10

Find the angle B in the following triangle.

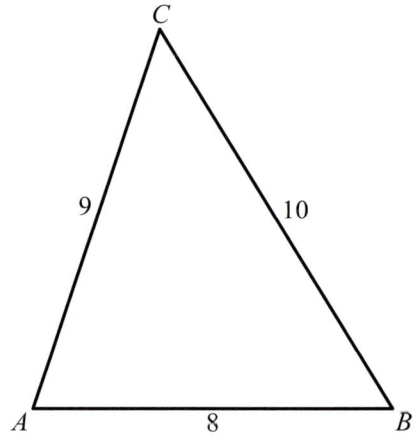

Use $\cos B = \dfrac{a^2 + c^2 - b^2}{2ac}$

By the cosine rule,
$$\cos B = \frac{10^2 + 8^2 - 9^2}{2(10)(8)}$$

$$B = \cos^{-1}\left(\frac{10^2 + 8^2 - 9^2}{2(10)(8)}\right) = 58.8°$$

You are the Researcher

There is also a rule called the tan rule which has fallen out of favour because all problems solved using it can be solved using a combination of the sine and cosine rules. However, it does have several interesting applications and proofs that you could research.

Be the Examiner 5.1

In triangle ABC, $AB = 7\,\text{cm}$, $AC = 12\,\text{cm}$, $A\hat{B}C = 60°$ and $A\hat{C}B = 25°$

Find the length of side BC.

Which is the correct solution? Identify the errors made in the incorrect solutions.

Solution 1	Solution 2	Solution 3
By cosine rule, $BC^2 = 7^2 + 12^2 - 2(7)(12)\cos 60$ $= 109$ So, $BC = 10.4\,\text{cm}$	By cosine rule, $BC^2 = 7^2 + 12^2 - 2(7)(12)\cos 25$ $= 40.7403$ So, $BC = 6.38\,\text{cm}$	By cosine rule, $BC^2 = 7^2 + 12^2 - 2(7)(12)\cos 95$ $= 207.642$ So, $BC = 14.4\,\text{cm}$

Area of a triangle

You already know you can use Area $= \dfrac{1}{2}bh$ to find the area of a triangle if you know the base and the perpendicular height.

Often, though, you will not know the perpendicular height. In that case there is an alternative formula.

Tip

In this formula, the angle you are interested in (C) is always between the two side lengths you are interested in (a and b).

KEY POINT 5.5

- Area $= \dfrac{1}{2}ab\sin C$

Proof 5.3

Prove that for a triangle Area $= \dfrac{1}{2}ab\sin C$.

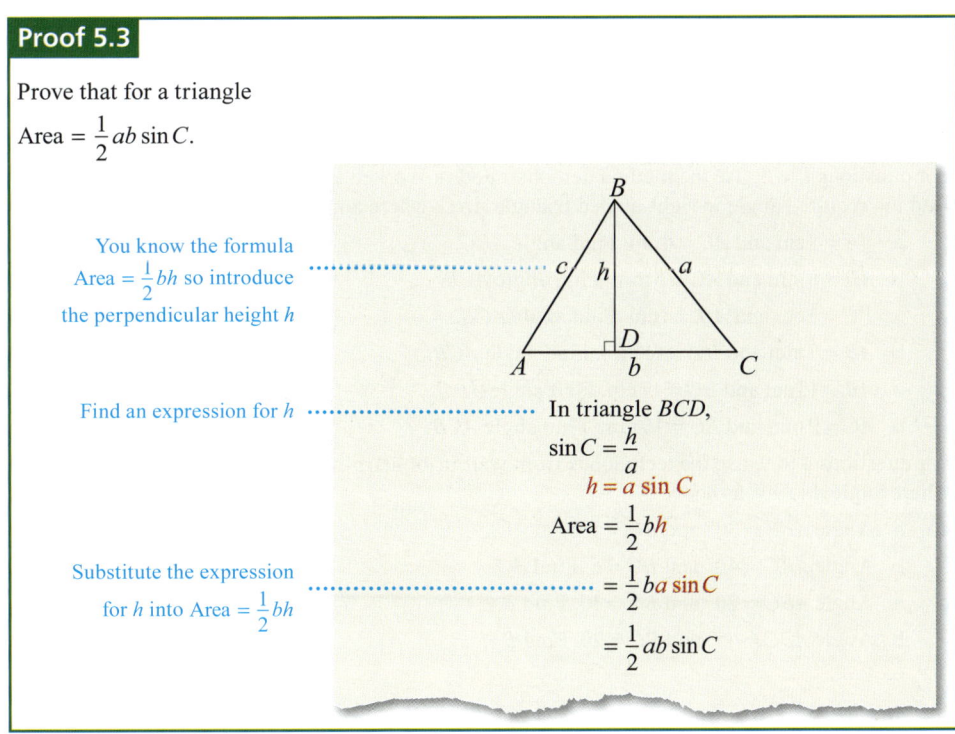

You know the formula Area $= \dfrac{1}{2}bh$ so introduce the perpendicular height h

Find an expression for h

In triangle BCD,
$\sin C = \dfrac{h}{a}$
$h = a\sin C$

Area $= \dfrac{1}{2}bh$

Substitute the expression for h into Area $= \dfrac{1}{2}bh$

$= \dfrac{1}{2}ba\sin C$

$= \dfrac{1}{2}ab\sin C$

WORKED EXAMPLE 5.11

Find the area of the triangle on the right.

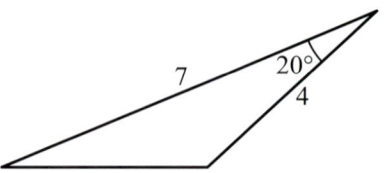

Use Area = $\frac{1}{2} ab \sin C$

Remember that the angle C will be between the side lengths a and b

Area = $\frac{1}{2}(7)(4)\sin 20$

= 4.79

WORKED EXAMPLE 5.12

The triangle on the right has area 15 cm².

Find the acute angle marked x.

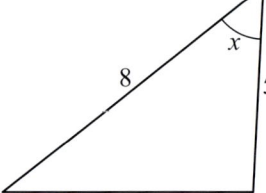

Use Area = $\frac{1}{2} ab \sin C$, with Area = 15

$15 = \frac{1}{2} \times 5 \times 8 \sin x$

Solve for x

$15 = 20 \sin x$

$\sin x = \frac{3}{4}$

$x = 48.6°$

Exercise 5B

For questions 1 to 3, use the method demonstrated in Worked Example 5.4 to find the required angle in right-angled triangle ABC, where angle A is 90°.

1. **a** $AB = 7$ cm and $AC = 4$ cm. Find angle ABC.
   b $AB = 8$ mm and $AC = 5$ mm. Find angle ACB.
2. **a** $BC = 5$ cm and $AB = 4$ cm. Find angle ACB.
   b $AB = 7$ mm and $BC = 11$ mm. Find angle ACB.
3. **a** $AB = 13$ cm and $BC = 17$ cm. Find angle ABC.
   b $AC = 9$ mm and $BC = 15$ mm. Find angle ACB.

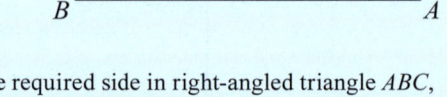

For questions 4 to 9, use the techniques from your prior learning to find the required side in right-angled triangle ABC, where angle A is 90° as above.

4. **a** Angle $ABC = 30°$ and $BC = 5$. Find AC.
   b Angle $ABC = 40°$ and $BC = 6$. Find AC.
5. **a** Angle $ABC = 50°$ and $AC = 6$. Find BC.
   b Angle $ABC = 60°$ and $AC = 10$. Find BC.
6. **a** Angle $ABC = 20°$ and $BC = 10$. Find BA.
   b Angle $ABC = 45°$ and $BC = 20$. Find BA.
7. **a** Angle $ABC = 55°$ and $AB = 10$. Find BC.
   b Angle $ABC = 70°$ and $AB = 9$. Find BC.

5B Rules of trigonometry

8   **a** Angle $ABC = 15°$ and $AB = 10$. Find AC.
    b Angle $ABC = 65°$ and $AB = 12$. Find AC.

9   **a** Angle $ABC = 2°$ and $AC = 12$. Find AB.
    b Angle $ABC = 80°$ and $AC = 24$. Find AB.

For questions 10 to 12, use the method demonstrated in Worked Example 5.5 to find the angle between the two lines.

10 **a** Find the angle between the positive x-axis and $y = 3x$.
    b Find the angle between the positive x-axis and $y = 5x$.

11 **a** Find the angle between the positive y-axis and $y = 3x$.
    b Find the angle between the positive y-axis and $y = \frac{7}{2}x$.

12 **a** Find the angle between the positive y-axis and $y = -\frac{5}{2}x$.
    b Find the angle between the positive y-axis and $y = -\frac{9}{2}x$.

For questions 13 to 15, use the method demonstrated in Worked Example 5.6 to find the angle between the two lines.

13 **a** Find the acute angle between the lines $y = 5x$ and $y = 3x$.
    b Find the acute angle between the lines $y = 2x$ and $y = \frac{7}{2}x$.

14 **a** Find the acute angle between the lines $y = 5x$ and $y = -3x$.
    b Find the acute angle between the lines $y = 2x$ and $y = -3x$.

15 **a** Find the acute angle between the lines $y = 5x - 3$ and $y = \frac{5}{2}x + 1$.
    b Find the acute angle between the lines $y = 9x - 6$ and $y = 4 - 7x$.

For questions 16 to 18, use the method demonstrated in Worked Example 5.7, applying the sine rule to find the required side length.

16 **a** In triangle ABC, angle $A = 45°$, angle $B = 30°$ and $a = 12$ cm. Find b.
    b In triangle ABC, angle $A = 60°$, angle $B = 45°$ and $a = 8$ mm. Find b.

17 **a** In triangle PQR, angle $P = 70°$, angle $Q = 40°$ and $QR = 5$ cm. Find PR.
    b In triangle PQR, angle $P = 60°$, angle $Q = 15°$ and $QR = 11$ mm. Find PR.

18 **a** In triangle ABC, angle $A = 23°$, angle $C = 72°$ and $a = 1.3$ cm. Find b.
    b In triangle ABC, angle $A = 39°$, angle $C = 74°$ and $a = 2.8$ mm. Find b.

For questions 19 to 21, use the method demonstrated in Worked Example 5.8, applying the sine rule to find the required angle.

19 **a** In triangle ABC, angle $A = 75°$. $a = 11$ cm and $b = 7$ cm. Find angle B.
    b In triangle ABC, angle $A = 82°$. $a = 9$ cm and $b = 7$ cm. Find angle B.

20 **a** In triangle PQR, angle $P = 104°$. $QR = 2.8$ cm and $PQ = 1.7$ cm. Find angle Q.
    b In triangle PQR, angle $P = 119°$. $QR = 13$ cm and $PQ = 7$ cm. Find angle Q.

21 **a** In triangle ABC, angle $A = 84°$. $a = 7.3$ cm and $c = 7.1$ cm. Find angle B.
    b In triangle ABC, angle $A = 70°$. $a = 89$ m and $c = 81$ m. Find angle B.

For questions 22 to 24, use the method demonstrated in Worked Example 5.9, applying the cosine rule to find the unknown side length x.

22 **a** In triangle ABC, angle $A = 60°$. $b = 4$ cm and $c = 7$ cm. Find side length a.
    b In triangle ABC, angle $A = 60°$. $b = 5$ mm and $c = 8$ mm. Find side length a.

23 **a** In triangle PQR, angle $P = 45°$. $PR = 7$ cm and $PQ = 6$ cm. Find side length QR.
    b In triangle PQR, angle $P = 50°$. $PR = 8$ mm and $PQ = 7$ mm. Find side length QR.

24 **a** In triangle ABC, angle $B = 70°$. $a = 5$ cm and $c = 4$ cm. Find side length b.
    b In triangle ABC, angle $C = 65°$. $a = 4$ mm and $b = 7$ mm. Find side length a.

For questions 25 to 27, use the method demonstrated in Worked Example 5.10, applying the cosine rule to find the unknown angle.

25   a   In triangle ABC, $a = 7$ cm, $b = 8$ cm and $c = 11$ cm. Find angle A.
     b   In triangle ABC, $a = 10$ cm, $b = 12$ cm and $c = 15$ cm. Find angle A.

26   a   In triangle PQR, $PQ = 4.2$ cm, $QR = 5.1$ cm and $PR = 11$ cm. Find angle Q.
     b   In triangle PQR, $PQ = 16$ cm, $QR = 13$ cm and $PR = 17$ cm. Find angle R.

27   a   In triangle ABC, $a = 5.7$ cm, $b = 8.1$ cm and $c = 6.6$ cm. Find angle C.
     b   In triangle ABC, $a = 1.4$ cm, $b = 2.5$ cm and $c = 1.9$ cm. Find angle B.

For questions 28 to 30, use the method demonstrated in Worked Example 5.11 to find the area of each triangle.

28   a   In triangle ABC, $a = 12$ cm, $b = 5$ cm and angle $C = 72°$.
     b   In triangle ABC, $a = 6$ mm, $b = 7$ mm and angle $C = 30°$.

29   a   In triangle PQR, $PR = 19$ cm, $QR = 17$ cm and angle $R = 52°$.
     b   In triangle PQR, $PR = 3.5$ mm, $QR = 2.1$ mm and angle $R = 28°$.

30   a   In triangle ABC, $a = 37$ cm, $c = 51$ cm and angle $B = 42°$.
     b   In triangle ABC, $b = 61$ mm, $c = 71$ mm and angle $A = 52°$.

For questions 31 to 33, use the method demonstrated in Worked Example 5.12 to find the unknown value.

31   a   In triangle ABC, $a = 15.5$ cm, $b = 14.7$ cm and the area is 90 cm^2. Find acute angle C.
     b   In triangle ABC, $a = 4.8$ mm, $b = 5.2$ mm and the area is 11 mm^2. Find acute angle C.

32   a   In triangle PQR, $PR = 12$ cm, angle $R = 40°$ and the area is 32 cm^2. Find length QR.
     b   In triangle PQR, $QR = 8$ mm, angle $R = 35°$ and the area is 12 mm^2. Find length PR.

33   a   In triangle ABC, $a = 5.2$ cm, angle $B = 55°$ and the area is 15 cm^2. Find length c.
     b   In triangle ABC, $a = 6.8$ cm, angle $B = 100°$ and the area is 28 mm^2. Find length c.

34   Find the length marked x in the diagram.

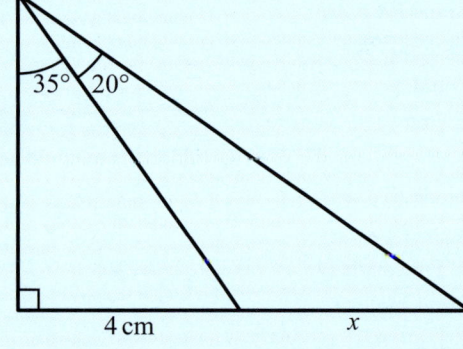

35   Find the angle marked θ in the diagram.

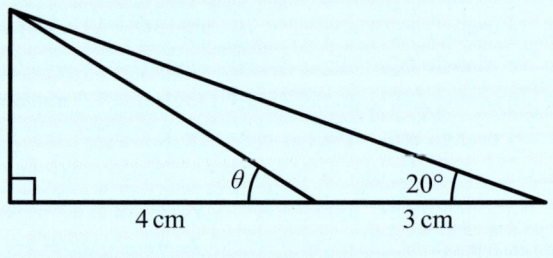

36   The diagram shows the line with equation $y = \frac{1}{2}x + 3$.

    a   Write down the coordinates of both axis intercepts.
    b   Find the size of the angle between the line and the x-axis.

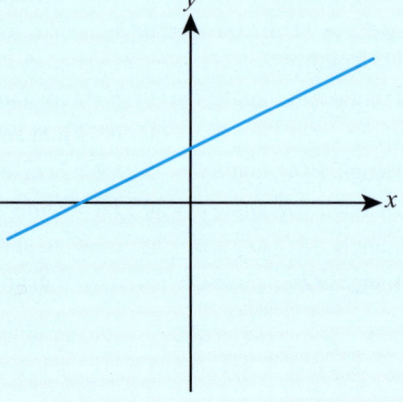

37 **a** Draw a line with equation $y = 3x - 1$, labelling both axis intercepts.
   b Find the size of the angle that the line makes with the x-axis.

38 Find the length marked a.

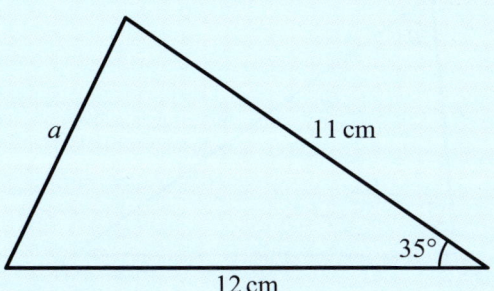

39 Find the missing angles in this triangle.

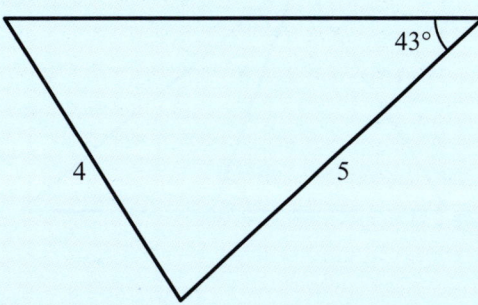

40 Find the size of the angle $C\hat{A}B$.

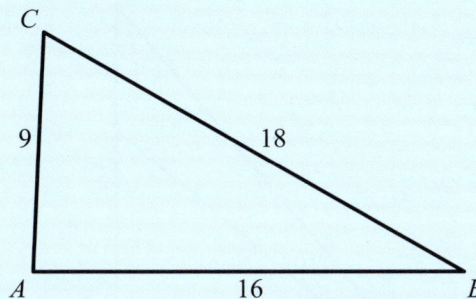

41 For the triangle shown in the diagram:

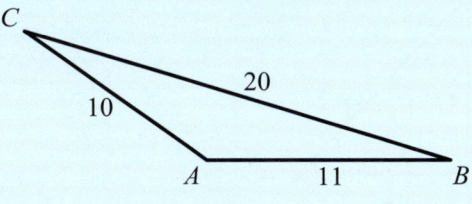

   a Find the size of angle $\hat{C}$.
   b Find the area of the triangle.

42 The area of this triangle is 241. Find the length of AB.

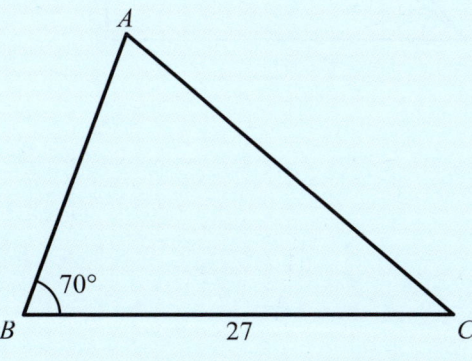

43 Find the length marked x and the angle marked θ.

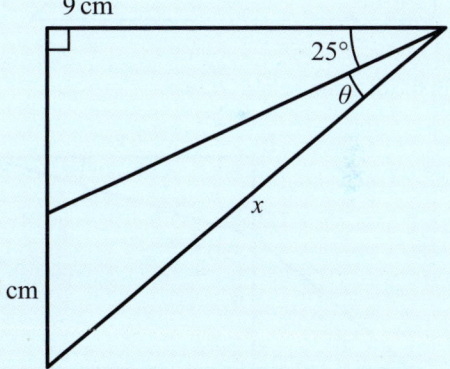

44 Find the acute angle that the line with equation $4x + 5y = 40$ makes with the x-axis.

45 **a** Show that the lines $y = 2x - 8$ and $y = \frac{1}{4}x - 1$ intersect on the x-axis.
   b Find the angle between the two lines.

46 Find the acute angle between the lines $2x - 5y = 7$ and $4x + y = 8$.

47 In triangle ABC, $A = 40°$, $B = 60°$ and $a = 12$ cm. Find the length of side b.

48 In triangle ABC, $A = 45°$, $b = 5$ cm and $c = 8$ cm. Find the length of side a.

49 In triangle ABC, the sides are $a = 4$ cm, $b = 6$ cm, $c = 8$ cm. Find angle A.

50 Triangle XYZ has $X = 66°$, $x = 10$ cm and $y = 8$ cm. Find angle Y.

51 Triangle PQR has $P = 102°$, $p = 7$ cm and $q = 6$ cm. Find angle R.

52 In triangle ABC, $B = 32°$, $C = 64°$ and $b = 3$ cm. Find side a.

53 Find the length of the side *BC*.

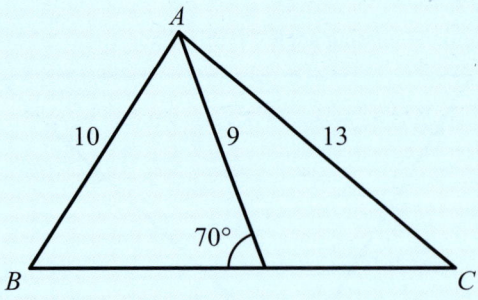

54 The area of this triangle is 26. θ is acute. Find the value of θ and the length of *AB*.

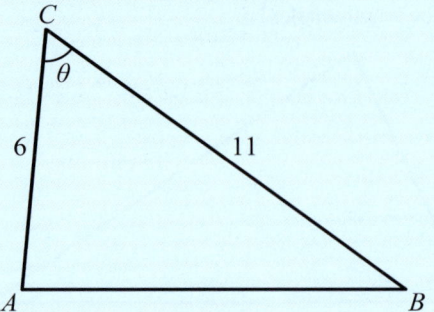

55 Find the value of *x*.

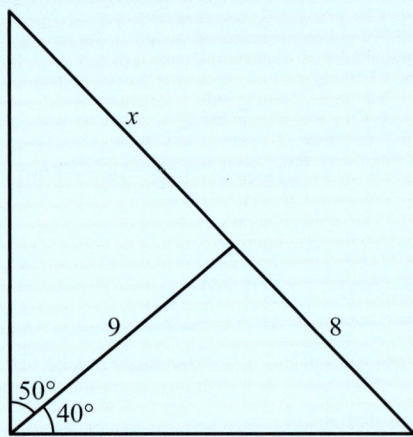

56 Express *h* in terms of *d*.

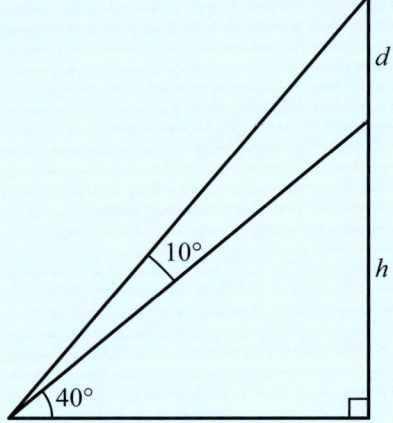

57 In the diagram below, $AC = 8$.

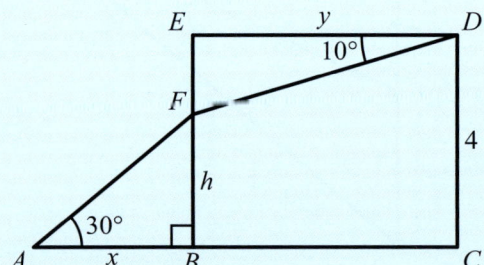

  a Express *x* and *y* in terms of *h*.
  b Hence find the value of *h*.

5C Applications of trigonometry

There are many different situations that can arise in which the rules of trigonometry can be applied. The best approach is to always draw a good diagram and look for appropriate triangles, especially right-angled triangles.

■ Angles between two intersecting lines in 3D shapes

The key idea when finding angles in 3D shapes is to look for useful triangles.

WORKED EXAMPLE 5.13

For the following cuboid,

a find the acute angle HAG

b find the acute angle between AG and EC.

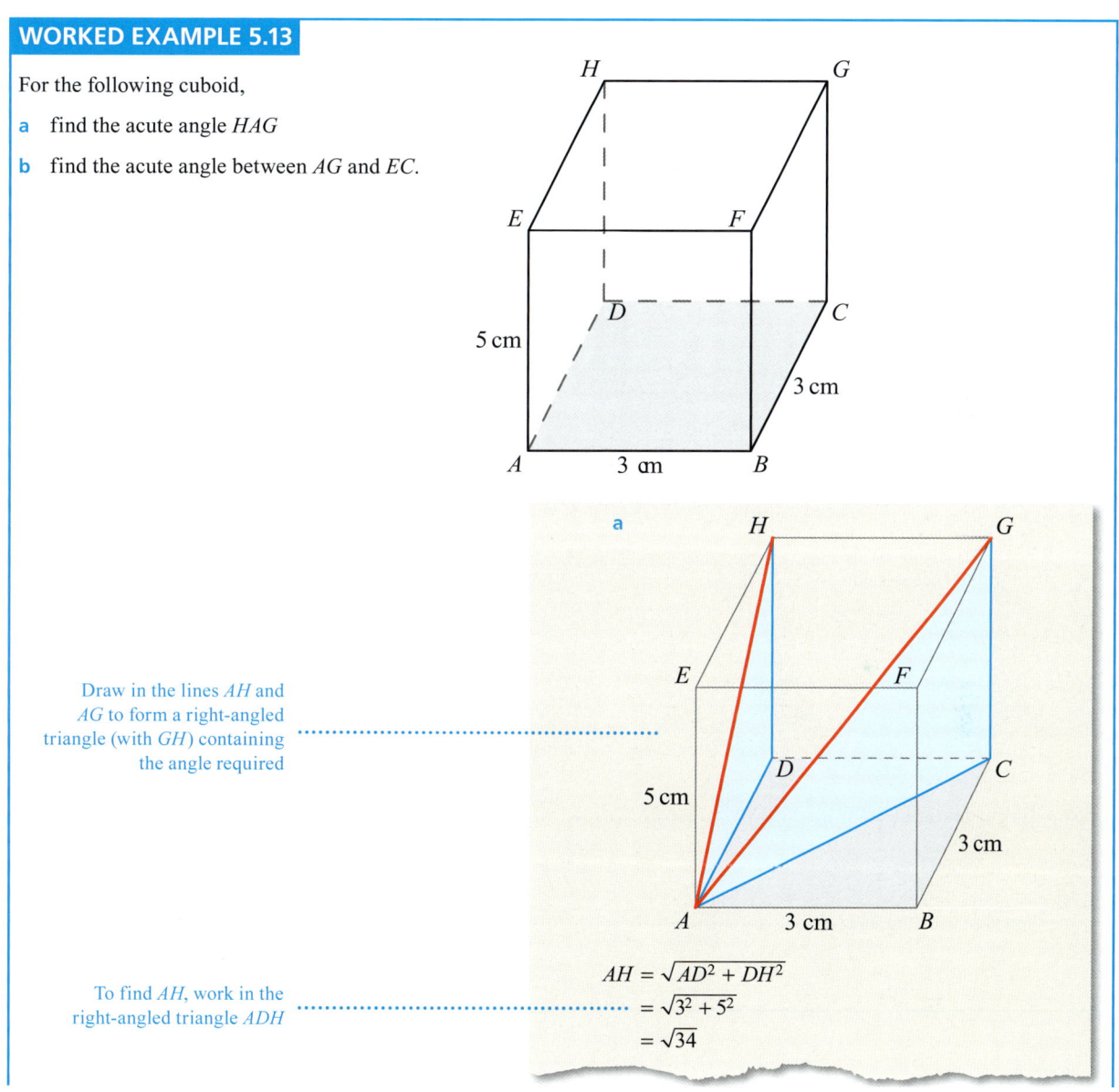

a   Draw in the lines AH and AG to form a right-angled triangle (with GH) containing the angle required

To find AH, work in the right-angled triangle ADH

$$AH = \sqrt{AD^2 + DH^2}$$
$$= \sqrt{3^2 + 5^2}$$
$$= \sqrt{34}$$

Now work in the right-angled triangle *AGH*

In triangle *AGH*,
$$\tan A = \frac{3}{\sqrt{34}}$$
$$A = \tan^{-1}\left(\frac{3}{\sqrt{34}}\right)$$
$$= 27.2°$$
∴ angle *HAG* = 27.2°

b
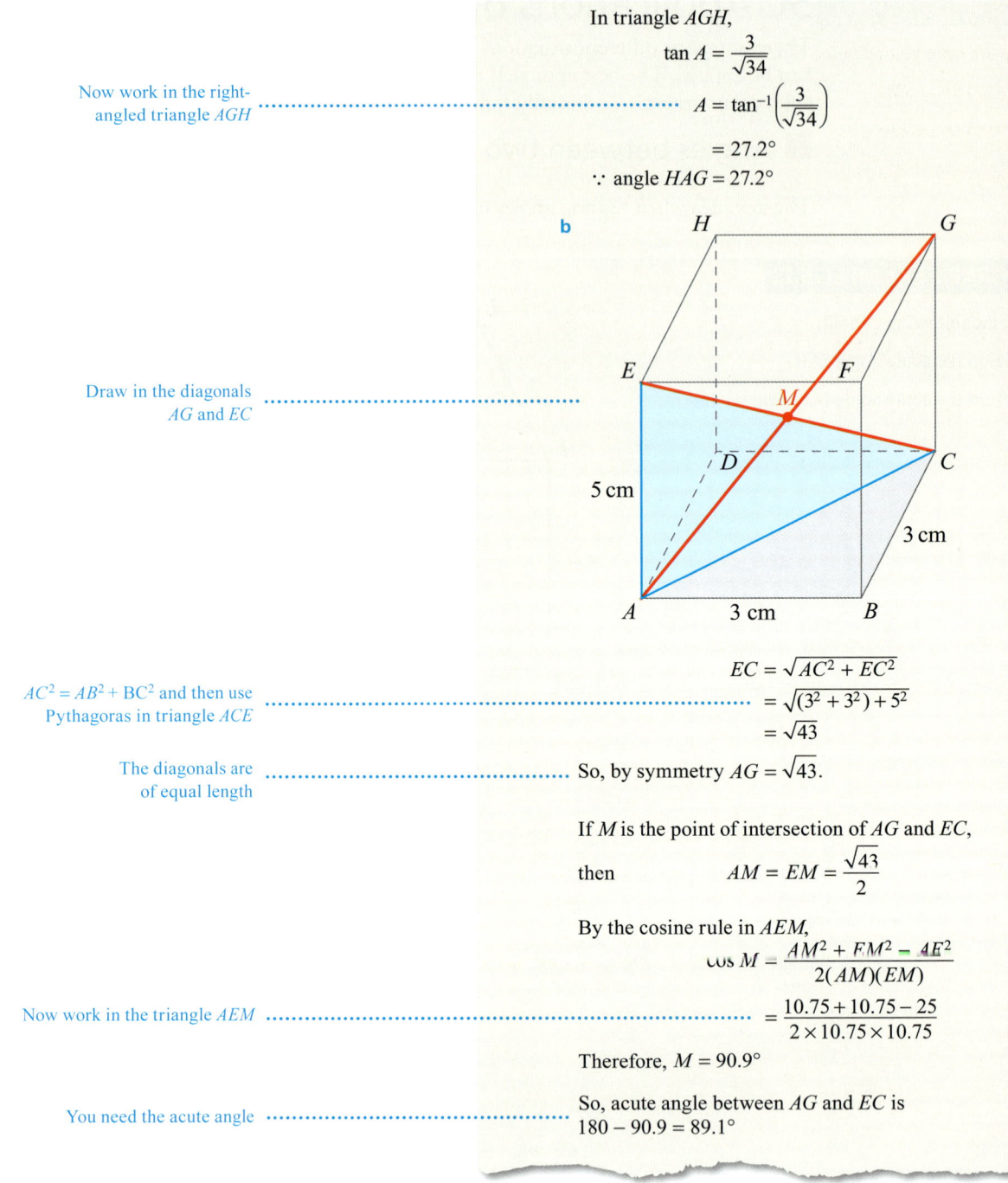

Draw in the diagonals *AG* and *EC*

$AC^2 = AB^2 + BC^2$ and then use Pythagoras in triangle *ACE*

$$EC = \sqrt{AC^2 + EC^2}$$
$$= \sqrt{(3^2 + 3^2) + 5^2}$$
$$= \sqrt{43}$$

The diagonals are of equal length

So, by symmetry $AG = \sqrt{43}$.

If *M* is the point of intersection of *AG* and *EC*, then
$$AM = EM = \frac{\sqrt{43}}{2}$$

By the cosine rule in *AEM*,
$$\cos M = \frac{AM^2 + EM^2 - AE^2}{2(AM)(EM)}$$

Now work in the triangle *AEM*

$$= \frac{10.75 + 10.75 - 25}{2 \times 10.75 \times 10.75}$$

Therefore, $M = 90.9°$

You need the acute angle

So, acute angle between *AG* and *EC* is $180 - 90.9 = 89.1°$

5C Applications of trigonometry

WORKED EXAMPLE 5.14

The diagram below on the right a square-based right pyramid.

a Find angle *EBC*.
b Find angle *EBD*.
c Find the height, *h*.

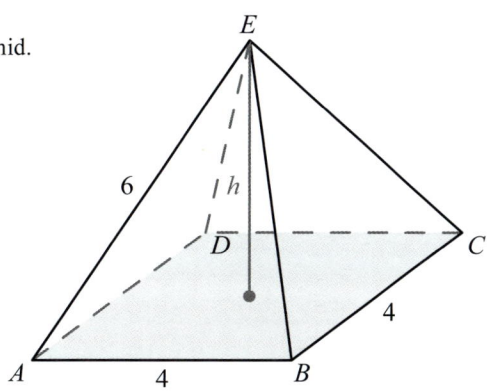

a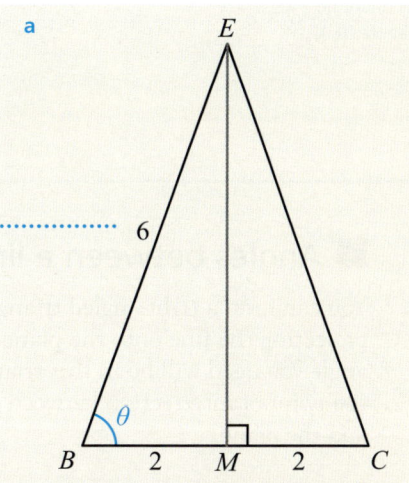

You could use the cosine rule in triangle *BCE*, but it is easier to put in the perpendicular from *E* and work in the right-angled triangle *BEM*

In triangle *BEM*,
$$\cos\theta = \frac{BM}{BE}$$
$$\theta = \cos^{-1}\left(\frac{2}{6}\right)$$
$$= 70.5°$$
∴ angle *EBC* = 70.5°

b Let *N* be the midpoint of *DB*, perpendicular from *E*.

Work first in the base plane and then in the triangle *BEN*

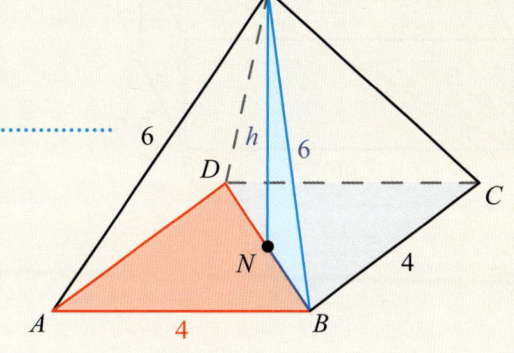

$BN = \frac{1}{2}BD$ ....

$$BD = \sqrt{AB^2 + AD^2}$$
$$= \sqrt{4^2 + 4^2}$$
$$= 4\sqrt{2}$$

So, $BN = 2\sqrt{2}$

In triangle BEN,
$$\cos\alpha = \frac{BN}{BE}$$
$$\alpha = \cos^{-1}\left(\frac{2\sqrt{2}}{6}\right)$$
$$= 61.9°$$

Use Pythagoras' theorem in triangle BEN ....

c $h = \sqrt{BN^2 + BE^2}$
$= \sqrt{8 + 36}$
$= 2\sqrt{11}$

■ Angles between a line and a plane in 3D shapes

You can form a right-angled triangle by projecting the line onto the plane. The angle you need will be in this triangle at the point of intersection between the line and the plane.

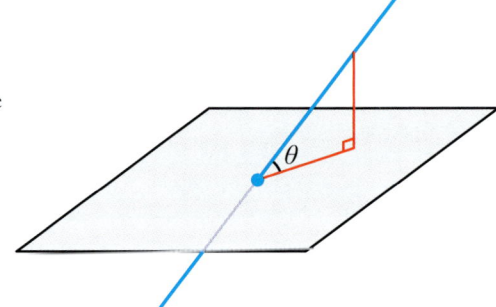

WORKED EXAMPLE 5.15

Find the angle between the line AG and the plane $CDHG$ in the diagram below.

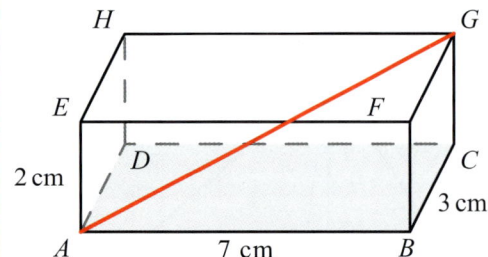

5C Applications of trigonometry

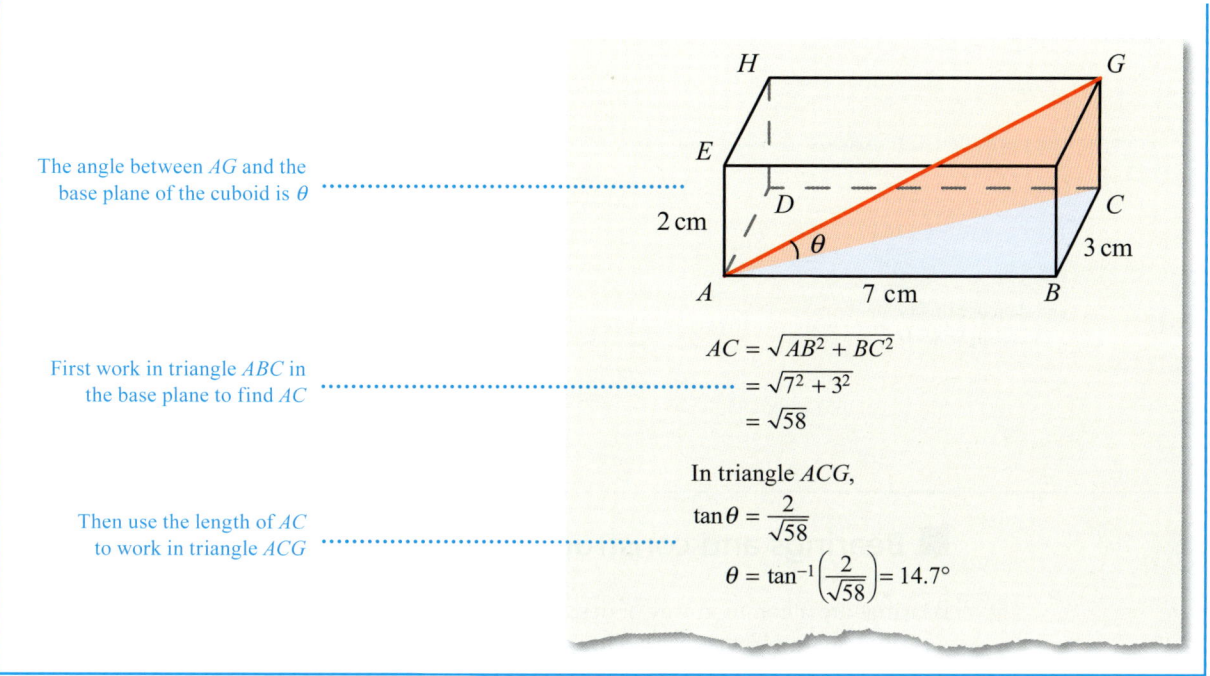

The angle between AG and the base plane of the cuboid is θ

First work in triangle ABC in the base plane to find AC

$AC = \sqrt{AB^2 + BC^2}$
$= \sqrt{7^2 + 3^2}$
$= \sqrt{58}$

Then use the length of AC to work in triangle ACG

In triangle ACG,
$\tan \theta = \dfrac{2}{\sqrt{58}}$

$\theta = \tan^{-1}\left(\dfrac{2}{\sqrt{58}}\right) = 14.7°$

WORKED EXAMPLE 5.16

Find the angle between the line AD and the plane ABC in the triangular-based right pyramid shown on the right.

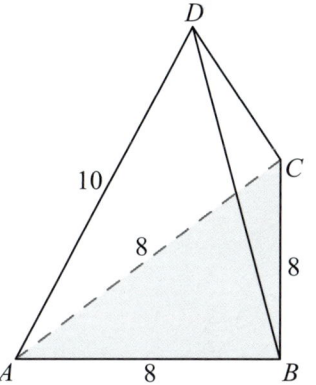

Let the perpendicular from D intersect the base plane ABC at M and let N be the midpoint of AB:

In the base plane, AM will bisect the angle at A (which is 60° as triangle ABC is equilateral)

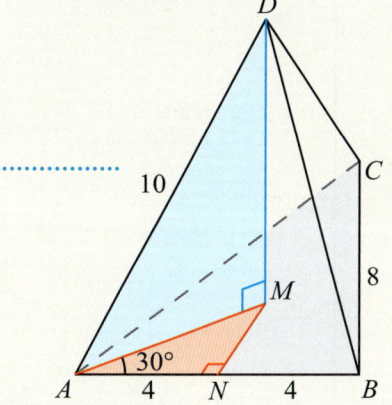

From the base plane, find AM

In triangle AMN,
$$\cos 30 = \frac{4}{AM}$$
$$AM = \frac{8\sqrt{3}}{3}$$

... and then use this in the triangle ADM

In triangle ADM,
$$\cos\theta = \frac{AM}{AD} = \frac{4\sqrt{3}}{10}$$
$$\theta = \cos^{-1}\left(\frac{4\sqrt{3}}{10}\right)$$
$$= 46.1°$$

■ Bearings and constructing diagrams from information

Bearings are a common way of describing the direction an object is travelling or the position of two objects relative to each other. They are angles measured clockwise from north.

> Bearings are one of a large number of mathematical ideas which originated from navigation. The word 'geometry' comes from the ancient Greek word for 'measuring the Earth'. How important is it for all ships and aeroplanes to use the same conventions when describing journeys?

WORKED EXAMPLE 5.17

A ship leaves a port on a bearing of 020° and travels 100 km. It unloads some of its cargo then travels on a bearing of 150° for 200 km to unload the rest of its cargo. Find

a the distance it must now travel to return to the original port

b the bearing it must travel on to return to the original port.

Start by drawing the situation described

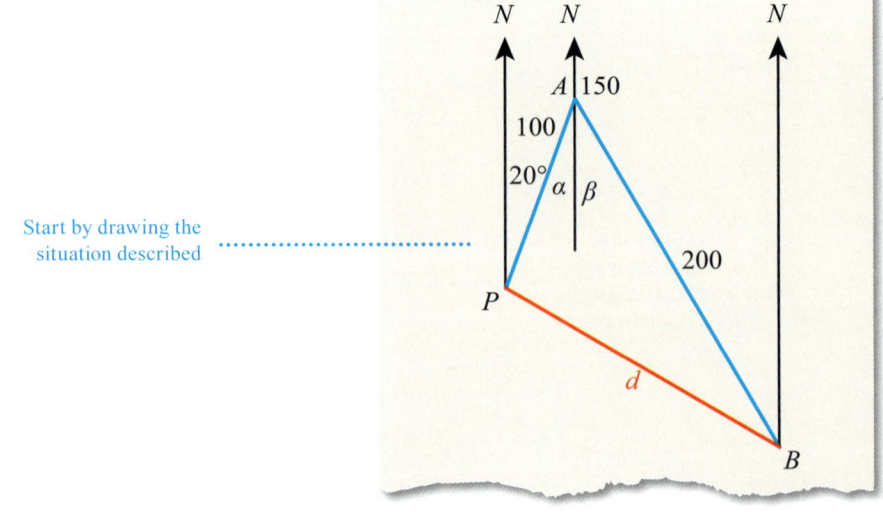

5C Applications of trigonometry

Split the angle PAB using a vertical through A so you can relate α and β to the information you have

a $\alpha = 20°$ by alternate angles.
$\beta = 180 - 150 = 30°$
So, $P\hat{A}B = 20 + 30 = 50°$

You can now work in the triangle PAB

By cosine rule in triangle PAB,
$d^2 = 100^2 + 200^2 - 2 \times 100 \times 200 \cos 50$
$= 24\,288.495\,61$
So,
$d = \sqrt{24\,288.495\,61}$
$= 155$ m

b By sine rule in PAB,
$\dfrac{\sin APB}{200} = \dfrac{\sin 50}{155.848}$

$APB = \sin^{-1}\left(\dfrac{\sin 50}{155.848} \times 200\right)$
$= 79.4°$

■ Angles of elevation and depression

An angle of elevation is an angle above the horizontal and an angle of depression is an angle below the horizontal. These are often used to describe the objects from a viewer's perspective.

WORKED EXAMPLE 5.18

When a man stands at a certain distance from a building, the angle of elevation of top of the building is 21.8°. When the man walks a further 50 m away in the same direction, the new angle of elevation is 11.3°.

If the measurements are being taken at the man's eye level, which is 1.8 m above the ground, find the height of the building.

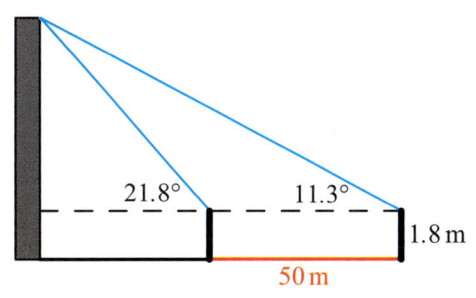

Work first in the non-right-angled triangle in order to find the length that is shared with the right-angled triangle

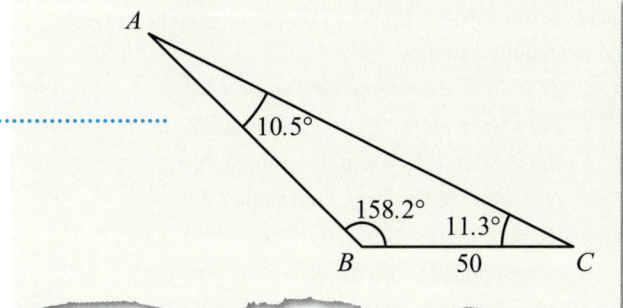

Angles on a straight line ⋯⋯ $A\hat{B}C = 180 - 21.8 = 158.2°$

Sum of angles in a triangle ⋯⋯ $B\hat{A}C = 180 - 158.2 - 11.3 = 10.5°$

By sine rule,
$$\frac{AB}{\sin 11.3} = \frac{50}{\sin 10.5}$$

You will need the sine rule ⋯⋯ $AB = \frac{50}{\sin 10.5} \times \sin 11.3$
$= 53.8\,\text{m}$

Now work in the right-angled triangle ⋯⋯

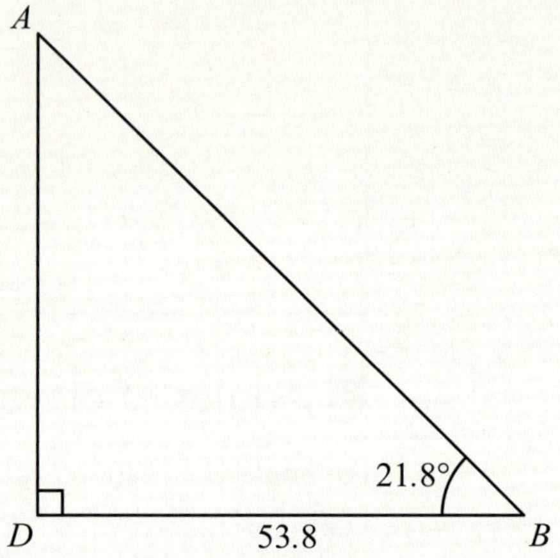

$$\sin 21.8 = \frac{AD}{53.7618}$$
$$AD = 20.0\,\text{m}$$

Therefore,

Add on the man's height to get the height of the building ⋯⋯ height of building = 20.0 + 1.8 = 21.8 m

Exercise 5C

Questions 1 to 5 refer to the cuboid shown on the right.
Use the method demonstrated in Worked Example 5.13 to find the required angles.

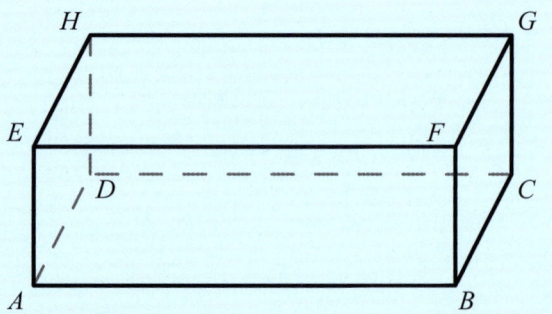

1 a $AC = 7$ cm, $AE = 5$ cm. Find angle ACE.
   b $FH = 6$ cm, $HD = 10$ cm. Find angle HFD.
2 a $BD = 10$ cm, $AE = 3$ cm. Find angle ACE.
   b $HC = 4$ cm, $CB = 5$ cm. Find angle CEB.
3 a $AE = 3$ cm, $AB = 4$ cm, $AD = 5$ cm. Find angle HBA.
   b $AD = 4$ cm, $DC = 10$ cm, $DH = 6$ cm. Find angle GDF.
4 a $AE = 3$ cm, $AB = 4$ cm, $AD = 5$ cm. Find angle HFC.
   b $AE = 6$ cm, $AB = 3$ cm, $AD = 10$ cm. Find angle GEB.
5 a $AE = 3$ cm, $AB = 4$ cm, $AD = 5$ cm. Find the acute angle between AG and HB.
   b $AE = 6$ cm, $AB = 3$ cm, $AD = 10$ cm. Find the acute angle between CE and AG.

5C Applications of trigonometry

Questions 6 to 9 refer to a right pyramid with rectangular base $ABCD$ and vertex M. O is the centre of the base so that OM is the height of the pyramid.

Use the method demonstrated in Worked Example 5.14 to find the required lengths and angles.

6. a $AC = 5$ cm, $OM = 6$ cm. Find angle ACM.
   b $AC = 9$ cm, $OM = 4$ cm. Find angle ACM.
7. a $AB = 5$ cm, $BC = 7$ cm, $OM = 5$ cm. Find angle ACM.
   b $AB = 4$ mm, $BC = 8$ mm, $OM = 5$ mm. Find angle ACM.
8. a $AB = BC = 12$ cm, $AM = 17$ cm. Find angle ABM.
   b $AB = BC = 11$ cm, $AM = 8$ cm. Find angle ABM.
9. a $AB = 6$ cm, $BC = 5$ cm, angle $MAO = 35°$. Find pyramid height OM.
   b $AB = 9$ m, $BC = 7$ m, $AM = 10$ m. Find pyramid height OM.

Questions 10 to 12 refer to a cuboid as labelled in questions 1–5.

Use the method demonstrated in Worked Example 5.15 to find the smallest angle between the diagonal AG and the base plane $ABCD$, given the following measurements.

10. a $BH = 15, DH = 9$
    b $BH = 25, DH = 17$
11. a $AB = 5, AD = 8, AE = 7$
    b $AB = 8, AD = 3, AE = 9$
12. a $AE = 6, AF = 7, AH = 11$
    b $AE = 1.2, AF = 2.1, CF = 4.3$

Questions 13 to 15 refer to a right pyramid as labelled in questions 6–9. Use the method demonstrated in Worked Example 5.15 to find the angle between the edge AM and plane $ABCD$ in each case.

13. a $AB = BC = 7$ cm, $OM = 5$ cm
    b $AB = BC = 7$ mm, $OM = 5$ mm
14. a $AB = 9$ cm, $BC = 4$ cm, $OM = 6$ cm
    b $AB = 11$ mm, $BC = 12$ mm, $OM = 15$ mm
15. a $AB = 3$ cm, $BC = 7$ cm, $AM = 4$ cm
    b $AB = 6$ mm, $BC = 8$ mm, $AM = 13$ mm

Question 16 refers to a right pyramid with triangular base ABC and vertex M. O is the centre of the base so that OM is the height of the pyramid.

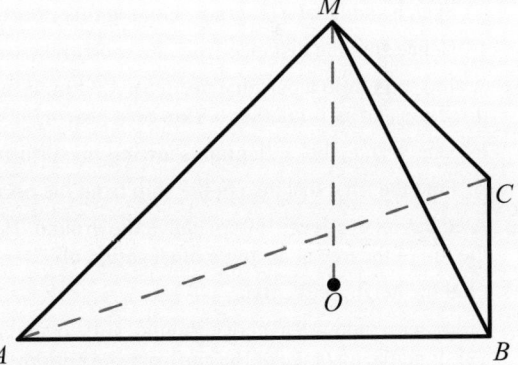

Use the method demonstrated in Worked Example 5.16 to find the angle between the edge AM and plane ABC in each case.

16. a ABC is an equilateral triangle with side length 5 cm, $OM = 4$ cm.
    b ABC is an equilateral triangle with side length 11 mm, $OM = 6$ mm.

In each of questions 17 to 19, a drone is sent on two legs of a journey. Use the method demonstrated in Worked Example 5.17 to calculate:
   i the distance
   ii the bearing on which the drone must travel to return to its original position.

17. a 2 km north and then 1.8 km on a bearing 145°.
    b 3.3 km south and then 2.1 km on a bearing 055°.
18. a 1.7 km east and then 2.3 km on a bearing 085°.
    b 6.8 km west and then 9.1 km on a bearing 035°.

19 a 2.2 km on a bearing 130° and then 1.8 km on a bearing 145°.
    b 13 km on a bearing 220° and then 17 km on a bearing 105°.
20 Viewed from 40 m away, a building has an angle of elevation of 55°. Find the height of the building.
21 For the cuboid shown in the diagram:
    a Find the length of *HB*.
    b Find the angle between *HB* and *BD*.

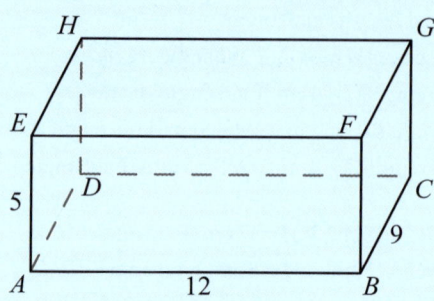

22 The diagram shows a square – based pyramid.
    a Draw a sketch of triangle *ACE*, labelling the lengths of all the sides.
    b Find the height of the pyramid.
    c Find the angle between *AE* and *EC*.

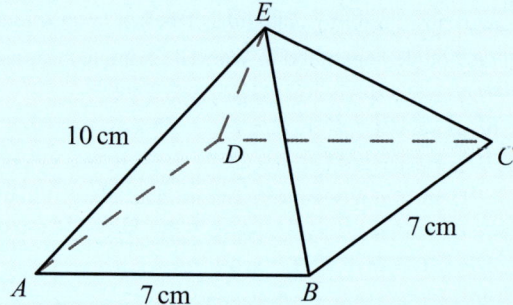

23 A cone has diameter 12 cm and vertical height 9 cm. Find the angle θ between the sloping edge and the base.

24 From his tent, Mario can see a tree 120 m away on a bearing of 056°. He can also see a rock that is due east of his tent and due south of the tree.
    a Sketch and label a diagram showing this information.
    b Hence find the distance from the rock to the tree.

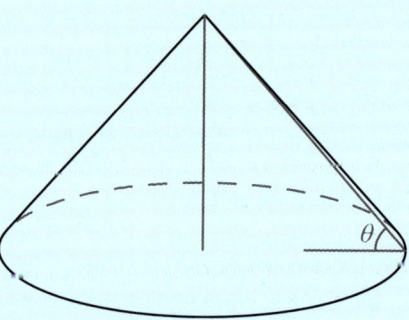

25 A ship sails from a port due north for 1.2 km. It then changes direction and sails another 0.8 km on a bearing of 037°.
    a Sketch and label a diagram showing this information.
    b Find the final distance of the ship from the port.

26 Alec has an eye-level 160 cm above the ground. He stands 6.5 m from a tree. He can see the top of the tree at an angle of elevation of 62°.
    Find the height of the tree.

27 From her window, 9 m above ground, Julia observes a car at an angle of depression of 12°. Find the distance of the car from the bottom of Julia's building.

28 After recording the angle of elevation of the top of a statue at an unknown distance from the statue's base, a student walks exactly 5 m directly away from the statue along horizontal ground and records a second angle of elevation. The two angles recorded are 17.7° and 12.0°. Find the height of the statue.

29 From the top of his lighthouse, the keeper observes two buoys, the nearest of which is directly to the east of the lighthouse and the second buoy 18 m south of the first. The surface of the water is still.

If the angle of depression to the first buoy is 42.5° and the angle of depression to the second buoy is 41.3°, find the height of the lighthouse above sea level, to the nearest metre.

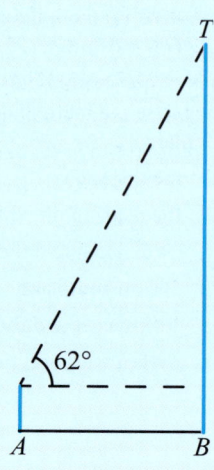

5C Applications of trigonometry

30. A cuboid ABCD with sides 4 cm, 7 cm and 9 cm is shown in the diagram.
    a Find the length of the diagonal AG.
    b Find the angle between the diagonal AG and the base ABCD.
    c Find the angle between AG and the side AB.

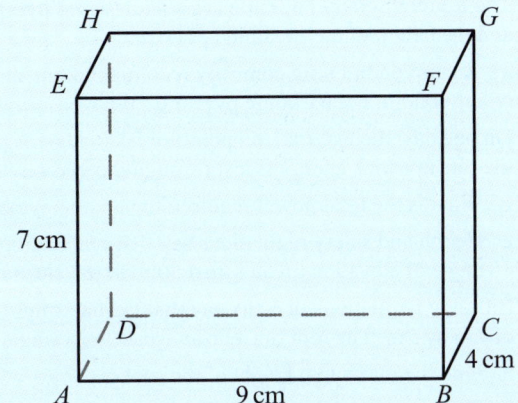

31. In a cuboid with base ABCD and upper surface EFGH, the three face diagonals have lengths AC = 13 m, AF = 7 m, CF = 11 m.
    a Find the length of the body diagonal AG, giving your answer to four significant figures.
    b Find the smallest angle between line AG and the base of the cuboid.

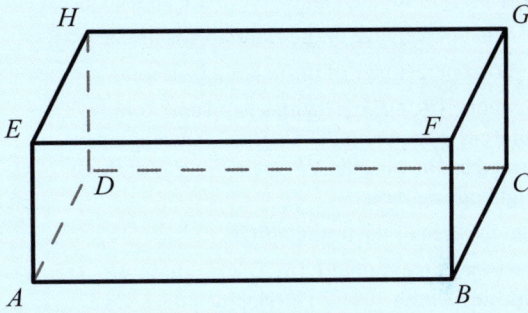

32. The cuboid ABCDEFGH is shown in the diagram on the right.
    a Find the lengths of AG and CE.
    b Find the acute angle between AG and CE.

33. A dog runs 220 m on a bearing of 042° and then a further 180 m on a bearing of 166°. Find the distance and the bearing on which the dog must run to return to the starting position.

34. A lighthouse is 2.5 km from the port on a bearing of 035°. An island is 1.3 km from the port on a bearing of 130°. Find the distance and the bearing of the lighthouse from the island.

35. The Great Pyramid of Giza has a square base of side 230 m. One of the sloping edges makes an angle of 42° with the base. Find the height of the pyramid.

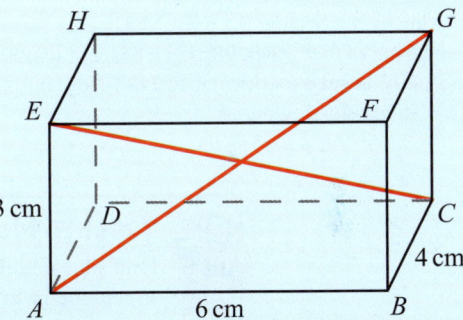

36. Ramiz stands at the point R, 19.5 m from the base B of a vertical tree. He can see the top of the tree, T, at an angle of elevation of 26°.
    a Find the height of the tree.

    Mia can see the top of the tree at an angle of elevation of 41°.

    b Find Mia's distance from the bottom of the tree.
    c Given that the distance between Mia and Ramiz is 14.7 m, find the size of the angle $R\hat{B}M$ correct to the nearest degree.

37. A visitor at an art gallery sits on the floor, 2.4 m from the wall, and looks up at a painting. He can see the bottom of the painting at an angle of elevation of 55° and the top of the painting at an angle of elevation of 72°. Find the height of the painting.

38 The Louvre pyramid in Paris is a square based pyramid made mainly of glass. The square base has sides of 34 m and the height of the pyramid is 21.6 m.

  a An air conditioning company recommend one unit per 1000 m³ of air volume. How many units are needed to air-condition the Louvre pyramid? What assumptions are you making?

On one day, the external temperature is 20 degrees below the required internal temperature. In these conditions, the rate at which energy is lost through the glass is 192 Watts per m².

  b What is the total power required to heat the pyramid to offset the energy lost through the glass?
  c Health and safety regulations say that scaffolding must be used to clean any glass building with a maximum angle of elevation greater than 50°. Do the cleaners need to use scaffolding? Justify your answer.

39 Building regulations in a city say that the maximum angle of elevation of a roof is 35°. A building has a footprint of 7 m by 5 m. The roof must be an isosceles triangle-based prism with a vertical line of symmetry.

  a Find the maximum height of the roof.
  b Find the maximum volume.

Only parts of the roof above 0.6 m are classed as usable.

  c What percentage of the floor area is usable?
  d What percentage of the volume is usable?

40 A cuboid *ABCDEFGH* with sides 4 cm, 7 cm and 9 cm is shown in the diagram. A triangle is formed by the diagonals *BG*, *GE* and *EB* of three neighbouring faces.

Find the area of the triangle *BGE*.

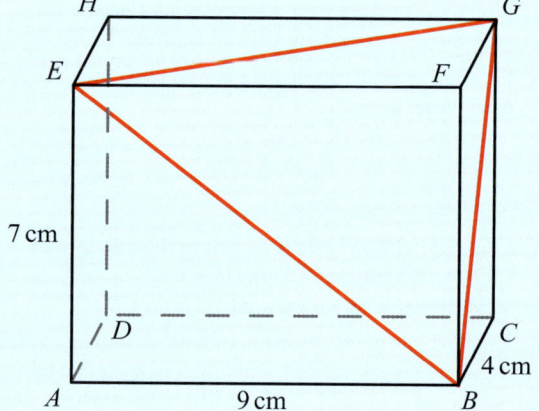

41 The base of a pyramid *VABCD* is a square *ABCD* with side length 20 cm. The sloping edges have length 23 cm. *M* is the midpoint of the edge *AB* and *N* is the midpoint of the edge *BC*. Find the size of the angle *MVN*.

42 Amy cycles around the park. She cycles 120 m on a bearing of 070°, then 90 m in a bearing of 150° and finally 110 m on a bearing of 250°. How far is she from her starting point?

TOOLKIT: Modelling

Find a local building and model it using some of the shapes you have met. Use trigonometry to estimate the dimensions of the shapes. Calculate the volume and surface area of the building, trying to keep track of the estimated size of any errors in your values.

Use these values to answer real world questions about the buildings such as:
- How much paint would be required to paint it?
- How much does it cost to maintain the building's temperature?
- How long would it take to clean the building?
- How energy efficient is the building?

See if you can compare your answers to available data on these values.

Checklist

- You should be able to find the volume and surface area of three-dimensional solids:

Shape	Volume	Surface area
Sphere of radius r	$\frac{4}{3}\pi r^3$	$4\pi r^2$
Cone of base radius r, height h and slant height l	$\frac{1}{3}\pi r^2 h$	$\pi r l + \pi r^2$
Pyramid of base area B and height h	$\frac{1}{3}Bh$	Area of triangular sides $+B$

- You should be able to find the angle between two intersecting lines in two dimensions.
- You should be able to use the sine rule to find side lengths and angles in non-right-angled triangles:
  - □ $\dfrac{a}{\sin A} = \dfrac{b}{\sin B} = \dfrac{c}{\sin C}$ or, equivalently,   □ $\dfrac{\sin A}{a} = \dfrac{\sin B}{b} = \dfrac{\sin C}{c}$
- You should be able to use the sine rule to find side lengths and angles in non-right-angled triangles:
  - □ $a^2 = b^2 + c^2 - 2bc\cos A$ or, equivalently,   □ $\cos A = \dfrac{b^2 + c^2 - a^2}{2bc}$
- You should be able to find the area of a triangle when you do not know the perpendicular height:
  - □ Area $= \dfrac{1}{2}ab\sin C$
- You should be able to find the angle between two intersecting lines in three-dimensional shapes.
- You should be able to find the angle between a line and a plane in three-dimensional shapes.
- You should be able to construct diagrams from given information.
- You should be able to use trigonometry in questions involving bearings.
- You should be able to use trigonometry in questions involving angles of elevation and depression.

■ Mixed Practice

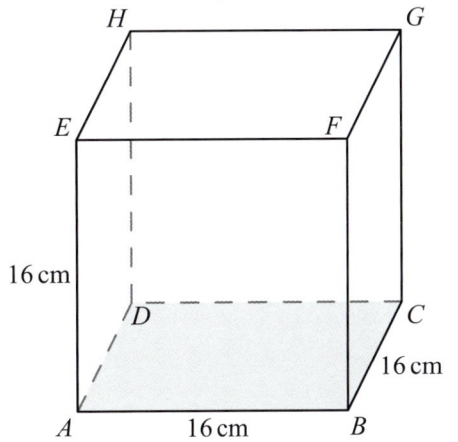

1 Viewed from 50 m away, a building has an angle of elevation of 35°. Find the height of the building.

2 The cube in the diagram has side 16 cm.
  a Find the lengths of AC and AG.
  b Draw a sketch of triangle ACG, labelling the lengths of all the sides.
  c Find the angle between AC and AG.

3 The base of a pyramid is a square of side 23 cm.
The angle between AC and AE is 56°.
  a Find the length of AC.
  b Find the height of the pyramid.
  c Find the length of AE.

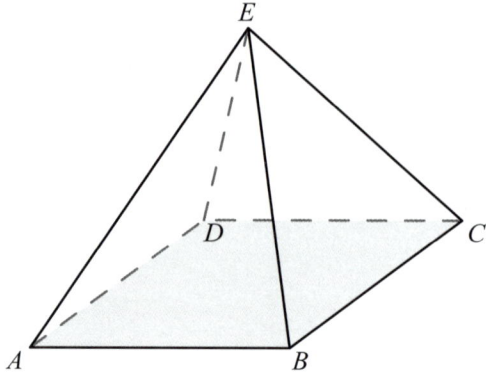

4 A cone has radius 5 cm and vertical height 12 cm.
Find the size of angle θ.

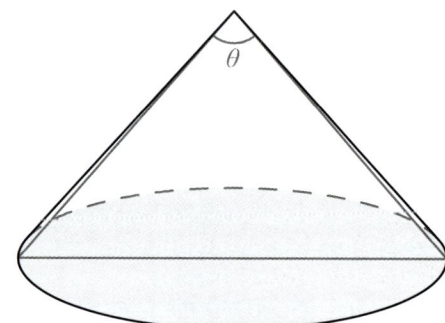

5 A and B are points on a straight line as shown on the graph on the right.
  a Write down the y-intercept of the line AB.
  b Calculate the gradient of the line AB.
The acute angle between the line AB and the x-axis is θ.
  c Calculate the size of θ.

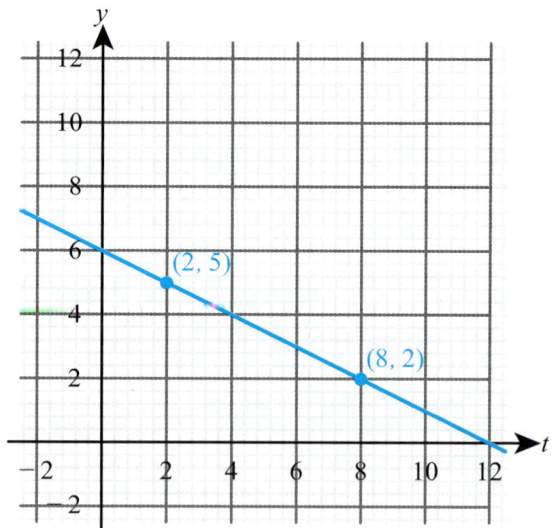

Mathematical Studies SL May 2009 Paper 1 TZ1 Q4

Mixed Practice

6 The quadrilateral ABCD has AB = 10 cm. AD = 12 cm and CD = 7 cm.
The size of angle ABC is 100° and the size of angle ACB is 50°.
   a Find the length of AC in centimetres.
   b Find the size of angle ADC.

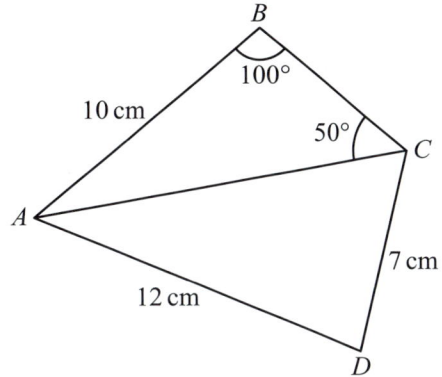

Mathematical Studies SL May 2010 Paper 1 TZ2 Q5

7 In triangle ABC, A = 50°, B = 70° and a = 10 cm. Find the length of side b.

8 In triangle ABC, A = 15°, b = 8 cm and c = 10 cm. Find the length of side a.

9 In triangle ABC, the sides are a = 3 cm, b = 5 cm and c = 7 cm. Find angle A.

10 Triangle XYZ has X = 42°, x = 15 cm and y = 12 cm. Find angle Y.

11 Triangle PQR has P = 120°, p = 9 cm and q = 4 cm. Find angle R.

12 In triangle ABC, B = 32°, C = 72° and b = 10 cm. Find side a.

13 After recording the angle of elevation of the top of a tower at an unknown distance from the tower's base, a student walks exactly 20 m directly away from the tower along horizontal ground and records a second angle of elevation. The two angles recorded are 47.7° and 38.2°. Find the height of the tower.

14 All that remains intact of an ancient castle is part of the keep wall and a single stone pillar some distance away. The base of the wall and the foot of the pillar are at equal elevations.

From the top of the keep wall, the tip of the pillar is at an angle of depression of 23.5° and the base of the pillar is at an angle of depression of 37.7°.

The wall is known to have a height of 41 m. Find the height of the pillar, to the nearest metre.

15 a Sketch the lines with equations $y = \frac{1}{3}x + 5$ and $y = 10 - x$, showing all the axis intercepts.
   b Find the coordinates of the point of intersection between the two lines.
   c Find the size of the acute angle between the two lines.

16 A square-based pyramid has height 26 cm. The angle between the height and one of the sloping edges is 35°. Find the volume of the pyramid.

17 The base of a cuboid ABCDEFGH is a square of side 6 cm. The height of the cuboid is 15 cm. M is the midpoint of the edge BC.
   a Find the angle between ME and the base ABCD.
   b Find the size of the angle HME.

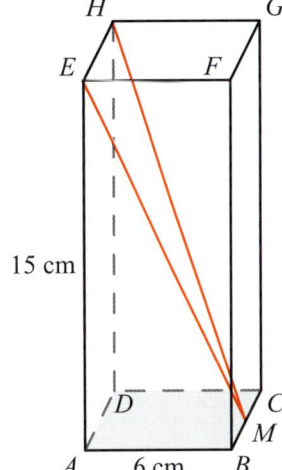

18 In triangle ABC, $AB = x$, $AC = 2x$, $BC = x + 4$ and $B\hat{A}C = 60°$. Find the value of x.

19 The area of this triangle is 84 units². Find the value of x.

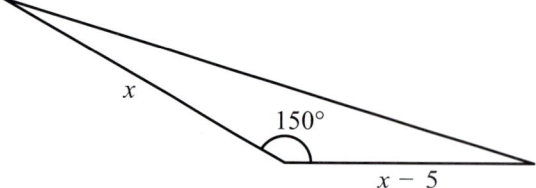

20 A 30 m tall tower and a vertical tree both stand on horizontal ground. From the top of the tower, the angle of depression of the bottom of the tree is 50°. From the bottom of the tower, the angle of elevation of the top of the tree is 35°. Find the height of the tree.

21 Tennis balls are sold in cylindrical tubes that contain four balls. The radius of each tennis ball is 3.15 cm and the radius of the tube is 3.2 cm. The length of the tube is 26 cm.
 a Find the volume of one tennis ball.
 b Calculate the volume of the empty space in the tube when four tennis balls have been placed in it.

Mathematical Studies SL May 2009 Paper 1 TZ1 Q13

22 The diagram shows a right triangular prism, $ABCDEF$, in which the face $ABCD$ is a square.

$AF = 8$ cm, $BF = 9.5$ cm, and angle BAF is 90°.
 a Calculate the length of AB.
 M is the midpoint of EF.
 b Calculate the length of BM.
 c Find the size of the angle between BM and the face $ADEF$.

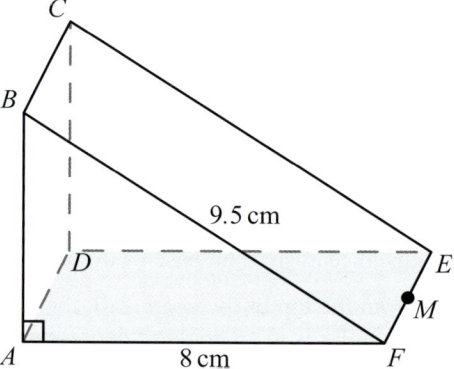

Mathematical Studies SL November 2012 Paper 1 Q12

23 Part A
The diagram on the right shows a square-based right pyramid. $ABCD$ is a square of side 10 cm. VX is the perpendicular height of 8 cm. M is the midpoint of BC.
 a Write down the length of XM.
 b Calculate the length of VM.
 c Calculate the angle between VM and $ABCD$.

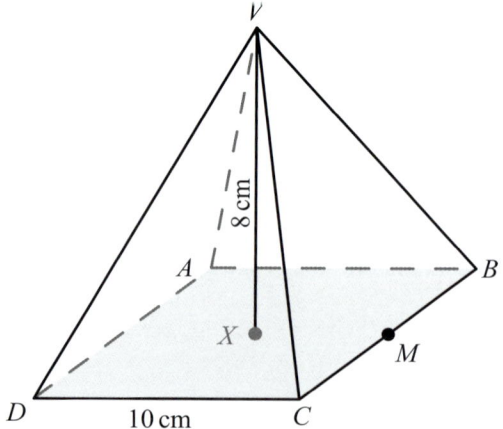

Mixed Practice

Part B

A path goes around a forest so that it forms the three sides of a triangle. The lengths of two sides are 550 m and 290 m. These two sides meet at an angle of 115°. A diagram is shown on the right.

a Calculate the length of the third side of the triangle. Give your answer correct to the nearest 10 m.

b Calculate the area enclosed by the path that goes around the forest.

Inside the forest a second path forms the three sides of another triangle named *ABC*. Angle *BAC* is 53°, *AC* is 180 m and *BC* is 230 m.

c Calculate the size of angle *ACB*.

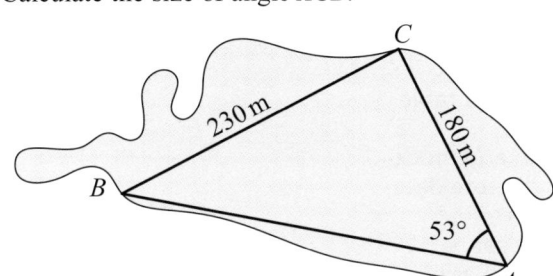

Mathematical Studies SL November 2009 Paper 2 Q1

24 The diagram shows an office tower of total height 126 metres. It consists of a square-based pyramid *VABCD* on top of a cuboid *ABCDPQRS*.

V is directly above the centre of the base of the office tower.

The length of the sloping edge *VC* is 22.5 metres and the angle that *VC* makes with the base *ABCD* (angle *VCA*) is 53.1°.

a i Write down the length of *VA* in metres.
   ii Sketch the triangle *VCA* showing clearly the length of *VC* and the size of angle *VCA*.
b Show that the height of the pyramid is 18.0 metres correct to 3 significant figures.
c Calculate the length of *AC* in metres.
d Show that the length of *BC* is 19.1 metres correct to 3 significant figures.
e Calculate the volume of the tower.

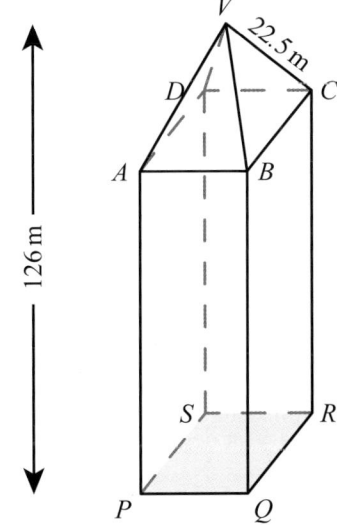

To calculate the cost of air conditioning, engineers must estimate the weight of air in the tower. They estimate that 90% of the volume of the tower is occupied by air and they know that 1 m³ of air weighs 1.2 kg.

f Calculate the weight of air in the tower.

Mathematical Studies SL May 2010 Paper 2 TZ1 Q4

25 Find the area of the triangle formed by the lines $y = 8 - x$, $2x - y = 10$ and $11x + 2y = 25$.

6 Core: Statistics

ESSENTIAL UNDERSTANDINGS

- Statistics is concerned with the collection, analysis and interpretation of data.
- Statistical representations and measures allow us to represent data in many different forms to aid interpretation.

In this chapter you will learn…

- about the concepts of population and sample
- about discrete and continuous data
- about potential bias in sampling
- about a range of sampling techniques and their effectiveness
- how to identify and interpret outliers
- about frequency distributions
- how to estimate the mean for a grouped frequency table
- how to find the modal class
- how to use your GDC to find the mean, median and mode
- how to use your GDC to find the quartiles of discrete data
- how to use your GDC to find standard deviation
- about the effect of constant changes on a data set
- how to construct and use statistical diagrams such as histograms, cumulative frequency graphs and box-and-whisker plots
- about scatter graphs and how to add a line of best fit
- how to calculate (using your GDC) and interpret a numerical measure of linear correlation
- how to use your GDC to find the equation of the line of best fit (the regression line)
- how to use the regression line to predict values not in the data set and how to interpret the coefficients of the regression line
- about piecewise linear models.

■ Figure 6.1 How can statistics be used to mislead?

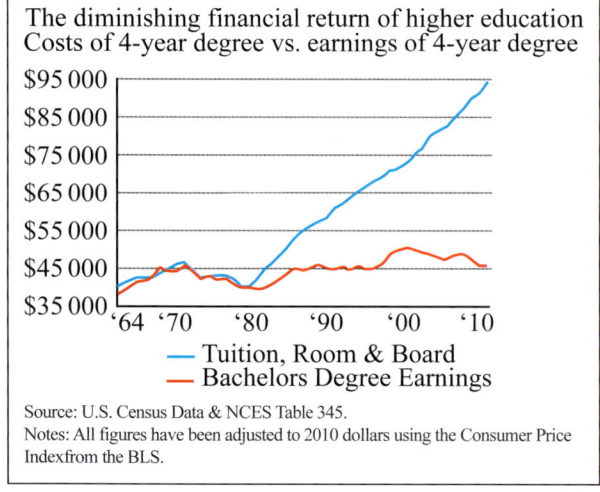

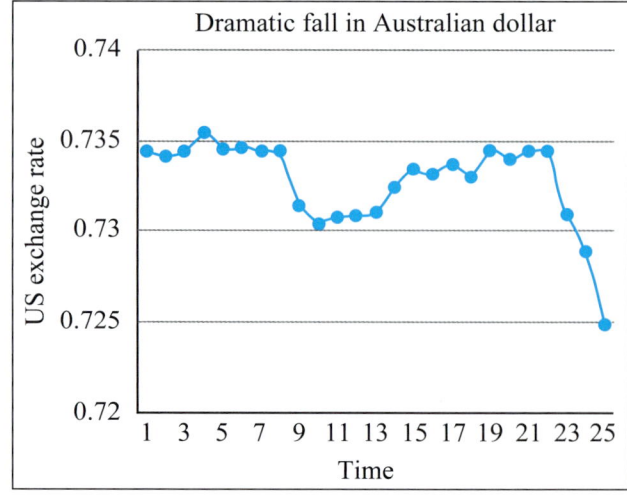

Core: Statistics

CONCEPTS

The following key concepts will be addressed in this chapter:
- Organizing, **representing**, analysing and interpreting data, and utilizing different statistical tools, facilitate prediction and drawing of conclusions.
- Different statistical techniques require justification and the identification of their limitations and **validity**.
- **Approximation** in data can approach the truth but may not always achieve it.
- Correlation and regression are powerful tools for identifying **patterns**.

PRIOR KNOWLEDGE

Before starting this chapter, you should already be able to complete the following:

1 For the data 1, 2, 2, 5, 6, 6, 6, 10, find:
   a the mean
   b the median
   c the mode
   d the range.
2 For the straight line with equation $y = 1.5x + 7$, state
   a the gradient
   b the coordinates of the point where the line crosses the y axis.

Starter Activity

Look at the graphs in Figure 6.1 and discuss what message they are trying to convey. What techniques do they use to persuade you?

Now look at this problem:

The annual salaries of people working in a small business (in thousands of £s) are:

10, 10, 10, 15, 15, 20, 20, 25, 30, 100.

a Work out the mean, median and mode.
b Which average would a union representative wish to use?
c Which would the owner wish to use?
d Which is the most meaningful?

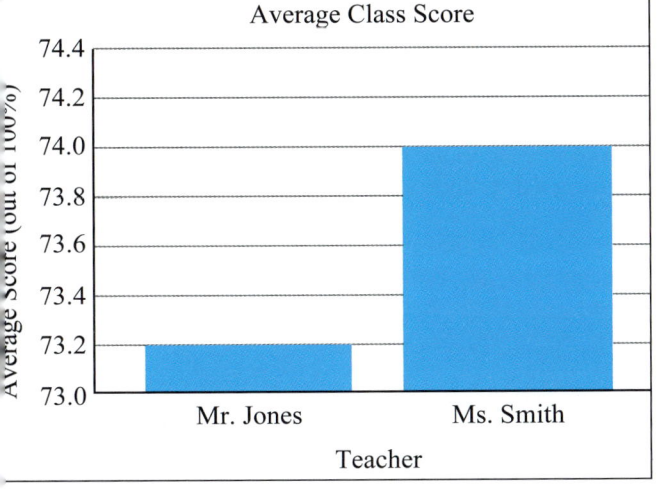

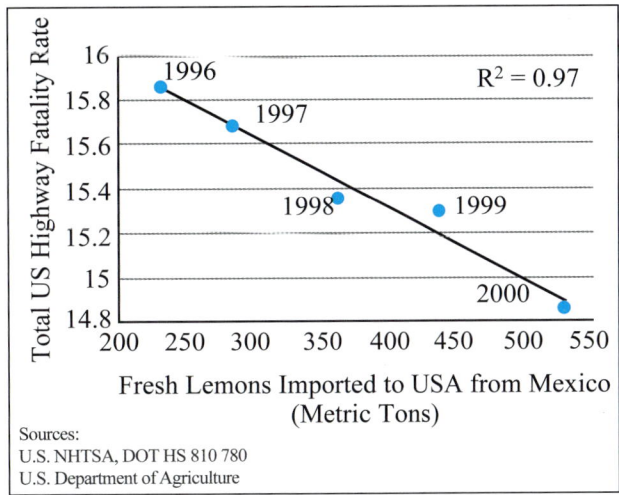

6A Sampling

■ Sampling concepts

If you want to find the average height of an adult female in the UK, one approach would be to measure the height of every person in this **population**, that is, every adult female in the UK. Of course, in reality, this is not practical, so instead you could estimate the population average by taking a **sample** of the adult female population. The average height of the sample will most likely differ from the population average, but if the sample is well chosen it should provide a reasonable estimate.

Data such as height, which can take any value, limited only by the accuracy of the instrument you are measuring with, are known as **continuous**.

Data that can take only distinct values, such as the number of people who are taller than 2 metres, are known as **discrete**.

> **WORKED EXAMPLE 6.1**
>
> Below is an extract from a report on workplace productivity:
>
> '492 employees from a single company participated in the study. Those who were on flexible hours showed a 5.3% increase in productivity. This suggests that all workers in the UK should have access to flexible hours.'
>
> Identify the sample and the population in the extract. From the information you are given, do you think it is reasonable to generalize from the sample to the population in this case?
>
>> The sample is the employees from the company who participated.
>>
>> The population is all workers in the UK.
>>
>> There is no indication that the sample of employees from this company is representative of the population of all workers in the UK, so it is not reasonable to generalize.

■ Reliability of data sources and bias in sampling

If we want a sample to provide a good estimate of a population value such as the mean, then the sample needs to be representative of the population. This means that the distribution of the values in the sample is roughly the same as in the whole population.

This will not be the case if the sampling procedure is **biased**. This does not necessarily mean an intent to get the wrong answer. It means using a sample which does not represent the population of interest. For example, if we wanted to find out about people's political views and decided to conduct a phone poll in the middle of a week day, our results might be skewed by getting a disproportionate number of retired people's responses.

CONCEPTS – VALIDITY

In 1948 the Chicago Tribune ran a telephone survey that suggested an overwhelming victory for Thomas Dewey in the US presidential election. They were so confident that they ran a newspaper with a headline announcing Dewey's win. The picture shows the actual winner, Harry Truman, holding this paper. Why did the poll get it so wrong? It was because in 1948 those people who had a phone were significantly better off than the average US citizen, so their sample was entirely unrepresentative! It is very important that when assessing any statistics you consider their **validity**. This means the extent to which you are answering the question you really want to answer. In this example, finding out what proportion of telephone owners would vote for Truman does not answer the question about what proportion of the electorate would vote for Truman.

TOOLKIT: Problem Solving

For each of the following situations, explain how the sampling procedure was biased. You may like to research these situations further.

- In 2013, Google used the number of people searching for flu related terms on the internet to predict the number of people who would go to the doctor to seek flu treatment. They overestimated the true number by 140%.
- In 1936, the Literary Digest used a poll of 10 million people with 2.4 million responses to predict the outcome of the presidential election in the USA. They got the result spectacularly wrong, but George Gallup predicted the correct result by asking 50 000 people.
- In 2014, an app used in Boston, USA, for reporting pot holes led to the least damaged roads being repaired.

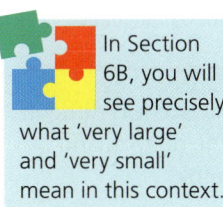

In Section 6B, you will see precisely what 'very large' and 'very small' mean in this context.

■ Interpretation of outliers

Any extreme value – either very large or very small – compared to the rest of the data set is said to be an **outlier**.

You should be aware of any outliers in a data set. Sometimes these will be perfectly valid data values, but at other times these might be errors and therefore should be discarded.

WORKED EXAMPLE 6.2

Alana was analysing data from questionnaires asking schools for the time in hours they spend each year teaching mathematics. Alana's statistical package flagged up one item as an outlier. All the rest of her data were between 120 and 200 hours. What would you suggest Alana should do if the outlier has the value:

a −175

b 4

c 240?

The number of hours of teaching cannot be negative so this must be an error	**a** This is likely to be a data entry error. She should check the original questionnaire and update to the correct value.
This is far too low to be a plausible response	**b** This is likely to be a school giving the number of hours per week rather than per year, but it could be a data entry error again. She should check the original questionnaire but if found not to be a data entry error then discard this value from the analysis.
This seems rather large but is not obviously an error	**c** This seems like an extreme value rather than an error, so it should not be discarded from the analysis.

Links to: Environmental Systems and Societies

The NASA Nimbus satellites collected data on the ozone layer from the early 1970s. Unfortunately, the data were processed by a program that automatically filtered outliers. That meant evidence of the huge 'hole' in the ozone layer above the Antarctic was effectively discarded. The hole was first reported by the British Antarctic Survey in 1985. The historic data were rerun without the outlier-filter and evidence of the hole was seen as far back as 1976. It is not always a good idea to remove outliers – sometimes they are the most important part of the data!

■ Sampling techniques and their effectiveness

There are several different methods for selecting samples, each of which has strengths and weaknesses.

Simple random sample

This is the type of sample most people have in mind when they talk about random samples.

> **KEY POINT 6.1**
>
> With simple random sampling, every possible sample (of a given size) has an equal chance of being selected.

This is good in theory and will produce an unbiased sample, but in practice it is difficult to do as you need a list of the entire population from which to select the sample, and then you need to actually obtain data from everyone in the chosen sample.

Convenience sampling

This approach avoids the difficulties of simple random sampling.

> **KEY POINT 6.2**
>
> With convenience sampling, respondents are chosen based upon their availability.

Tip

Don't think that just because a sampling method is unbiased the chosen sample will necessarily be representative of the population. It might just so happen that by chance the chosen sample contains only extreme values.

This does not produce a random sample, but if the sample size is large enough it can still provide useful information. However, it can introduce bias if the group consists of very similar members and, as such, results may not be generalizable to the population.

Systematic sampling

This requires a list of all participants ordered in some way.

> **KEY POINT 6.3**
>
> With systematic sampling, participants are taken at regular intervals from a list of the population.

This may be more practical than using, say, a random number generator to select a sample, but you still need a list of the entire population and it is less random than simple random sampling due to the fact that selections are no longer independent.

Stratified sampling

Another random sampling method is stratified sampling.

> **KEY POINT 6.4**
>
> With stratified sampling the population is split into groups based on factors relevant to the research, then a random sample from each group is taken in proportion to the size of that group.

This produces a sample representative of the population over the factors identified, but again you need a list of the entire population, this time with additional information about each member so as to identify those with particular characteristics.

Quota sampling

This is a common alternative to stratified sampling.

> **KEY POINT 6.5**
>
> With quota sampling, the population is split into groups based on factors relevant to the research, then convenience sampling from each group is used until a required number of participants are found.

This produces a sample representative of the population for the factors identified, but the convenience sampling element means it can introduce bias if the group consists of very similar members and as such results may not be generalizable to the population.

> **WORKED EXAMPLE 6.3**
>
> In a survey, researchers questioned shoppers in a shopping mall until they had responses from 100 males and 100 females in the 18–35 age range. Name the sampling technique and state one advantage and one disadvantage compared to a simple random sample.
>
>> Quota sampling.
>>
>> One advantage over simple random sampling is that quota sampling is practical. To take a simple random sample, researchers would have to know in advance who was going to the shopping mall that day.
>>
>> One disadvantage is that the shoppers who are prepared to stop and talk to the researchers may not be representative of all shoppers.

> **WORKED EXAMPLE 6.4**
>
> A school has 60% boys and 40% girls. Describe how a stratified sample of 50 students could be formed, reflecting the gender balance of the school.
>
> 60% of the sample of 50 need to be boys .... Number of boys in sample $= \frac{60}{100} \times 50 = 30$
>
> The remainder of the sample of 50 must be girls .... So, number of girls in sample $= 20$

> **CONCEPTS – APPROXIMATION**
>
> Most advanced statistics is about trying to make some inference about a population based on a sample. For example, if the mean IQ of a sample of 50 students in a school is 120 we might think this is a good estimate of the mean IQ of all students in that school. Although this is not perfect, if the sample was representative then this is a good **approximation** to the true value. With further study of statistics you might even be able to suggest how far from the true value this might be.

Exercise 6A

For questions 1 to 4, use the method demonstrated in Worked Example 6.4 to determine the number of each type of participant or item required for a stratified sample.

1. a In a wildlife park, 30% of big cats are lions and 70% are tigers. Select 30 big cats.
   b Of ice creams sold in a café, 45% are strawberry and 55% chocolate. Select 20 ice creams.
2. a A school is attended by 120 boys and 80 girls. Select 40 pupils.
   b There are 84 pupils studying Maths HL and 126 pupils studying Maths SL. Select 45 pupils.
3. a 25% of pupils at school play football, 35% play hockey and 40% play basketball. Select 40 pupils.
   b 30% of fish caught are cod, 45% are haddock and 25% are mackerel. Select a sample of 20 fish.
4. a A manufacturer produced 240 chairs, 90 tables and 40 beds. Select a sample of 37 pieces of furniture.
   b A park has 64 oak trees, 56 willows trees and 32 chestnut trees. Select 19 trees.
5. Anke wants to find out the proportion of households in Germany who have a pet. For her investigation, she decides to ask her friends from school, which is located in the centre of a large city, whether their family owns a pet.
   a What is the relevant population for Anke's investigation?
   b Name the sampling method that Anke is using.
   c State one reason why Anke's sample may not be representative of the population.
6. Leonie wants to collect information on the length of time pupils at her school spend on homework each evening. She thinks that this depends on the school year, so her sample should contain some pupils from each year group.
   a What information does Leonie need in order to be able to select a stratified sample?

   Leonie decides to ask pupils in the lunch queue until she has responses from at least 10 pupils in each year group.

   b Name this sampling method.
   c Having collected and analysed the data, Leonie found two outliers. For each value, suggest whether it should be kept or discarded.
      i   10 minutes
      ii 20 hours
7. A student wants to conduct an investigation into attitudes to environmental issues among the residents of his village. He decides to talk to the first 20 people who arrive at his local bus stop.
   a Name this sampling technique.
   b State the population relevant to his investigation.

c Give one reason why a sample obtained in this way may not be representative of the whole population.

d Explain why it would be difficult to obtain a simple random sample in this situation.

8 Joel obtains a random sample of 20 pupils from a college in order to conduct a survey. He finds that his sample contains the following numbers of students of different ages:

Age	16	17	18	19
Number	6	0	7	7

Mingshan says: 'There are no 17-year-olds in the sample, so your sampling procedure must have been biased.'

Comment on Mingshan's statement.

9 A manufacturer wants to test the lifetime, in hours, of its light bulbs.

a Are the data they need discrete or continuous?

b Explain why they need to test a sample, rather than the whole population.

c Each lightbulb produced is given a serial number. Explain how to obtain a systematic sample consisting of 5% of all lightbulbs produced.

10 A shop owner wants to find out what proportion of the scarves she sells are bought by women. She thinks that this may depend on the colour of scarves, so she records the gender of the customers who bought the first 10 of each colour of scarf.

a Give the name of this type of sampling.

b Explain why a stratified sample would be more appropriate.

The shop owner knows that of all the scarves she sells, 30% are red, 30% are green, 25% are blue and 15% are white. She has a large set of historical sales records which identify the gender of the purchaser and are sorted by colour of scarf sold. She does not want to look through all of them, so she will take a sample of 40 records.

c How many scarves of each colour should be included to make this a stratified sample?

11 A park ranger wants to estimate the proportion of adult animals in a wildlife park that are suffering from a particular disease. She believes that of the adults of these species present in the park, 20% are deer, 30% are tigers, 40% are wolves and 10% are zebras. She decides to observe animals until she has recorded 10 deer, 15 tigers, 20 wolves and 5 zebras.

a State the name of this sampling method.

b Why might this sampling method be better than a convenience sample of the first 50 animals she encounters?

c Explain why a simple random sample may be difficult to obtain in this situation.

12 Shakir wants to find out the average height of students in his school. He decides to use his friends from the basketball team as a sample.

a Is Shakir collecting discrete or continuous data?

b Identify one possible source of bias in his sample.

Shakir decides to change his sampling technique. He obtains an alphabetical list of all students in the school and selects every 10th student for his sample.

c Name this sampling technique.

d Explain why this does not produce a simple random sample.

13 The table shows the number of cats, dogs and fish kept as pets by a group of children.

Cat	Dog	Fish
27	43	30

A sample of 20 pets is required, where the type of pet may be a relevant factor in the investigation.

Find the number of each type of pet that should be included in a stratified sample.

14 Dan needs to select a sample of 20 children from a school for his investigation into the amount of time they spend playing computer games. He thinks that age and gender are relevant factors and so chooses a stratified sample. The table shows the number of pupils of each relevant age and gender at the school.

Gender/Age	12	13	14
Boys	40	52	50
Girls	0	37	21

   a Create a similar table showing how many pupils from each group should be selected for the sample.

   b Do you think that it would be reasonable to generalize the results from Dan's investigation to all 12- to 14-year-olds in the country?

15 A zoologist wants to investigate the distribution of the number of spots on ladybirds. She prepares the following table to record her data:

Number of spots	2	7	10
Number of ladybirds			

   a Are the data she is collecting discrete or continuous?
   b Would it be possible to collect the data for the whole population of ladybirds in a particular field?
   c The zoologist collects her data by counting the number of spots on the first 100 ladybirds she catches. State the name for this sampling procedure.
   b Give one reason why the results from this sample may not be generalizable to the whole population.

16 Ayesha selects a sample of six children from a primary school and measures their heights in centimetres. She finds that five of the children are taller than the national average for their age.

   a Comment whether each of the following could be a sensible conclusion for Ayesha to draw:
      i The children at this school are taller than average.
      ii The sampling methods must have been biased.
      iii This just happens to be an unusual sample.
   b If instead, Ayesha took a sample of 60 children and found that 50 were taller than the national average, how would your answer to a change?
   c In a larger sample, Ayesha identified two values as outliers: –32 cm and 155 cm. For each of the values, comment whether it should be kept or discarded from further analysis.

17 In many situations, researchers want to find out about something which people do not want to admit – for example, criminal activity. Randomized Response Theory (RRT) is one method which allows researchers to estimate the proportion of a population with the trait without ever knowing if an individual has that trait.

   a Research subjects take one of three cards at random and are asked to follow the instructions on the card. The cards have the following text:

   Card 1: Say yes.

   Card 2: Say no.

   Card 3: Say yes if you have ever taken illegal drugs, otherwise say no.

   If 24 out of 60 research subjects say 'Yes' estimate the percentage of the population who have taken illegal drugs.

   b A sample of students were shown two statements.

   Statement 1: I have cheated on a test.

   Statement 2: I have never cheated on a test

   They were asked to secretly roll a dice and if they got a 6 state honestly say true or false about the first statement, otherwise say true or false honestly about the second statement.
      i If 20% of students have cheated on a test, how many out of 120 students would be expected to answer 'True'?
      ii If 48 out of 120 students say 'True' estimate the percentage of the students who have cheated on a test.

18 Ecologists wanted to estimate the number of cod in the North Sea. They captured 50 000 cod and humanely labelled them. Six months later they captured a sample of 40 000 cod and 20 are found to be labelled.

  a Estimate the number of cod in the North Sea.

  b State two assumptions required for your calculation in part **a**.

TOK Links

If you do an internet search for 'How many adult cod are in the North Sea', you might get a surprising answer – lots of sources say that there are only 100. This is based on a research paper which used the threshold for 'adulthood' for a different species of cod. Can you find any other examples when the definitions of terms in statistical arguments are poorly defined? What criteria do you use when judging the authenticity of information you get on the internet? Is it possible to remove misleading information from common knowledge?

6B Summarizing data

■ Frequency distributions

A frequency distribution is a table showing all possible values a variable can take and the number of times the variable takes each of those values (the frequency).

Often the data will be presented in groups (or classes). For continuous data, these groups must cover all possible data values in the range, so there can be no gaps between the classes.

WORKED EXAMPLE 6.5

The following table shows the times, recorded to the nearest minute, taken by students to complete an IQ test.

Time, t (minutes)	1	2	3	4	5	6	7	8	9	10
Frequency	1	2	5	12	16	14	10	4	2	1

a How many students took the test?

b Copy and complete the following table to summarize this information:

Time, t (minutes)	$0.5 \leq t < 3.5$	$3.5 \leq t < 6.5$	$6.5 \leq t < 10.5$
Frequency			

Sum all the frequencies .... **a** $1 + 2 + 5 + 12 + 16 + 14 + 10 + 4 + 2 + 1 = 67$

The interval $0.5 \leq t < 3.5$ .... **b**
includes anyone who
took 1, 2 or 3 min

Time, t (minutes)	$0.5 \leq t < 3.5$	$0.5 \leq t < 3.5$	$0.5 \leq t < 3.5$
Frequency	8	42	17

The interval $3.5 \leq t < 6.5$
includes anyone who
took 4, 5, or 6 min

The interval $6.5 \leq t < 10.5$
includes anyone who
took 7, 8, 9 or 10 min

■ Measures of central tendency

It is very useful to have one number that represents the whole data set, and so it makes sense that this measures the centre of the data set. Such a value is known as an average.

You already know that there are three commonly used measures for the average: the mean (often given the symbol $\bar{x}$), the median and the mode. Usually you will just find these from your GDC.

WORKED EXAMPLE 6.6

For the data set 1, 1, 3, 6, 8 find:

a the mean b the median c the mode.

Enter the list of data in the GDC and then read off the mean ($\bar{x}$)... **a** mean = 3.6

```
1-Variable
x̄      =3.6
Σx     =18
Σx²    =106
xσn    =2.87054001
xσn-1  =3.2093613
n      =5                ↓
```

... the median (med) and the mode (mod) ..... **b** median = 2
 c mode = 1

```
1-Variable
minX  =1                 ↑
Q1    =1
Med   =2
Q3    =7
maxX  =8
Mod   =1                 ↓
```

CONCEPTS – REPRESENTATION

Is it useful to **represent** a set of data using a single number? What information is lost when we do this?

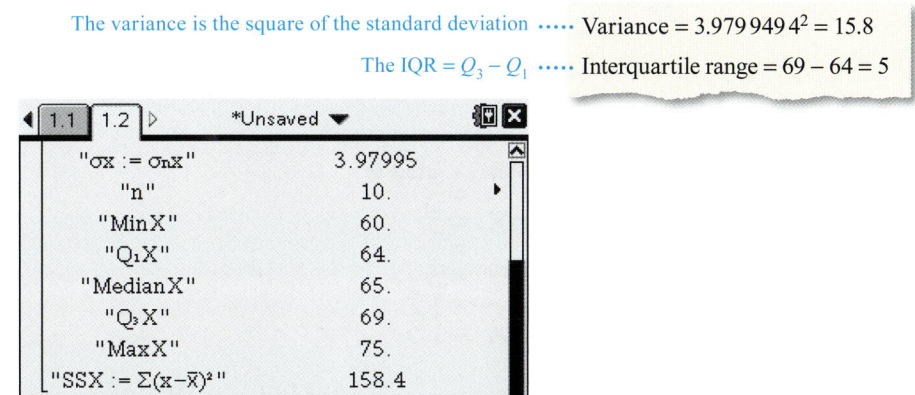

The variance is the square of the standard deviation ····· Variance = $3.9799494^2 = 15.8$

The IQR = $Q_3 - Q_1$ ····· Interquartile range = $69 - 64 = 5$

You are the Researcher

You might like to investigate why calculators provide two different symbols for standard deviation.

CONCEPTS – VALIDITY

Suppose I wanted to compare the spread of lengths of two species of snake. A sample of species A has an IQR of 16 cm and a sample of species B has a standard deviation of 12 cm. Can I say which species has a greater spread? The answer is no, for two reasons:

- I cannot directly compare two different measures of spread.
- I cannot assume that just because the spread of one sample is larger than another that this will also be the case for the two populations.

Both of these points are issues of **validity**. You need to consider the validity of your statistical analysis as well as the sampling process.

Identifying outliers

One way of deciding when a particularly large or small value qualifies as an outlier is to see whether it is far enough above the upper quartile or below the lower quartile when compared to the general spread of the data values.

The exact definition you need for this course is:

KEY POINT 6.8

The data value x is an outlier if $x < Q_1 - 1.5(Q_3 - Q_1)$ or $x > Q_3 + 1.5(Q_3 - Q_1)$.

WORKED EXAMPLE 6.13

A set of data has lower quartile 60 and upper quartile 70. Find the range of values for which data would be flagged as outliers.

x is an outlier if

$x < Q_1 - 1.5(Q_3 - Q_1)$ ················· $x < 60 - 1.5(70 - 60)$

$x < 45$

... or if

$x > Q_3 + 1.5(Q_3 - Q_1)$ ················· $x > 70 + 1.5(70 - 60)$

$x > 85$

x is an outlier if $x < 45$ or $x > 85$

6B Summarizing data

Tip

There are slightly different methods for working out the quartiles, so you may find that the values from your GDC differ from those obtained by hand. The values from your GDC will always be acceptable though.

WORKED EXAMPLE 6.11

Find the upper and lower quartiles of 3, 3, 5, 5, 12, 15.

The lower quartile is Q_1 and the upper quartile Q_3 .... Lower quartile = 3
Upper quartile = 12

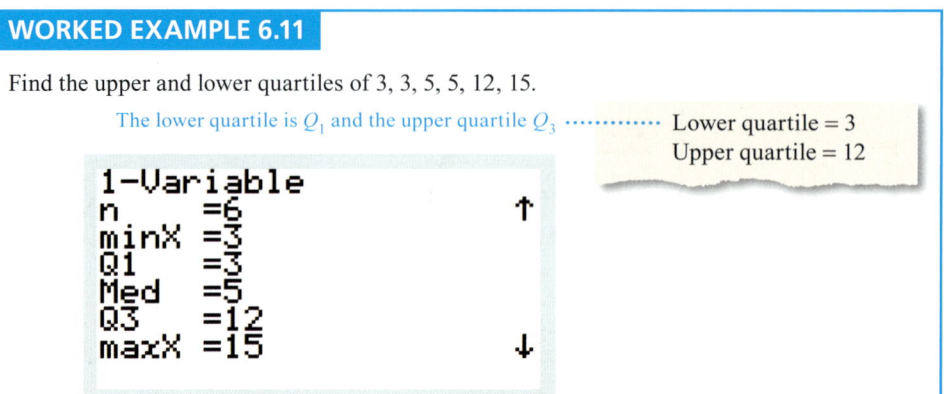

■ Measures of dispersion (interquartile range, standard deviations and variance)

Once you have a measure of the centre of the data set (the average), it is useful to have a measure of how far the rest of the data set is from that central value.

This distance from the average is known as the dispersion (or spread) of the data, and like averages there are several measures of dispersion.

You are already familiar with the **range** as a measure of dispersion. An adjusted version of this is the **interquartile range** (often abbreviated to IQR), which measures the distance between the upper and lower **quartiles** (that is, the width of the central half of the data set).

KEY POINT 6.7

- $IQR = Q_3 - Q_1$

The **standard deviation** (often denoted by σ) is another measure of dispersion, which can be thought of as the mean distance of each point from the mean. The **variance** is the square of the standard deviation (σ^2).

You only need to be able to find the standard deviation using your GDC.

WORKED EXAMPLE 6.12

In a quality control process, eggs are weighed and the following 10 masses, in grams, are found:
64, 65, 68, 64, 65, 70, 75, 60, 64, 69.

Find the standard deviation, variance and interquartile range of the data.

The standard deviation in this screenshot is $x\sigma_n$, although it is displayed differently in differently models .... Standard deviation = 3.98

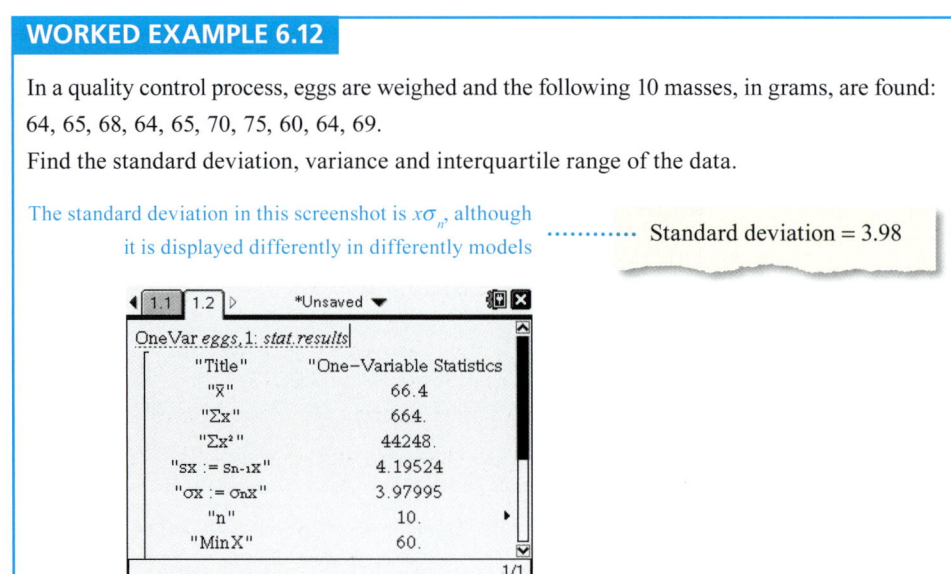

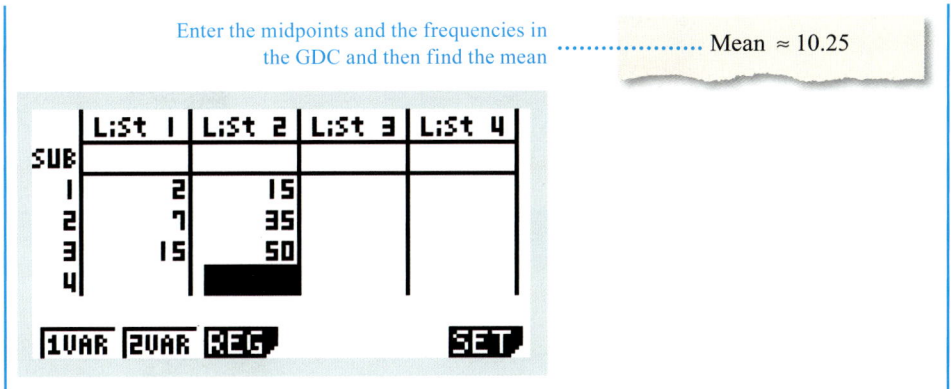

Enter the midpoints and the frequencies in the GDC and then find the mean

Mean ≈ 10.25

Modal class

It is not possible to find the mode from grouped data as we do not have any information on individual data values, so the best we can do is identify the modal group – this is simply the group that has the highest frequency.

> **You are the Researcher**
>
> This definition is fine if all the groups are have the same width, but if that is not the case you have to be careful that a group is not labelled as a modal class just because it is wider than other groups. This leads on to an idea called frequency density which is used in some types of histograms and leads naturally on to the ideas of continuous random variables.

> **WORKED EXAMPLE 6.10**
>
> Find the modal class in the table below:
>
> | x | $10 \leq x < 14$ | $14 \leq x < 18$ | $18 \leq x < 22$ |
> |---|---|---|---|
> | **Frequency** | 6 | 14 | 22 |
>
> The modal class is the group with the highest frequency
>
> Modal class is $18 \leq x < 22$

There are many websites which tell you statistics about different countries. What are the benefits of sharing and analysing data from different countries? If you know the average wage in the US, or the average life expectancy in India, what does that tell you about that country?

Quartiles of discrete data

Once ordered from lowest to highest, the lower quartile (often abbreviated to Q_1) is the value that is one-quarter of the way through the data set and the upper quartile (or Q_3) is the value that is three-quarters of the way through the data set.

So, the quartiles (together with the median, which can be written as Q_2), divide the data set into four parts.

6B Summarizing data

You should also be able to use the formula for the mean of n items:

> **KEY POINT 6.6**
>
> - $\bar{x} = \dfrac{\sum x}{n}$

WORKED EXAMPLE 6.7

If the mean of 1, 2, 4, 5, a, $2a$ is 8, find the value of a.

Use the formula $\bar{x} = \dfrac{\sum x}{n}$ .... $\dfrac{1+2+4+5+a+2a}{6} = 8$

Solve for x .... $12 + 3a = 48$

$3a = 36$

$a = 12$

The formula can also be applied to frequency tables.

WORKED EXAMPLE 6.8

Given that the mean of the following data set is 4.2, find y.

Data value	1	2.5	6	y
Frequency	4	5	8	3

You can use the formula $\bar{x} = \dfrac{\sum x}{n}$ again. The data value 2.5 occurs 5 times, so it will contribute 2.5×5 to the total when you are summing the data values

.... $\dfrac{(1 \times 4) + (2.5 \times 5) + (6 \times 8) + 3y}{20} = 4.2$

The total frequency is $4 + 5 + 8 + 3 = 20$

$\dfrac{64.5 + 3y}{20} = 4.2$

Solve for x .... $64.5 + 3y = 84$

$3y = 19.5$

$y = 6.5$

Estimation of the mean from grouped data

If you are dealing with grouped data you can find an estimate for the mean by replacing each group with its midpoint.

WORKED EXAMPLE 6.9

For the following frequency table, estimate the mean value.

x	$0 \leqslant x < 4$	$4 \leqslant x < 10$	$10 \leqslant x < 20$
Frequency	15	35	50

Re-write the table replacing each group with the value at its midpoint ....

Midpoint	2	7	15
Frequency	15	35	50

For example, the midpoint of the interval $0 \leqslant x < 4$ is $\dfrac{0+4}{2} = 2$

6B Summarizing data

■ Effect of constant changes on the original data

If you add a constant to every value in a data set you will cause the average to change by that same value, but this shift will not affect the spread of the data.

> **KEY POINT 6.9**
>
> Adding a constant, k, to every data value will:
> - change the mean, median and mode by k
> - not change the standard deviation or interquartile range.

> **WORKED EXAMPLE 6.14**
>
> The mean of a set of data is 12 and the standard deviation is 15. If 100 is added to every data value, what would be the new mean and standard deviation?
>
> Adding 100 to every value will increase the mean by 100 ······· New mean = $12 + 100 = 112$
>
> The standard deviation is unaffected ······· New standard deviation = 15

You could also multiply every value in the data set by some factor. Again, this will alter the average by the same factor but this time it will also affect the spread of the data too, since the data set will be stretched.

> **KEY POINT 6.10**
>
> Multiplying every data value by a positive constant, k, will:
> - multiply the mean, median and mode by k
> - multiply the standard deviation and interquartile range by k.

> **WORKED EXAMPLE 6.15**
>
> The median of a set of data is 2.4 and the interquartile range is 3.6. If every data item is halved, find the new median and interquartile range.
>
> Halving every value will halve both ······· New median = $\frac{2.4}{2} = 1.2$
> the median and the interquartile range
>
> New interquartile range = $\frac{3.6}{2} = 1.8$

Exercise 6B

For questions 1 and 2, use the method demonstrated in Worked Example 6.5 to complete the grouped frequency tables summarizing the data in each question. State also the total number of observations.

1 The times taken by members of an athletics team running a sprint race are given below, to the nearest second.

Time t (sec)	15	16	17	18	19	20	21	22	23
Frequency	2	1	7	5	3	0	1	0	1

Copy and complete each table.

a

Time t (sec)	$14.5 \leqslant t < 16.5$		$18.5 \leqslant t < 20.5$	
Frequency				

b

Time t (sec)		$17.5 \leq t < 20.5$	
Frequency			

2 Some books from two parts of a library are sampled, and the number of pages in each book is recorded, rounding down to the nearest hundred.

Number of pages n		200	300	400	500	600	700	800
Frequency	Historical fiction	0	2	2	13	12	7	4
	Romantic fiction	2	7	12	8	5	4	2

Copy and complete the tables, representing the actual number of pages in the sampled books.

a Historical fiction:

Number of pages n			
Frequency	4	25	

b Romantic fiction:

Number of pages			
Frequency		9	11

For questions 3 and 4, use the method demonstrated in Worked Example 6.6 to find the mean, median and mode for each set of data.

3 a 1, 1, 2, 3, 5, 7, 8
   b 2, 4, 5, 6, 6, 7, 9

4 a 5, 3, 2, 6, 3, 5, 4, 3, 8, 8
   b 12, 24, 25, 24, 33, 36, 24, 55, 55, 24

For questions 5 and 6, use the method demonstrated in Worked Example 6.7 to find the value of x.

5 a The mean of 2, 5, 8, x, $2x$ is 9
   b The mean of 1, 2, 4, 6, x, $x+3$ is 5

6 a The mean of x, $x+2$, $3x$, $4x+1$ is 8.5
   b The mean of $x-3$, $2x$, $2x+1$, $3x-2$ is 11.2

For questions 7 and 8, use the method demonstrated in Worked Example 6.8 to find the value of y.

7 a Mean = 5.6

Data value	2	4	7	y
Frequency	6	10	8	6

   b Mean = 34.95

Data value	27	31	40	y
Frequency	5	8	4	3

8 a Mean = 5.71

Data value	y	5.2	6.1	7.3
Frequency	26	32	18	24

   b Mean = 2.925

Data value	y	2.7	3.5	3.8
Frequency	5	7	6	2

For questions 9 and 10, use the methods demonstrated in Worked Examples 6.9 and 6.10 to:

i   estimate the mean
ii find the modal group.

9 a

x	$0 \leq x < 10$	$10 \leq x < 20$	$20 \leq x < 30$
Frequency	15	12	8

   b

x	$5 \leq x < 15$	$15 \leq x < 25$	$25 \leq x < 35$
Frequency	13	14	15

10 a

x	$10.5 \leq x < 12.5$	$12.5 \leq x < 14.5$	$14.5 \leq x < 16.5$
Frequency	17	23	18

   b

x	$3.5 \leq x < 7.5$	$7.5 \leq x < 11.5$	$11.5 \leq x < 15.5$
Frequency	23	31	40

6B Summarizing data

For questions 11 and 12, use the method demonstrated in Worked Example 6.9 to estimate the mean.

11 a

x	$0 \leq x < 5$	$5 \leq x < 12$	$12 \leq x < 20$
Frequency	6	8	11

b

x	$2 \leq x < 5$	$5 \leq x < 6$	$6 \leq x < 9$
Frequency	12	9	5

12 a

x	$2.5 \leq x < 3.5$	$3.5 \leq x < 5.5$	$5.5 \leq x < 10.5$
Frequency	23	31	27

b

x	$10.5 \leq x < 20.5$	$20.5 \leq x < 40.5$	$40.5 \leq x < 45.5$
Frequency	14	11	26

For questions 13 and 14, use the method demonstrated in Worked Example 6.13 to identify the range of values which would be flagged as outliers.

13 a Lower quartile = 50, upper quartile = 80
    b Lower quartile = 16, upper quartile = 20

14 a $Q_1 = 6.5$, $Q_3 = 10.5$
    b $Q_1 = 33.5$, $Q_3 = 45.5$

For questions 15 and 16, use the method demonstrated in Worked Example 6.14 to find the new measures of average and spread.

15 a The mean of a data set is 34 and the standard deviation is 8. Then 12 is added to every data value.
    b The mean of a data set is 162 and the standard deviation is 18. Then 100 is subtracted from every data value.

16 a A data set has median 75 and interquartile range 13. Every data item is decreased by 50.
    b A data set has median 4.5 and interquartile range 2.1. Every data item is decreased by 4.

For questions 17 and 18, use the method demonstrated in Worked Example 6.15 to find the new measures of average and spread.

17 a The median of a data set is 36 and the interquartile range is 18. Every data item is multiplied by 10.
    b The median of a data set is 50 and the interquartile range is 26. Every data item is divided by 10.

18 a A mean of a data set is 16 and the standard deviation is 6. Every data item is halved.
    b A mean of a data set is 8.2 and the standard deviation is 0.6. Every data item is multiplied by 5.

19 The times (in minutes) taken by a group of children to complete a puzzle were rounded to the nearest minute and recorded in the frequency table:

Time (min)	4	5	6	7	8
Frequency	5	12	12	8	11

  a How many children were in the group?
  b How many children took more than 6.5 minutes to complete the puzzle?
  c Find the mean and standard deviation of the times.

20 The masses of some kittens (in kg) are recorded in the following grouped frequency table:

Mass (kg)	$0.8 \leq m < 1.0$	$1.0 \leq m < 1.2$	$1.2 \leq m < 1.4$	$1.4 \leq m < 1.6$	$1.6 \leq m < 1.8$
Frequency	12	19	22	27	19

  a How many kittens have a mass less than 1.2 kg?
  b State the modal group.
  c Estimate the mean mass of the kittens.

21 The table shows the times (in seconds) of a group of school pupils in a 100 m race.

Time (s)	$11.5 \leq t < 13.5$	$13.5 \leq t < 15.5$	$15.5 \leq t < 17.5$	$17.5 \leq t < 19.5$	$19.5 \leq t < 21.5$
Frequency	1	3	12	8	4

  a How many pupils took part in the race?
  b State the modal group of the times.
  c Estimate the mean time. Why is this only an estimate?

22 The lengths of songs on the latest album by a particular artist, in minutes is recorded as:

3.5, 4, 5, 4.5, 6, 3.5, 4, 5.5, 3, 4.5.

Find:

a the median song length

b the interquartile range of the lengths.

The lengths of songs on an album by another artist have the median of 5.5 and interquartile range of 1.

c Write two comments comparing the lengths of songs by the two artists.

23 For the set of data:

3, 6, 1, 3, 2, 11, 3, 6, 8, 4, 5

a find the median

b calculate the interquartile range.

c Does the data set contain any outliers? Justify your answer with clear calculations.

24 The frequency table shows foot lengths of a group of adults, rounded to the nearest cm.

Length (cm)	21	22	23	24	25	26	27	28	29
Frequency	4	7	12	8	11	11	8	4	2

a How many adults are in the group?

b What is the modal foot length?

c Find the mean foot length.

d Complete the grouped frequency table:

Length (cm)	$20.5 \leq l < 23.5$	$23.5 \leq l < 26.5$	
Frequency			

e How many adults have foot length above 23.5 cm?

f Does the modal group include the mode?

25 The mean of the values 2, 5, $a+2$, $2a$ and $3a+1$ is 17.

a Find the value of a.

b Find the standard deviation of the values.

26 The mean of the values $2x$, $x+1$, $3x$, $4x-3$, x and $x-1$ is 10.5.

a Find the value of x.

b Find the variance of the values.

27 A group of 12 students obtained a mean mark of 67.5 on a test. Another group of 10 students obtained a mean mark of 59.3 on the same test. Find the overall mean mark for all 22 students.

28 Lucy must sit five papers for an exam. In order to pass her Diploma, she must score an average of at least 60 marks. In the first four papers Lucy scored 72, 55, 63 and 48 marks. How many marks does she need in the final paper in order to pass her Diploma?

29 A set of data is summarized in a frequency table:

Data values	2	5	7	a	$a+2$
Frequency	5	8	13	14	10

a Find the mean of the data in terms of a.

b Given that the mean is 8.02, find the value of a.

6B Summarizing data

30 The shoe sizes of a group of children are summarized in a frequency table:

Shoe size	4.5	5	6.5	7	x
Frequency	4	12	8	4	2

a Given that the mean shoe size is 5.9, find the value of x.
b Find the median and quartiles of the shoe sizes.
c Hence determine whether the value x is an outlier.

31 The weekly wages of the employees of a small company are:
£215, £340, £275, £410, £960
a Find the mean and standard deviation of the wages.
b Would the mean or the median be a better representation of the average wage for this company?
c The wages are converted into US dollars, with £1 = \$1.31. Find the new mean and standard deviation.

32 a Find the mean and standard deviation of 1, 3 and 8.
b Hence write down the mean and standard deviation of 2003, 2009 and 2024.

33 The temperature measured in degrees Fahrenheit (F) can be converted to degrees Celsius (C) using the formula:
$$C = \frac{5}{9}F - \frac{160}{9}$$
The average January temperature in Austin, Texas is 51 °F with the standard deviation of 3.6 °F Find the mean and standard deviation of the temperatures in °C.

34 The mean distance between stations on a certain trail line is 5.7 miles and the variance of the distances is 4.6 miles². Given that 1 mile = 1.61 km, find the mean and variance of the distances in kilometres.

35 A data set has median 25 and interquartile range 14. Every data value is multiplied by −3. Find the new median and interquartile range.

36 The marks of 11 students on a test are:
35, 35, 37, 39, 42, 42, 42, 43, 45, 46, m
where $m > 46$. The marks are all integers.
a Find the median and the interquartile range of the marks.
b Find the smallest value of m for which this data set has an outlier.

37 Juan needs to take six different tests as part of his job application. Each test is scored out of the same total number of marks. He needs an average of at least 70% in order to be invited for an interview. After the first four tests his average is 68%. What average score does he need in the final two tests?

38 The table summarizes the History grades of a group of students.

Grade	3	4	5	6	7
Frequency	1	8	15	p	4

a Given that the mean grade was 5.25, find the value of p.
b Find the standard deviation of the grades.

39 The frequency table below shows 50 pieces of data with a mean of 5.34. Find the values of p and q.

Value	3	4	5	6	7	8
Frequency	5	10	13	11	p	q

40 Three positive integers a, b and c, where $a < b < c$, are such that their median is 26, their mean is 25 and their range is 11. Find the value of c.

41 The numbers a, b, 1, 2, 3 have got a median of 3 and a mean of 4. Find the largest possible range.

42 The positive integers 1, 3, 4, 10, 10, 16, x, y have a median of 6 and mean of 7. Find all possible values of x and y if $x < y$.

6C Presenting data

An alternative to summarizing a data set by numerical values such as an average and a measure of dispersion is to represent the data graphically.

There are several ways this can be done.

■ Histograms

A histogram uses the height of each bar to represent the frequency of each group from a frequency table.

While it looks much like a bar chart, the difference is that the horizontal scale is continuous for a histogram, whereas there will be gaps between the bars in a bar chart, which is used for discrete data.

> **WORKED EXAMPLE 6.16**
>
> Using technology, or otherwise, plot a histogram for the following data.
>
> | x | $10 < x \leqslant 20$ | $20 < x \leqslant 30$ | $30 < x \leqslant 40$ | $40 < x \leqslant 50$ |
> |---|---|---|---|---|
> | **Frequency** | 16 | 10 | 18 | 14 |
>
> The width of each bar is the width of the corresponding group in the frequency table and the height of each bar is the frequency of that group

6C Presenting data

> **WORKED EXAMPLE 6.17**
>
> For the histogram on the right estimate the number of items which have x values between 15 and 35.
>
>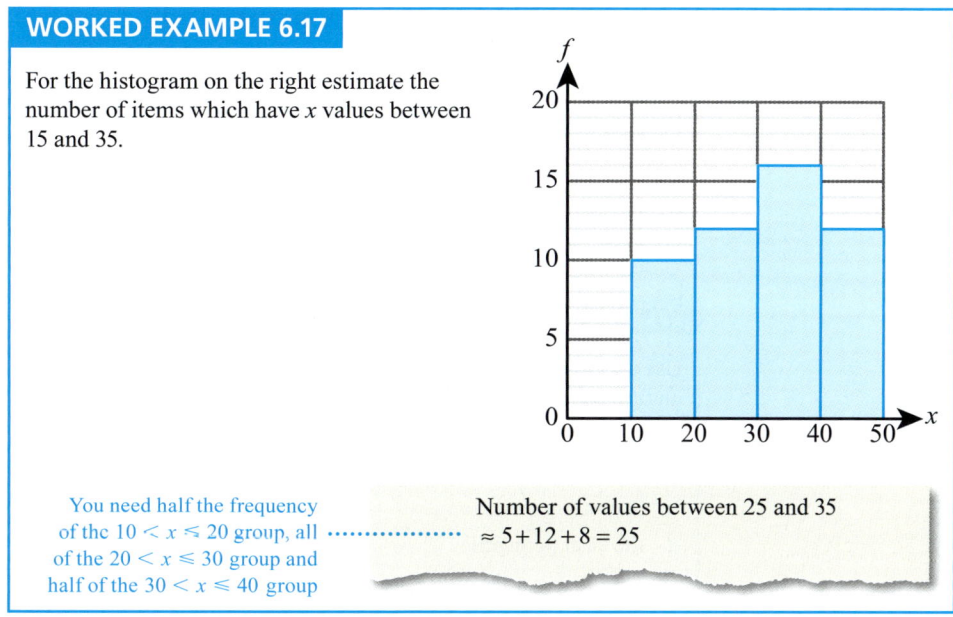
>
> You need half the frequency of the $10 < x \leqslant 20$ group, all of the $20 < x \leqslant 30$ group and half of the $30 < x \leqslant 40$ group
>
> Number of values between 25 and 35
> $\approx 5 + 12 + 8 = 25$

■ Cumulative frequency graphs

The cumulative frequency is the number of values that are less than or equal to a given point in the data set.

A cumulative frequency graph is a plot of cumulative frequency (on the y-axis) against the data values (on the x-axis). When working from a grouped frequency table, take the x value at the upper boundary.

> **WORKED EXAMPLE 6.18**
>
> Construct a cumulative frequency diagram for the following data.
>
> | x | $10 < x \leqslant 20$ | $20 < x \leqslant 30$ | $30 < x \leqslant 40$ | $40 < x \leqslant 50$ |
> |---|---|---|---|---|
> | **Frequency** | 10 | 12 | 16 | 12 |
>
> The x values in the cumulative frequency table are the upper boundaries of each group
>
> | x | 23 | 30 | 40 | 50 |
> |---|---|---|---|---|
> | **Cumulative frequency** | 10 | 12 | 38 | 50 |
>
> Plot the cumulative frequency values against the x values in the cumulative frequency table
>
>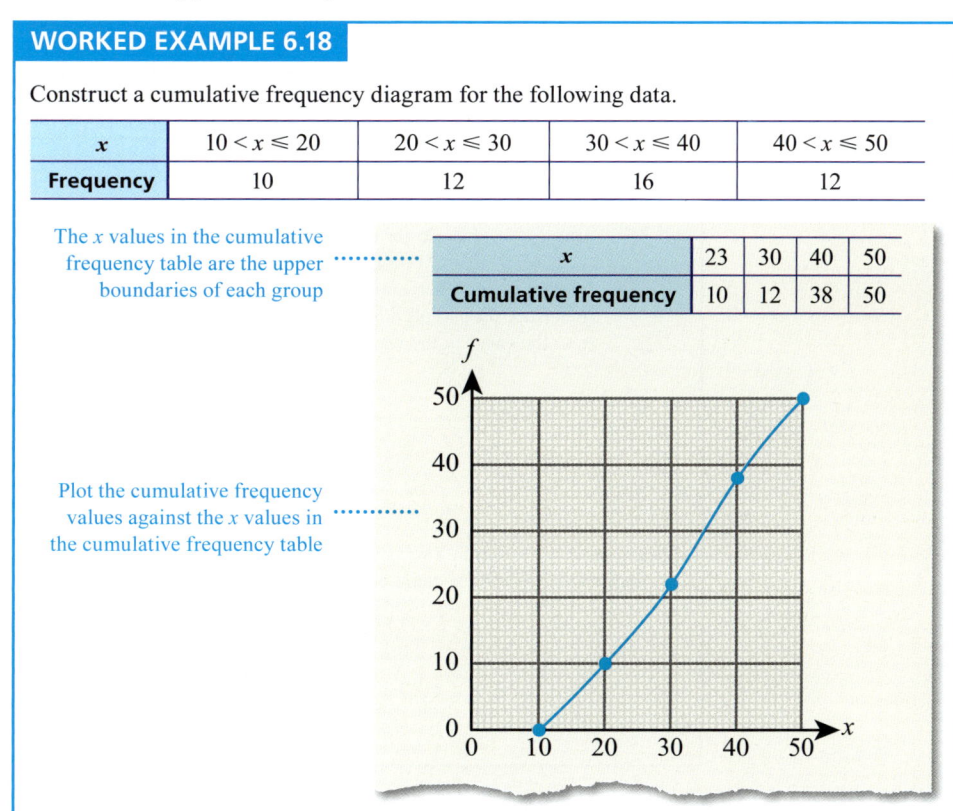

Using cumulative frequency graphs to find medians, quartiles, percentiles, range and interquartile range

One of the main purposes of a cumulative frequency graph is to estimate the median and interquartile range.

The median and quartiles are particular examples of percentiles, which are points that are a particular percentage of the way through the data set. The median is the 50th percentile, and the lower quartile the 25th percentile.

WORKED EXAMPLE 6.19

Use the cumulative frequency graph on the right to estimate:

a the median
b the range
c the interquartile range
d the 90th percentile.

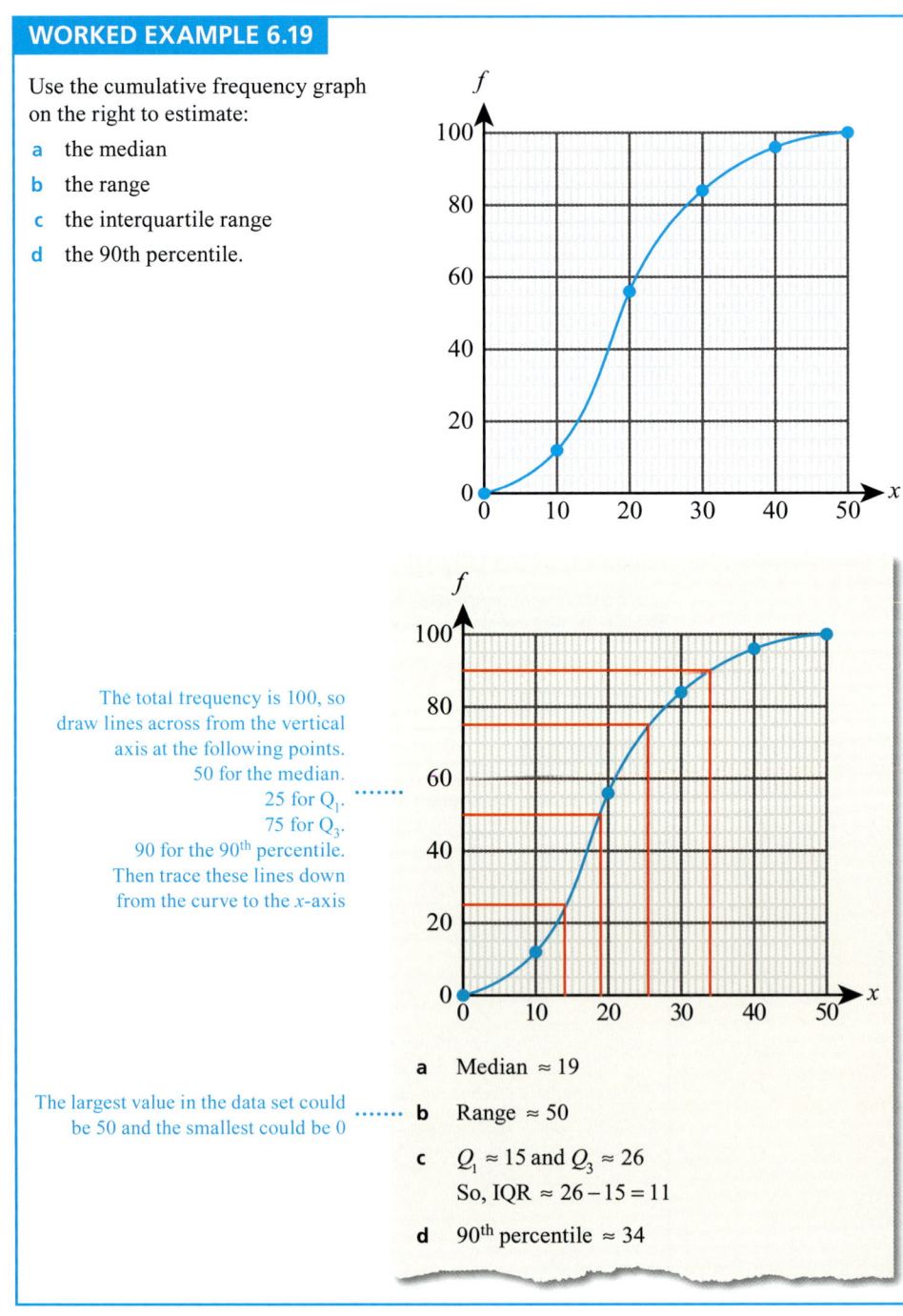

The total frequency is 100, so draw lines across from the vertical axis at the following points.
50 for the median.
25 for Q_1.
75 for Q_3.
90 for the 90th percentile.
Then trace these lines down from the curve to the x-axis

The largest value in the data set could be 50 and the smallest could be 0

a Median ≈ 19

b Range ≈ 50

c $Q_1 \approx 15$ and $Q_3 \approx 26$
So, IQR ≈ 26 − 15 = 11

d 90th percentile ≈ 34

Box-and-whisker diagrams

A box-and-whisker diagram represents five key pieces of information about a data set:
- the smallest value
- the lower quartile
- the median
- the upper quartile
- the largest value.

This excludes outliers – they are represented with crosses.

WORKED EXAMPLE 6.20

Construct a box-and-whisker diagram for the following data:

3, 6, 7, 7, 7, 9, 9, 11, 15, 20.

Use the GDC to find Q_1, Q_3 and the median

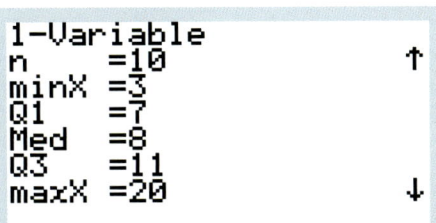

Check to see if there are any outliers

x is an outlier if
$$x < 7 - 1.5(11-7)$$
$$x < 1$$
or if
$$x > 11 + 1.5(11-7)$$
$$x > 17$$
So, $x = 20$ is an outlier

The minimum value is 3. As 20 is an outlier, you need to mark it with a cross

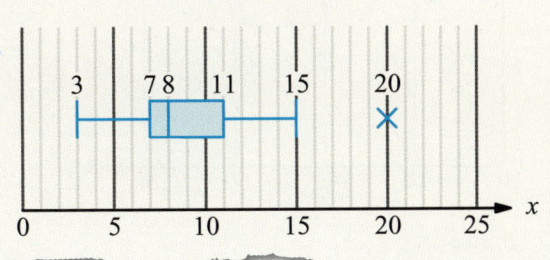

 You can use normal box-and-whisker plots to check whether data might plausibly be normally distributed. You will do this in Chapter 8.

CONCEPTS – REPRESENTATION

Is a picture worth a thousand words? Do you learn more about data by **representing** them as a diagram or summary statistics? Which is easier to manipulate to persuade your intended audience? Are any statistics entirely unbiased?

Exercise 6C

For questions 1 and 2, use technology or the method demonstrated in Worked Example 16.16 to present the data in a histogram.

1 a

x	$5 \leqslant x < 10$	$10 \leqslant x < 15$	$15 \leqslant x < 20$	$20 \leqslant x < 25$
Frequency	5	10	12	8

b

x	$0 \leqslant x < 20$	$20 \leqslant x < 40$	$40 \leqslant x < 60$	$60 \leqslant x < 80$
Frequency	18	15	6	13

2 a

h	$1.2 \leqslant h < 2.5$	$2.5 \leqslant h < 3.8$	$3.8 \leqslant h < 5.1$	$5.1 \leqslant h < 6.4$	$6.4 \leqslant h < 7.7$
Frequency	5	0	12	10	4

b

t	$30 \leqslant t < 36$	$36 \leqslant t < 42$	$42 \leqslant t < 48$	$48 \leqslant t < 54$	$54 \leqslant t < 60$
Frequency	25	31	40	62	35

In each of the questions 3 and 4, a data set is presented as a histogram. Use the method demonstrated in Worked Example 6.17 to estimate the number of data items in the given range.

3

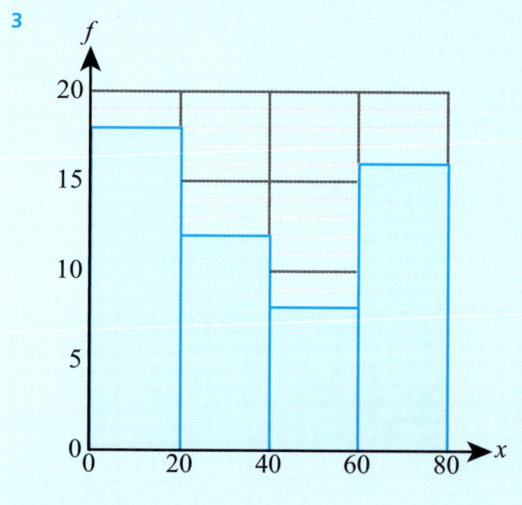

a $20 \leqslant x < 50$
b $30 \leqslant x < 80$

4

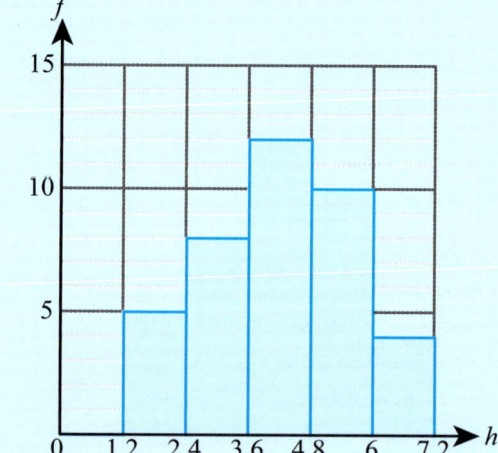

a $2.7 \leqslant h < 5.4$
b $4.0 \leqslant h < 4.4$

For questions 5 and 6, use technology or the method demonstrated in Worked Example 6.18 to construct a cumulative frequency diagram for the data in the table.

5 a

x	$5 \leqslant x \leqslant 10$	$10 \leqslant x < 15$	$15 \leqslant x < 20$	$20 \leqslant x < 25$
Frequency	5	10	12	8

b

x	$0 \leqslant x < 20$	$20 \leqslant x < 40$	$40 \leqslant x < 60$	$60 \leqslant x < 80$
Frequency	18	12	8	16

6C Presenting data

6 a

h	1.2 ≤ h < 2.5	2.5 ≤ h < 3.8	3.8 ≤ h < 5.1	5.1 ≤ h < 6.4	6.4 ≤ h < 7.7
Frequency	10	8	4	0	5

b

t	30 ≤ t < 36	36 ≤ t < 42	42 ≤ t < 48	48 ≤ t < 54	54 ≤ t < 60
Frequency	20	0	12	12	25

For question 7, use the method demonstrated in Worked Example 6.19 to find:
  i   the median
  ii the range
  iii the interquartile range
  iv the 90th percentile.

7 a

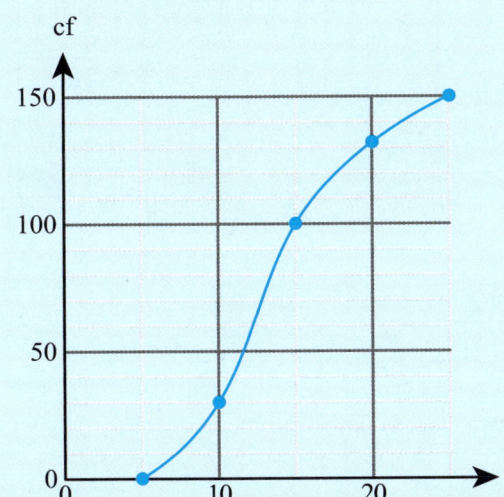

b

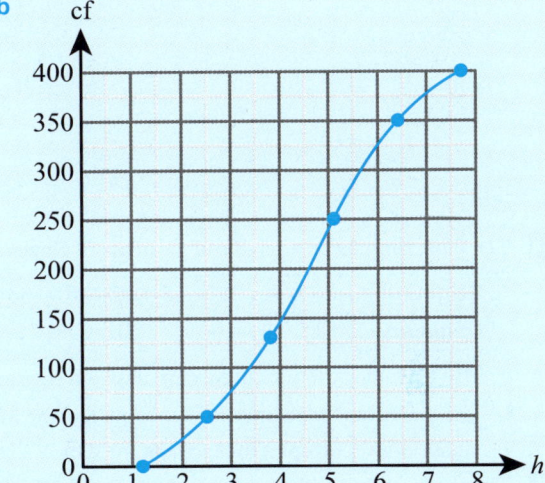

For questions 8 and 9, use the method demonstrated in Worked Example 6.20 to construct a box-and-whisker diagram for the given data.

8 a 3, 5, 5, 6, 8, 9, 9
   b 11, 13, 15, 15, 16, 18, 19, 21, 24, 24

9 a 11, 13, 15, 15, 16, 18, 19, 21, 25, 32
   b 4, 13, 14, 16, 17, 18, 18, 19, 19

10 The length of fossils found at a geological dig is summarized in the following table.

Length (cm)	0 < l ≤ 4	4 < l ≤ 8	8 < l ≤ 12	12 < l ≤ 16	16 < l ≤ 20
Frequency	10	15	15	20	20

  a Using technology, draw a histogram to represent this set of data.
  b Estimate the number of fossils which were longer than 15 cm.

11 The masses of some apples are summarized in the following table.

Mass (g)	80 ≤ m < 100	100 ≤ m < 120	120 ≤ m < 140	140 ≤ m < 160	160 ≤ m < 180
Frequency	16	14	25	32	18

  a How many apples were weighed?
  b Using technology, draw a histogram to represent the data.
  c What percentage of apples had masses between 110 g and 150 g?

12 The maximum daily temperatures in a particular town, over a period of one year, are summarized in the table.

Temperature (°C)	5 ≤ t < 10	10 ≤ t < 15	15 ≤ t < 20	20 ≤ t < 25	25 ≤ t < 30
Frequency	43	98	126	67	21

  a Use technology to draw a cumulative frequency graph for this data.
  b Use your graph to estimate:
   i the median temperature
   ii the interquartile range.

13 The histogram shows the times taken by a group of pupils to complete a writing task.

a How many pupils completed the task?
b Estimate the percentage of pupils who took more than 22 minutes.
c Complete the grouped frequency table.

Time (min)	$5 \leq t < 10$			
Frequency	7			

d Hence estimate the mean time taken to complete the task.

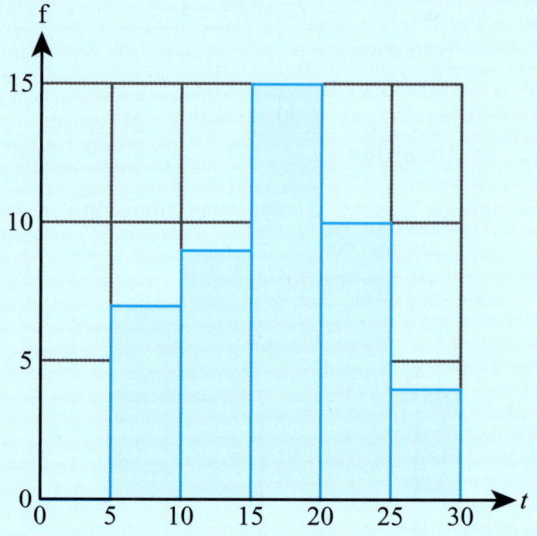

14 The table summarizes heights of a sample of flowers.

Height (cm)	$20 \leq h < 25$	$25 \leq h < 30$	$30 \leq h < 35$	$35 \leq h < 40$	$40 \leq h < 45$
Frequency	16	21	34	18	6

a Use technology to produce a cumulative frequency graph for the data.
b Use your graph to estimate the median and the interquartile range of the heights.
c Hence draw a box plot to represent the data.

15 The number of candidates taking Mathematics SL at a sample of schools is recorded in the grouped frequency table.

Number of candidates	Number of schools
$10 < n \leq 30$	34
$30 < n \leq 50$	51
$50 < n \leq 70$	36
$70 < n \leq 90$	18
$90 < n \leq 110$	7

a Draw a cumulative frequency graph for the data.
b Estimate the number of schools with more than 60 Mathematics SL candidates.
c Find the median and interquartile range of the data.
d Hence draw a box plot to show the number of Mathematics SL candidates.

The box plot below shows the number of candidates taking History SL at the same group of schools.

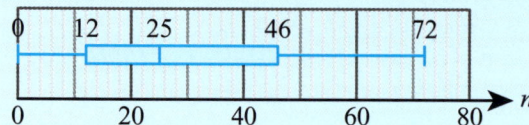

e Write two comments comparing the number of candidates taking Mathematics SL and History SL at this group of schools.

6C Presenting data

16 The histogram summarizes heights of children in a school.

Use the histogram to estimate the mean height, correct to the nearest centimetre.

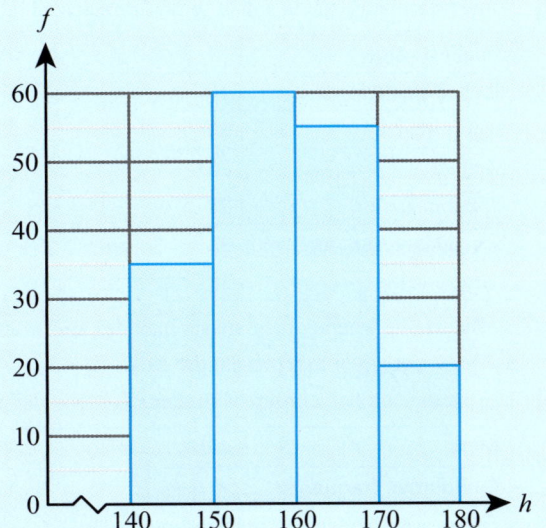

17 The cumulative frequency graph summarizes the examination scores at a particular school.

a How many students scored below 73?

b The pass mark for the examination was 55. What percentage of students passed?

c Find the 60th percentile of the scores.

d Estimate the median and quartiles of the scores. Hence produce a box plot for the data. (You may assume that there are no outliers.)

e The box plot below summarizes the scores on the same exam from a different school. Make two comments comparing the scores at the two schools.

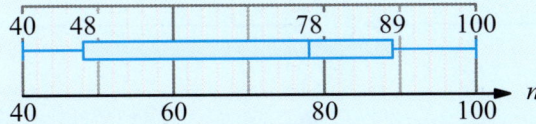
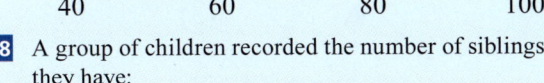

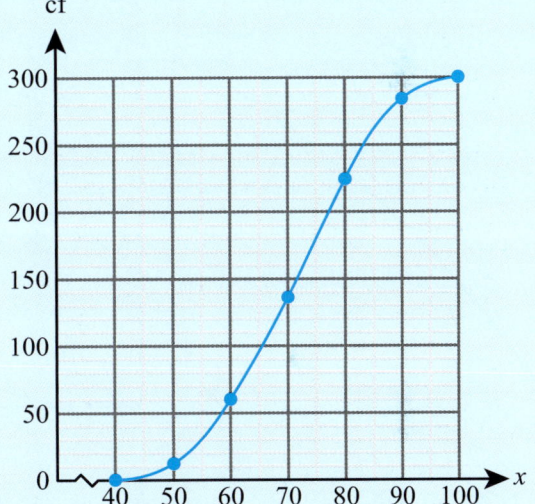

18 A group of children recorded the number of siblings they have:
0, 2, 1, 3, 2, 1, 1, 0, 2, 3, 2, 2, 1, 4, 7.

a Find the median and quartiles of the data.

b Calculate the interquartile range.

c Determine whether this data set contains any outliers.

d Draw the box plot for the data.

19 Some information about a data set is summarized in this table.

Smallest value	7
Second smallest value	11
Lower quartile	20
Median	25
Upper quartile	28
Second largest value	32
Largest value	38

a Determine whether the data set includes any outliers.

b Draw a box plot to represent the data.

20 The exam marks of a group of students are recorded in a cumulative frequency table.

Mark	≤ 20	≤ 40	≤ 60	≤ 80	≤ 100
Cumulative frequency	6	45	125	196	247

a Represent the data in the histogram.

b Estimate the mean of the marks.

21 Match each histogram with the cumulative frequency diagram drawn from the same data.

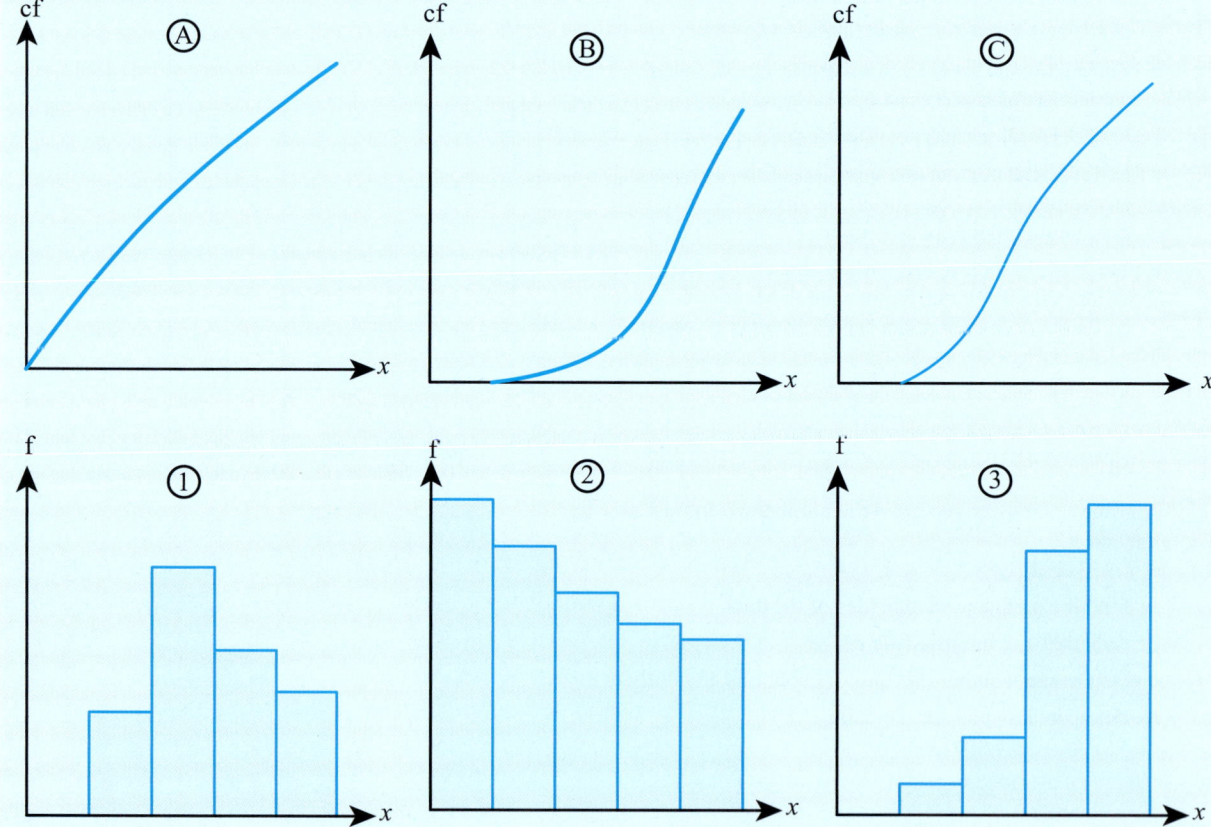

6D Correlation and regression

So far, we have only looked at one variable at a time, but sometimes we want to investigate whether there is a relationship between two variables, such as a person's height and mass. These types of data are known as bivariate.

■ Scatter diagrams and lines of best fit

A good starting point for investigating a possible relationship between two variables is to plot one variable against the other on a scatter graph.

If it looks like there is a linear relationship, then we can plot a line of best fit.

> **KEY POINT 6.11**
>
> The line of best fit will always pass through the mean point $(\bar{x}, \bar{y})$.

WORKED EXAMPLE 6.21

For the data below:

a find the mean point

b create a scatter diagram, showing the mean point

c draw a line of best fit by eye.

x	1	2	3	4	4	5	6	6	7	9
y	0	2	4	4	5	6	7	10	9	10

Use the GDC to find $\bar{x}$ and $\bar{y}$: ..... a From GDC,

$(\bar{x}, \bar{y}) = (4.7, 5.7)$

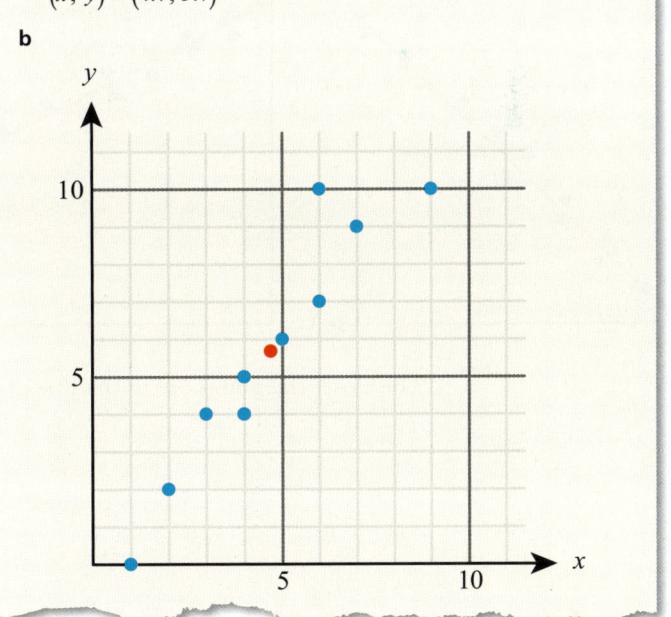

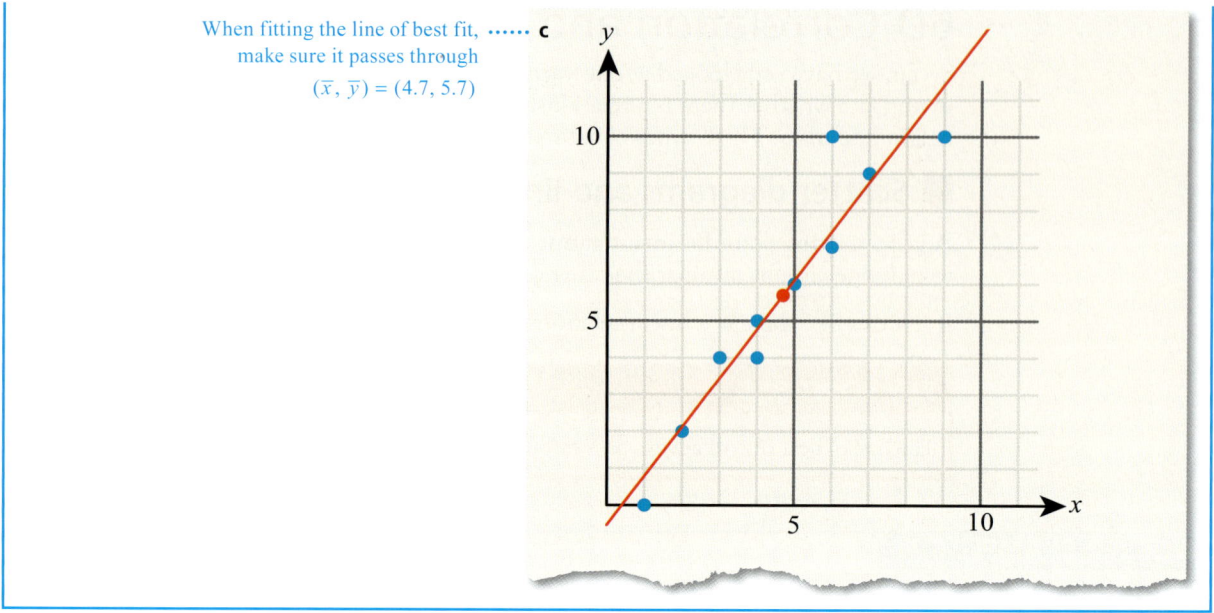

When fitting the line of best fit, make sure it passes through $(\bar{x}, \bar{y}) = (4.7, 5.7)$

■ Linear correlation of bivariate data

The extent to which two variables are related is called correlation. In this course we will focus on linear correlation:

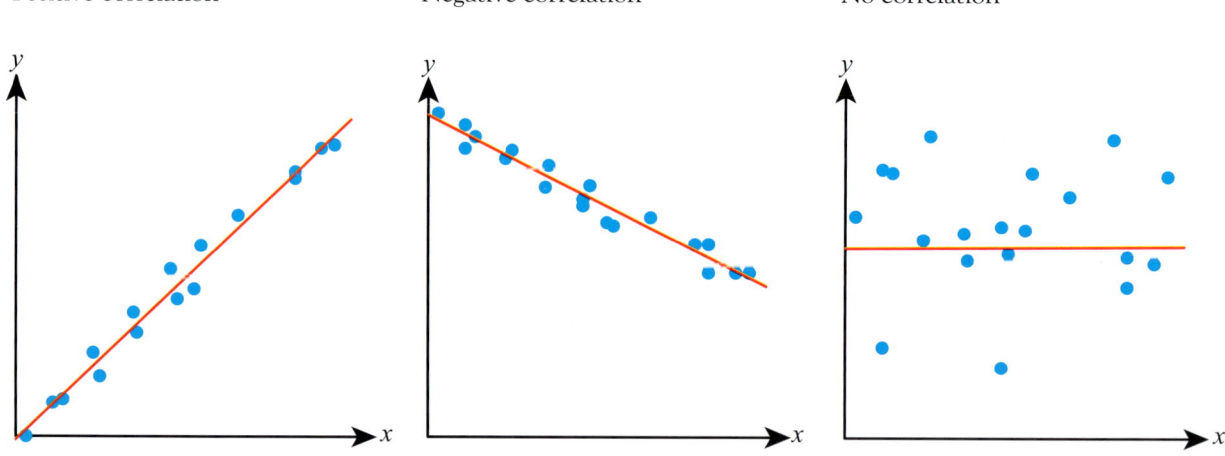

Positive correlation   Negative correlation   No correlation

The closer the points are to the line of best fit, the stronger the correlation.

6D Correlation and regression

WORKED EXAMPLE 6.22

Describe the linear correlation in the scatter graph below.

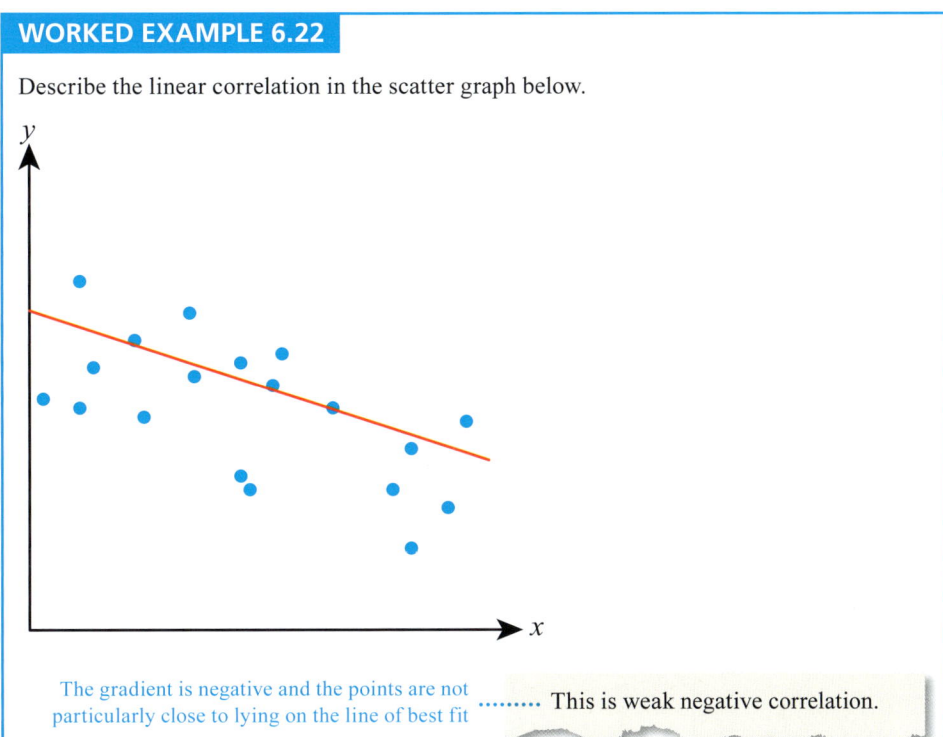

The gradient is negative and the points are not particularly close to lying on the line of best fit .... This is weak negative correlation.

Pearson's product-moment correlation coefficient, r

Rather than just describing the correlation as 'strong positive' or 'weak negative', etc, there is a numerical value called the Pearson's product-moment correlation coefficient (given the letter r) that can be used to represent the strength of linear correlation. You need to be able to find r using your calculator.

KEY POINT 6.12

The Pearson's product-moment correlation coefficient, r, is such that $-1 \leqslant r \leqslant 1$.
- $r \approx 1$ means strong positive linear correlation
- $r \approx 0$ means no linear correlation
- $r \approx -1$ means strong negative linear correlation

Tip

Just because r is close to 1 or to -1, this strong correlation does not necessarily mean that one variable causes a change in the other. The correlation might just be a coincidence or be due to a third hidden variable.

While it seems clear that a value of r such as 0.95 gives good evidence of positive linear correlation, it might not be so clear whether a value such as 0.55 does.

In fact, this depends on how many data points we have – the fewer points there are, the nearer to 1 the value of r needs to be to give good evidence of positive linear correlation (and the nearer to -1 for negative correlation).

The 'cut-off' values at which the value of r provides significant evidence of positive (or negative) correlation are known as critical values.

WORKED EXAMPLE 6.23

a For the data below, use technology to find the value of Pearson's product-moment correlation coefficient.

x	4	5	4	8	9	10	11	8	6	0
y	9	4	5	9	8	1	2	5	8	11

b Interpret qualitatively what this suggests.

A table suggests that the critical value of r when there are 10 pieces of data is 0.576.

c What does this mean for your data?

Use your GDC to find r: ......

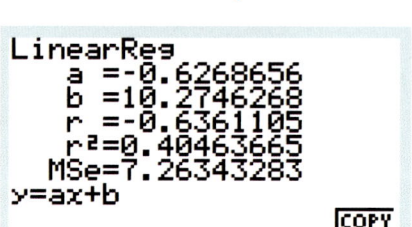

a From GDC,
$r = -0.636$

b There is weak negative correlation, that is, a general trend that as x increases, y decreases.

The critical value given will always be positive, but this same value can be used ...... to check for negative correlation

c $r < -0.576$, so there is significant evidence of negative correlation.

 ■ Equation of the regression line of y on x

Once there is evidence of linear correlation, you would like to know what the precise nature of the linear relationship is between the variables – that is, you would like to know the equation of the line of best fit. This is known as the regression line.

WORKED EXAMPLE 6.24

For the data below, find the equation of the regression line using technology.

x	1	2	3	4	4	5	6	6	7	9
y	0	2	4	4	5	6	7	10	9	10

The GDC will give you the values of the coefficients a and b in the straight-line equation $y = ax + b$

Regression line is
$y = 1.33x - 0.534$

```
LinearReg
    a =1.32629558
    b =-0.5335892
    r =0.94742821
    r²=0.89762022
  MSe=1.30662188
y=ax+b
                    COPY
```

6D Correlation and regression

TOOLKIT: Problem Solving

For each of the following data sets, find **a** the mean of x and y, **b** the standard deviation of x and y, **c** the correlation coefficient, and **d** the equation of the regression line. What does this suggest about the data sets?

I		II		III		IV	
x	y	x	y	x	y	x	y
10.0	8.04	10.0	9.14	10.0	7.46	8.0	6.58
8.0	6.95	8.0	8.14	8.0	6.77	8.0	5.76
13.0	7.58	13.0	8.74	13.0	12.74	8.0	7.71
9.0	8.81	9.0	8.77	9.0	7.11	8.0	8.84
11.0	8.33	11.0	9.26	11.0	7.81	8.0	8.47
14.0	9.96	14.0	8.10	14.0	8.84	8.0	7.04
6.0	7.24	6.0	6.13	6.0	6.08	8.0	5.25
4.0	4.26	4.0	3.10	4.0	5.39	19.0	12.50
12.0	10.84	12.0	9.13	12.0	8.15	8.0	5.56
7.0	4.82	7.0	7.26	7.0	6.42	8.0	7.91
5.0	5.68	5.0	4.74	5.0	5.73	8.0	6.89

Now use technology to create scatterplots of each data set.

These data sets are called Anscombe's quartet. They highlight the importance of visualizing your data.

■ Use of the regression line for prediction purposes

Once you have found the equation of the regression line, you can use it to predict the y value for a given x value. If the x value you use in the equation is within the range of the data set (interpolation) then the prediction can be considered reliable, but if the y value is beyond the data set (extrapolation), the prediction should be treated with caution as there is no guarantee that the relationship continues beyond the observed values.

WORKED EXAMPLE 6.25

Based on data with x values between 10 and 20 and a correlation coefficient of 0.93, a regression line is formed: $y = 10.2x - 5.3$.

a Use this line to predict the value of y when:
  i $x = 3$
  ii $x = 13$.
b Comment on the reliability of your answers in **a**.

Substitute each value of x into the regression line

a i $y = 10.2(3) - 5.3 = 25.3$
  ii $y = 10.2(13) - 5.3 = 127$

b The prediction when $x = 13$ can be considered reliable as 13 is within the range of known x values.

The prediction when $x = 3$, however, cannot be considered reliable, as the relationship has had to be extrapolated significantly beyond the range of given data to make this prediction.

CONCEPTS – APPROXIMATION AND PATTERNS

When you are using a regression line you should not assume that your prediction is totally accurate. There will always be natural variation around the line of best fit. However, even though it is only giving us an **approximate** prediction, it can help us to see the underlying **patterns** in the data.

■ Interpreting the meaning of regression parameters

The parameters a and b in the regression line $y = ax + b$ are just the gradient and y-intercept respectively, which means they can be easily interpreted in context.

WORKED EXAMPLE 6.26

Data for the number of flu cases (f, in tens of thousands) in a country against the amount spent on promoting a vaccination program (p, in millions) are given by:

$f = 12.4 - 2.1p$.

Interpret, in context, the meaning of

a 12.4

b 2.

in this equation.

12.4 is the y-intercept of the line, which corresponds to the value of y when $x = 0$. Remember here that f is in tens of thousands

a If no money were spent on promoting a vaccination program then the number of flu cases would be $12.4 \times 10^5 = 1\,240\,000$.

The gradient of the line is -2.1, which means that for an increase of 1 unit in x there is a decrease of 2.1 units in y

b For every extra million spent on promoting a vaccination program, the number of flu cases decreases by $2.1 \times 10^5 = 210\,000$.

CONCEPTS – PATTERNS

Especially when there are large amounts of data and many variables, it can be difficult to spot **patterns** visually. Correlation provides an objective way to assess whether there are any underlying linear trends. One of the most useful parts of regression turns out to be deciding whether the underlying regression has a non-zero gradient, as this tells you whether the variables you are considering are actually related in some way.

■ Piecewise linear models

In some cases, a scatter graph might show that the same regression line is not a good fit for all the data, but that two (or more) different regression lines do fit very well with different parts of the data set. A model like this is known as piecewise linear.

6D Correlation and regression

WORKED EXAMPLE 6.27

a Use technology to plot a scatter diagram of the data below.

x	0	1	3	3	4	5	6	6	7	8	9	9
y	1	2	4	3	4.5	5	10	11	12	15	17	18

b Use your scatter diagram to separate the data into two distinct linear regions. Form a piecewise linear model to fit to the data.

c Use your model to estimate the value of y when x is 7.5.

Rather than fit a single line of best fit to the data, it is clear that there are two distinct regions, so add a line of best fit separately for each of these .... **a**

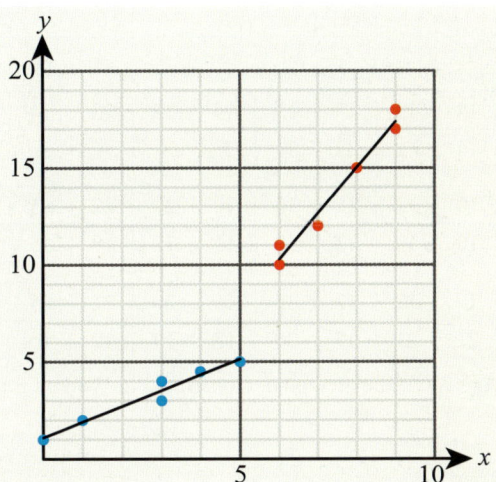

Use your GDC to find the equation of each regression line: .... **b** For $0 \leq x \leq 5$,
$y = 0.808x + 1.10$

For $6 \leq x \leq 9$,
$y = 2.37x - 3.93$

For $x = 7.5$, use the second regression line. .... **c** When $x = 7.5$,
$y = 2.37(7.5) - 3.93 = 13.8$

Exercise 6D

For questions 1 to 3, use the method demonstrated in Worked Example 6.21 to create a scatter diagram, plot the mean point and add a line of best fit by eye.

1 a

x	3	4	6	7	9	10	12	13
y	2	3	5	4	7	6	5	7

b

x	1	2	2	3	5	7	8	8
y	−3	0	1	−2	−1	2	4	3

2 a

x	1	1	2	3	5	6	6	8	9
y	6	7	5	4	5	4	2	1	2

b

x	14	13	15	19	19	22	23	24	22
y	14	12	12	11	9	9	8	6	5

3 a

x	10	12	12	15	16	20	22	25	26	28	31	34
y	10	10	13	10	16	12	19	15	20	17	21	21

b

x	10	12	12	15	16	20	22	25	26	28	31	34
y	21	21	17	20	15	19	12	16	10	13	10	10

For questions 4 to 7, use the method demonstrated in Worked Example 6.22 to describe the linear correlation in the scatter graph.

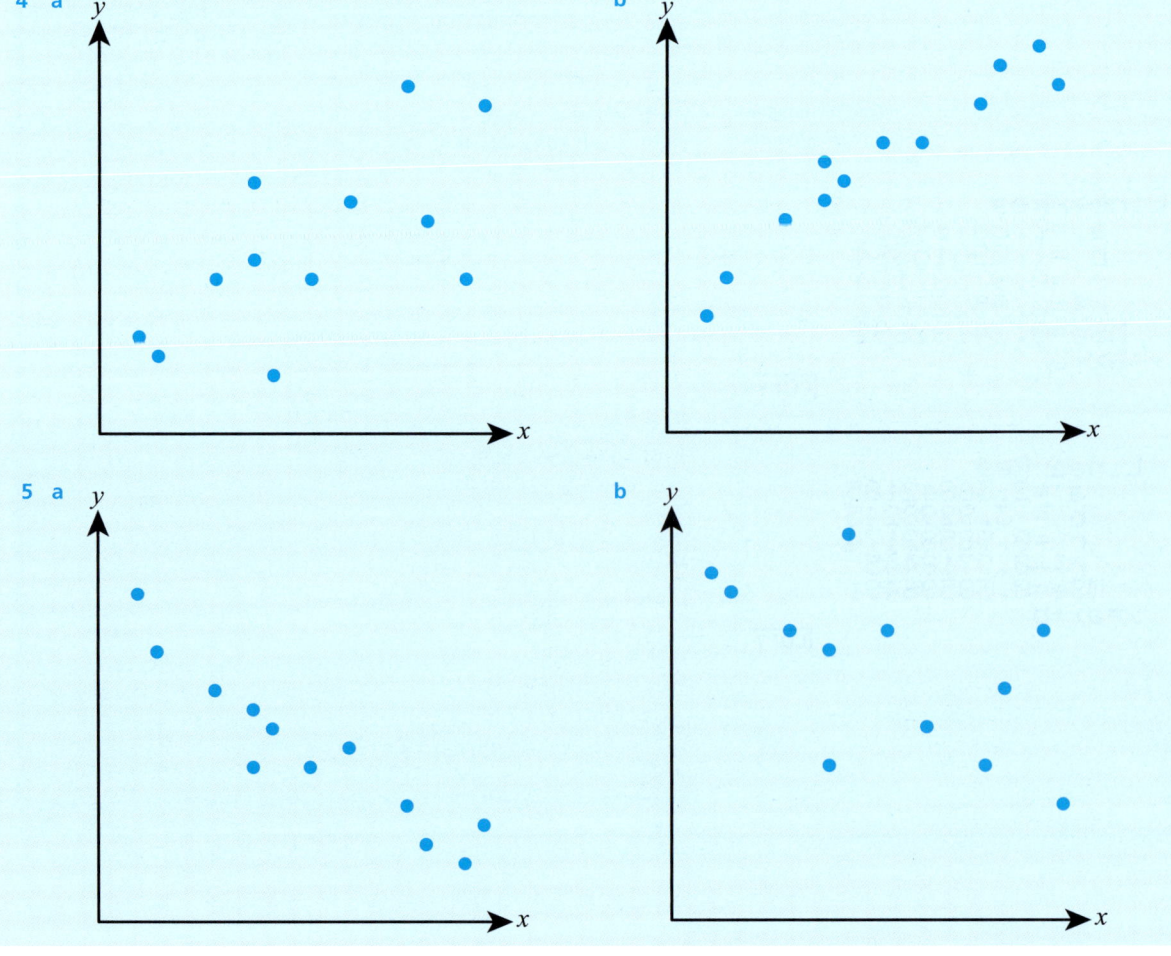

6D Correlation and regression

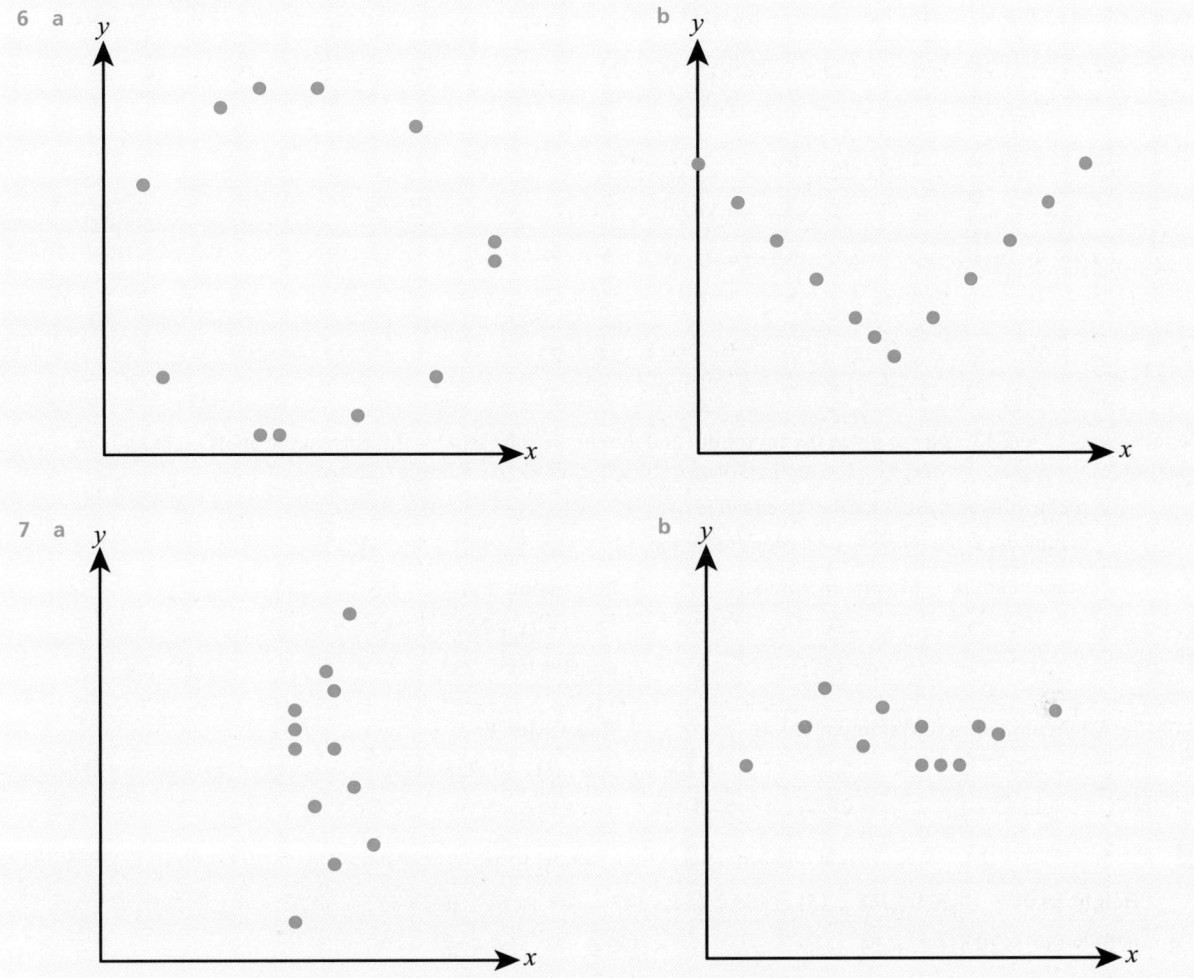

6 a
b
7 a
b

For questions 8 and 9, use the method demonstrated in Worked Example 6.23 to:
  i   find the value of the Pearson's product-moment correlation coefficient
  ii interpret qualitatively what the correlation coefficient suggests
  iii use the appropriate critical value from this table to decide whether the correlation is significant.

Number of data pairs	6	7	8	9	10
Critical value	0.621	0.584	0.549	0.521	0.497

8 a

x	3	4	6	7	9	10	12	13
y	2	3	5	4	7	6	5	7

b

x	1	1	2	3	5	6	6	8	9
y	6	7	5	4	5	4	2	1	2

9 a

x	3	7	2	5	4	9
y	6	10	7	8	9	8

b

x	14	13	15	19	19	22	23
y	14	10	12	11	9	9	10

For questions 10 and 11, use the method demonstrated in Worked Example 6.24 to find the equation of the regression line.

10 a

x	3	4	6	7	9	10	12	13
y	2	3	5	4	7	6	5	7

b

x	1	2	2	3	5	7	8	8
y	−3	0	1	−2	−1	2	4	3

11 a

x	1	1	2	3	5	6	6	8	9
y	6	7	5	4	5	4	2	1	2

b

x	14	13	15	19	19	22	23	24	22
y	14	12	12	11	9	9	8	6	5

For questions 12 and 13, you are given the minimum and maximum data values, the correlation coefficient and the equation of the regression line. Use the method demonstrated in Worked Example 6.25 to:

   i   predict the value of y for the given value of x

   ii comment on the reliability of your prediction.

12

	Minimum value	Maximum value	r	Regression line	x
a	10	20	0.973	$y = 1.62x - 7.31$	18
b	1	7	−0.875	$y = -0.625x + 1.37$	9

13

	Minimum value	Maximum value	r	Regression line	x
a	20	50	0.154	$y = 2.71x + 0.325$	27
b	12	27	−0.054	$y = 4.12x - 2.75$	22

14 The table shows the data for height and arm length for a sample of 10 15-year-olds.

Height (cm)	154	148	151	165	154	147	172	156	168	152
Arm length (cm)	65	63	58	71	59	65	75	62	61	61

  a Plot the data on a scatter graph.
  b Describe the correlation between the height and arm length.
  c Find the mean height and mean arm length. Add the corresponding point to your graph.
  d Draw the line of best fit by eye.
  e Use your line of best fit to predict the arm length of a 15-year-old whose height is 150 cm.
  f Comment on whether or not it would be appropriate to use your line to predict the arm length for
     i   a 15-year-old whose height is 150 cm
     ii   a 15-year-old whose height is 192 cm
     iii a 72-year-old whose height is 150 cm.

15 The table shows the distance from the nearest train station and the average house price for seven villages.

Distance (km)	0.8	1.2	2.5	3.7	4.1	5.5	7.4
Average house price (000s $)	240	185	220	196	187	156	162

  a Plot the data on a scatter graph.
  b Describe the correlation between the distance and the average house price.
  c Find the mean distance and the mean house price.
  d Draw a line of best fit on your graph.
  e A new village is to be developed 6.7 km from a train station. Predict the average house price in the new village.

16 Match the scatter diagrams with the following values of r:
A: $r = 0.98$ B: $r = -0.34$ C: $r = -0.93$ D: $r = 0.58$

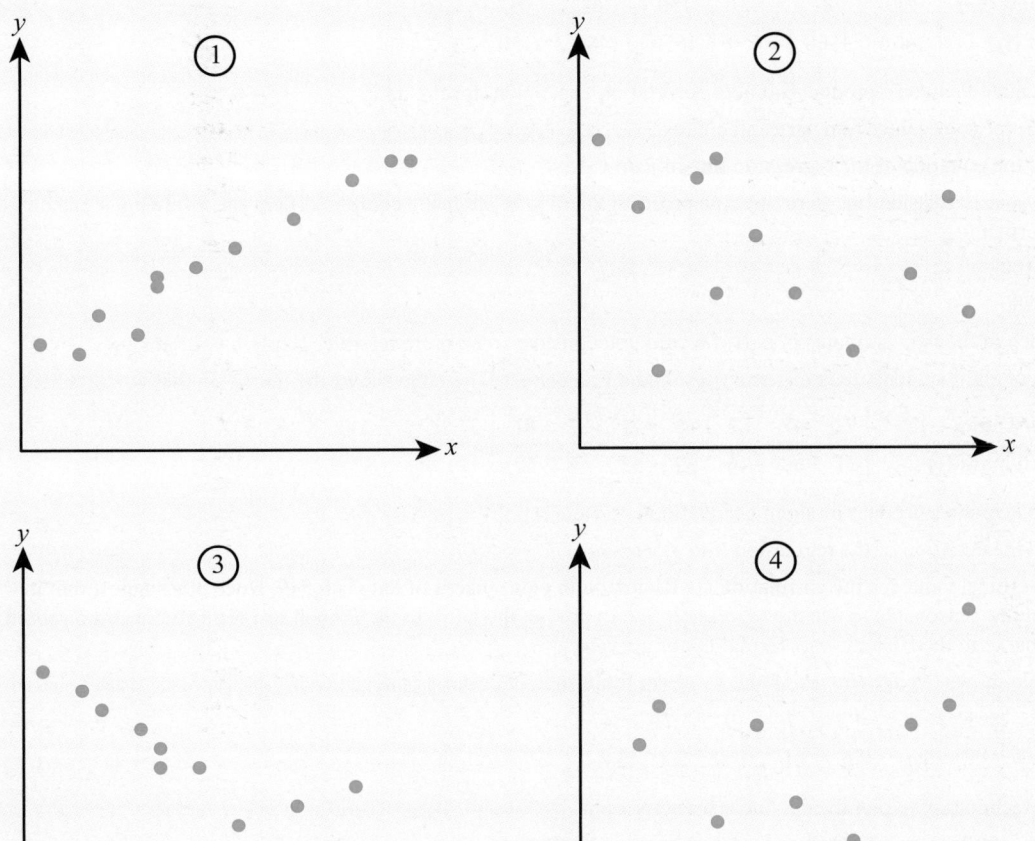

17 A group of 11 students took a maths test and a chemistry test. Their marks on the two tests are shown in the table.

Maths test mark	63	81	57	72	93	47	61	82	65	83	71
Chemistry test mark	40	57	46	51	60	37	52	48	33	48	51

a Find Pearson's product-moment correlation coefficient between the marks on the two tests.

b Find the equation of the regression line of y on x, where x is the mark on the first test and y is the mark on the second test.

c Another student scored 68 marks on the maths test but missed the chemistry test. Use your regression line to estimate the mark they would have got on the chemistry test.

d Asher says: 'The strong correlation proves that by getting better at maths you would also get better at chemistry'. Do you agree with this statement?

18 A small company records the amount spent on advertising and the profit they made the following month. The results are summarized in the table:

Amount (x)	120	90	65	150	80	95
Profit (y)	1600	1300	450	1650	1480	1150

  a Calculate Pearson's product-moment correlation coefficient for the data.
  b Interpret your value from part **a** in context.
  c Find the equation of the regression line of y on x.
  d Use your regression line to predict the profit in the months after the company spent the following sums on advertising:
    i $100
    ii $200.
  e Which of the two estimates in part **d** would you consider to be more reliable? Explain your answer.

19 A class of eight students took a History test and a French test. The table shows the marks on the two tests.

History mark (x)	72	47	82	65	71	83	81	57
French mark (y)	51	60	46	37	52	48	57	41

  a Find Pearson's product-moment correlation coefficient.
  b Find the equation of the regression line of y on x.
  c The critical value for the correlation coefficient with eight pieces of data is 0.549. Does this suggest that it would be reasonable to use the regression line to predict the mark in the French test for a student who scored 75 marks in the History test? Explain your answer.

20 The table shows the age (in years) and value (in thousands of dollars) of seven cars.

Age	3	5	12	7	9	7	4
Value	3.2	1.1	0.6	3.1	1.8	2.4	3.4

  a Use technology to plot the data on a scatter graph.
  b Calculate Pearson's product-moment correlation coefficient.
  c The critical value of the correlation coefficient for seven pieces of data is 0.584. What does this suggest about your data?

21 The values of two variables, measured for nine different items, are shown in the table.

x	6	3	5	8	7	5	3	6	9
y	31	35	52	22	26	26	41	42	31

  a Use technology to plot the data on a scatter graph.
  b Describe the correlation suggested by the graph.
  c Calculate Pearson's product-moment correlation coefficient.
  d The critical value of the correlation coefficient for nine items is 0.521. Do the data suggest that there is significant correlation between the two variables?

22 Daniel wants to investigate whether there is any relationship between head circumference (x cm) and arm length (y cm). He collects data from six of his friends and finds that the value of Pearson's product-moment correlation coefficient is 0.741 and the equation of the regression line is $y = 1.13x - 17.6$. The head circumferences ranged from 48.5 cm to 53.4 cm and the mean head circumference was 51.7 cm.

  a Describe the correlation between the head circumference and arm length.
  b Daniel uses his regression line to estimate the arm length of a friend whose head circumference is 49.2 cm. Should this estimate be considered reliable? Explain your answer.
  c Find the mean arm length for this sample.

6D Correlation and regression

23 Theo investigates whether spending more time practising his spellings leads to better marks in weekly spelling tests. Each week he records the length of time spent practising and his spelling test mark (out of 20).

Time (t min)	17	5	10	7	25	14	20
Mark (m/20)	20	8	6	12	19	16	18

a Find Pearson's product-moment correlation coefficient and describe the correlation between the time spent practising and the test mark.

b Find the equation of the regression line of m on t.

c Interpret the gradient and the intercept of your regression line.

24 A company records its advertising budget, x-thousand euros, and its profit, y-thousand euros, over a number of years. They find that the correlation coefficient between the two variables is 0.825 and that the equation of the regression line is $y = 3.25x + 13.8$.

a Describe, in context, the relationship between the advertising budget and the profit.

b Does the value of the correlation coefficient show that increasing the advertising budget will lead to an increase in profit?

c Interpret the values

  i 3.25 and

  ii 138

in the equation of the regression line.

25 The data below show the sales of ice creams at different maximum daily temperatures.

Temp (°C)	Sales
9	12
11	9
11	14
13	11
14	9
14	13
15	10
16	12
21	32
23	32

Temp (°C)	Sales
25	36
26	34
26	37
27	41
29	40
30	41
31	44
32	42
32	45
34	50

a Using technology, or otherwise, create a scatter diagram illustrating the data.

b Suggest what the two regions in your graph might represent.

c Describe the correlation between the temperature and the number of ice creams for each of the two regions.

d Use an appropriate regression line to estimate the ice cream sales on a day when the maximum temperature is 28°C.

26 Alessia collected height and mass measurements for a sample of children and adults in the school playground.

h (cm)	125	119	152	131	175	121	135	164	158	127
m (kg)	23	18	52	29	65	20	31	58	52	25

a Plot the data on a scatter graph.

b Explain why the data contain two linear regions.

c Find the midpoint for each of the two regions and add two regression lines to your scatter graph.

d Use the appropriate regression line to predict the mass of a child whose height is 120 cm.

27 The table shows pairs of data values (x, y).

x	20	10	12	15	7	19	29	24	25	16
y	17	11	12	6	9	16	22	19	20	15

a Find the upper and lower quartiles for each of x and y.
b Hence show that there are no outliers in the x data values or the y data values.
c Using technology, or otherwise, plot the data on a scatter graph.
d Which point on the scatter graph would you describe as an outlier?
e Add a line of best fit to your graph
   i excluding the outlier,
   ii with the outlier included.

TOOLKIT: Modelling

Take a population where you can find the 'true' population statistics – for example, data on the age of pupils in your school, or the incomes in your country. Then try to collect data using different sampling methods. How close do your sample statistics come to the 'true' values. Is there a difference in the accuracy of the mean, median and standard deviation?

Checklist

- You should be able to understand and work with the concepts of population and sample.
- You should understand when data are discrete or continuous.
- You should be able to identify potential sources of bias in sampling.
- You should be able to understand and evaluate a range of sampling techniques:
  - With simple random sampling, every possible sample (of a given size) has an equal chance of being selected.
  - With convenience sampling, respondents are chosen based upon their availability.
  - With systematic sampling, participants are taken at regular intervals from a list of the population.
  - With stratified sampling, the population is split into groups based on factors relevant to the research, then a random sample from each group is taken in proportion to the size of that group.
  - With quota sampling, the population is split into groups based on factors relevant to the research, then convenience sampling from each group is used until a required number of participants are found.
- You should be able to use your GDC to find the mean, median, mode and range.
- You should be able to use the formula for the mean of data:
  - $\bar{x} = \dfrac{\sum x}{n}$
- You should be able to use your GDC to find the quartiles of discrete data, and know that the interquartile (IQR) is $IQR = Q_3 - Q_1$.
- You should be able to use your calculator to find the standard deviation.
- You should understand the effect of constant changes to a data set.
- Adding a constant, k, to every data value will:
  - change the mean, median and mode by k
  - not change the standard deviation or interquartile range.
- Multiplying every data value by a constant, k, will:
  - multiply the mean, median and mode by k
  - multiply the standard deviation and interquartile range by k.
- You should know that a data value x is an outlier if $x < Q_1 - 1.5(Q_3 - Q_1)$ or $x < Q_3 + 1.5(Q_3 - Q_1)$.
- You should be able to construct and use statistical diagrams such as histograms, cumulative frequency graphs and box-and-whisker plots.
- You should be able to draw scatter graphs and add a line of best fit.
  - The line of best fit will always pass through the mean point $(\bar{x}, \bar{y})$.
- You should be able to use your GDC to find Pearson's product-moment correlation coefficient and interpret the value:
  - the Pearson's product-moment correlation coefficient, r, is such that $-1 \leq r \leq 1$
  - $r = 1$ means strong positive linear correlation
  - $r = 0$ means no linear correlation
  - $r = -1$ means strong negative linear correlation.
- You should be able to use your GDC to find the regression line, use the regression line to predict values not in the given data set, and interpret the regression coefficients.
- You should be able to fit a piecewise linear model when relevant.

Mixed Practice

1 A librarian is investigating the number of books borrowed from the school library over a period of 10 weeks. She decided to select a sample of 10 days and record the number of books borrowed on that day.
  a She first suggests selecting a day at random and then selecting every seventh day after that.
   i State the name of this sampling technique.
   ii Identify one possible source of bias in this sample.
  b The librarian changes her mind and selects a simple random sample of 10 days instead.
   i Explain what is meant by a simple random sample in this context.
   ii State one advantage of a simple random sample compared to the sampling method from part **a**.
  c For the days in the sample, the numbers of books borrowed were:
  17, 16, 21, 16, 19, 20, 18, 11, 22, 14
  Find
   i the range of the data
   ii the mean number of books borrowed per day
   iii the standard deviation of the data.

2 The table shows the maximum temperature ($T°C$) and the number of cold drinks (n) sold by a small shop on a random sample of nine summer days.

T	21	28	19	21	32	22	27	18	30
n	20	37	21	18	35	25	31	17	38

  a Using technology, or otherwise, plot the data on a scatter graph.
  b Describe the relationship between the temperature and the sales of cold drinks.
  c Find the equation of the regression line of n on T.
  d Use your regression line to estimate the number of cold drinks sold on the day when the maximum temperature is $26°C$.

3 The masses of 50 cats are summarized in the grouped frequency table:

Mass (kg)	$1.2 \leq m < 1.6$	$1.6 \leq m < 2.0$	$2.0 \leq m < 2.4$	$2.4 \leq m < 2.8$	$2.8 \leq m < 3.2$
Frequency	4	10	8	16	12

  a Use this table to estimate the mean mass of a cat in this sample. Explain why your answer is only an estimate.
  b Use technology to create a cumulative frequency graph.
  c Use your graph to find the median and the interquartile range of the masses.
  d Create a box plot to represent the data. You may assume that there are no outliers.

4 A survey was carried out on a road to determine the number of passengers in each car (excluding the driver). The table shows the results of the survey.

Number of passengers	0	1	2	3	4
Number of cars	37	23	36	15	9

  a State whether the data are discrete or continuous.
  b Write down the mode.
  c Use your GDC to find
   i the mean number of passengers per car
   ii the median number of passengers per car
   iii the standard deviation.

Mathematical Studies SL November 2013 Paper 1 Q1

Mixed Practice

5 Two groups of 40 students were asked how many books they have read in the last two months. The results for **the first group** are shown in the following table.

Number of books read	Frequency
2	5
3	8
4	13
5	7
6	4
7	2
8	1

The quartiles for these results are 3 and 5.
  a Write down the value of the median for these results.
  b Draw a box-and-whisker diagram for these results.

The results for **the second group** of 40 students are shown in the following box-and-whisker diagram.

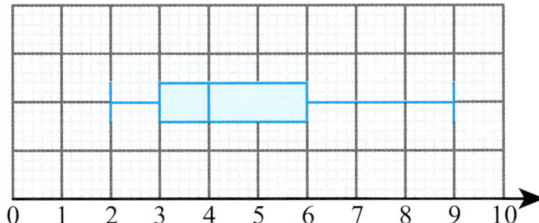

Number of books read

  c Estimate the number of students **in the second group** who have read at least 6 books.

Mathematical Studies SL May 2015 Paper 1 TZ2 Q4

6 The following table shows the Diploma score x and university entrance mark y for seven IB Diploma students.

Diploma score (x)	28	30	27	31	32	25	27
University entrance mark (y)	73.9	78.1	70.2	82.2	85.5	62.7	69.4

  a Find the correlation coefficient.

The relationship can be modelled by the regression line with equation $y = ax + b$.
  b Write down the value of a and b.

Rita scored a total of 26 in her IB Diploma.
  c Use your regression line to estimate Rita's university entrance mark.

Mathematics SL November 2014 Paper 2 Q2

7 A student recorded, over a period of several months, the amount of time he waited in the queue for lunch at the college canteen. He summarized the results in this cumulative frequency table.

Time (minutes)	⩽ 2	⩽ 4	⩽ 6	⩽ 8	⩽ 10	⩽ 12
Cumulative frequency	4	9	16	37	45	48

  a Draw a cumulative frequency graph for the data.
  b Use your graph to estimate
    i the median
    ii the interquartile range of the times
    iii the 90th percentile.
  c Complete the grouped frequency table:

Time (min)	$0 < t \leqslant 2$	$0 < t \leqslant 4$				
Frequency						

  d Estimate the mean waiting time.

8 The number of customers visiting a shop is recorded over a period of 12 days:
26, 33, 28, 47, 52, 45, 93, 61, 37, 55, 57, 34
  a Find the median and the quartiles.
  b Determine whether there are any outliers.
  c Draw a box-and-whisker diagram for the data.

9 The heights of a group of 7 children were recorded to the nearest centimetre:
127, 119, 112, 123, 122, 126, 118
  a Find the mean and the variance of the heights.
  b Each child stands on a 35-centimetre-high stool. Find the mean and variance of their heights.

10 Theo is keeping a record of his travel expenses. The cost of each journey is $15 plus $3.45 per kilometre. The mean length of Theo's journeys is 11.6 km and the standard deviation of the lengths is 12.5 km. Find the mean and standard deviation of his cost per journey.

11 The frequency table summarizes data from a sample with mean 1.6. Find the value of x.

x	0	1	2	3
Frequency	5	6	8	x

Mixed Practice

12 A scientist measured a sample of 12 adult crabs found on a beach, measuring their shell length (s) and mass (m).

Shell length (cm)	Mass (g)	Shell length (cm)	Mass (g)
7.1	165	5.9	143
8.1	256	8.4	190
8.5	194	9.2	208
6.0	150	5.1	194
9.0	275	6.3	217
5.3	204	9.1	268

a Using technology or otherwise plot a scatter diagram to illustrate the scientist's results.
b The scientist later realized that the beach contains two species of crab – the Lesser European Crab and the Giant European Crab. Her research suggests that the Giant European crab tends to be heavier than similar-sized Lesser European Crabs. Find the equation for a regression line for the mass of the Giant European crab if its shell length is known.
c Find the correlation coefficient for the data for the Giant European crab and comment on your result.
d Estimate the mass of a Giant European Crab with shell length 8 cm.
e Juvenile Giant European Crabs have a shell length of between 2 and 4 cm. Estimate the possible masses of these crabs and comment on the reliability of your results.

13 The one hour distances, in miles, covered by runners before (x) and after (y) going on a new training program are recorded.

The correlation between these two distances is found to be 0.84. The regression line is $y = 1.2x + 2$.
a Describe the significance of
  i the intercept of the regression line being positive
  ii the gradient of the regression line being greater than 1.
b If the previous mean distance is 8 miles, find the new mean.

Their trainer wants to have the data in km. To convert miles to km all the distances are multiplied by 1.6. The new variables in km are X and Y.
c Find
  i the correlation between X and Y
  ii the regression line connecting X and Y.

7 Core: Probability

ESSENTIAL UNDERSTANDINGS

- Probability enables us to quantify the likelihood of events occurring and so evaluate risk.
- Both statistics and probability provide important representations which enable us to make predictions, valid comparisons and informed decisions.
- These fields have power and limitations and should be applied with care and critically questioned, in detail, to differentiate between the theoretical and the empirical/observed.

In this chapter you will learn…

- how to find experimental and theoretical probabilities
- how to find probabilities of complementary events
- how to find the expected number of occurrences
- how to use Venn diagrams to find probabilities
- how to use tree diagrams to find probabilities
- how to use sample space diagrams to find probabilities
- how to calculate probabilities from tables of events
- a formula for finding probabilities of combined events
- how to find conditional probabilities
- about mutually exclusive events
- about independent events.

CONCEPTS

The following key concepts will be addressed in this chapter:
- **Modelling** and finding structure in seemingly random events facilitates prediction.

■ Figure 7.1 What are the chances of a particular outcome in each case?

Core: Probability

> **PRIOR KNOWLEDGE**
>
> Before starting this chapter, you should already be able to complete the following:
>
> 1 For the sets $A = \{2, 3, 5, 6, 8\}$ and $B = \{4, 5, 7, 8, 9\}$, find the sets
>   a $A \cup B$ b $A \cap B$
>
> 2 Out of a class of 20 students, 8 play the piano, 5 play the violin and 2 play both.
>
>   Draw a Venn diagram to represent this information.
>
> 3 A bag contains six red balls and four blue balls. Two balls are removed at random without replacement.
>
>   Draw a tree diagram to represent this information.

Probability is a branch of mathematics that deals with events that depend on chance. There are many circumstances in which you cannot be certain of the outcome, but just because you do not know for sure what will happen, does not mean you cannot say anything useful. For example, probability theory allows predictions to be made about the likelihood of an earthquake occurring or a disease spreading. Understanding and evaluating risk is crucial in a range of disciplines from science to insurance.

Starter Activity

Look at the images in Figure 7.1. Discuss what you can say about the outcomes in each case.

Now look at this problem:

Flip a coin 10 times and record the number of heads and the number of tails. Could you predict what the outcome would be?

Do you have to get five heads and five tails to conclude that the coin is fair?

> **LEARNER PROFILE – Principled**
> Are mathematicians responsible for the applications of their work? Some major mathematical breakthroughs have been inspired by and used in warfare, code-breaking and gambling. Should such areas of research be banned?

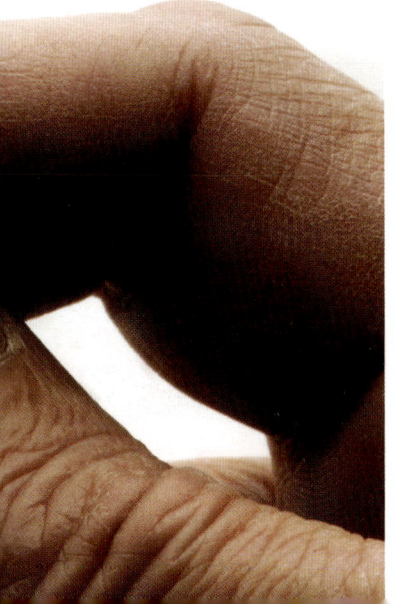

7A Introduction to probability

■ Concepts in probability

If you roll a standard fair dice, there are six possible equally likely **outcomes** – 1, 2, 3, 4, 5 and 6 – each having probability $\frac{1}{6}$ of occurring.

You might be interested in a particular combination of these outcomes, such as rolling an even number on the dice. A combination of outcomes is known as an **event**. The probability of an event, A, happening is denoted by P(A). This is a number between 0 and 1 inclusive which measures how likely the event is.

You might wish to repeat the roll of the dice several times. Each repetition of the process is known as a **trial**. In practice you often won't know for sure what the probability of any given outcome occurring is, so you have to use experimental data to estimate the probability. One way of estimating the probability of an event is to use the **relative frequency**. This is the fraction of trials which are favourable to an event.

> **You are the Researcher**
>
> Probability is the most common way of representing uncertain events, but there are others you might want to investigate, such as likelihood, odds or possibility theory. They all have different uses and different rules.

> **WORKED EXAMPLE 7.1**
>
> A random sample of 80 collectable card packets was opened; 23 were found to contain a special rare card.
>
> Another card packet was purchased. Estimate the probability that it contains a special rare card.
>
> You know from the sample data that 23 out of 80 packets contained the rare card so this relative frequency is the best estimate you have of the probability of getting a rare card
>
> P(pack contains rare card) = $\frac{23}{80}$

TOK Links

How reliable is using information about the past when predicting the future?

■ Theoretical approach to probability

If you know that all outcomes are equally likely then you can work out probabilities theoretically, rather than needing to approximate them from experimental evidence. To do this you need to list all possible outcomes.

The set of all possible outcomes is called the **sample space**, which is denoted by U. For example, the sample space for rolling a standard dice is 1, 2, 3, 4, 5, 6.

Using the notation $n(A)$ to mean the number of outcomes in the event A we have the following result:

> **KEY POINT 7.1**
>
> If all outcomes are equally likely, then P(A) = $\frac{n(A)}{n(U)}$.

WORKED EXAMPLE 7.2

A fair standard six-sided dice is rolled.

Find the probability that the outcome is a prime number.

Use $P(A) = \frac{n(A)}{n(U)}$ — 2, 3 and 5 are prime

$P(\text{prime}) = \frac{3}{6}$

There are six outcomes in the sample space and three of these are prime

$= \frac{1}{2}$

■ Complementary events

The complement of the event A (denoted by A') occurs when the event A does not happen. Since an event either happens or doesn't, the sum of an event and its complement must be 1.

KEY POINT 7.2

- $P(A') = 1 - P(A)$

WORKED EXAMPLE 7.3

The probability of no buses arriving in a 10-minute interval is 0.4.

Find the probability of at least one bus arriving in that 10-minute interval.

At least one means every possible outcome except no buses

$P(\text{at least 1 bus}) = 1 - 0.4$
$= 0.6$

■ Expected number of occurrences

If you know the probability of an event happening, then you can use this to estimate how many of the total number of trials will result in that outcome.

KEY POINT 7.3

Expected number of occurrences of $A = P(A) \times n$, where n is the number of trials.

WORKED EXAMPLE 7.4

An archer has a probability of 0.3 of hitting the bullseye.

If he takes 12 shots in a competition, find the expected number of bullseyes.

Use Expected number $= P(A) \times n$

Expected number of bullseyes $= 0.3 \times 12$
$= 3.6$

Tip

The expected number does not need to be a whole number.

 **TOOLKIT: Problem Solving**

An actuarial study tracked 100 000 people born in the UK in 1940. By 2010, 84 210 were still alive.

a Use the data to estimate the probability of a person in the UK living to age 70.

b An insurance company uses this study to price the cost of a life insurance policy. Give two reasons why the answer to **a** might be an underestimate of the probability of a new client living to 70.

c A 21-year-old takes out a policy which pays out £100 000 on death up to the age of 70, but nothing thereafter. The expected total amount that the 21-year-old will pay is £22 000. Estimate the expected profit of the company on this policy.

d Why might the 21-year-old take out a policy on which they expect to make a loss?

Exercise 7A

For questions 1 and 2, use the method demonstrated in Worked Example 7.1 to estimate the probability from given information.

1 a Out of a random sample of 40 children at a school, 12 were found to have a pet. Estimate the probability that a child at this school has a pet.

   b A random sample of 60 flowers was collected from a meadow and 8 were found to have six petals. Find the probability that a flower from this meadow has six petals.

2 a A biased dice is rolled 90 times and 18 sixes were obtained. Find the probability of rolling a six on this dice.

   b Daniel took 25 shots at a target and hit it 15 times. Find the probability that he hits the target next time.

For questions 3 to 5, use the method demonstrated in Worked Example 7.2 to find the probabilities.

3 A standard six-sided dice is rolled. Find the probability that the outcome is

   a an even number   b a multiple of 3.

4 A card is picked at random from a standard pack of 52 cards. Find the probability that the selected card is

   a a diamond   b an ace.

5 A card is picked at random from a standard pack of 52 cards. Find the probability that the selected card is

   a a red 10   b a black picture card (jack, queen or king).

For questions 6 to 11, use the method demonstrated in Worked Example 7.3.

6 a The probability that Elsa does not meet any of her friends on the way to school is 0.06. Find the probability that she meets at least one friend on the way to school.

   b The probability that Asher does not score any goals in a football match is 0.45. Find the probability that he scores at least one goal.

7 a The probability that a biased coin comes up heads is $\frac{9}{20}$. Find the probability that it comes up tails.

   b The probability of rolling an even number on a biased dice is $\frac{17}{40}$. Find the probability of rolling an odd number.

8 a The probability that Daniel is late for school on at least one day during a week is 0.15. Find the probability that he is on time every day.

   b Theo either walks or cycles to school. The probability that he cycles at least one day a week is 0.87. Find the probability that he walks every day during a particular week.

9 a When a biased coin is flipped 10 times, the probability of getting more than six tails is $\frac{73}{120}$. What is the probability of getting at most six tails?

   b When a biased coin is flipped six times, the probability of getting fewer than three heads is $\frac{7}{48}$. What is the probability of getting at least three heads?

7A Introduction to probability

10 a The probability that it rains on at least four days in a week is 0.27. Find the probability that it rains on no more than three days.

   b The probability that it is sunny on at most two days in a week is 0.34. Find the probability that it is sunny on no fewer than three days.

11 a The probability that an arrow lands within 20 cm of the target is 0.56. Find the probability that it lands more than 20 cm away.

   b The probability that a parachutist lands within 20 m of the landing spot is 0.89. Find the probability that she lands more than 20 m away.

For questions 12 to 15, use the method demonstrated in Worked Example 7.4.

12 a The probability that a packet of crisps contains a toy is 0.2. Find the expected number of toys in 20 packets of crisps.

   b The probability that a footballer scores a penalty is 0.9. If he takes 30 penalties find the expected number of goals scored.

13 a The probability that it rains on any particular day is $\frac{2}{5}$. Find the expected number of rainy days in a 30-day month.

   b The probability that a pack of collectable cards contains a particular rare card is $\frac{3}{70}$. If I buy 140 packs of cards, how many copies of this rare card should I expect to find?

14 a The probability that a biased coin comes up heads is 0.12. Find the expected number of heads when the coin is tossed 40 times.

   b The probability that a biased dice lands on a 4 is 0.15. Find the expected number of 4s when the dice is rolled 50 times.

15 a The probability that Adam is late for work is 0.08. Find the number of days he is expected to be late in a 20-day period.

   b The probability that Sasha forgets her homework is 0.15. Find the number of days she is expected to forget her homework in a 10-day period.

16 A clinical trial involved 350 participants. 26 of them experienced some negative side-effects from the drug.

   a Estimate the probability that the drug produces a negative side-effect.

   b A larger trial is conducted with 900 participants. How many of them are expected to experience negative side-effects?

17 A fair eight-sided dice has sides numbered 1, 1, 1, 2, 2, 3, 4, 5. The dice is rolled 30 times. Estimate the number of times it will land on a 2.

18 A card is selected from a standard pack of 52 cards and returned to the pack. This is repeated 19 more times. Find the expected number of selected cards which are not a diamond.

19 A standard fair n-sided dice is rolled 400 times and the expected number of 1s is 50. What is the value of n?

20 When rolling a biased dice 100 times, the expected number of odd outcomes is three times the expected number of even outcomes. Find the probability of an odd outcome.

21 An actuarial study finds that in a sample of 2000 18-year-old drivers, 124 are involved in an accident. The average cost of such an accident to the insurance company is $15 000. If the company wants to make a 20% profit on policies, what should they charge 18-year-old drivers?

22 An alternative measure of probability is called odds. If the probability of an event is p, then the odds are defined as $\frac{p}{1-p}$.

   a If an event has a probability of 0.6, find its odds.

   b If an event has odds of 4, find its probability.

23 An urn contains red balls, green balls and blue balls. The ratio of red balls to green balls is 1:3. The ratio of green balls to blue balls is 1:4. A random ball is drawn from the urn. What is the probability that it is a red ball?

24 A computer scientist uses a method called Monte Carlo sampling to estimate the value of π. He uses a random number generator to generate 100 points at random inside a 1 by 1 square. Of these points, 78 are inside the largest circle which can be drawn in the square. What would be the estimate of π she would form on the basis of this sample?

7B Probability techniques

■ Venn diagrams

You are already familiar with the idea of using Venn diagrams to represent information about two or more sets. Here you will see how they can be used to calculate probabilities.

When filling in a Venn diagram it is a good idea to start in the middle and work outwards.

WORKED EXAMPLE 7.5

In a school year of 60 students, 32 study Geography and 40 study Biology. If 18 study both of these subjects, find the probability of a randomly selected student studying neither Geography nor Biology.

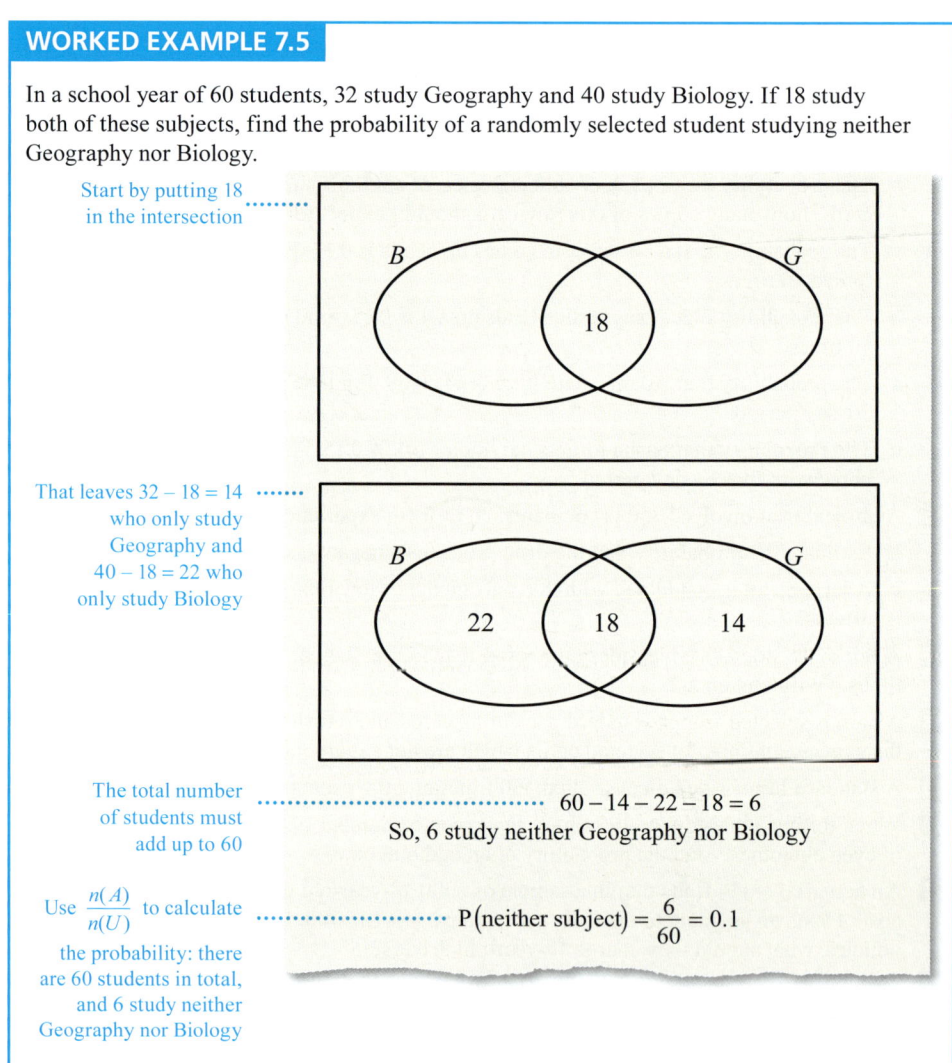

Start by putting 18 in the intersection

That leaves $32 - 18 = 14$ who only study Geography and $40 - 18 = 22$ who only study Biology

The total number of students must add up to 60

$60 - 14 - 22 - 18 = 6$
So, 6 study neither Geography nor Biology

Use $\dfrac{n(A)}{n(U)}$ to calculate the probability: there are 60 students in total, and 6 study neither Geography nor Biology

$P(\text{neither subject}) = \dfrac{6}{60} = 0.1$

■ Tree diagrams

Another technique you are familiar with is tree diagrams. These are used when you have one event followed by another, particularly when the second event depends on the first.

7B Probability techniques

WORKED EXAMPLE 7.6

In a city, the probability of it raining is $\frac{1}{4}$. If it rains, then the probability of Anya travelling by car is $\frac{2}{3}$. If it is not raining, then the probability of Anya travelling by car is $\frac{1}{2}$.

Find the overall probability of Anya travelling by car.

Let R be the event 'it rains' and C the event 'travels by car'

Multiply the probabilities as you work through the tree diagram from left to right

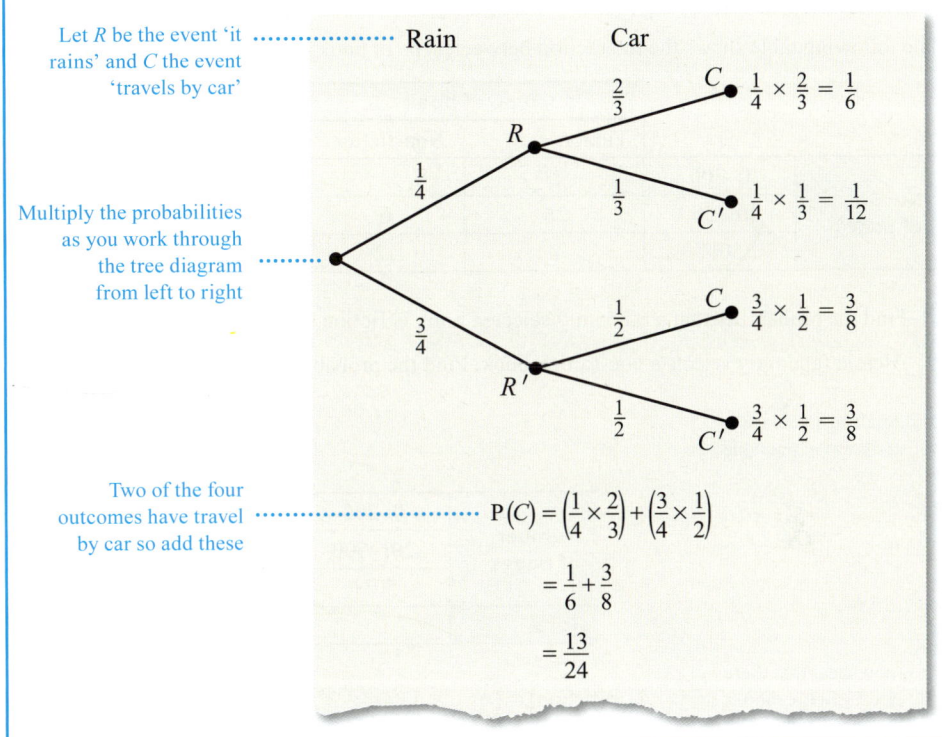

Two of the four outcomes have travel by car so add these

$$P(C) = \left(\frac{1}{4} \times \frac{2}{3}\right) + \left(\frac{3}{4} \times \frac{1}{2}\right)$$

$$= \frac{1}{6} + \frac{3}{8}$$

$$= \frac{13}{24}$$

■ Sample space diagrams

Sometimes the easiest way of finding probabilities is to record all possible outcomes in a sample space diagram.

WORKED EXAMPLE 7.7

Two fair standard four-sided dice are rolled. Find the probability that the sum is larger than 6.

Draw a sample space diagram consisting of all possible sums of the outcomes on the two dice

		Dice 1			
		1	2	3	4
Dice 2	1	2	3	4	5
	2	3	4	5	6
	3	4	5	6	7
	4	5	6	7	8

3 outcomes in the sample space are greater than 6 out of 16 in total

$$P(\text{sum} > 6) = \frac{3}{16}$$

■ Tables of outcomes

Information may be given in a table. In this case it is a good idea to start by finding the totals of the rows and columns.

> **WORKED EXAMPLE 7.8**
>
> The following table shows the interaction between type of book and size of book in a random sample of library books.
>
> | | | Type of book | |
> | --- | --- | --- | --- |
> | | | Fiction | Non-fiction |
> | **Number of pages** | 0–200 | 23 | 3 |
> | | 201–500 | 28 | 6 |
> | | 501+ | 8 | 12 |
>
> **a** Find the probability that a randomly selected book is fiction.
>
> **b** Alessia randomly selects a non-fiction book. Find the probability that it has 0–200 pages.
>
> Add totals to all columns and rows of the table
>
> | | | Type of book | | Total |
> | --- | --- | --- | --- | --- |
> | | | Fiction | Non-fiction | |
> | **Number of pages** | 0–200 | 23 | 3 | 26 |
> | | 201–500 | 28 | 6 | 34 |
> | | 501+ | 8 | 12 | 20 |
> | **Total** | | 59 | 21 | 80 |
>
> It is now clear that there are 59 fiction books out of a total of 80
>
> **a** $P(\text{fiction}) = \dfrac{59}{80}$
>
> All the non-fiction books are in the highlighted column of the table
>
> **b** $P(0\text{–}200 \text{ pages given it is non-fiction}) = \dfrac{3}{21} = \dfrac{1}{7}$

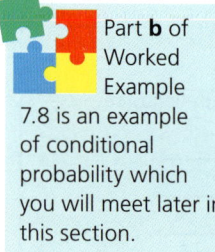

Part **b** of Worked Example 7.8 is an example of conditional probability which you will meet later in this section.

■ Combined events

If you want the probability of either event A or event B occurring, or both occurring, then you can think of this from a Venn diagram as being the probability of being in the union of these two sets ($A \cup B$).

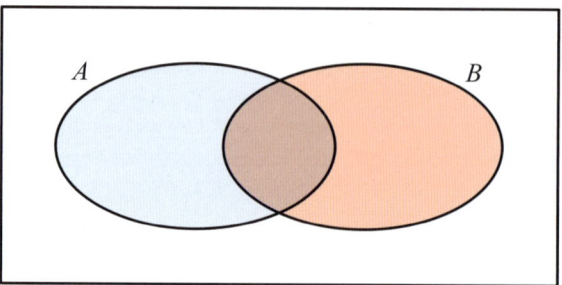

7B Probability techniques

Simply adding the probability of being in A to the probability of being in B will count the intersection $(A \cap B)$ twice so you need to then subtract one lot of the intersection.

> **KEY POINT 7.4**
> - $P(A \cup B) = P(A) + P(B) - P(A \cap B)$

> **WORKED EXAMPLE 7.9**
>
> If $P(A) = 0.4$, $P(B) = 0.3$ and l $P(A \cup B) = 0.5$, find $P(A \cap B)$.
>
> Use $P(A \cup B) = P(A) + P(B) - P(A \cap B)$ .... $P(A \cup B) = P(A) + P(B) - P(A \cap B)$
> and rearrange to find $P(A \cap B)$ .... $0.5 = 0.4 + 0.3 - P(A \cap B)$
> .... $P(A \cap B) = 0.2$

■ Mutually exclusive events

Two events are mutually exclusive if they cannot both happen at the same time, for example, throwing a 1 and throwing a 6 on a dice. In other words, the probability of two mutually exclusive events happening together is zero.

Tip

The second formula in Key Point 7.5 follows directly from the formula $P(A \cup B) = P(A) + P(B) - P(A \cap B)$ with $P(A \cap B) = 0$.

> **KEY POINT 7.5**
>
> If events A and B are mutually exclusive, then:
> - $P(A \cap B) = 0$
>
> or equivalently
> - $P(A \cup B) = P(A) + P(B)$

> **WORKED EXAMPLE 7.10**
>
> If two events A and B are mutually exclusive and $P(A) = 0.5$ and $P(A \cup B) = 0.8$ find $P(B)$.
>
> Use $P(A \cup B) = P(A) + P(B)$ .... Since A and B are mutually exclusive,
> and rearrange to find $P(B)$ .... $P(A \cup B) = P(A) + P(B)$
> .... $0.8 = 0.5 + P(B)$
> .... $P(B) = 0.3$

■ Conditional probability

If the probability of event A happening is dependent on a previous event, B, then the probability is said to be conditional. This can be written as $P(A|B)$, which means the probability of A happening given that B has happened.

> **WORKED EXAMPLE 7.11**
>
> A pack of 52 cards contains 13 hearts. If the first card taken from the pack (without replacement) is a heart, what is the probability that the second card will also be a heart.
>
> If the first card is a heart then there are .... $P(\text{heart 2nd}|\text{heart 1st}) = \dfrac{12}{51} = \dfrac{4}{17}$
> now 12 hearts left out of 51 cards

WORKED EXAMPLE 7.12

The probability of a randomly chosen student studying only French is 0.4. The probability of a student studying French and Spanish is 0.1.

Find the probability of a student studying Spanish given that they study French.

When the situation is a little more complex, it is a good idea to draw a Venn diagram

Given that the student studies French means that you can ignore anything outside the French set

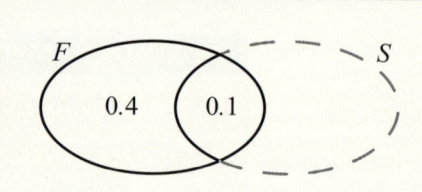

The remaining probability of speaking Spanish is 0.1 out of a remaining total of 0.5

$$P(\text{Spanish}|\text{French}) = \frac{0.1}{0.5}$$
$$= 0.2$$

You can always do conditional probability questions using the Venn diagram method demonstrated in Worked Example 7.12, but this method does suggest a general formula:
$$P(A|B) = \frac{P(A \cap B)}{P(B)}.$$

■ Independent events

Two events are independent if one doesn't affect the other; in other words if $P(A|B) = P(A)$. Substituting this into the above formula and rearranging gives the following convenient relationship for independent events:

KEY POINT 7.6

If events A and B are independent, then:
- $P(A \cap B) = P(A) \times P(B)$

WORKED EXAMPLE 7.13

The probability of a randomly chosen individual in a society being male is 0.52. The probability of the individual being of Irish origin is 0.18.

If these two characteristics are independent, find the probability that a randomly chosen individual is a male of Irish origin.

Being a male of Irish origin means being male and Irish. As these two characteristics are independent you can multiply the probabilities

$$P(\text{male} \cap \text{Irish}) = P(\text{male}) \times P(\text{Irish})$$
$$= 0.52 \times 0.18$$
$$= 0.0936$$

CONCEPTS – MODELLING

In reality, there are often links between events which mean that they are not independent. For example, there might be a small difference in genetic or social factors which mean that an individual of Irish origin is more or less likely to be male than other members of the same population. We often make assumptions that such links are small so independence is valid, but you should be aware that you are making these assumptions.

Be the Examiner 7.1

A bag contains 10 balls labelled uniquely with the numbers 1 to 10. Hanna picks a ball at random.

What is the probability she gets a prime number or an even number?

Which is the correct solution? Identify the errors made in the incorrect solutions.

Solution 1	Solution 2	Solution 3
$P(\text{prime}) = \frac{4}{10}$	$P(\text{prime}) = \frac{4}{10}$	$P(\text{prime}) = \frac{4}{10}$
$P(\text{even}) = \frac{5}{10}$	$P(\text{even}) = \frac{5}{10}$	$P(\text{even}) = \frac{5}{10}$
So, $P(\text{prime or even}) = \frac{4}{10} + \frac{5}{10}$ $= \frac{9}{10}$	$P(\text{prime and even}) = \frac{1}{10}$ So, $P(\text{prime or even}) = \frac{4}{10} + \frac{5}{10} - \frac{1}{10}$ $= \frac{8}{10}$	$P(\text{prime and even}) = \frac{4}{10} \times \frac{5}{10} = \frac{1}{5}$ So, $P(\text{prime or even}) = \frac{4}{10} + \frac{5}{10} - \frac{1}{5}$ $= \frac{7}{10}$

Exercise 7B

For questions 1 to 7, use the method demonstrated in Worked Example 7.5 to draw a Venn diagram and then find the required probability.

1 Out of 25 students in a class, 18 speak French, 9 speak German and 5 speak both languages. Find the probability that a randomly chosen student speaks

   a neither language
   b German but not French.

2 In a group of 30 children, 12 like apples, 18 like bananas and 10 like both fruits. Find the probability that a randomly chosen child

   a likes bananas but not apples
   b likes neither fruit.

3 In a school of 150 pupils, 80 pupils play basketball, 45 play both hockey and basketball and 25 play neither sport. Find the probability that a randomly selected pupil

   a plays basketball but not hockey
   b plays hockey but not basketball.

4 In a certain school, 80 pupils study both Biology and History, 35 study neither subject, 45 study only Biology and 60 study only History. Find the probability that a randomly selected pupil

   a studies at least one of the two subjects
   b studies History.

5 Out of 35 shops in a town, 28 sell candy, 18 sell ice cream and 12 sell both ice cream and candy. Find the probability that a randomly selected shop sells

   a ice cream but not candy
   b candy but not ice cream.

6 At a certain college, the probability that a student studies Geography is 0.4, the probability that they study Spanish is 0.7 and the probability that they study both subjects is 0.2. Find the probability that a randomly selected student

  a studies neither subject
  b studies Spanish but not Geography.

7 In a given school, the probability that a child plays the piano is 0.4. The probability that they play both the piano and the violin is 0.08, and the probability that they play neither instrument is 0.3. Find the probability that a randomly selected child

  a plays the violin but not the piano
  b plays the piano but not the violin.

For questions 8 to 12, use the method demonstrated in Worked Example 7.6 to draw a tree diagram and then find the required probability.

8 Every day, Rahul either drives to work or takes the bus. The probability that he drives is 0.3. If he drives to work the probability that he is late is 0.05. If he takes the bus the probability that he is late is 0.2. Find the probability that Rahul is

  a late for work
  b not late for work.

9 The probability that the school canteen serves pizza for lunch is 0.3. If there is pizza, the probability that they also serve fries is 0.8. If there is no pizza, the probability that they serve fries is 0.5. Find the probability that on a randomly selected day

  a there are no fries
  b there are fries.

10 Sophia rolls a fair six-sided dice. If she gets a 6, she tosses a fair coin. If she does not get a 6, she tosses a biased coin which has the probability 0.6 of showing tails. Find the probability that the coin shows

  a heads
  b tails.

11 A bag contains 12 red balls and 18 blue balls. Two balls are taken out at random, one after the other, without replacement. Find the probability that

  a the two balls are the same colour
  b the second ball is blue.

12 The probability that I revise for a test is 0.8. If I revise, the probability that I pass the test is 0.9. If I do not revise, the probability that I pass is 0.7. Find the probability that

  a I revise and pass the test
  b I do not revise but pass the test.

For questions 13 to 17, use the method demonstrated in Worked Example 7.7 to create a sample space diagram showing all possible outcomes, and use it to find the required probability.

13 Two fair six-sided dice are rolled, and the scores are added. Find the probability that the total score is

  a equal to 6
  b greater than 9.

14 Two fair six-sided dice are rolled. Find the probability that

  a the two scores are equal
  b the second score is larger than the first.

15 Two fair coins are tossed. Find the probability that

  a both show tails
  b the two outcomes are different.

16 A family has two children, born independently. Find the probability that they are

  a a boy and a girl
  b two girls.

17 Write out all possible arrangements of the letters T, A, I, L. Hence find the probability that, when those four letters are arranged in a random order,

  a the 'word' ends with two consonants
  b the two vowels are next to each other.

For questions 18 to 21, use the method demonstrated in Worked Example 7.8 to find the required probability.

18 The table shows the number of students at a school taking different science subjects at Standard Level and Higher Level.

	HL	SL
Biology	32	18
Chemistry	13	25
Physics	9	12

Find the probability that a randomly selected student

  a studies Chemistry
  b takes a Standard Level Science subject.

19 The favourite colour and favourite fruit for a group of children are shown in the table.

	Apple	Banana	Strawberry
Red	12	23	18
Blue	16	18	9
Green	3	9	6

A child is selected at random. Find the probability that
   a their favourite colour is green
   b their favourite fruit is banana.

20 The table shows information on hair colour and eye colour of a group of people.

	Blond hair	Brown hair
Blue eyes	23	18
Brown eyes	16	20

Find the probability that a randomly selected person
   a has blue eyes and brown hair
   b has brown eyes and blond hair.

21 The table shows information on some items of clothing in a shop.

	Black	White	Red
Trousers	26	8	9
Skirts	30	12	23
Tops	14	26	15

Find the probability that
   a a randomly selected skirt is red
   b a randomly selected white item is a top.

For questions 22 to 24, use the method demonstrated in Worked Example 7.9.

22 a If $P(A) = 0.3$, $P(B) = 0.8$ and $P(A \cup B) = 0.9$, find $P(A \cap B)$.
   b If $P(A) = 0.8$, $P(B) = 0.2$ and $P(A \cup B) = 0.7$, find $P(A \cap B)$.

23 a If $P(A) = 0.6$, $P(B) = 0.4$ and $P(A \cap B) = 0.2$, find $P(A \cup B)$.
   b If $P(A) = 0.8$, $P(B) = 0.5$ and $P(A \cap B) = 0.4$, find $(A \cup B)$.

24 a If $P(A) = 0.4$, $P(A \cap B) = 0.3$ and $P(A \cup B) = 0.6$, find $P(B)$.
   b If $P(A) = 0.7$, $P(A \cap B) = 0.5$ and $P(A \cup B) = 0.9$, find $P(B)$.

For questions 25 and 26, use the method demonstrated in Worked Example 7.10.

25 a If two events A and B are mutually exclusive and $P(A) = 0.6$ and $P(A \cup B) = 0.8$, find $P(B)$.
   b If two events A and B are mutually exclusive and $P(A) = 0.4$ and $P(A \cup B) = 0.5$, find $P(B)$.

26 a Events A and B are mutually exclusive, $P(A) = \frac{1}{3}$ and $P(B) = \frac{2}{5}$. Find $P(A \cup B)$.
   b Events A and B are mutually exclusive, $P(A) = \frac{3}{4}$ and $P(B) = \frac{1}{10}$. Find $P(A \cup B)$.

For questions 27 to 30, use the method demonstrated in Worked Example 7.11 to find conditional probabilities.

27 A bag contains 20 red counters and 30 blue counters. Two counters are taken out at random. Given that the first counter is red, find the probability that the second counter is
   a red
   b blue.

28 Two cards are taken out of a standard pack of 52 cards. Given that the first card is a red 10, find the probability that the second card is
   a red
   b a 10.

29 The favourite colour and favourite fruit for a group of children are shown in the table.

	Apple	Banana	Strawberry
Red	12	23	18
Blue	16	18	9
Green	3	9	6

Find the probability that
   a a child's favourite colour is green, given that their favourite fruit is apples
   b a child's favourite fruit is strawberries, given that their favourite colour is red.

30 The table shows information on hair colour and eye colour of a group of people.

	Blond hair	Brown hair
Blue eyes	23	18
Brown eyes	16	20

Find the probability that a randomly selected person
   a has brown eyes, given that they have blond hair
   b has blond hair, given that they have brown eyes.

For questions 31 to 37, use the method demonstrated in Worked Example 7.12 to draw a Venn diagram and find the required conditional probability.

31 Out of 25 students in a class, 18 speak French, 9 speak German and 5 speak both languages. Find the probability that a randomly chosen student speaks
   a German, given that they speak French
   b French, given that they speak German.

32 In a group of 30 children, 12 like apples, 18 like bananas and 10 like both fruits. Find the probability that a randomly chosen child
   a likes bananas, given that they like apples
   b likes apples, given that they like bananas.

33 In a school of 150 pupils, 80 pupils play basketball, 45 play both hockey and basketball and 25 play neither sport. Find the probability that a randomly selected pupil
   a plays basketball, given that they do not play hockey
   b plays hockey, given that they do not play basketball.

34 In a certain school, 80 pupils study both Biology and History, 35 study neither subject, 45 study only Biology and 60 study only History. Find the probability that a randomly selected pupil
   a studies Biology, given that they study at least one of the two subjects
   b studies History, given that they study at least one of the two subjects.

35 Out of 35 shops in a town, 28 sell candy, 18 sell ice cream and 12 sell both ice cream and candy. Find the probability that a randomly selected shop
   a sells neither, given that it does not sell candy.
   b sells neither, given that it does not sell ice cream.

36 At a certain college, the probability that a student studies Geography is 0.4, the probability that they study Spanish is 0.7 and the probability that they study both subjects is 0.2. Find the probability that a randomly selected student
   a studies both subjects, given that they study Geography
   b studies both subjects, given that they study Spanish.

37 In a school, the probability that a child plays the piano is 0.4. The probability that they play both the piano and the violin is 0.08, and the probability that they play neither instrument is 0.3. Find the probability that a randomly selected child
   a plays the violin, given that they play at least one instrument
   b plays the piano, given that they play at least one instrument.

7B Probability techniques

For questions 38 and 39, use the method demonstrated in Worked Example 7.13 to find the combined probability of two independent events.

38   a   The probability that a child has a sibling is 0.7. The probability that a child has a pet is 0.2. If having a sibling and having a pet are independent, find the probability that a child has both a sibling and a pet.

   b   The probability that a child likes bananas is 0.8. The probability that their favourite colour is green is 0.3. If these two characteristics are independent, find the probability that a child's favourite colour is green and they like bananas.

39   a   My laptop and my phone can run out of charge independently of each other. The probability that my laptop is out of charge is 0.2 and the probability that they are both out of charge is 0.06. Find the probability that my phone is out of charge.

   b   The probability that my bus is late is 0.2. The probability that it rains and that my bus is late is 0.08. If the two events are independent, find the probability that it rains.

40   Two fair dice are rolled and the score is the product of the two numbers.

   a   Draw a sample space diagram showing all possible outcomes.

   b   If this were done 180 times, how many times would you expect the score to be at least 20?

41   A fair four-sided dice (numbered 1 to 4) and a fair eight-sided dice (numbered 1 to 8) are rolled and the outcomes are added together. Find the probability that the total is less than 10.

42   A survey of 180 parents in a primary school found that 80 are in favour of building a new library, 100 are in favour of a new dining hall and 40 are in favour of both.

   a   Draw a Venn diagram to represent this information.

   b   Find the probability that a randomly selected parent is in favour of neither a new library, nor a new dining hall.

   c   Find the probability that a randomly selected parent is in favour of a new library, given that they are not in favour of a new dining hall.

43   A bag contains 12 green balls and 18 yellow balls. Two balls are taken out at random. Find the probability that they are different colours.

44   The table summarizes the information on a pupil's punctuality, depending on the weather.

	Raining	Not raining
On time	35	42
Late	5	6

   a   Find

      i   P(late|it is raining)   ii   P(late|it is not raining)

   b   Hence, determine whether being late is independent of the weather. Justify your answer.

45   Asher has 22 toys: 12 are teddies, 5 are green teddies, and 8 are neither green nor teddies.

   a   Draw a Venn diagram showing this information and find the number of toys which are green but not teddies.

   b   Find the probability that a randomly selected toy is green.

   c   Given that a toy is green, find the probability that it is a teddy.

46   Out of 130 students at a college, 50 play tennis, 45 play badminton and 12 play both sports.

   a   How many students play neither sport?

   b   Find the probability that a randomly selected student plays at least one of the two sports.

   c   Given that a student plays exactly one of the two sports, find the probability that it is badminton.

47   A football league consists of 18 teams. My team has a 20% chance of winning against a higher-ranking team and a 70% chance of winning against a lower-ranking team. If my team is currently ranked fifth in the league, what it is the probability that we win the next match against a randomly selected team?

48   An office has two photocopiers. One of them is broken with probability 0.2 and the other with probability 0.3. If a photocopier is selected at random, what it the probability that it is working?

49 Theo either walks, takes the bus or cycles to school, each with equal probability. The probability that he is late is 0.06, if he walks, it is 0.2 if he takes the bus and 0.1 if he cycles. Using a tree diagram, or otherwise, find the probability that
   a Theo takes the bus to school and is late
   b Theo is on time for school.

50 Three fair six-sided dice are thrown and the score is the sum of the three outcomes. Find the probability that the score is equal to 5.

51 The table shows information on hair colour and eye colour of a group of adults.

	Blue eyes	Brown eyes	Green eyes
Blond hair	26	34	19
Brown hair	51	25	32

Find the probability that a randomly selected adult has
   a blue eyes
   b blond hair
   c both blue eyes and blond hair.

52 Events A and B satisfy: $P(A) = 0.6$, $P(A \cap B) = 0.2$ and $P(A \cup B) = 0.9$. Find $P(B)$.

53 Events A and B satisfy: $P(A) = 0.7$, $P(A \cup B) = 0.9$ and $P(B') = 0.3$. Find $P(A \cap B)$.

54 Let A and B be events such that $P(A) = \frac{2}{5}$, $P(B|A) = \frac{1}{2}$ and $P(A \cup B) = \frac{3}{4}$.
   a Using a Venn diagram, or otherwise, find $P(A \cap B)$.
   b Find $P(B)$.

55 A box contains 40 red balls and 30 green balls. Two balls are taken out at random, one after another without replacement.
   a Given that the first ball is red, find the probability that the second ball is also red.
   b What is the probability that both balls are red?
   c Is it more likely that the two balls are the same colour or different colour?

56 The probability that it rains on any given day is 0.12, independently of any other day. Find the probability that
   a it rains on at least one of two consecutive days
   b it rains on three consecutive days.

57 In a group of 100 people, 27 speak German, 30 speak Mandarin, 40 speak Spanish, 3 speak all three languages. 12 speak only German, 15 speak only Mandarin and 8 speak Spanish and Mandarin but not German.
   a Draw a Venn diagram showing this information, and fill in the missing regions.
   b Find the probability that a randomly selected person speaks only Spanish.
   c Given that a person speaks Spanish, find the probability that they also speak Mandarin.

58 In a class of 30 pupils, 22 have a phone, 17 have a smart watch and 15 have a tablet. 11 have both a smart watch and a phone, 11 have a phone and a tablet and 10 have a tablet and a smart watch. Everyone has at least one of those items.
   a Let x be the number of pupils who have all three items. Draw a Venn diagram showing this information.
   b Use your Venn diagram to show that $x + 22 = 30$.
   c Hence find the probability that a randomly selected pupil has all three items.
   d Given that a pupil does not have a tablet, find the probability that they have a phone.
   e Given that a pupil has exactly two of the items, find the probability that they are a smart watch and a tablet.

59 a A mother has two children. If one of them is a boy, what is the probability that they are both boys?
   b A mother has two children. If the older one is a boy, what is the probability that they are both boys?

TOOLKIT: Problem Solving

In a 1950s US TV show, a contestant played the following game. He is shown three doors and told that there is a car behind one of them and goats behind the other two. He is asked to select a door at random. The show host (who knows what is behind each door) then opens one of the other two doors to reveal a goat. The contestant is then given the option to switch to the third door.

a Does your intuition suggest that the contestant should stick, switch or that it does not matter?

b If the car is behind door 1 and the contestant initially selects door 2, explain why they will definitely win if they switch.

c The table shows all possible positions of the car and all possible initial selections. For each combination it shows whether the contestant will win or lose if they switch.

Complete the table.

		Car behind door		
		1	2	3
Door initially selected	1	L		
	2	W		
	3			

d Hence decide whether the contestant should switch or stick with the original door.

e How does your intuition change if you exaggerate the problem in the following way?

A contestant is shown 100 doors. He selects door 1, then the host opens all the other doors except door 83, revealing goats behind all of them. Should he stick with door 1 or switch to door 83?

Exaggerating is a very powerful problem-solving technique!

Checklist

- You should be able to find experimental and theoretical probabilities. If all outcomes are equally likely then:

$$P(A) = \frac{n(A)}{n(U)}$$

- You should be able to find probabilities of complementary events: $P(A') = 1 - P(A)$

- You should be able to find the expected number of occurrences of an event:
  - Expected number of occurrences of A is $P(A) \times n$, where n is the number of trials.

- You should be able to use Venn diagrams to find probabilities.

- You should be able to use tree diagrams to find probabilities.

- You should be able to use sample space diagrams to find probabilities.

- You should be able to calculate probabilities from tables of events.

- You should be able to use the formula for probabilities of combined events:
  - $P(A \cup B) = P(A) + P(B) - (A \cap B)$

- You should be able to find conditional probabilities.
- You should know about mutually exclusive events:
  - If events A and B are mutually exclusive, then $P(A \cap B) = 0$ or, equivalently,
  - $P(A \cup B) = P(A) + P(B)$
- You should know about independent events:
  - If the events A and B are independent, then $P(A \cap B) = P(A) \times P(B)$

■ Mixed Practice

1 In a clinical trial, a drug is found to have a positive effect on 128 out of 200 participants. A second trial is conducted, involving 650 participants. On how many of the participants is the drug expected to have a positive effect?

2 The table shows the genders and fruit preferences of a group of children.

	Boy	Girl
Apples	16	21
Bananas	32	14
Strawberries	11	21

Find the probability that
a a randomly selected child is a girl
b a randomly selected child is a girl who prefers apples
c a randomly selected boy prefers bananas
d a randomly selected child is a girl, given that they prefer strawberries.

3 Every break time at school, Daniel randomly chooses whether to play football or basketball. The probability that he chooses football is $\frac{2}{3}$. If he plays football, the probability that he scores is $\frac{1}{5}$. If he plays basketball, the probability that he scores is $\frac{3}{4}$.

Using a tree diagram, find the probability that Daniel
a plays football and scores
b does not score.

4 The Venn diagram shows two sets:
$T = \{$multiples of 3 between 1 and 20$\}$
$F = \{$multiples of 5 between 1 and 20$\}$
a Copy the diagram and place the numbers from 1 to 20 (inclusive) in the correct region of the Venn diagram.
b Find the probability that a number is
  i a multiple of 5
  ii a multiple of 5 given that it is not a multiple of 3.

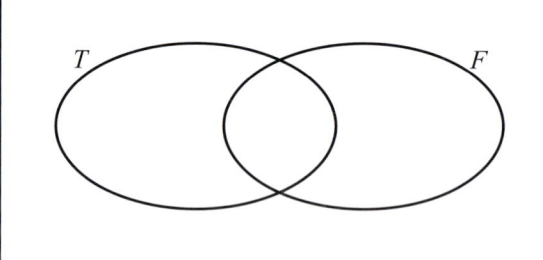

5 In a certain school, the probability that a student studies Geography is 0.6, the probability that a student studies French is 0.4 and the probability that they study neither subject is 0.2.
a Find the probability that a student studies both subjects.
b Given that a student does not study French, what is the probability that they study Geography?

6 Events A and B satisfy: $P(A) = 0.6$, $P(B) = 0.3$, $P(A \cup B) = 0.72$.
a Find $P(A \cap B)$.
b Determine whether A and B are independent.

Mixed Practice

7 Events A and B are independent with $P(A) = 0.6$ and $P(B) = 0.8$. Find $P(A \cup B)$.

8 Events A and B satisfy: $P(A) = 0.7$, $P(B) = 0.8$ and $P(A \cap B) = 0.6$
 a Draw a Venn diagram showing events A and B.
 b Find $P(A \mid B)$.
 c Find $P(B \mid A')$.

9 Elsa has a biased coin with the probability $\frac{1}{3}$ of showing heads, a fair six-sided dice (numbered 1 to 6) and a fair four-sided dice (numbered 1 to 4). She tosses the coin once. If it comes up heads, she rolls the six-sided dice, and if comes up tails she rolls the four-sided dice.
 a Find the probability that the dice shows
  i a '1'
  ii a '6'.
 b Find the probability that the number on the dice is a multiple of 3.

10 An integer is chosen at random from the first 10 000 positive integers. Find the probability that it is
 a a multiple of 7
 b a multiple of 9
 c a multiple of at least one of 7 and 9.

11 A fair coin is tossed three times.
 a Copy and complete the tree diagram showing all possible outcomes.

Find the probability that
 b the coin shows tails all three times
 c the coin shows heads at least once
 d the coin shows heads exactly twice.

12 There are six black and eight white counters in a box. Asher takes out two counters without replacement. Elsa takes out one counter, returns it to the box and then takes another counter. Who has the larger probability of selecting one black and one white counter?

13 A drawer contains eight red socks, six white socks and five black socks. Two socks are taken out at random. What is the probability that they are the same colour?

14 A pack of cards in a game contains eight red, six blue and ten green cards. You are dealt two cards at random.
 a Given that the first card is blue, write down the probability that the second card is red.
 b Find the probability that you get at least one red card.

15 Alan's laundry basket contains two green, three red and seven black socks.

He selects one sock from the laundry basket at random.
 a Write down the probability that the sock is red.

Alan returns the sock to the laundry basket and selects two socks at random.
 b Find the probability that the first sock he selects is green and the second sock is black.

Alan returns the socks to the laundry basket and again selects two socks at random.
 c Find the probability that he selects two socks of the same colour.

Mathematical Studies SL May 2013 Paper 1 TZ2 Q10

16 100 students at IB College were asked whether they study Music (*M*), Chemistry (*C*) or Economics (*E*) with the following results.

10 study all three

15 study Music and Chemistry

17 study Music and Economics

12 study Chemistry and Economics

11 study Music **only**

6 study Chemistry **only**

  a Draw a Venn diagram to represent the information above.
  b Write down the number of students who study Music but not Economics.

There are 22 Economics students **in total**.

  c **i** Calculate the number of students who study Economics only.
     ii Find the number of students who study none of these three subjects.

A student is chosen at random from the 100 that were asked above.
  d Find the probability that this student
     i studies Economics
     ii studies Music and Chemistry but not Economics
     iii does not study either Music or Economics
     iv does not study Music given that the student does not study Economics.

Mathematical Studies SL May 2013 Paper 2 TZ1 Q2

17 A bag contains 10 red sweets and *n* yellow sweets. Two sweets are selected at random. The probability that the two sweets are the same colour is $\frac{1}{2}$.
  a Show that $n^2 - 21n + 90 = 0$.
  b Hence find the possible values of *n*.

18 At a large school, students are required to learn at least one language, Spanish or French. It is known that 75% of the students learn Spanish, and 40% learn French.
  a Find the percentage of students who learn **both** Spanish and French.
  b Find the percentage of students who learn Spanish, but not French.

At this school, 52% of the students are girls, and 85% of the girls learn Spanish.
  c A student is chosen at random. Let *G* be the event that the student is a girl, and let *S* be the event that the student learns Spanish.
     i Find P $(G \cap S)$.
     ii Show that *G* and *S* are **not** independent.
  d A boy is chosen at random. Find the probability that he learns Spanish.

Mathematics SL November 2012 Paper 2 Q10

19 A group of 100 customers in a restaurant are asked which fruits they like from a choice of mangoes, bananas and kiwi fruits. The results are as follows.

15 like all three fruits

22 like mangoes and bananas

33 like mangoes and kiwi fruits

27 like bananas and kiwi fruits

8 like none of these three fruits

x like **only** mangoes

a **Copy** the following Venn diagram and correctly insert all values from the above information.

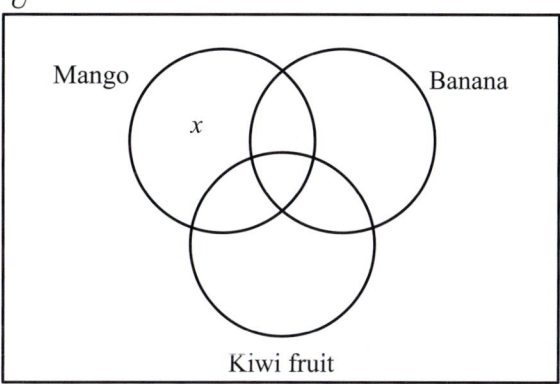

The number of customers that like **only** mangoes is equal to the number of customers that like **only** kiwi fruits. This number is half of the number of customers that like **only** bananas.

b Complete your Venn diagram from part **a** with this additional information **in terms of x**.

c Find the value of x.

d Write down the number of customers who like
  i mangoes;
  ii mangoes or bananas.

e A customer is chosen at random from the 100 customers. Find the probability that this customer
  i likes none of the three fruits;
  ii likes only two of the fruits;
  iii likes all three fruits given that the customer likes mangoes and bananas.

f Two customers are chosen at random from the 100 customers. Find the probability that the two customers like none of the three fruits.

Mathematical Studies SL May 2015 Paper 2 TZ2 Q1

8 Core: Probability distributions

ESSENTIAL UNDERSTANDINGS

- Probability enables us to quantify the likelihood of events occurring and to evaluate risk.
- Both statistics and probability provide important representations which enable us to make predictions, valid comparisons and informed decisions.
- These fields have power and limitations and should be applied with care and critically questioned, in detail, to differentiate between the theoretical and the empirical/observed.
- Probability theory allows us to make informed choices, to evaluate risk, and to make predictions about seemingly random events.

In this chapter you will learn…
- about discrete random variables and their probability distributions
- how to find the expected value of a discrete random variable and apply this to real-life contexts
- the circumstances under which the binomial distribution is an appropriate model
- how to find the mean and variance of the binomial distribution
- about the key properties of the normal distribution
- how to carry out normal probability calculation and inverse normal calculations on your GDC.

CONCEPTS

The following key concepts will be addressed in this chapter:
- **Modelling** and finding structure in seemingly random events facilitate prediction.
- Different probability distributions provide a **representation** of the relationship between the theory and reality, allowing us to make predictions about what might happen.

■ **Figure 8.1** How are the different variables represented distributed?

Core: Probability distributions

> **PRIOR KNOWLEDGE**
>
> Before starting this chapter, you should already be able to complete the following:
> 1. A bag contains six red and four blue balls. Two balls are removed at random without replacement. Find:
>   a the probability of getting two red balls
>   b the probability of getting one red ball.
> 2. Draw a box-and-whisker diagram for the following data: 2, 6, 8, 9, 10, 12.

In some cases, you are equally as likely to see an outcome at the lower or upper end of the possible range as in the middle; for example, getting a 1 or a 6 on a fair dice is just as likely as getting a 3 or 4. However, often the smallest or largest values are far less likely than those around the average; for example, scores in an exam tend to be grouped around the average mark, with scores of 0% or 100% rather less likely!

Starter Activity

Look at the images in Figure 8.1. Discuss the variables that these extreme cases represent. If you had to draw a histogram representing these variables what would it look like?

Now look at this problem:

If you roll a fair dice 10 times, what do you expect the mean score to be?

LEARNER PROFILE – Caring
Game theory has combined with ecology to explain the 'mathematics of being nice'. There are some good reasons why some societies are advantaged by working together. Is any action totally altruistic?

8A Discrete random variables

Concept of discrete random variables and their probability distributions

A **discrete random variable** (usually denoted by a capital letter such as X) is a variable whose discrete output depends on chance, for example, 'the score on a fair dice'. In this case we might write a statement such as $P(X = 2) = \frac{1}{6}$, which just means 'the probability that the score on the dice is 2 is $\frac{1}{6}$'.

The sample space of a random variable, together with the probabilities of each outcome, is called the probability distribution of the variable. This is often displayed in a table.

> **WORKED EXAMPLE 8.1**
>
> A drawer has five black socks and three red socks. Two socks are removed without replacement. Find the probability distribution of X, the number of red socks removed.
>
> A tree diagram is a useful way of working out the probabilities
>
>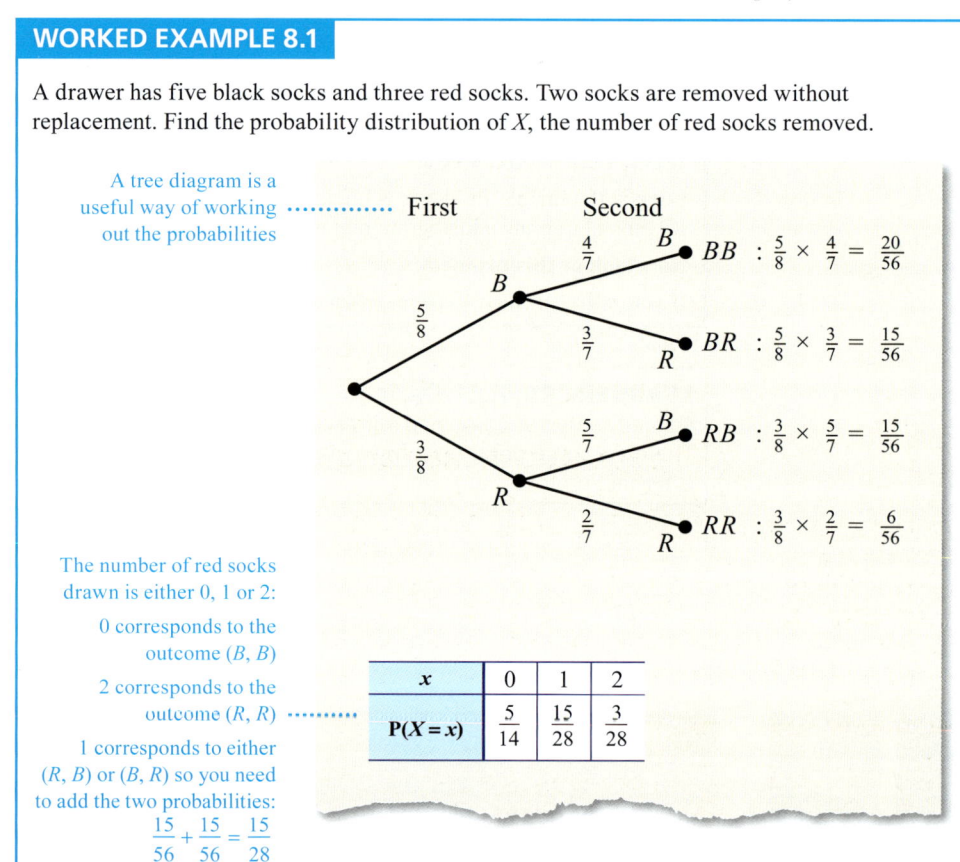
>
> The number of red socks drawn is either 0, 1 or 2:
>
> 0 corresponds to the outcome (B, B)
>
> 2 corresponds to the outcome (R, R)
>
> 1 corresponds to either (R, B) or (B, R) so you need to add the two probabilities: $\frac{15}{56} + \frac{15}{56} = \frac{15}{28}$
>
> | x | 0 | 1 | 2 |
> |---|---|---|---|
> | $P(X=x)$ | $\frac{5}{14}$ | $\frac{15}{28}$ | $\frac{3}{28}$ |

Once you had found $P(B, B)$ and $P(R, R)$ from the tree diagram, you didn't actually need to work out $P(B, R) + P(R, B)$ as well because you knew that the sum of all possible outcomes on the right of the tree diagram must be 1.

This leads to a very important property of probability distributions.

Tip

Key Point 8.1 just says that the sum of all probabilities must be 1.

> **KEY POINT 8.1**
>
> In any discrete probability distribution:
>
> - $\sum_{x} P(X = x) = 1$

8A Discrete random variables

WORKED EXAMPLE 8.2

The table below shows the probability distribution of Y, the number of cracks observed in a 1 km region of a sewer pipe.

y	0	1	2	>2
$P(Y = y)$	0.45	0.3	0.15	k

a Find the value of k.

b Find $P(Y \leq 1)$.

c Find $P(Y = 0 | Y \leq 1)$.

The sum of the probabilities must be 1

a $\quad 0.45 + 0.3 + 0.15 + k = 1$
$\quad\quad k = 0.1$

If $Y \leq 1$ it can be either 0 or 1

b $\quad P(Y \leq 1) = P(Y = 0) + P(Y = 1)$
$\quad\quad = 0.45 + 0.3$
$\quad\quad = 0.75$

Given $Y \leq 1$ means that we only use this part of the table:

y	0	1
$P(Y = y)$	0.45	0.3

c $\quad P(Y = 0 | Y \leq 1) = \dfrac{0.45}{0.75}$
$\quad\quad = 0.6$

Expected value for discrete data

If you rolled a fair dice enough times you would expect to get a mean score of 3.5 (the mean of 1, 2, 3, 4, 5, 6). This is the theoretical mean or the expected value of the random variable, but in any given number of rolls, you may well get a different outcome for the average.

The expected value, $E(X)$, of a discrete random variable (also referred to as the mean, μ) is the average value you would get if you carried out infinitely-many repetitions.

Tip

Note that the expected value/mean of a random variable is not necessarily the most likely single outcome – that would be the mode.

KEY POINT 8.2

For a discrete random variable, X:

$$E(X) = \sum_{x} x \, P(X = x)$$

CONCEPTS – REPRESENTATION

Using a single value to **represent** the expected outcome will always involve a loss of information. We are not suggesting that this will always be the outcome, or even that it is the most likely outcome. It really only represents what we would predict is the mean outcome if we were allowed an infinite number of observations of the random variable. Whenever we use probability to represent real-life situations, we should be aware of the inherent uncertainty in the predictions. However, if the number of observations is large enough, our predictions can become unerringly accurate. This underpins much of the statistical interpretations of economics and physics. Even though the behaviour of a single molecule of water going down a waterfall or a single customer in a market is difficult to predict, the overall behaviour of groups is remarkably predictable.

WORKED EXAMPLE 8.3

The random variable Y has the probability distribution $P(Y = y) = 0.125(x^2 - 7x + 14)$ for $y \in \{2, 3, 4\}$

Find the expected value, $E(Y = y)$.

Evaluate the formula at $y = 1, 2$ and 3 to create a probability distribution table

y	2	3	4
$P(Y=y)$	0.5	0.25	0.25

Use $E(Y) = \sum_y y\, P(Y = y)$

$E(Y) = (2 \times 0.5) + (3 \times 0.25) + (4 \times 0.25)$
$= 2.75$

Applications

There are many applications of discrete random variables. They are used to model things from the creation of muon particles in nuclear physics to the number of infected individuals in an epidemic. One particular example you need to know in the IB is the application to games of chance. The definition of a fair game you need to use is one where the expected gain for each participant is zero.

WORKED EXAMPLE 8.4

Jenny and Hiroki play a game. Hiroki pays Jenny $4 to play, then rolls a fair standard six-sided dice. He gets back the score on the dice in dollars.

Is this a fair game?

Define the random variable so it is clear what you are referring to

Let the random variable X be the score on the dice.

Use $E(X) = \sum_x x\, P(X = x)$ to find the expected score

$E(X) = \left(1 \times \frac{1}{6}\right) + \left(2 \times \frac{1}{6}\right) + \left(3 \times \frac{1}{6}\right) + \left(4 \times \frac{1}{6}\right) + \left(5 \times \frac{1}{6}\right) + \left(6 \times \frac{1}{6}\right)$
$= 3.5$

This is the amount Hiroki expects to receive

So his expected payout is $3.50

But he pays out $4 to play

A player would expect to lose $0.50 when playing this game so it is not fair.

TOK Links

What is meant by a fair game? Within IB maths we use the definition that the expected gain of each player is zero, but there are other possibilities. The expected gain of a player in a lottery is negative, but all the rules are clearly displayed and everybody is free to choose whether or not to play. There are often issues when commonly used words such as 'fair' are used in a precise mathematical context because different people may apply subtly different prior meanings. Language can be both a conduit and a barrier to knowledge.

Is 'expected gain' a useful criterion for deciding on the fairness of a game? What role should mathematics take in judging the ethics of a situation? Should mathematicians be held responsible for unethical applications of their work?

8A Discrete random variables

Exercise 8A

For questions 1 to 6, use the method demonstrated in Worked Example 8.1 to find the probability distribution of X.

1. **a** A bag contains three blue counters and four yellow counters. Two counters are removed without replacement and X is the number of blue counters removed.
   b Adam has seven blue shirts and eight purple shirts in the wardrobe. He picks two shirts without looking and X is the number of blue shirts selected.
2. **a** A fair coin is tossed twice and X is the number of heads.
   b A fair coin is tossed twice and X is the number of tails.
3. **a** A fair coin is tossed three times, X is the number of tails.
   b A fair coin is tossed three times, X is the number of heads.
4. **a** A fair six-sided dice is rolled once and X is the number shown on the dice.
   b A fair eight-sided dice is rolled once and X is the number shown on the dice.
5. **a** A fair six-sided dice is rolled three times and X is the number of 5s.
   b A fair eight-sided dice is rolled three times and X is the number of 1s.
6. **a** The probability that it rains on any given day is 0.2. X is the number of rainy days in a two-day weekend.
   b The probability that I go to the gym on any given day is 0.7. X is the number of days I go to the gym during a two-day weekend.

For questions 7 to 9, use the method demonstrated in Worked Example 8.2 to find the value of k and then find the required probabilities.

7. **a**

y	0	1	2	>2
$P(Y=y)$	0.32	0.45	0.06	k

   i $P(Y \leq 1)$
   ii $P(Y = 0 | Y \leq 1)$

   b

y	0	1	2	>2
$P(Y=y)$	0.12	0.37	0.25	k

   i $P(Y \leq 1)$
   ii $P(Y = 0 | Y \leq 1)$

8. **a**

x	0	1	2	3	4
$P(X=x)$	0.1	k	0.3	0.3	0.1

   i $P(X > 2)$
   ii $P(X = 3 | X > 2)$

   b

x	0	1	2	3	4
$P(X=x)$	0.06	0.15	k	0.35	0.08

   i $P(X > 2)$
   ii $P(X = 3 | X > 2)$

9. **a**

z	2	3	5	7	11
$P(Z=z)$	0.1	k	$3k$	0.3	0.1

i $P(Z < 7)$

  ii $P(Z \geq 3 \mid Z < 7)$

b

z	1	4	9	16	25
P(Z = z)	0.1	k	0.2	2k	0.1

  i $P(Z \leq 16)$

  ii $P(Z > 4 \mid Z \leq 16)$

For questions 10 to 12, use the method demonstrated in Worked Example 8.3 to find the expected value of the random variable with the given probability distribution.

10 a

y	1	2	3
P(Y = y)	0.4	0.1	0.5

  b

y	1	2	3
P(Y = y)	0.4	0.3	0.3

11 a

x	0	1	2	3	4
P(X = x)	0.1	0.2	0.3	0.3	0.1

  b

x	0	1	2	3	4
P(X = x)	0.06	0.15	0.36	0.35	0.08

12 a

z	2	3	5	7	11
P(Z = z)	0.1	0.2	0.2	0.3	0.1

  b

z	1	4	9	16	25
P(Z = z)	0.1	0.4	0.2	0.2	0.1

13 A discrete random variable X has the probability distribution given in this table.

x	1	2	3	4
P(X = x)	0.2	0.1	0.3	k

  a Find the value of k. b Find $P(X \geq 3)$. c Find $E(X)$.

14 A probability distribution of Y is given in the table.

y	1	3	6	10
P(Y = y)	0.1	0.3	k	2k

  a Find the value of k. b Find $P(Y < 6)$. c Find $E(Y)$.

15 A bag contains eight red sweets and six yellow sweets. Two sweets are selected at random. Let X be the number of yellow sweets.

  a Find the probability distribution of X.
  b Find the expected number of yellow sweets.

16 Two cards are selected at random from a standard pack of 52 cards.
  a Find the probability that both cards are hearts.
  b Find the probability distribution of H, the number of hearts drawn.
  c Find the expected number of hearts.

17 Simon and Olivia play the following game:

Simon gives Olivia £5. Olivia then tosses a fair coin. If the coin comes up heads she gives Simon £7; if it comes up tails she gives him £2.

Determine whether this is a fair game.

18 Maria and Shinji play the following game: Shinji selects a card at random from a standard pack of 52 cards. If the card is a diamond, he gives Maria £3; otherwise Maria gives him £n.

Find the value of n which makes this a fair game.

19 A game stall offers the following game: You toss three fair coins. You receive the number of dollars equal to the number of tails.

How much should the stall charge for one game in order to make the game fair?

20 A fair coin is tossed three times.
  a Draw a tree diagram showing all possible outcomes.
  b Find the probability that exactly two heads are rolled.

Let X be the number of heads.
  c Find the probability distribution of X.
  d Find the expected number of heads.

21 A fair four-sided dice (with faces numbered 1 to 4) is rolled twice and X is the sum of the two scores.
  a Find the probability distribution of X.
  b Find the expected total score.

22 A discrete random variable has the probability distribution given by $P(X=x) = \frac{1}{15}(x+3)$ for $x \in \{1, 2, 3\}$.
  a Write down $P(X=3)$.
  b Find the expected value of X.

23 A discrete random variable Y has the probability distribution given by $P(Y=y) = k(y-1)$ for $y \in \{4, 5, 6\}$.
  a Find the value of k.   b Find $P(Y \geq 5)$.   c Calculate $E(Y)$.

24 A probability distribution is given by $P(X=x) = \frac{c}{x}$ for $x = 1, 2, 3, 4$.
  a Find the value of c.   b Find $P(X=2 \mid X \leq 3)$.   c Calculate $E(X)$.

25 The discrete random variable X has the probability distribution given in the table.

x	1	2	3	4
$P(X=x)$	a	0.2	0.3	b

Given that $E(X) = 2.4$, find the values of a and b.

8B Binomial distribution

There are certain probability distributions that arise as the result of commonly occurring circumstances. One of these 'standard' distributions is the binomial distribution.

> **KEY POINT 8.3**
>
> The binomial distribution occurs when the following conditions are satisfied:
> - There is a fixed number of trials.
> - Each trial has one of two possible outcomes ('success' or 'failure').
> - The trials are independent of each other.
> - The probability of success is the same in each trial.

If a random variable X has the binomial distribution with n trials and a constant probability of success p, we write $X \sim B(n, p)$.

> **WORKED EXAMPLE 8.5**
>
> A drawer contains five black and three red socks. Two socks are pulled out of the drawer without replacement and X is the number of black socks removed.
>
> Does this situation satisfy the standard conditions for X to be modelled by a binomial distribution?
>
> There is a fixed number of trials (2), each trial has two possible outcomes (black or red) ..... No, since the probability of success is not the same in each trial.

> **You are the Researcher**
>
> There are many standard probability distributions that you might like to research, all with interesting properties. In Worked Example 8.5, X is actually modelled by a distribution called the hypergeometric distribution.

 You can calculate probabilities from the binomial distribution with your GDC.

> **WORKED EXAMPLE 8.6**
>
> The number of bullseyes hit by an archer in a competition is modelled by a binomial distribution with 10 trials and a probability 0.3 of success. Find the probability that the archer hits
>
> a exactly three bullseyes
> b at least three bullseyes.
>
> Define the random variable you are going to use. ..... Let X be the number of bullseyes.
>
> X is binomial with $n = 10$ and $p = 0.3$ ..... $X \sim B(10, 0.3)$
>
> Use your GDC, making sure you are using the option which just finds the probability of a single value of X: ..... **a** $P(X = 3) = 0.267$
>
> ```
> Binomial P.D
> Data     :Variable
> x        :3
> Numtrial :10
> p        :0.3
> Save Res :None
> Execute
> CALC
> ```
>
> Use your GDC, making sure you are using the option which just finds the probability of a being less than or equal to a value of X: ..... **b** $P(X \geq 3) = 1 - P(X \leq 2)$
> $= 1 - 0.383$
> $= 0.617$
>
> ```
> Binomial C.D
> Data     :Variable
> x        :2
> Numtrial :10
> p        :0.3
> Save Res :None
> Execute
> CALC
> ```

> **CONCEPTS – MODELLING**
>
> In the example above, it is unlikely that the conditions for the binomial are perfectly met. For example, there might be different wind conditions for the different shots. Missing the previous shot might make the archer lose confidence or perhaps become more focused. However, even though the **model** is not perfect it does not mean it is useless, but the reporting of the results should reflect how certain you are in the assumptions.

Mean and variance of the binomial distribution

Being a standard distribution, you do not have to work out the expected value (mean) of a binomial distribution as you would for a general discrete probability distribution using the method from Section 8A.

Instead, there is a formula you can use (derived from the general method), and a similar formula for the variance as well.

> **KEY POINT 8.4**
>
> If $X \sim B(n, p)$ then:
>
> - $E(X) = np$
> - $Var(X) = np(1 - p)$

> **WORKED EXAMPLE 8.7**
>
> Dima guesses randomly the answers to a multiple-choice quiz. There are 20 questions and each question has five possible answers. If X is the number of questions Dima answers correctly, find
>
> **a** the mean of X
>
> **b** the standard deviation of X.
>
> X is binomial with $n = 20$
> and $p = \frac{1}{5} = 0.2$ .... $X \sim B(20, 0.2)$
>
> Use $E(X) = np$ .... **a** $E(X) = 20 \times 0.2$
> $= 4$
>
> Use $Var(X) = np(1 - p)$ .... **b** $Var(X) = 20 \times 0.2 \times (1 - 0.2)$
> $= 3.2$
>
> Standard deviation is the
> square root of the variance .... So, standard deviation $= 1.79$

Be the Examiner 8.1

Mia tosses a far coin 10 times.

What is the probability she gets more than four but fewer than eight heads?

Which is the correct solution? Identify the errors made in the incorrect solutions.

Solution 1	Solution 2	Solution 3
X is number of heads	X is number of heads	X is number of heads
$X \sim B(10, 0.5)$	$X \sim B(10, 0.5)$	$X \sim B(10, 0.5)$
$P(4 < X < 8) = P(X \leq 7) - P(X \leq 3)$	$P(4 < X < 8) = P(X \leq 8) - P(X \leq 3)$	$P(4 < X < 8) = P(X \leq 7) - P(X \leq 4)$
$= 0.9453 - 0.1719$	$= 0.9893 - 0.1719$	$= 0.9453 - 0.3770$
$= 0.773$	$= 0.817$	$= 0.568$

Exercise 8B

For questions 1 to 5, use the method demonstrated in Worked Example 8.5 to decide whether the random variable X can be modelled by a binomial distribution. If not, state which of the conditions are not met. If yes, identify the distribution in the form $X \sim B(n, p)$.

1. **a** A fair coin is tossed 30 times and X is the number of heads.
   b A fair six-sided dice is rolled 45 times and X is the number of 3s.
2. **a** A fair six-sided dice is rolled until it shows a 6 and X is the number of 4s rolled up until that point.
   b A fair coin is tossed until it shows tails and X is the number of tosses up until that point.
3. **a** Dan has eight white shirts and five blue shirts. He selects four shirts at random and X is the number of white shirts.
   b Amy has seven apples and six oranges. She selects three pieces of fruit at random and X is the number of apples.
4. **a** In a large population, it is known that 12% of people carry a particular gene. 50 people are selected at random and X is the number of people who carry the gene.
   b It is known that 23% of people in a large city have blue eyes. 40 people are selected at random and X is the number of people with blue eyes.
5. **a** In a particular town, the probability that a child takes the bus to school is 0.4. A family has five children and X is the number of children in the family who take the bus to school.
   b The probability that a child goes bowling on a Sunday is 0.15. X is the number of friends, out of a group of 10, who will go bowling next Sunday.

For questions 6 to 12, use the method demonstrated in Worked Example 8.6 to find the required binomial probability.

In questions 6 to 8, $X \sim B(20, 0.6)$

6. **a** $P(X = 11)$ **b** $P(X = 12)$
7. **a** $P(X \leq 12)$ **b** $P(X \leq 14)$
8. **a** $P(X > 11)$ **b** $P(X > 13)$

In questions 9 to 12, $X \sim B\left(16, \frac{1}{3}\right)$

9. **a** $P(X = 4)$ **b** $P(X = 7)$
10. **a** $P(X \geq 5)$ **b** $P(X \geq 8)$
11. **a** $P(3 < X \leq 5)$ **b** $P(2 < X \leq 4)$
12. **a** $P(4 \leq X < 7)$ **b** $P(3 \leq X < 5)$

8B Binomial distribution

For questions 13 to 15, an archer has a probability of 0.7 of hitting the target. If they take 12 shots, find the probability that they hit the target:

13 a exactly six times   b exactly eight times
14 a at least seven times   b at least six times
15 a fewer than eight times   b fewer than four times.

For questions 16 to 18, use the method demonstrated in Worked Example 8.7 to find the mean and standard deviation of X.

16 a A multiple-choice quiz has 25 questions and each question has four possible answers. Elsa guesses the answers randomly and X is the number of questions she answers correctly.
    b A multiple-choice quiz has 30 questions and each question has three possible answers. Asher guesses the answer randomly and X is the number of questions he answers correctly.

17 a A fair coin is tossed 10 times and X is the number of heads.
    b A fair coin is tossed 20 times and X is the number of tails.

18 a A fair dice is rolled 30 times and X is the number of 6s.
    b A fair dice is rolled 20 times and X is the number of 4s.

19 A fair six-sided dice is rolled 10 times. Let X be the number of 6s.
    a Write down the probability distribution of X.
    b Find the probability that exactly two 6s are rolled.
    c Find the probability that more than two 6s are rolled.

20 A footballer takes 12 penalties. The probability that he scores on any shot is 0.85, independently of any other shots.
    a Find the probability that he will score exactly 10 times.
    b Find the probability that he will score at least 10 times.
    c Find the expected number of times he will score.

21 The random variable Y has distribution $B(20, 0.6)$. Find
    a $P(Y = 11)$ b $P(Y > 9)$ c $P(7 \leq Y < 10)$ d the variance of Y.

22 A multiple-choice test consists of 25 questions, each with five possible answers. Daniel guesses answers at random.
    a Find the probability that Daniel gets fewer than 10 correct answers.
    b Find the expected number of correct answers.
    c Find the probability that Daniel gets more than the expected number of correct answers.

23 It is known that 1.2% of people in a country suffer from a cold at any time. A company has 80 employees.
    a What assumptions must be made in order to model the number of employees suffering from a cold by a binomial distribution?
    b Suggest why some of the assumptions might not be satisfied.

    Now assume that the conditions for a binomial distribution are satisfied. Find the probability that on a particular day,
    c exactly three employees suffer from a cold
    d more than three employees suffer from a cold.

24 A fair six-sided dice is rolled 20 times. What is the probability of rolling more than the expected number of sixes?

25 Eggs are sold in boxes of six. The probability that an egg is broken is 0.06, independently of other eggs.
    a Find the probability that a randomly selected box contains at least one broken egg.

    A customer will return a box to the shop if it contains at least one broken egg. If a customer buys 10 boxes of eggs, find the probability that
    b she returns at least one box
    c she returns more than two boxes.

26 An archer has the probability of 0.7 of hitting the target. A round of a competition consists of 10 shots.
    a Find the probability that the archer hits the target at least seven times in one round.
    b A competition consists of five rounds. Find the probability that the archer hits the target at least seven times in at least three rounds of the competition.

27 Daniel has a biased dice with probability p of rolling a 6. He rolls the dice 10 times and records the number of sixes. He repeats this a large number of times and finds that the average number of 6s from 10 rolls is 2.7. Find the probability that in the next set of 10 rolls he gets more than four 6s.

28 A random variable X follows binomial distribution B (n, p). It is known that the mean of X is 36 and the standard deviation of X is 3. Find the probability that X takes its mean value.

29 A machine produces electronic components that are packaged into packs of 10. The probability that a component is defective is 0.003, independently of all other components.

  a Find the probability that at least one of the components in the pack is defective.

The manufacturer uses the following quality control procedure to check large batches of boxes:

A pack of 10 is selected at random from the batch. If the pack contains at least one defective component, then another pack is selected from the same batch. If that pack contains at least one defective component, then the whole batch is rejected; otherwise the whole batch is accepted.

  b Find the probability that a batch is rejected.

  c Suggest a reason why the assumption of independence might not hold.

8C The normal distribution

Properties of the normal distribution

So far, we have been dealing with discrete distributions where all the outcomes and probabilities can be listed. With a continuous variable this is not possible, so instead continuous distributions are often represented graphically, with probabilities being given by the area under specific regions of the curve.

One of the most widely occurring distributions is the normal distribution, which is symmetrical with the most likely outcomes around a central value and increasingly unlikely outcomes farther away from the centre:

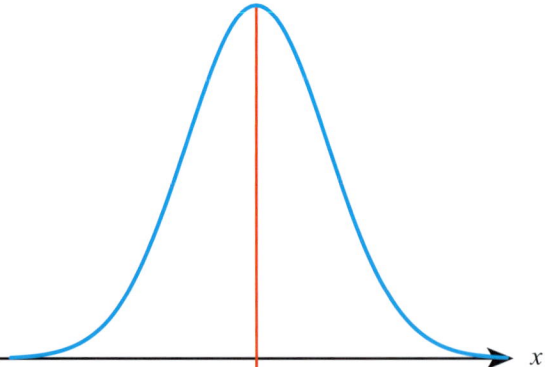

Typical examples of random variables that are modelled by a normal distribution include people's height, mass and scores in an IQ test.

To specify any particular normal distribution, you need to know the central value (its mean, μ) and how spread out the curve is (its standard deviation, σ). We then write $X \sim N(\mu, \sigma^2)$.

8C The normal distribution

> **You are the Researcher**
>
> The normal distribution is not just any curve with a peak in the middle. The basic normal distribution N(0,1) has the formula $\frac{1}{\sqrt{2\pi}}e^{-\frac{x^2}{2}}$. You might want to research how this formula was derived by mathematicians such as De Moivre and Gauss.
>
> There are several other similar shaped curves – for example, the Cauchy distribution has formula $\frac{1}{\pi(1+x^2)}$. You might want to research its uses.

It will always be the case that approximately:

68% of the data lie between $\mu \pm \sigma$

95% of the data lie between $\mu \pm 2\sigma$

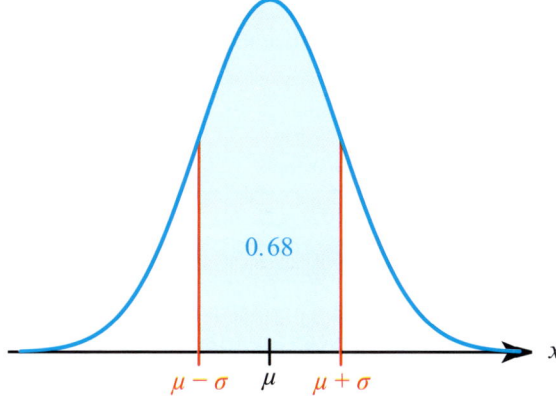
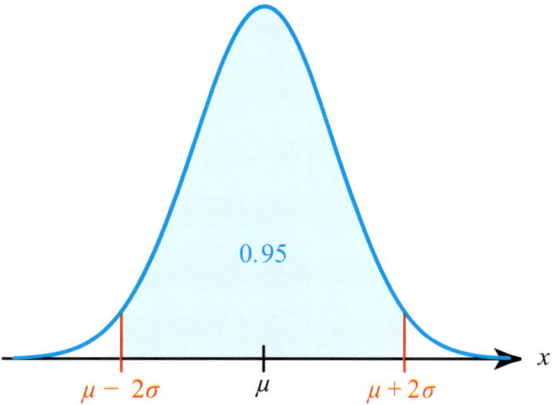

99.7% of the data lie between $\mu \pm 3\sigma$

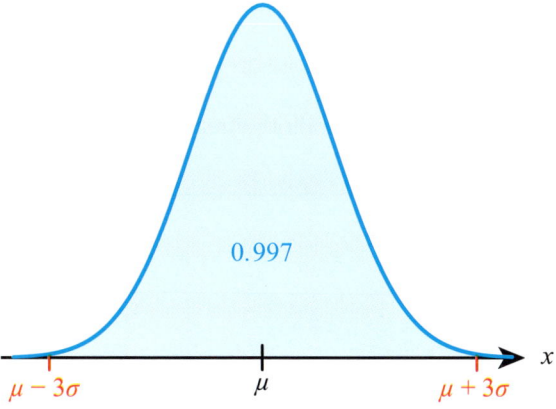

WORKED EXAMPLE 8.8

A sample of 20 students was asked about the number of siblings they had. The results are shown below.

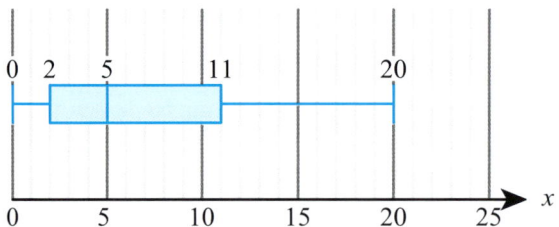

State two reasons why a normal distribution would be a poor model for the number of siblings of students in this population.

The normal distribution is symmetric ····· The sample suggests that the distribution is not symmetric.

The normal distribution is continuous ····· The number of siblings is discrete and not continuous.

CONCEPTS – MODELLING

There are many real-world situations where we assume that a normal distribution is an appropriate **model**. There are various more formal tests we can use to see whether data are likely to be taken from a normal distribution – for example, Q–Q plots or the Shapiro–Wilk test. Having ways to check assumptions is a key part of modelling. However, even if the data are not perfectly normal, there are many situations where we say that they are good enough. For example, if discrete data have enough levels, we often say that they are approximately normal.

Normal probability calculations

As with the binomial distribution, you can calculate probabilities directly from your GDC.

Tip

With a continuous variable we usually only use statements such as $P(X < k)$, rather than $P(X \leq k)$. In fact, both mean exactly the same – this is not the case for discrete variables!

WORKED EXAMPLE 8.9

If $X \sim N(10, 25)$ find $P(X > 20)$.

Use the GDC in Normal C.D. mode to find the probability of X taking a range of values. For the upper value just choose a very large number: ·········· $P(X > 20) = 0.0228$

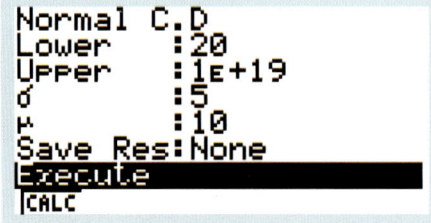

■ Inverse normal calculations

You may be given a particular probability and asked to find the corresponding value of X. Again, this can be done on your GDC, this time using the Inverse Normal distribution.

> **WORKED EXAMPLE 8.10**
>
> The waist circumference of adult males is believed to follow a normal distribution with mean 92 cm and standard deviation 4.5 cm. A trouser manufacturer wants to make sure that their smallest size does not exclude more than 5% of the population.
>
> Find the smallest size waist circumference their trousers should fit.
>
> Use the Inverse Normal function on your GDC. $X \sim N(92, 4.5^2)$
> The tail is to the left as you want $P(X < x)$
>
> $P(X < x) = 0.05$
>
> $x = 84.6$

Exercise 8C

For questions 1 to 8, use the method demonstrated in Worked Example 8.9 to find the required probability.

In questions 1 and 2, $X \sim N(10, 25)$.

1 a $P(X > 15)$ b $P(X > 12)$
2 a $P(X < 13)$ b $P(X < 7)$

In questions 3 and 4, $Y \sim N(25, 12)$.

3 a $P(18 < Y < 23)$ b $P(23 < Y < 28)$
4 a $P(Y > 24.8)$ b $P(Y > 27.4)$

In questions 5 and 6, T has a normal distribution with mean 12.5 and standard deviation 4.6.

5 a $P(T < 11.2)$ b $P(T < 13.1)$
6 a $P(11.6 < T < 14.5)$ b $P(8.7 < T < 12.0)$

In questions 7 and 8, M has a normal distribution with mean 62.4 and variance 8.7.

7 a $P(M > 65)$ b $P(M > 60)$
8 a $P(61.5 < M < 62.5)$ b $P(62.5 < M < 63.5)$

For questions 9 to 12, use the method demonstrated in Worked Example 8.10 to find the value of x.

In questions 9 and 10, $X \sim N(10, 25)$.

9 a $P(X < x) = 0.75$ b $P(X < x) = 0.26$
10 a $P(X > x) = 0.65$ b $P(X > x) = 0.12$

In questions 11 and 12, $X \sim N(38, 18)$.

11 a $P(X < x) = 0.05$ b $P(X < x) = 0.97$
12 a $P(X > x) = 0.99$ b $P(X > x) = 0.02$

13 The box plot on the right summarizes the ages of pupils at a school. Does the box plot suggest that the ages are normally-distributed? Give one reason to explain your answer.

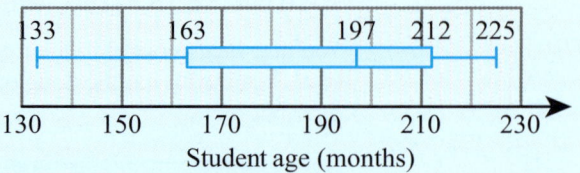

14 The heights of trees in a forest can be modelled by a normal distribution with mean 8.7 m and standard deviation 2.3 m. Elsa wants to find the probability that a tree is between 5 and 10 metres tall.

   a Sketch a normal distribution curve and shade the area representing the required probability.
   b Find the probability that a randomly chosen tree is between 5 and 10 metres tall.

15 The battery life of a particular model of a phone can be modelled by a normal distribution with mean 56 hours and standard deviation 8 hours. Find the probability that a randomly selected phone has battery life

   a between 50 and 60 hours
   b more than 72 hours.

16 The heights of trees in a forest are modelled by a normal distribution with mean 17.2 m and standard deviation 6.3 m. Find the probability that a randomly selected tree

   a has height between 15 m and 20 m
   b is taller than 20 m.

17 Charlotte's times for running a 400 m race are normally distributed with mean 62.3 seconds and standard deviation 4.5 seconds.

   a What is the probability that Charlotte's time in a randomly selected race is over 65 seconds?
   b Charlotte ran 38 races this season. What is the expected number of races in which she ran over 65 seconds?
   c In order to qualify for the national championships, Charlotte needs to run under 59.7 seconds. What is the probability that she will qualify with a single race?

18 The times taken by a group of children to complete a puzzle can be modelled by a normal distribution with mean 4.5 minutes and variance 2.25 minutes^2.

   a What is the probability that a randomly selected child takes more than 7 minutes to complete the puzzle?
   b Find the length of time such that 90% of the children take less than this time to complete the puzzle.

19 The daily amount of screen time among teenagers can be modelled by a normal distribution with mean 4.2 hours and standard deviation 1.3 hours.

   a What percentage of teenagers get more than 6 hours of screen time per day?
   b Let T be the time such that 5% of teenagers get more screen time than this. Find the value of T.
   c In a group of 350 teenagers, how many would you expect to get less than 3 hours of screen time per day?

20 In a long jump competition, the distances achieved by a particular age group can be modelled by a normal distribution with mean 5.2 m and variance 0.6 m^2.

   a In a group of 30 competitors from this age group, how many would you expect to jump further than 6 m?
   b In a large competition, the top 5% of participants will qualify for the next round. Estimate the qualifying distance.

21 Among 17-year-olds, the times taken to run 100 m are normally distributed with mean 14.3 s and variance 2.2 s^2. In a large competition, the top 15% of participants will qualify for the next round. Estimate the required qualifying time.

22 A random variable follows a normal distribution with mean 12 and variance 25. Find

   a the upper quartile
   b the interquartile range of the distribution.

23 A random variable follows a normal distribution with mean 3.6 and variance 1.44. Find the 80th percentile of the distribution.

Checklist

24 Given that $X \sim N(17, 3.2^2)$, find the value of k such that $P(15 < X < k) = 0.62$.

25 Given that $Y \sim N(13.2, 5.1^2)$, find the value of c such that $P(c < Y < 17.3) = 0.14$.

26 Percentage scores on a test, taken by a large number of people, are found to have a mean of 35 with a standard deviation of 20. Explain why this suggests that the scores cannot be modelled by a normal distribution.

27 The box plot shows the distribution of masses of a large number of children in a primary school.

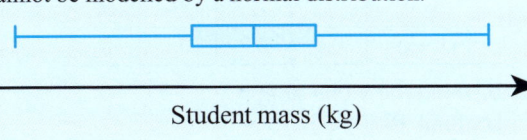

Student mass (kg)

  a Identify one feature of the box plot which suggests that a normal distribution might be a good model for the masses.

  b The mean mass of the children is 36 kg and the standard deviation is 8.5 kg. Copy and complete the box plot by adding the relevant numbers. You may assume that the end points are the minimum and maximum values which are not outliers.

28 The reaction time of a sprinter follows a normal distribution with mean 0.2 seconds and standard deviation 0.1 seconds.

  a Find the probability of getting a negative reaction time. Explain in this context why this might be a plausible outcome.

  A false start is declared if the reaction time is less than 0.1 seconds.

  b Find the probability of the sprinter getting a false start.

  c In 10 races, find the probability that the sprinter gets a false start more than once.

  d What assumptions did you have to make in part **c**? How realistic are these assumptions?

29 A farmer has chickens that produce eggs with masses that are normally distributed with mean 60 g and standard deviation 5 g. Eggs with a mass less than 53 g cannot be sold. Eggs with a mass between 53 g and 63 g are sold for 12 cents and eggs with a mass above 63 g are sold for 16 cents. If the farmer's hens produce 6000 eggs each week, what is the farmer's expected income from the eggs.

Checklist

- You should be able to work with discrete random variables and their probability distributions. In any discrete probability distribution:
  - $\sum_x P(X = x)$

- You should be able to find the expected value of a discrete random variable and apply this to real-life contexts. For a discrete random variable, X:
  - $E(X) = \sum_x x \, P(X = x)$

- You should know the conditions that need to be satisfied for the binomial distribution to be appropriate:
  - There is a fixed number of trials.
  - Each trial has one of two possible outcomes ('success' or 'failure').
  - The trials are independent of each other.
  - The probability of success is the same in each trial.

- You should be able to find the mean and variance of the binomial distribution. If $X \sim B(n, p)$ then:
  - $E(X) = np$
  - $Var(X) = np(1-p)$

- You should know the key properties of the normal distribution.

- You should be able to carry out normal probability calculations and inverse normal calculations on your GDC.

■ Mixed Practice

1 Random variable X has the probability distribution given in the table.

x	1	2	3	4
$P(X=x)$	0.2	0.2	0.1	k

    a Find the value of k.
    b Find $P(X \geq 3)$.
    c Find $E(X)$.

2 A fair six-sided dice is rolled twelve times. Find the probability of getting
    a exactly two 16s
    b more than two 1s.

3 Lengths of films are distributed normally with mean 96 minutes and standard deviation 12 minutes. Find the probability that a randomly selected film is
    a between 100 and 120 minutes long
    b more than 105 minutes long.

4 Scores on a test are normally distributed with mean 150 and standard deviation 30. What score is needed to be in the top 1.5% of the population?

5 A factory making plates knows that, on average, 2.1% of its plates are defective. Find the probability that in a random sample of 20 plates, at least one is defective.

6 Daniel and Alessia play the following game. They roll a fair six-sided dice. If the dice shows an even number, Daniel gives Alessia $1. If it shows a 1, Alessia gives Daniel $1.50; otherwise Alessia gives Daniel $0.50.
    a Complete the table showing possible outcomes of the game for Alessia.

Outcome	$1	−$0.50	−$1.50
Probability			

    b Determine whether the game is fair.

7 A spinner with four sectors, labelled 1, 3, 6 and N (where $N > 6$) is used in a game. The probabilities of each number are shown in the table.

Number	1	3	6	N
Probability	$\frac{1}{2}$	$\frac{1}{5}$	$\frac{1}{5}$	$\frac{1}{10}$

A player pays three counters to play a game, spins the spinner once and receives the number of counters equal to the number shown on the spinner. Find the value of N so that the game is fair.

Mixed Practice

8 A random variable X is distributed normally with a mean of 20 and variance 9.
   a Find $P(X \leq 24.5)$.
   b Let $P(X \leq k) = 0.85$.
      i Represent this information on a copy of the following diagram.

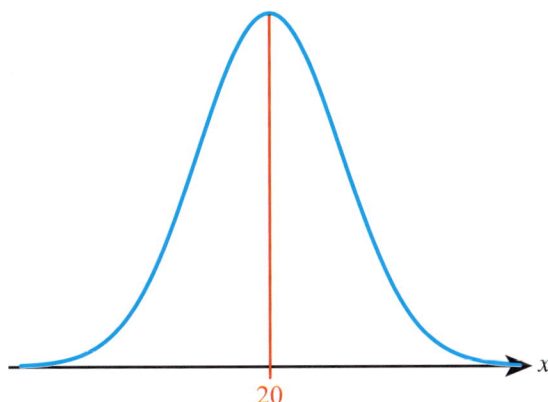

   ii Find the value of k.

   Mathematics SL May 2011 Paper 2 TZ1 Q4

9 The random variable X has the distribution $B(30, p)$. Given that $E(X) = 10$, find
   a the value of p;
   b $P(X = 10)$;
   c $P(X \geq 15)$.

   Mathematics HL May 2012 Paper 2 TZ2 Q2

10 The mean mass of apples is 110 g with standard deviation 12.2 g.
   a What is the probability that an apple has a mass of less than 100 g?
   b Apples are packed in bags of six. Find the probability that more than one apple in a bag has a mass of less than 100 g.

11 Masses of eggs are normally distributed with mean 63 g and standard deviation 6.8 g. Eggs with a mass over 73 g are classified as 'very large'.
   a Find the probability that a randomly selected box of six eggs contains at least one very large egg.
   b What is more likely: a box of six eggs contains exactly one very large egg, or that a box of 12 eggs contains exactly two very large eggs?

12 Bob competes in the long jump. The lengths of his jumps are normally distributed with mean 7.35 m and variance 0.64 m^2.
   a Find the probability that Bob jumps over 7.65 m.
   b In a particular competition, a jump of 7.65 m is required to qualify for the final. Bob makes three jumps. Find the probability that he will qualify for the final.

13 In an athletics club, the 100 m times of all the runners follow the same normal distribution with mean 15.2 s and standard deviation 1.6 s. Eight of the runners have a practice race. Heidi runs the time of 13.8 s. What is the probability that she wins the race?

14 The amount of paracetamol in a tablet is distributed normally, with mean 500 mg and variance 6400 mg^2. The minimum dose required for relieving a headache is 380 g. Find the probability that, in a trial of 25 randomly selected participants, more than two receive less than the required dose.

15 A dice is biased and the probability distribution of the score is given in the following table.

Score	1	2	3	4	5	6
Probability	0.15	0.25	0.08	0.17	0.15	0.20

    a Find the expected score for this dice.
    b The dice is rolled twice. Find the probability of getting one 5 and one 6.
    c The dice is rolled 10 times. Find the probability of getting at least two 1s.

16 A biased four-sided dice is used in a game. The probabilities of different outomes on the dice are given in this table.

Value	1	2	3	4
Probability	$\frac{1}{3}$	$\frac{1}{4}$	a	b

A player pays three counters to play the game, rolls the dice once and receives the number of counters equal to the number shown on the dice. Given that the player expects to lose one counter when he plays the game two times, find the value of a.

17 The heights of pupils at a school can be modelled by a normal distribution with mean 148 cm and standard deviation 8 cm.
    a Find the interquartile range of the distribution.
    b What percentage of the pupils at the school should be considered outliers?

18 The mean time taken to complete a test is 10 minutes and the standard deviation is 5 minutes. Explain why a normal distribution would be an inappropriate model for the time taken.

19 Quality control requires that no more than 2.5% of bottles of water contain less than the labelled volume. If a manufacturer produces bottles containing a volume of water following a normal distribution with mean value of 330 ml and standard deviation 5 ml, what should the labelled volume be, given to the nearest whole number of millilitres?

20 Jan plays a game where she tosses two fair six sided dice. She wins a prize if the sum of her scores is 5.
    a Jan tosses the two dice once. Find the probability that she wins a prize.
    b Jan tosses the two dice eight times. Find the probability that she wins three prizes.

Mathematics SL May 2010 Paper 2 TZ2 Q3

21 The time taken for a student to complete a task is normally distributed with a mean of 20 minutes and a standard deviation of 1.25 minutes.
    a A student is selected at random. Find the probability that the student completes the task in less than 21.8 minutes.
    b The probability that a student takes between k and 21.8 minutes is 0.3. Find the value of k.

Mathematics SL November 2013 Paper 2 Q6

22 The weight, W, of bags of rice follows a normal distribution with mean 1000 g and standard deviation 4 g.
    a Find the probability that a bag of rice chosen at random weighs between 990 g and 1004 g.

    95% of the bags of rice weigh less than k grams.
    b Find the value of k.

    For a bag of rice chosen at random, $P(1000 - a < W < 1000 + a) = 0.9$.
    c Find the value of a.

Mathematical Studies May 2015 Paper 1 TZ1 Q13

23 A test has five questions. To pass the test, at least three of the questions must be answered correctly. The probability that Mark answers a question correctly is $\frac{1}{5}$. Let X be the number of questions that Mark answers correctly.

   a i Find $E(X)$.

   ii Find the probability that Mark passes the test.

   Bill also takes the test. Let Y be the number of questions that Bill answers correctly. The following table is the probability distribution for Y.

y	0	1	2	3	4	5
$P(Y=y)$	0.67	0.05	$a+2b$	$a-b$	$2a+b$	0.04

   b i Show that $4a + 2b = 0.24$.

   ii Given that $E(Y) = 1$, find a and b.

   c Find which student is more likely to pass the test.

   Mathematics SL November 2010 Paper 2 Q9

24 The distances thrown by Josie in an athletics competition is modelled by a normal distribution with mean 40 m and standard deviation 5 m. Any distance less than 40 m gets 0 points. Any distance between 40 m and 46 m gets 1 point. Any distance above 46 m gets 4 points.

   a Find the expected number of points Josie gets if she throws

   i once

   ii twice.

   b What assumptions have you made in **a ii**? Comment on how realistic these assumptions are.

25 When a fair six-sided dice is rolled n times, the probability of getting no 6s is 0.194, correct to three significant figures. Find the value of n.

26 Find the smallest number of times that a fair coin must be tossed so that the probability of getting no heads is smaller than 0.001.

27 The probability of obtaining 'tails' when a biased coin is tossed is 0.57. The coin is tossed 10 times. Find the probability of obtaining

   a at least four tails

   b the fourth tail on the 10th toss.

   Mathematics SL May 2012 Paper 2 TZ1 Q7

28 A group of 100 people are asked about their birthdays. Find the expected number of dates on which no people have a birthday. You may assume that there are 365 days in a year and that people's birthdays are equally likely to be on any given date.

29 A private dining chef sends out invitations to an exclusive dinner club. From experience he knows that only 50% of those invited turn up. He can only accommodate four guests. On the first four guests he makes $50 profit per guest; however, if more than four guests turn up he has to turn the additional guests away, giving them a voucher allowing them to have their next dinner for free, costing him $100 per voucher.

   a Assuming that responses are independent, show that his expected profit if he invites five people is $120 to 3 significant figures.

   b How many invitations should he send out?

9 Core: Differentiation

ESSENTIAL UNDERSTANDINGS

- Differentiation describes the rate of change between two variables. Understanding these rates of change allows us to model, interpret and analyse real-world problems and situations.
- Differentiation helps us understand the behaviour of functions and allows us to interpret features of their graphs.

In this chapter you will learn…

- how to estimate the value of a limit from a table or graph
- informally about the gradient of a curve as a limit
- about different notation for the derivative
- how to identify intervals on which functions are increasing or decreasing
- how to interpret graphically $f'(x) > 0$, $f'(x) = 0$, $f'(x) < 0$
- how to differentiate functions of the form $f(x) = ax^n + bx^m$, where $n, m \in \mathbb{Z}$
- how to find the equations of the tangent and normal to a function at a given point.

CONCEPTS

The following key concepts will be addressed in this chapter:
- The derivative may be represented physically as a rate of **change** and geometrically as the gradient or slope function.
- Differentiation allows you to find a **relationship** between a function and its gradient.
- This relationship can have a graphical or algebraic **representation**.

LEARNER PROFILE – Knowledgeable

What types of mathematical knowledge are valued by society? There are few jobs where you will be using differentiation daily, so knowing lots of facts is not really the most important outcome of a mathematics education. However, there are lots of skills fundamental to mathematics which are desirable in everyday life: the ability to think logically, communicate precisely and pay attention to small details.

■ Figure 9.1 How can we model the changes represented in these photos?

Core: Differentiation

> **PRIOR KNOWLEDGE**
>
> Before starting this chapter, you should already be able to complete the following:
> 1. Write $\dfrac{x-2}{4x^2}$ in the form $ax^n + bx^m$.
> 2. Find the equation of the straight line through the point $(-2, 5)$ with gradient 3.
> 3. Find the gradient of the straight line perpendicular to $x - 2y + 4 = 0$.

Differentiation is the process of establishing the rate of change of the y-coordinate of a graph when the x-coordinate is changed. You are used to doing this for a straight line, by finding the gradient. In that case the rate of change is the same no matter where you are on the graph, but for non-linear graphs this is not the case – the rate of change is different at different values of x.

Differentiation has wide-ranging applications, from calculating velocity and acceleration in physics, to the rate of a reaction in chemistry, to determining the optimal price of a quantity in economics.

Starter Activity

Look at the pictures in Figure 9.1. In small groups, describe what examples of change you can see. Can you think of any more examples?

Now look at this problem:

In a drag race, the displacement, d metres, of a car along a straight road is modelled by $d = 4t^2$, where t is the time in seconds.

Find the average speed of the car:

a in the first 5 seconds

b between 1 and 5 seconds

c between 4 and 5 seconds.

What do you think the reading on the car's speedometer is at 5 seconds?

9A Limits and derivatives

Introduction to the concept of a limit

Suppose we are not sure what the value of f(*a*) is. If f(*x*) gets closer and closer to a particular value when *x* gets closer and closer to *a*, then it is said to have a limit. For example, in the Starter Activity problem you might have found that the speed of the car gets closer and closer to $40\,\mathrm{m\,s^{-1}}$.

You might reasonably ask why we do not just calculate f(*a*) directly. The main situation when we cannot do this is if, when we try to find f(*a*), we have to do a division by zero. If only the denominator is zero then the answer is undefined; however, if both the numerator and denominator are zero then it is possible that it has a limit.

TOK Links

Some mathematicians have tried to create a new number system where division by zero is defined – often as infinity. These systems have usually been rejected within mathematical communities. Who should decide whether these new number systems are acceptable, and what criteria should they use?

WORKED EXAMPLE 9.1

Use technology to suggest the limit of $\dfrac{\sin x}{\left(\dfrac{\pi x}{180}\right)}$ as *x* tends to zero, where *x* is in degrees.

We can use a spreadsheet to look at what happens when *x* gets very small. We could also use a graphical calculator to sketch the function ....

x	$\dfrac{\sin x}{\left(\dfrac{\pi x}{180}\right)}$
10	0.994 931
5	0.998 731
1	0.999 949
0.1	0.999 999
0.01	1

We can then interpret the spreadsheet. Notice that even though the spreadsheet has an output of 1, it does not mean that the value of the function here is exactly 1. It just means that it is 1 within the degree of accuracy the spreadsheet is using .... As *x* gets closer to 0, the function gets closer to 1, so this is the limit.

Links to: Physics

The function $\dfrac{\sin x}{x}$ turns out to be very important in optics, where it is used to give the amplitude of the light hitting a screen in a double slit experiment. This investigation of limits can be used to prove that the central line is brightest.

■ The derivative interpreted as a gradient function and as a rate of change

One of the main uses of limits in calculus is in finding the **gradient of a curve.**

KEY POINT 9.1

The gradient of a curve at a point is the gradient of the **tangent to the curve** at that point.

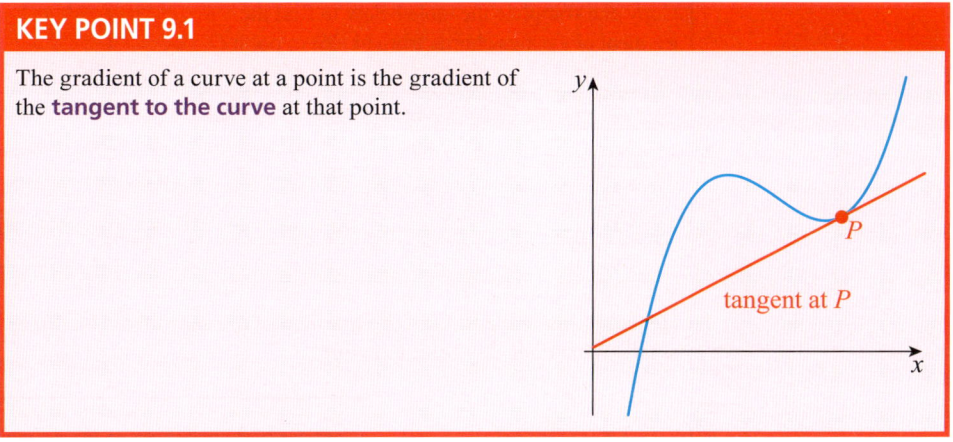

We could try to draw tangents on a graph and measure their gradients, but if you try this you will see that it is quite hard to do accurately. There is one important insight that helps us to calculate the gradient of the tangent.

KEY POINT 9.2

The gradient of a tangent at a point is the limit of the gradient of smaller and smaller chords from that point.

In Worked Example 9.2 the gradient of the chord was written as $\frac{\Delta y}{\Delta x}$ where 'Δ' is the Greek equivalent of 'd'. The German philosopher and mathematician Gottfried Leibniz came up with the notation of replacing 'Δ' for small changes with 'd' for infinitesimally small changes, that is, the limit as the change tends towards zero.

■ **Figure 9.2** Gottfried Leibniz

WORKED EXAMPLE 9.2

On the graph $y = x^2$ a chord is drawn from $P(4, 16)$ to the point $Q(x, y)$.

Copy and complete the table below to find the gradient of various chords from $(4, 16)$ to point Q.

x	y	Δx	Δy	Gradient of PQ
5	25	1	9	9
4.5				
4.1				
4.01				

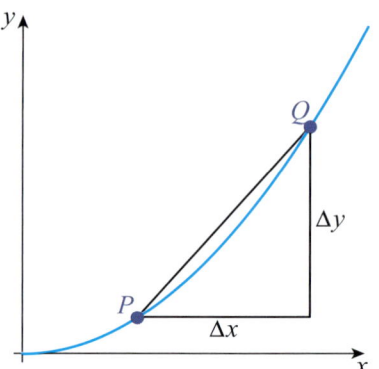

Hence, estimate the limit of the gradient of the chord from $(4, 16)$ as the chord gets very small. What does this limit tell you?

Fill in the table using:

$y = x^2$

$\Delta x = x - 4$

$\Delta y = y - 16$

Gradient $= \dfrac{\Delta y}{\Delta x}$

x	y	Δx	Δy	Gradient of PQ
5	25	1	9	9
4.5	20.25	0.5	4.25	8.5
4.1	16.81	0.1	0.81	8.1
4.01	16.0801	0.01	0.0801	8.01

Look at what the gradient is tending to

The gradient of the chord is tending to 8.
This suggests that the gradient of the curve at $(4, 16)$ is 8.

TOK Links

This approach of gathering evidence and suggesting a result is really a method of scientific induction rather than mathematical proof. There is a formalization of this method called 'differentiation from first principles' which does prove these results; however, some people find scientific induction gives them a better understanding of what is going on. Is proof always the best way to explain something? Does explanation mean different things in maths and science?

Links to: Physics

The notation $f'(x)$ is called Lagrange's notation. There is another notation popular in physics called Newton's notation, where $\dot{x}$ is used to represent $\dfrac{dx}{dt}$.

The process of finding the gradient of a curve is called differentiation. Each point of the curve can have a different value for the gradient. The function which gives the gradient at any point on a curve is called the **derivative, slope function** or **gradient function**. There are several different notations for the derivative. If the curve is expressed as y in terms of x, we would write the derivative as $\dfrac{dy}{dx}$. However, there is nothing special about y and x here. If there was a graph of variable G against variable t then the derivative would be $\dfrac{dG}{dt}$.

If the function being differentiated is $f(x)$ then the derivative is $f'(x)$. Much of the rest of this chapter will look at how we can calculate the derivative of a function, but it is also important to think about why we want to do this. As well as representing gradients of graphs, derivatives have another interpretation. If the graph is y against x then the gradient tells you the instantaneous amount y is increasing for each unit that x increases. This is called the rate of change of y with respect to x. In many real-life situations, we have ideas about the rates of change of variables, which allows us to create models called differential equations.

TOOLKIT: Modelling

There are many textbooks about solving differential equations, but increasingly technology can be used to solve them. One of the main skills modern mathematicians need is creating differential equations which capture all the important features of a real-world scenario.

WORKED EXAMPLE 9.3

The rate of growth of a bacterial population of size N in a Petri dish over time t is proportional to the population size. Write an equation that describes this information.

Interpret the rate of growth as a derivative. We are looking at N changing with respect to t	$\dfrac{dN}{dt} \propto N$
With direct proportion we can turn it into an equation with an unknown constant factor	$\dfrac{dN}{dt} = kN$

CONCEPTS – CHANGE

It is easy to think that 'rate of change' means **change** over time, but it could be the rate at which the height of a ball changes as the distance travelled changes, or the rate at which side-effects increase as the dose of a medicine increases.

Ancient Greek mathematicians came very close to 'discovering' calculus. Not realizing that the issue of zero divided by zero could be resolved by considering limits was one of the main hurdles they failed to overcome. They were aware of many difficulties concerned with dealing correctly with limits, the most famous of which are called Zeno's paradoxes.

Exercise 9A

In this exercise all angles are in degrees.

For questions 1 to 5, use the method demonstrated in Worked Example 9.1 to suggest the limit of the following functions as x tends to zero.

1. a $\dfrac{3x}{5x}$
   b $\dfrac{7x^2}{2x}$

2. a $\dfrac{3x + 5x^2}{2x}$
   b $\dfrac{x^2 + 3x}{2x^2 + x}$

3. a $\dfrac{\sin 2x}{\left(\dfrac{\pi x}{180}\right)}$
   b $\dfrac{\sin 3x}{\left(\dfrac{\pi x}{180}\right)}$

4. a $\dfrac{\tan x}{\left(\dfrac{\pi x}{180}\right)}$
   b $\dfrac{\tan 2x}{\left(\dfrac{\pi x}{90}\right)}$

5. a $\dfrac{2^x - 1}{x}$
   b $\dfrac{3^x - 1}{x}$

For questions 6 to 10, use a graphical calculator to sketch the graph of the function and hence suggest the limit as x tends to zero.

6. a $\dfrac{x}{5x}$
   b $\dfrac{7x^2}{10x}$

7. a $\dfrac{x + x^3}{2x}$
   b $\dfrac{x^2 + x}{x^2 + 2x}$

8. a $\dfrac{\sin 5x}{\left(\dfrac{\pi x}{36}\right)}$
   b $\dfrac{\sin 10x}{\left(\dfrac{\pi x}{180}\right)}$

9. a $\dfrac{\cos x - 1}{\left(\dfrac{\pi x}{180}\right)^2}$
   b $\dfrac{\cos x - 1}{\left(\dfrac{\pi x}{180}\right)}$

10. a $\dfrac{5^x - 1}{2x}$
    b $\dfrac{4^x - 1}{4x}$

For questions 11 to 15, use the method demonstrated in Worked Example 9.2 to suggest a value for the gradient of the tangent to the given curve at point P.

11. a The curve is $y = x^2$ and P is $(0, 0)$.

x	y	Δx	Δy	Gradient of PQ
1	1	1	1	1
0.5				
0.1				
0.01				

b The curve is $y = x^2$ and P is $(2, 4)$.

x	y	Δx	Δy	Gradient of PQ
3	9	1	5	5
2.5				
2.1				
2.01				

12. a The curve is $y = 2x^3$ and P is $(1, 2)$.

x	y	Δx	Δy	Gradient of PQ
2	16	1	14	14
1.5				
1.1				
1.01				

b The curve is $y = 3x^4$ and P is $(0, 0)$.

x	y	Δx	Δy	Gradient of PQ
1	3	1	3	3
0.5				
0.1				
0.01				

13. a The curve is $y = \sqrt{x}$ and P is $(1, 1)$.

x	y	Δx	Δy	Gradient of PQ
2	1.414	1	0.414	0.414
1.5				
1.1				
1.01				

b The curve is $y = \sqrt[3]{x}$ and P is $(8, 2)$.

x	y	Δx	Δy	Gradient of PQ
9	2.0801	1	0.0801	0.0801
8.5				
8.1				
8.01				

9A Limits and derivatives

14 a The curve is $y = \dfrac{1}{x}$ and P is $(1, 1)$.

x	y	Δx	Δy	Gradient of PQ
2	0.5	1	−0.5	−0.5
1.5				
1.1				
1.01				

b The curve is $y = \dfrac{1}{x^2}$ and P is $(1, 1)$.

x	y	Δx	Δy	Gradient of PQ
2	0.25	1	−0.75	−3
1.5				
1.1				
1.01				

15 a The curve is $y = \ln x$ and P is $(1, 0)$.

x	y	Δx	Δy	Gradient of PQ
2	0.693	1	0.693	0.693
1.5				
1.1				
1.01				

b The curve is $y = 2^x$ and P is $(0, 1)$.

x	y	Δx	Δy	Gradient of PQ
1	2	1	1	1
0.5				
0.1				
0.01				

For questions 16 to 20, write the given expressions as a derivative.

16 a The rate of change of z as v changes.
   b The rate of change of a as b changes.
17 a The rate of change of p with respect to t.
   b The rate of change of b with respect to x.
18 a How fast y changes when n is changed.
   b How fast t changes when f is changed.
19 a How quickly the height of an aeroplane (h) changes over time (t).
   b How quickly the weight of water (w) changes as the volume (v) changes.
20 a The rate of increase in the average wage (w) as the unemployment rate (u) increases.
   b The rate of increase in the rate of reaction (R) as the temperature (T) increases.

For questions 21 to 26, use the method demonstrated in Worked Example 9.3 to write the given information as an equation.

21 a The gradient of the graph of y against x at any point is equal to the y-coordinate.
   b The gradient of the graph of y against x at any point is equal to half of the x-coordinate.
22 a The gradient of the graph of s against t is proportional to the square of the t-coordinate.
   b The gradient of the graph of q against p is proportional to the square root of the q-coordinate.
23 a The growth rate of a plant of size P over time t is proportional to the amount of light it receives, $L(P)$.
   b The rate at which companies' profits (P) increase over time (t) is proportional to the amount spent on advertising, $A(P)$.
24 a The rate of increase of the height (h) of a road with respect to the distance along the road (x) equals the reciprocal of the height.
   b The rate at which distance away from the origin (r) of a curve (r) increases with respect to the angle (θ) with the positive x-axis is a constant.
25 a The rate at which $s(t)$ is increasing as t increases is 7.
   b The rate at which $q(x)$ increases with respect to x equals $7x$.
26 a The rate of change of acidity (pH) of a solution as the temperature changes (T) is a constant.
   b In economics, the marginal cost is the amount the cost of producing each item (C) increases by as the number of items (n) increases. In an economic model the marginal cost is proportional to the number of items produced below 1000.

27 The rate of increase of voltage (V) with respect to time (t) through an electrical component is equal to $V + 1$.
   a Express this information as an equation involving a derivative.
   b Find the gradient of the graph of V against t when $V = 2$.

28 Use technology to suggest the value of the limit of $\dfrac{\sin x^2}{\left(\dfrac{\pi x}{180}\right)^2}$ as x tends to zero.

29 Suggest the value of the limit of $\dfrac{\ln x}{x - 1}$ as x tends to 1.

30 a Find the gradient of the chord connecting $(1, 1)$ to (x, x^2).
   b Find the limit of the gradient of this chord as x tends towards 1. What is the significance of this value?

31 Use technology to suggest a value for the limit of $\sqrt{x^2 + 6x} - x$ as x gets very large.

32 The rate of change of x with respect to y is given by $3x$. Find the gradient on the graph of y against x when $x = 2$.

You are the Researcher

There are many models in natural sciences and social sciences which are described in terms of derivatives. Examples you could research include models such as the SIR model for epidemics, the von Bertalanffy model for animal growth or the Black–Scholes model for financial instruments. However, none of these models are perfect and current research involves looking at their assumptions and how they can be improved.

9B Graphical interpretation of derivatives

A good way to get an understanding of derivatives before getting too involved in algebra, is to focus on how they relate to the graph of the original function.

Increasing and decreasing functions

TOOLKIT: Modelling

You might see models in the news reporting things like: 'The government predicts that the national debt will rise to a value of 1.3 trillion then start falling'. Five years later, it turns out that the debt actually rose to 1.5 trillion and then started falling. Does this mean the model was wrong? Mathematical modellers know that there are quantitative results from a model – that is, numerical predictions – which are very uncertain (but they are often the things which make the headlines!). Often the more useful thing to take away from a model is a more qualitative result, such as 'the debt will rise then fall'. Try to find a model in the news and look for qualitative and quantitative results associated with it – which do you think is more useful?

When we are sketching a curve we often do not care whether the gradient at a point is 5 or 6 – all that matters is that it we get the direction roughly right. There is some terminology which we use to describe this.

> **KEY POINT 9.3**
>
> If when x increases f(x) increases (so that f'(x) > 0), the function is *increasing*.
>
> If when x increases f(x) decreases (so that f'(x) < 0), the function is *decreasing*.

> **TOK Links**
>
> Do you think that this terminology is obvious? Perhaps you are from a country where text is read from left to right. If you are used to the reading the other way, would that change the natural definition? If you look at a cross section of a mountain, is it obvious which way is uphill and which way is downhill? Some of the hardest assumptions to check are the ones you do not know you are making.

A function is not necessarily always increasing or decreasing. We might have to identify intervals in which it is increasing or decreasing.

> **WORKED EXAMPLE 9.4**
>
> For the graph on the right, use inequalities in x to describe the regions in which the function is increasing.
>
>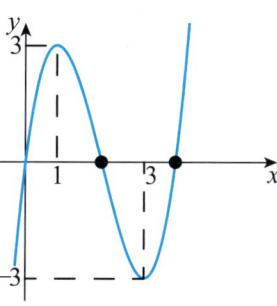
>
> The function starts off increasing until it gets to the point where the gradient is zero (at $x - 1$). The gradient ········· $x<1$ or $x>3$
> is then negative until it hits zero again at $x - 3$

We can use ideas of increasing and decreasing functions to sketch the derivative function of a given graph.

> **WORKED EXAMPLE 9.5**
>
> Sketch the derivative of this function.
>
>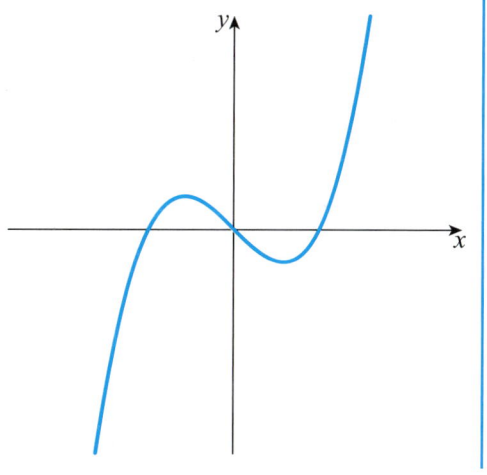

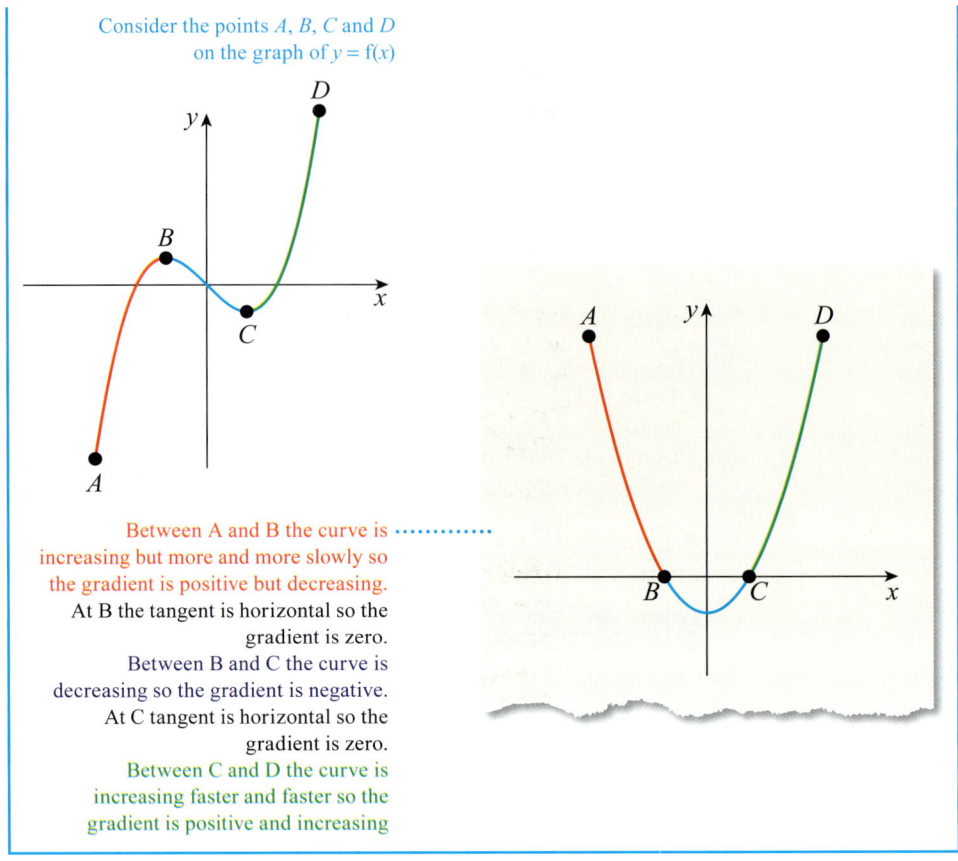

Consider the points A, B, C and D on the graph of $y = f(x)$

Between A and B the curve is increasing but more and more slowly so the gradient is positive but decreasing.
At B the tangent is horizontal so the gradient is zero.
Between B and C the curve is decreasing so the gradient is negative.
At C tangent is horizontal so the gradient is zero.
Between C and D the curve is increasing faster and faster so the gradient is positive and increasing

■ Graphical interpretation of $f'(x) > 0$, $f'(x) = 0$ and $f'(x) < 0$

One very important skill in mathematics is the ability to think backwards. As well as looking at graphs of functions and finding out about their derivatives, you might be given a graph of the derivative and have to make inferences about the original function.

WORKED EXAMPLE 9.6

The graph on the right shows $y = f'(x)$. Use inequalities to describe the region where $f(x)$ is increasing.

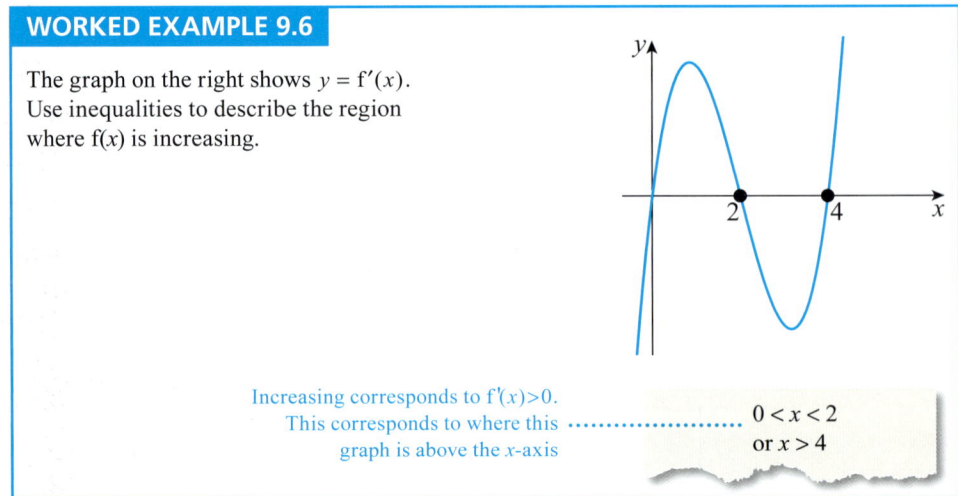

Increasing corresponds to $f'(x) > 0$.
This corresponds to where this graph is above the x-axis

$0 < x < 2$
or $x > 4$

9B Graphical interpretation of derivatives

If we have information about the derivative, that can also help us to sketch the original function.

WORKED EXAMPLE 9.7

The graph shows the derivative of a function. Sketch a possible graph of the original function.

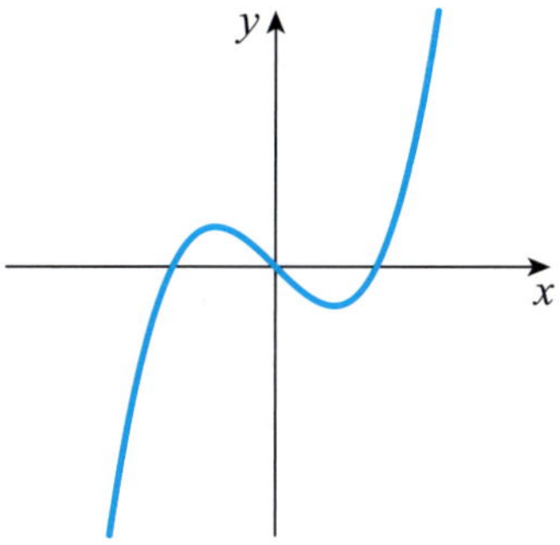

Consider the points A, B, C, D and E on the graph of $y = f'(x)$

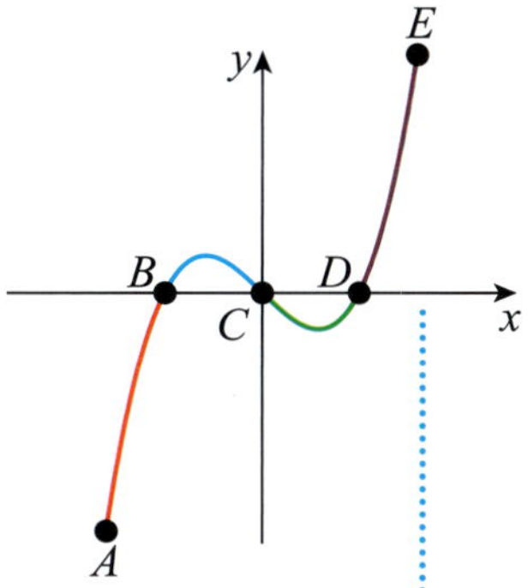

Between A and B the gradient is negative so the curve is decreasing.
At B the gradient is zero so the tangent is horizontal.
Between B and C the gradient is positive so the curve is increasing.
At C the gradient is zero so the tangent is horizontal.
Between C and D the gradient is negative so the curve is decreasing.
At D the gradient is zero so the tangent is horizontal.
Between D and E the gradient is positive so the curve is increasing

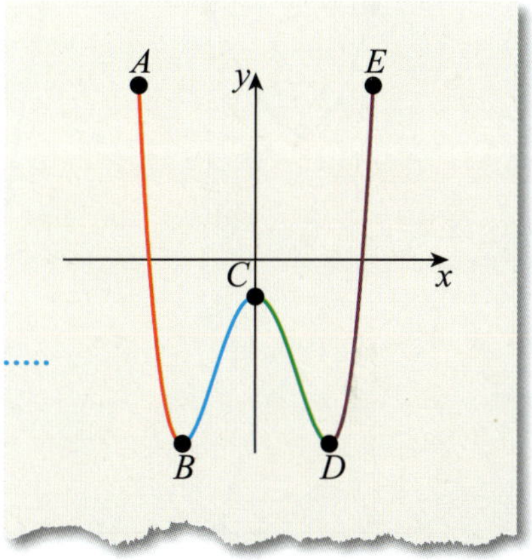

Note that in Worked Example 9.7 there was more than one possible graph that could have been drawn, depending on where the sketch started.

In Chapter 10 you will learn more about this ambiguity when you learn how to 'undo' differentiation.

Exercise 9B

For questions 1 to 3, use the method demonstrated in Worked Example 9.4 to write an inequality to describe where the function f(x) is increasing. The graphs all represent $y = f(x)$.

1 a 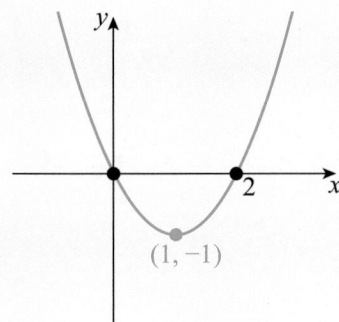   b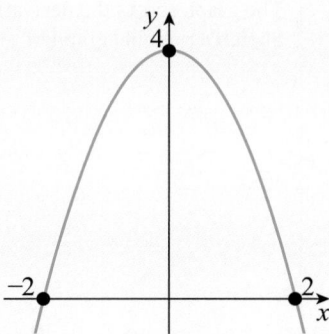

2 a 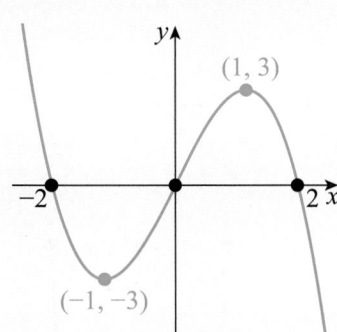   b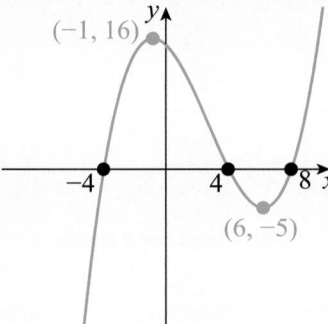

3 a 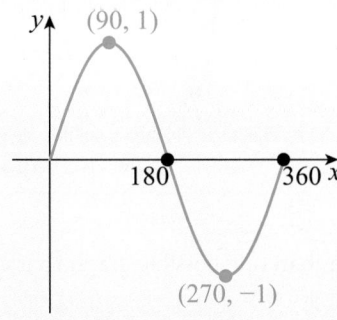   b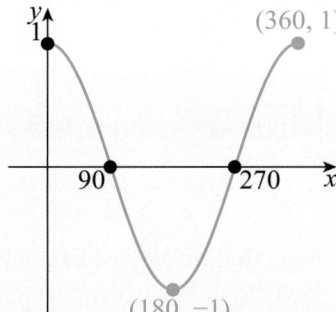

For questions 4 to 9, use the method demonstrated in Worked Example 9.5 to sketch the graph of the derivative of the function shown, labelling any intercepts with the *x*-axis.

4 a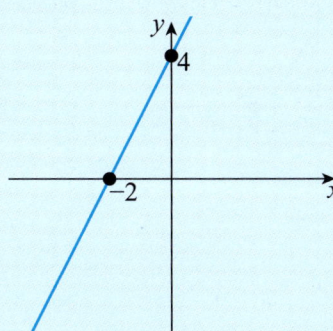

b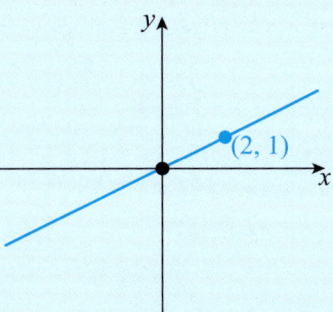

5 a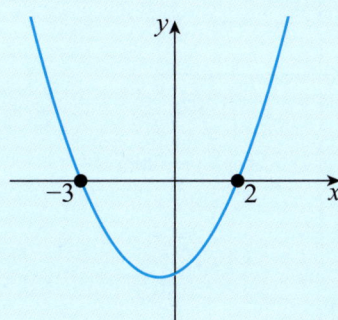

b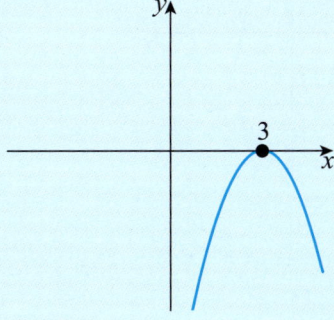

6 a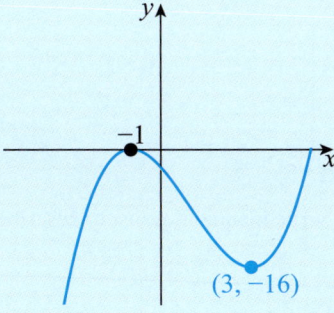

b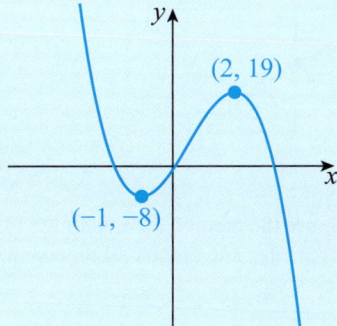

7 a 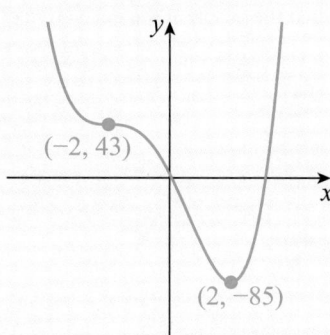 b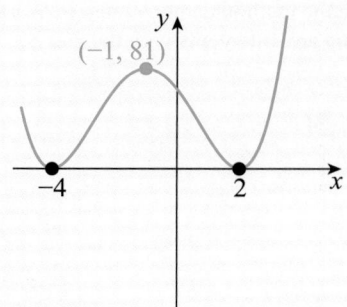

8 a 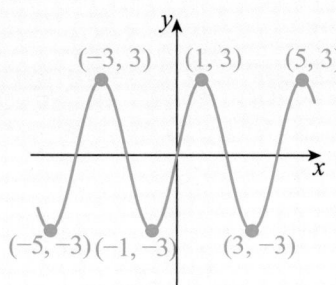 b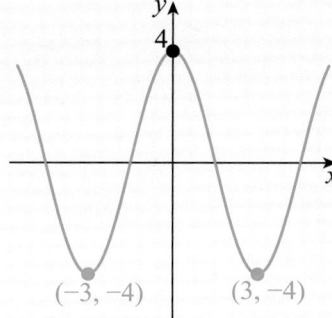

9 a 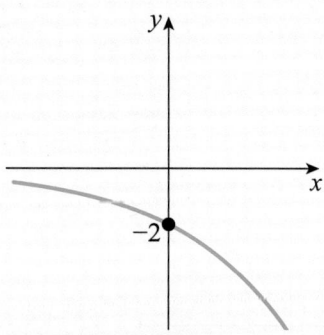 b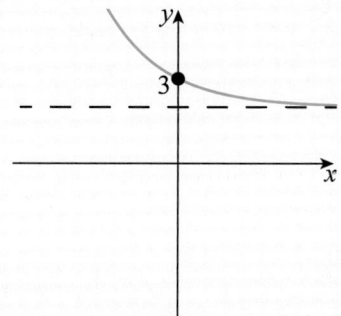

For questions 10 to 12, use the method demonstrated in Worked Example 9.6 to write inequalities in x to describe where the function $f(x)$ is increasing. The graphs all represent $y = f'(x)$.

10 a 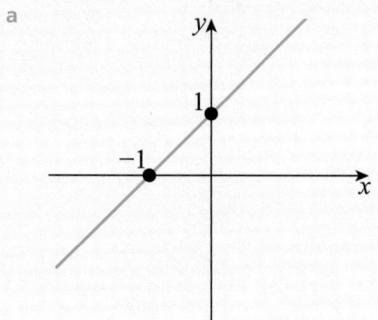 b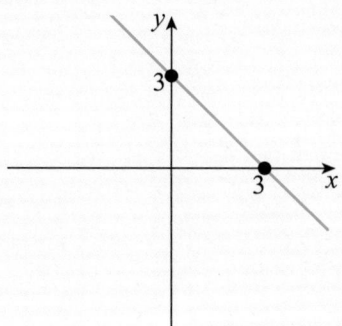

9B Graphical interpretation of derivatives

11 a   b

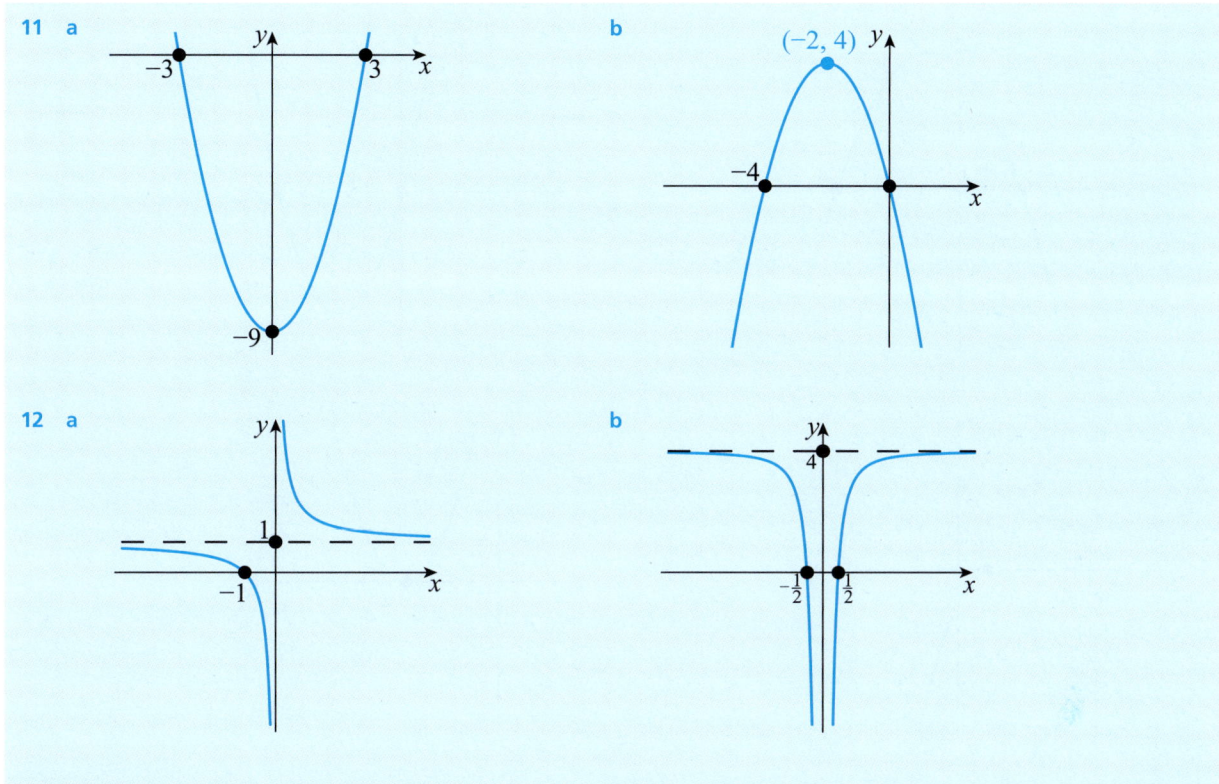

12 a   b

For questions 13 to 18, use the method demonstrated in Worked Example 9.7 to sketch a possible graph for $y = f(x)$, given that the graph shows $y = f'(x)$. Mark on any points on $y = f(x)$ where the gradient of the tangent is zero.

13 a   b

14 a   b

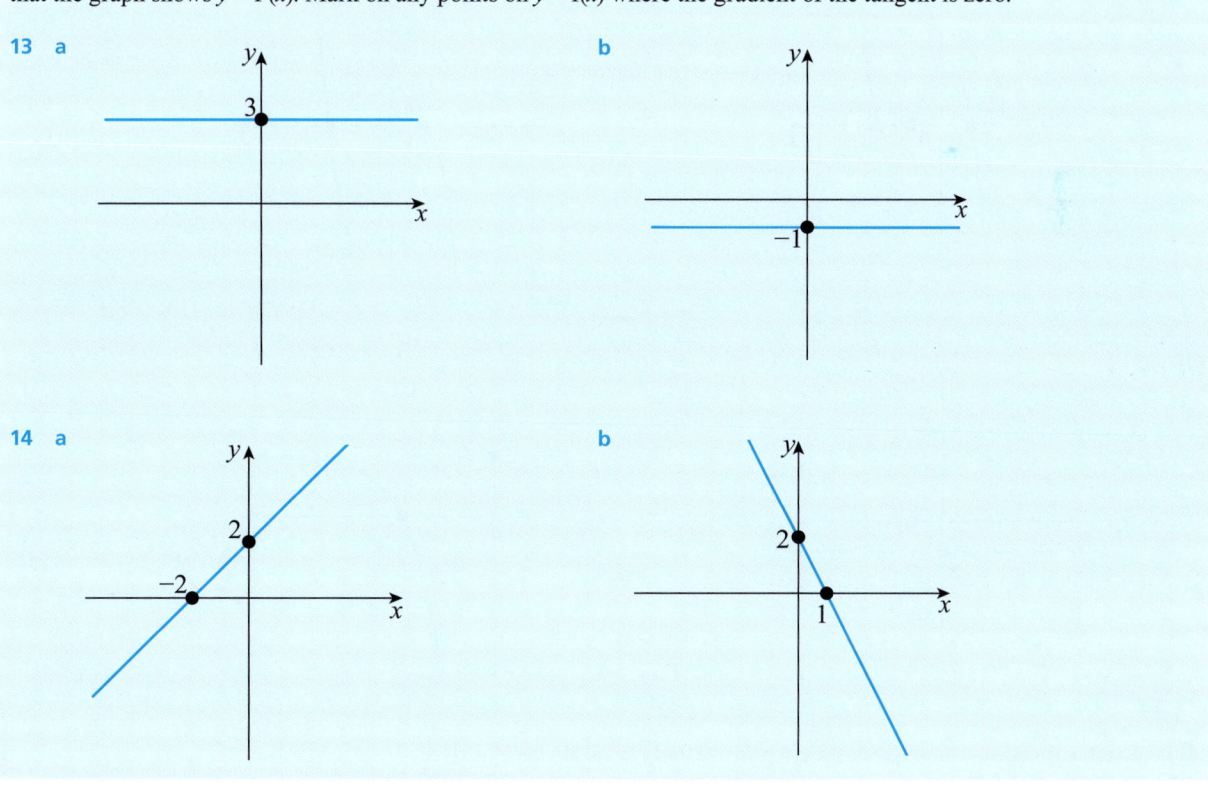

15 a 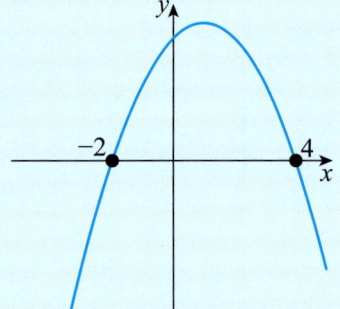 **b**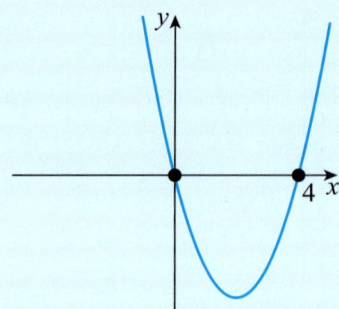

16 a 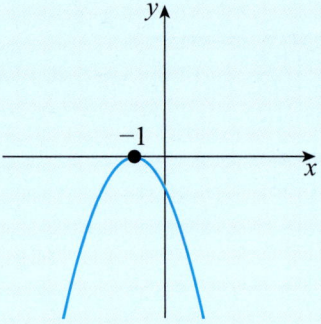 **b**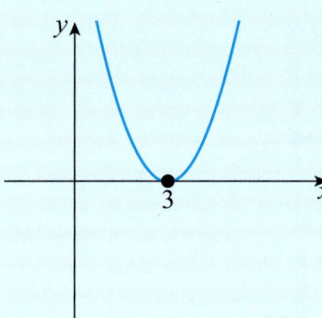

17 a 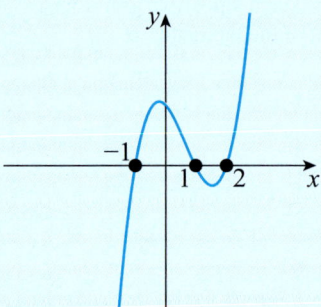 **b**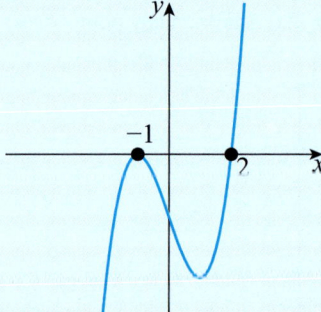

18 a 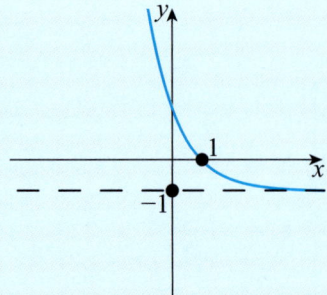 **b**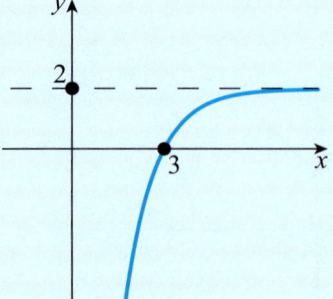

19 Sketch a function where f(x) is always positive and f'(x) is always positive.
20 Sketch a function where f(x) is always increasing but f'(x) is always decreasing.

21 For the following graph:

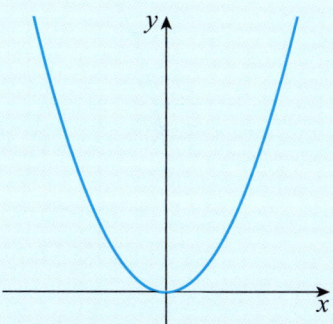

 a If the graph represents $y = f(x)$, sketch a graph of $y = f'(x)$.
 b If the graph represents $y = f'(x)$, sketch two possible graphs of $y = f(x)$.

22 For the following graph:

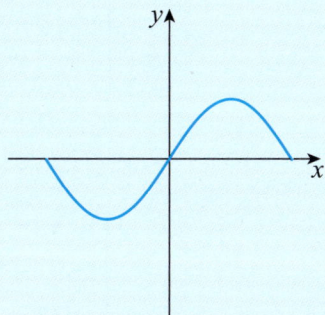

 a If the graph represents $y = f(x)$, sketch a graph of $y = f'(x)$.
 b If the graph represents $y = f'(x)$, sketch a possible graph of $y = f(x)$.
 c Explain why the answer to part (b) is not unique.

23 a Use technology to sketch the graph of $y = f(x)$ where $x^2(x^2 - 1)$.
 b Find the interval in which $f(x) > 0$.
 c Find the interval in which $f'(x) < 0$.

24 For the graphs below put them into pairs of a function and its derivative.

 a 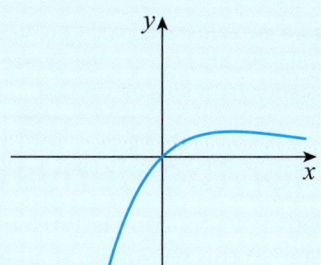   b 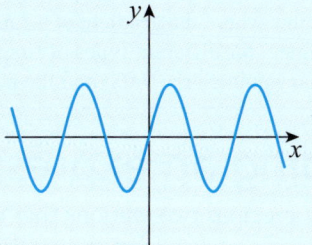   c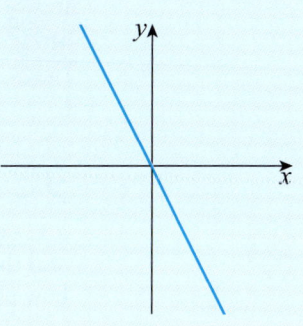

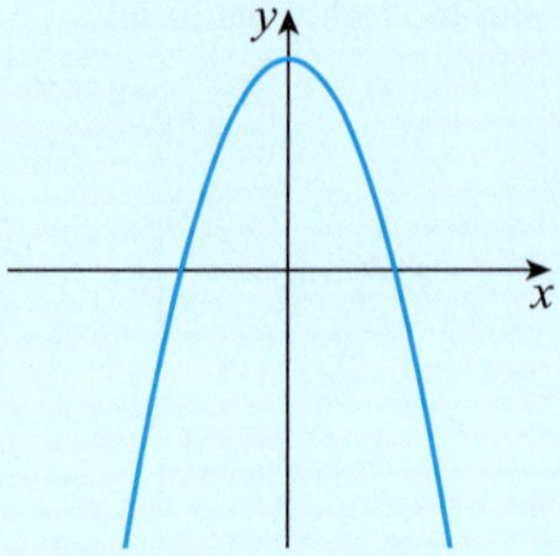

25 For the graphs below put them into pairs of a function and its derivative.

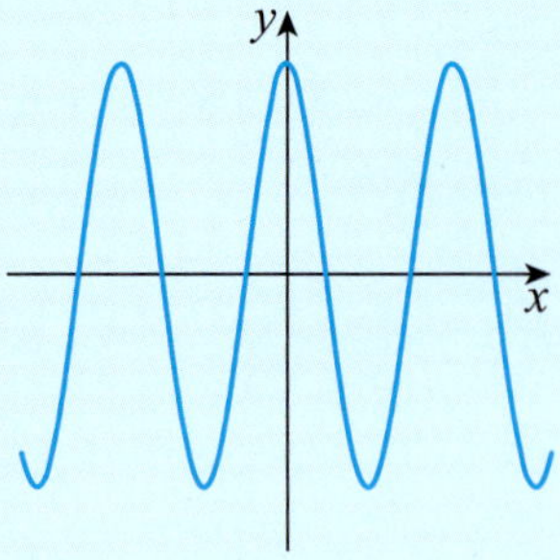

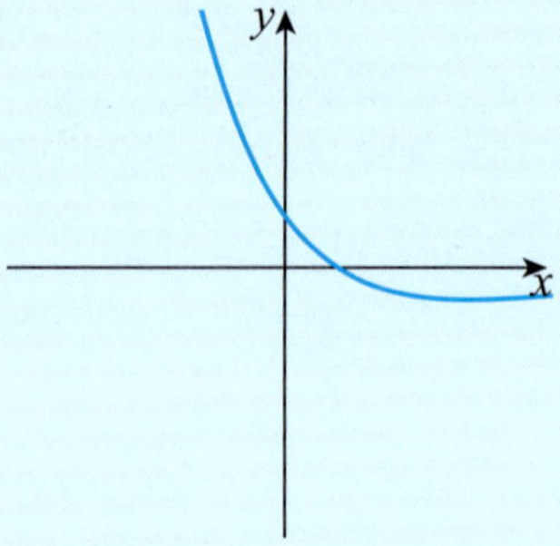

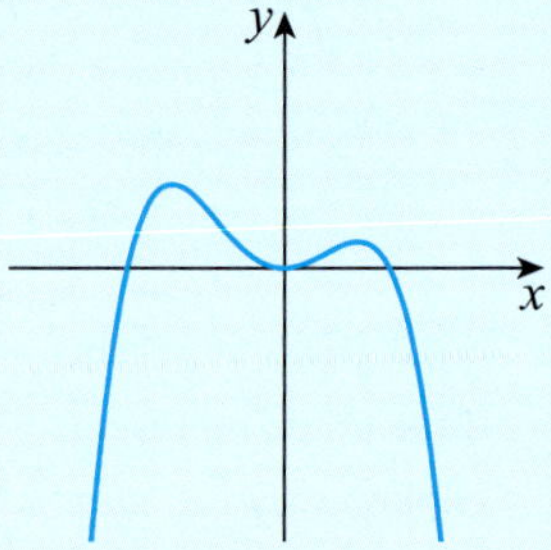

TOOLKIT: Problem Solving

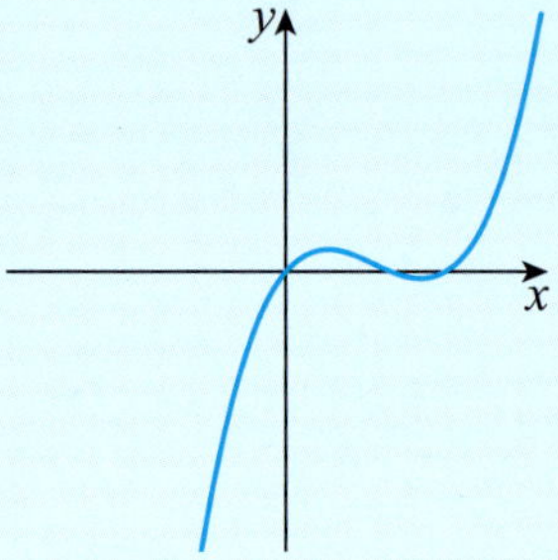

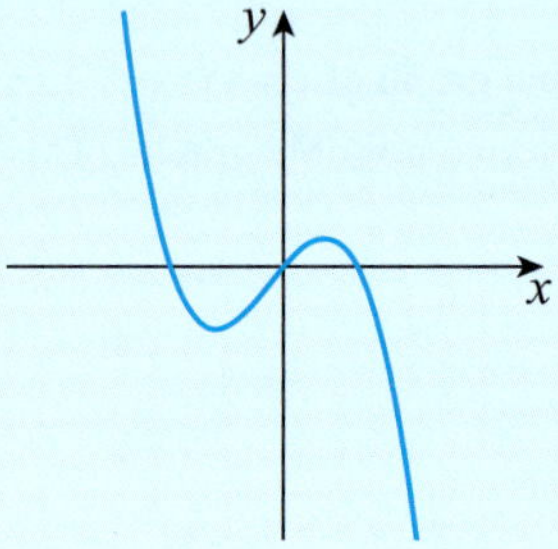

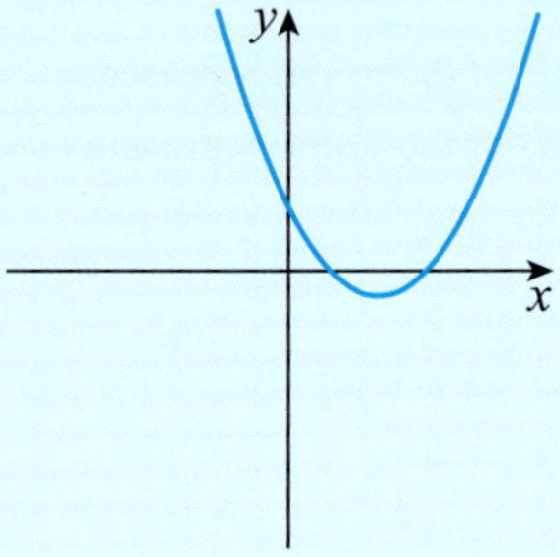

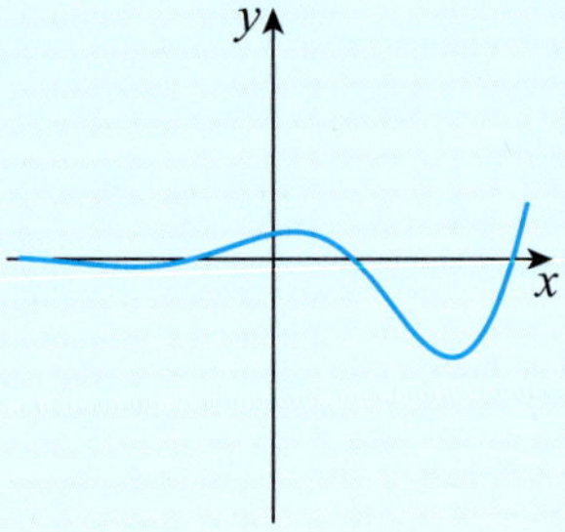

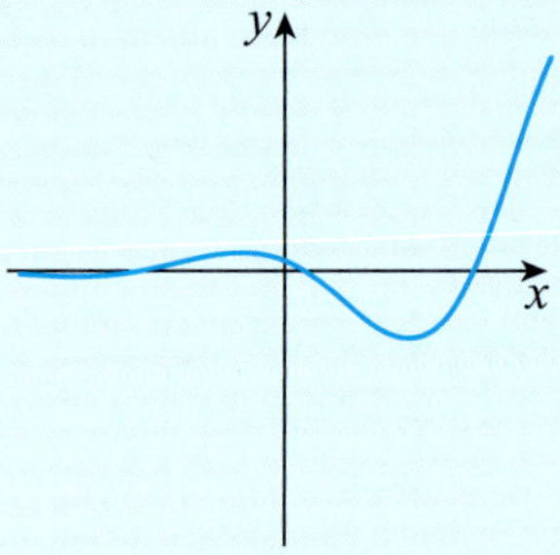

The points where the gradient of a function is zero are called stationary points. Investigate, for different types of function, the relationship between the number of stationary points on $y = f(x)$ and the number of stationary points on $y = f'(x)$.

9C Finding an expression for the derivative

You saw in the previous two sections that the derivative of a function is also a function, and that we could sketch it from the original function. In this section we shall look at how to find an expression for the derivative of some types of function.

The derivative of $f(x) = ax^n$

If a function is of the form $f(x) = x^n$, there is a rule for finding the derivative. This can be proved using the idea of limits, but for now we shall just quote it:

If $f(x) = x^n$, where $n \in \mathbb{Z}$, then $f'(x) = nx^{n-1}$.

WORKED EXAMPLE 9.8

Given $f(x) = x^3$, find $f'(x)$.

Use $f'(x) = nx^{n-1}$ .... $f'(x) = 3x^{3-1} = 3x^2$

WORKED EXAMPLE 9.9

A curve has equation $y = x^{-4}$. Find $\dfrac{dy}{dx}$.

Use $\dfrac{dy}{dx} = nx^{n-1}$ .... $\dfrac{dy}{dx} = -4x^{-4-1} = -4x^{-5}$

This basic rule is unaffected by multiplication by a constant.

KEY POINT 9.4

If $f(x) = ax^n$, where a is a real constant and $n \in \mathbb{Z}$, then $f'(x) = anx^{n-1}$.

WORKED EXAMPLE 9.10

Find the derivative of $f(t) = 2t^5$.

Differentiate t^5 and then multiply by 2 .... $f'(t) = 2 \times 5t^4 = 10t^4$

You will often need to use the laws of indices before you can differentiate.

See Chapter 1 if you need to review the rules of indices.

WORKED EXAMPLE 9.11

Differentiate $y = \dfrac{6}{x^2}$.

Use the laws of indices to rewrite in the form ax^n .... $y = \dfrac{6}{x^2} = 6x^{-2}$

Then differentiate x^{-2} and multiply by 6 .... $y' = 6(-2)x^{-3} = -12x^{-3}$

■ The derivative of $f(x) = ax^n + bx^m + \ldots$

Key Point 9.4 can be extended to deal with a sum of functions of the form ax^n. In this case you just differentiate each term in turn.

> **KEY POINT 9.5**
>
> If $f(x) = ax^n + bx^m + \ldots$, where a and b are real constants and $n, m \in \mathbb{Z}$, then
> $f'(x) = anx^{n-1} + bmx^{m-1} + \ldots$

Tip

Worked Example 9.12 shows that the derivative of ax will always be a and the derivative of a will always be zero. You can use these results without having to show the working each time.

WORKED EXAMPLE 9.12

Find $\dfrac{dy}{dx}$ for $y = 7x - 2$.

Write $7x$ as $7x^1$ and -2 as $-2x^0$ ········ $f(x) = 7x^1 - 2x^0$

Differentiate each term separately ········ $f'(x) = 7 \times 1x^{1-1} - 2 \times 0x^{0-1}$
$= 7x^0 - 0 = 7$

None of the Key Points in this section provide a rule for differentiating products or quotients of functions. So, before you can differentiate these, you will need to convert them into terms of the form ax^n.

WORKED EXAMPLE 9.13

If a is a constant, find the rate of change of $y = (x + 3)(x - a)$ with respect to x.

Expand the brackets ········ $y = (x + 3)(x - a)$
$= x^2 + 3x - ax - 3a$

Finding the rate of change is another way of saying differentiate. The constant a can be treated just like any other numerical constant ········ $\dfrac{dy}{dx} = 2x + 3 - a$

WORKED EXAMPLE 9.14

Differentiate $f(x) = \dfrac{x^2 - 6}{2x}$.

$f(x) = \dfrac{x^2 - 6}{2x}$

Use the laws of indices to rewrite as a sum of functions of the form ax^n ········ $= \dfrac{x^2}{2x} - \dfrac{6}{2x}$

$= \dfrac{1}{2}x - 3x^{-1}$

Then differentiate ········ $f'(x) = \dfrac{1}{2} - 3(-1)x^{-2}$

$= \dfrac{1}{2} + 3x^{-2}$

9C Finding an expression for the derivative

Be the Examiner 9.1

Given $y = \dfrac{6x^2 - 5x}{x^3}$, find $\dfrac{dy}{dx}$.

Which is the correct solution? Identify the errors made in the incorrect solutions.

Solution 1	Solution 2	Solution 3
$y = \dfrac{6x^2 - 5x}{x^3}$ $y' = \dfrac{12x - 5}{3x^2}$	$y = \dfrac{6x^2 - 5x}{x^3}$ $= 6x^{-1} - 5x^{-2}$ $y' = -6x^{-2} + 10x^{-3}$	$y = \dfrac{6x^2 - 5x}{x^3}$ $= x^{-3}(6x^2 - 5x)$ $y' = -3x^{-4}(12x - 5)$

You can use the rules of differentiation from this section together with the idea of increasing and decreasing functions from Section 9B.

WORKED EXAMPLE 9.15

Find the interval of values of x for which the function $f(x) = 5x^2 - 12x + 3$ is decreasing.

For any question about the increasing or decreasing of functions, start by finding f'(x) .... $f'(x) = 10x - 12$

A decreasing function has negative gradient .... $f'(x) < 0$
$10x - 12 < 0$
$x < 1.2$

Be the Examiner 9.2

Is the function $f(x) = 2x^3 - 3x$ increasing or decreasing at $x = 1$?

Which is the correct solution? Identify the errors made in the incorrect solutions.

Solution 1	Solution 2	Solution 3
$f(1) = 2 \times 1^3 - 3 \times 1$ $= 2 - 3$ $= -1 < 0$ $\therefore$ f is decreasing.	$f'(x) = 6x^2 - 3$ $f'(1) = 6 \times 1^2 - 3$ $= 6 - 3$ $= 3 > 0$ $\therefore$ f is increasing.	$f(0) = 0$ and $f(2) = 10$ $\therefore$ f is increasing

Exercise 9C

For questions 1 to 4, use the method demonstrated in Worked Examples 9.8 and 9.9 to find the derivative of the given function or graph.

1. a $f(x) = x^4$
   b $g(x) = x^6$
2. a $h(u) = u^{-1}$
   b $z(t) = t^{-4}$
3. a $y = x^8$
   b $p = q$
4. a $z = t^{-5}$
   b $s = r^{-10}$

For questions 5 to 7, use the method demonstrated in Worked Example 9.10 to find the derivative of the given function or graph.

5. a $f(x) = -4x$
   b $g(x) = 7x^2$
6. a $y = 6x$
   b $y = 3x^5$
7. a $y = 7$
   b $y = -6$

For questions 8 to 11, use the method demonstrated in Worked Example 9.11 to find the derivative of the given function or graph.

8   a $g(x) = \dfrac{1}{x}$ 9   a $z = \dfrac{3}{x^2}$ 10   a $y = \dfrac{1}{4x^4}$ 11   a $f(x) = -\dfrac{3}{2x}$

     b $h(x) = \dfrac{1}{x^3}$ b $y = -\dfrac{10}{t^5}$ b $y = \dfrac{1}{5x^5}$ b $f(x) = -\dfrac{5}{3x^3}$

For questions 12 to 15, use the method demonstrated in Worked Example 9.12 to find the derivative of the given function or graph.

12   a $f(x) = x^2 - 4x$ 13   a $y = 3x^3 - 5x^2 + 7x - 2$ 14   a $y = \dfrac{1}{2}x^4$ 15   a $y = 3x - \dfrac{1}{4}x^3$

     b $g(x) = 2x^2 - 5x + 3$ b $y = -x^4 + 6x^2 - 2x$ b $y = -\dfrac{3}{4}x^6$ b $y = 3 - 2x^3 + \dfrac{1}{4}x^4$

For questions 16 to 18, use the method demonstrated in Worked Example 9.13 to expand the brackets and hence find the derivative of the given function.

16   a $f(x) = x^3(2x - 5)$ 17   a $g(x) = (x + 3)(x - 1)$ 18   a $h(x) = \left(1 + \dfrac{1}{x}\right)^2$

     b $f(x) = x(x^2 + 3x - 9)$ b $f(x) = (x - 2)(x + 1)$ b $g(x) = \left(2x - \dfrac{3}{x}\right)^2$

For questions 19 to 25, use the method demonstrated in Worked Example 9.13 to find the rate of change of y with respect to x when a, b and c are constants.

19   a $y = ax + b$ 20   a $y = ax^2 + (3 - a)x$ 21   a $y = x^2 + a^2$ 22   a $y = x^{2a} + a^{2a}$

     b $y = ax^2 + bx + c$ b $y = x^3 + b^2x$ b $y = a^2x^3 + a^2b^3$ b $y = x^{-a} - x^{-b} + ab$

23   a $y = 7ax - \dfrac{3b}{x^2}$ 24   a $y = (ax)^2$ 25   a $y = (x + a)(x + b)$

     b $y = 5b^2x^2 + \dfrac{3a}{cx}$ b $y = (3ax)^2$ b $y = (ax + b)(bx + a)$

For questions 26 to 28, use the method demonstrated in Worked Example 9.14 to differentiate the expression with respect to x.

26   a $\dfrac{x + 10}{x}$ 27   a $\dfrac{2 + x}{x^7}$ 28   a $\dfrac{x^2 + 3x^3}{2x}$

     b $\dfrac{2x - 5}{x}$ b $\dfrac{1 + 2x}{x^3}$ b $\dfrac{7x^5 + 2x^9}{4x}$

For questions 29 to 30, use the method demonstrated in Worked Example 9.15 to find the interval of values of x for which the given function is decreasing.

29   a $f(x) = 3x^2 - 6x + 2$ 30   a $f(x) = -x^2 + 12x - 9$

     b $f(x) = x^2 + 8x + 10$ b $f(x) = 10 - x^2$

31 Differentiate $d = 6t - \dfrac{4}{t}$ with respect to t.

32 Find the rate of change of $q = m + 2m^{-1}$ as m varies.

33 A model for the energy of a gas particle (E) at a temperature T suggests that $E = \dfrac{3}{2}kT$ where k is a constant. Find the rate at which E increases with increasing temperature.

34 Find the interval in which $x^2 - x$ is an increasing function.

35 Find the interval in which $x^2 + bx + c$ is an increasing function, given that b and c are constants.

36 If $x + y = 8$ find $\dfrac{dy}{dx}$.

37 If $x^3 + y = x$ find $\dfrac{dy}{dx}$.

38 In an astronomical model of gravity it is believed that potential energy (V) depends on distance from a star (r) by the rule:

$V = \dfrac{k}{r}$

where k is a constant.

The force is defined as the rate of change of the potential energy with respect to distance.

a Find an expression for the force in terms of k and r.
b Find an expression for the force in terms of V and k.
c The star has two orbiting planets – Alpha and Omega. At a particular time Alpha lies exactly midway between the star and Omega. At this time, find the value of the ratio:

$\dfrac{\text{Force on Alpha}}{\text{Force on Omega}}$

39 A model for the reading age of a book (A) when the average sentence length is L suggests that $A = 2 + qL + qL^2$.

a Find the rate of change of A with respect to L.
b Explain why it might be expected that A is an increasing function. What constraint does this place on q?

40 Show that the function $f(x) = 4x^3 + 7x - 2$ is increasing for all x.

9D Tangents and normals at a given point and their equations

■ Calculating the gradient at a given point

To calculate the gradient of a function at any particular point, you simply substitute the value of x into the expression for the derivative.

> **WORKED EXAMPLE 9.16**
>
> Find the gradient of the graph of $y = 5x^3$ at the point where $x = 2$.
>
> The gradient is given by the derivative
> $\dfrac{dy}{dx} = 15x^2$
>
> Substitute the given value for x
> When $x = 2$,
> $\dfrac{dy}{dx} = 15 \times 2^2 = 60$
>
> So, the gradient is 60

If you know the gradient of a graph at a particular point, you can find the value of x at that point.

WORKED EXAMPLE 9.17

Find the values of x for which the tangent to the graph of $y = x^3 + 6x^2 + 2$ has the gradient 15.

The gradient is given by the derivative

$$\frac{dy}{dx} = 3x^2 + 12x$$

You are told that $\frac{dy}{dx} = 15$

$$3x^2 + 12x = 15$$
$$x^2 + 4x - 5 = 0$$
$$(x+5)(x-1) = 0$$
$$x = -5 \text{ or } 1$$

■ Analytic approach to finding the equation of a tangent

You already know from Section 9A that a tangent to a function at a given point is just a straight line passing through that point with the same gradient as the function at that point. Using the rules of differentiation in Section 9C, you can now find equations of tangents.

> See Chapter 3 for a reminder of how to find the equation of a straight line from its gradient and a point on the line.

KEY POINT 9.6

The equation of the tangent to the curve $y = f(x)$ at the point where $x = a$ is given by $y - y_1 = m(x - x_1)$, where:
- $m = f'(a)$
- $x_1 = a$
- $y_1 = f(a)$

WORKED EXAMPLE 9.18

Find the equation of the tangent to the curve $y = x^2 + 4x^{-1} - 5$ at the point where $x = 2$.

Give your answer in the form $ax + by + c = 0$.

Find the gradient of the tangent by finding the value of y' when $x = 2$

$$y' = 2x - 4x^{-2}$$
When $x = 2$,
$$y' = 2 \times 2 - 4 \times 2^{-2}$$
$$= 4 - 1$$

m is often used to denote the value of the gradient at a particular point

$$\therefore m = 3$$

Find the value of y when $x = 2$

When $x = 2$,
$$y = 2^2 + 4 \times 2^{-1} - 5$$
$$= 1$$

Substitute into $y - y_1 = m(x - x_1)$

So, equation of tangent is:
$$y - 1 = 3(x - 2)$$
$$3x - y - 5 = 0$$

9D Tangents and normals at a given point and their equations

> **CONCEPTS – RELATIONSHIPS AND REPRESENTATION**
>
> There are lots of **relationships** going on when we are doing calculus:
> - the x-coordinate of the curve and the y-coordinate of a point on the curve.
> - the x-coordinate of the tangent and the y-coordinate of a point on the tangent.
> - the x-coordinate of the curve and the value of the derivative at that point.
> - the relationship between the graph of the curve the graph of the derivative.
>
> It is very easy to use the same letters to **represent** different things and get very confused. Make sure when you write down an expression you know exactly what all your xs and ys represent.

You might have to interpret given information about the tangent and use it to find an unknown.

WORKED EXAMPLE 9.19

The tangent at a point on the curve $y = x^2 + 4$ passes through $(0, 0)$.

Find the possible coordinates of the point.

You need to find the equation of the tangent at the unknown point	Let the point have coordinates (p, q).
As the point lies on the curve, (p, q) must satisfy $y = x^2 + 4$	Then $q = p^2 + 4$
The gradient of the tangent is given by $\frac{dy}{dx}$ with $x = p$	$\frac{dy}{dx} = 2x$ When $x = p$, $\frac{dy}{dx} = 2p$ $\therefore m = 2x$
Write down the equation of the tangent	Equation of the tangent is: $y - q = 2p(x - p)$ $y - (p^2 + 4) = 2p(x - p)$
The tangent passes through the origin, so set $x = 0$ and $y = 0$ in the equation	Since the tangent passes through $(0, 0)$: $0 - (p^2 + 4) = 2p(0 - p)$ $-p^2 - 4 = -2p^2$ $p^2 = 4$ Hence, $p = 2$ or -2
Now use $q = p^2 + 1$ to find the corresponding values of q	When $p = 2$, $q = 8$ When $p = -2$, $q = 8$ So, the coordinates are $(2, 8)$ or $(-2, 8)$

■ Analytic approach to finding the equation of a normal

The **normal to a curve** at a given point is a straight line which crosses the curve at that point and is perpendicular to the tangent at that point.

You can use the fact that for perpendicular lines $m_1 m_2 = -1$ to find the gradient of the normal.

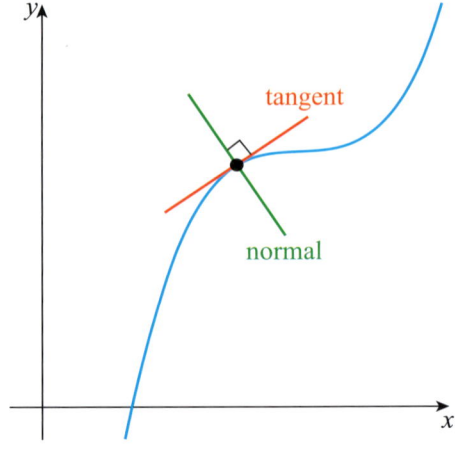

KEY POINT 9.7

The equation of the normal to the curve $y = f(x)$ at the point where $x = a$ is given by

$$y - y_1 = m(x - x_1),$$

where:
- $m = -\dfrac{1}{f'(a)}$
- $x_1 = a$
- $y_1 = f(a)$

We will often denote the gradient of the tangent by m_t and the gradient of the normal by m_n, so that $m_n = -\dfrac{1}{m_t}$.

WORKED EXAMPLE 9.20

Find the equation of the normal to the curve $y = x^2 + 4x^{-1} - 5$ at the point where $x = 2$ (from Worked Example 9.18).

Give your answer in the form $ax + by + c = 0$, where a, b and c are integers.

From Worked Example 9.18, you know that the gradient of the tangent at $x = 2$ is 3 .... At $x = 2$, gradient of tangent is $m_t = 3$

Find the gradient of the normal using $m_t m_n = -1$ ....

So, gradient of normal is

$$m_n = -\frac{1}{3}.$$

Substitute into $y - y_1 = m(x - x_1)$. You know from Worked Example 9.18 that $y_1 = 1$ ....

Hence, equation of normal is:

$$y - 1 = -\frac{1}{3}(x - 2)$$
$$3y - 3 = -x + 2$$
$$x + 3y - 5 = 0$$

Using technology to find the equation of tangents and normals

In some cases you will just be expected to use your calculator to find the gradient at a specific point, rather than needing to know how to differentiate the function.

Tip

Don't be put off if you are given an unfamiliar function that you don't know how to differentiate. In such cases you will be expected to use your calculator to find the value of the derivative at the relevant point.

WORKED EXAMPLE 9.21

Using technology, find the equation of the tangent to the curve $y = 4\ln x$ at the point where $x = 1$.

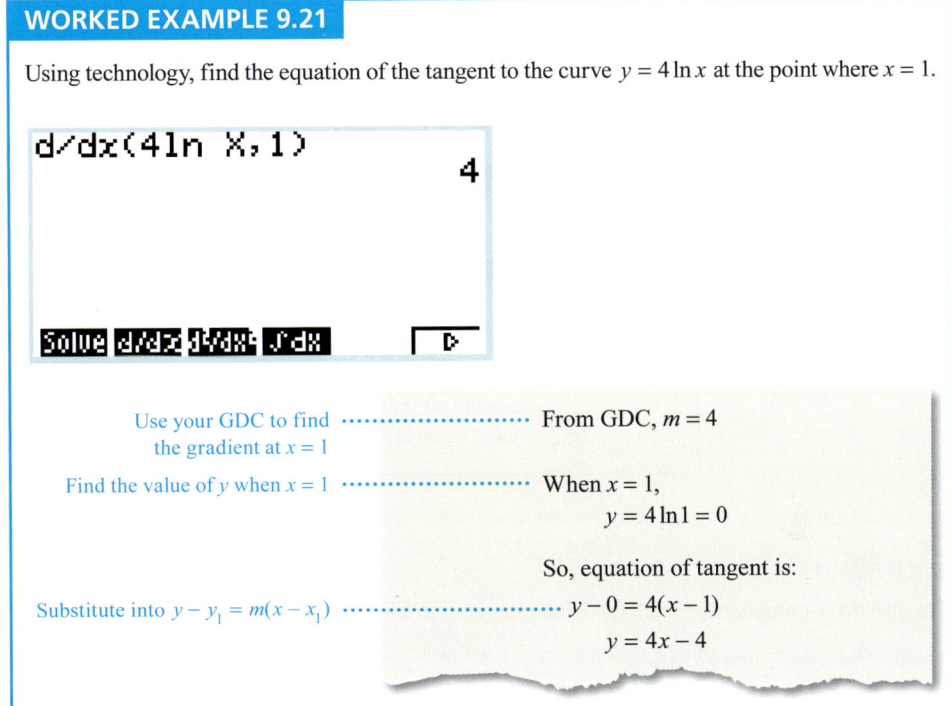

Use your GDC to find the gradient at $x = 1$	From GDC, $m = 4$
Find the value of y when $x = 1$	When $x = 1$, $y = 4\ln 1 = 0$
	So, equation of tangent is:
Substitute into $y - y_1 = m(x - x_1)$	$y - 0 = 4(x - 1)$ $y = 4x - 4$

Exactly the same approach as in Worked Example 9.21 can be used to find the equation of a normal. Start as before by using your GDC to find the gradient at the given point, then use $m_1 m_2 = -1$ to find the gradient of the normal.

A slightly more advanced calculator trick is to get the calculator to plot the derivative of the function. This is particularly useful when being asked to find the *x*-value of a point with a given gradient.

WORKED EXAMPLE 9.22

Find the x-coordinate of the point on the graph of xe^x with gradient 2.

This is not a function you know how to differentiate, so you need to get the calculator to sketch $y = f'(x)$. You also need to add the line $y = 2$ and find where these two graphs intersect.

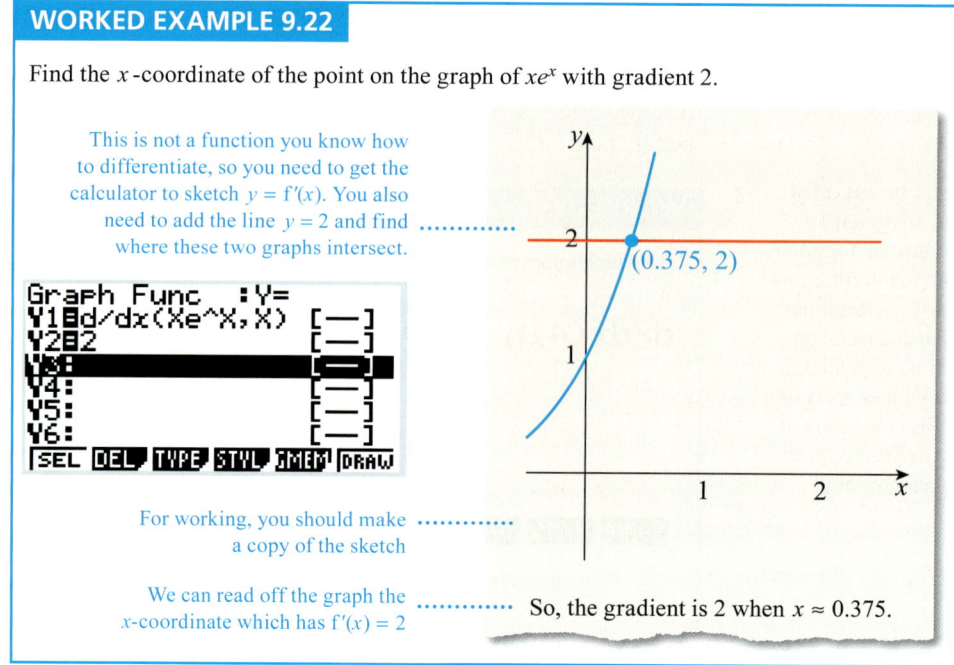

For working, you should make a copy of the sketch

We can read off the graph the x-coordinate which has $f'(x) = 2$ .... So, the gradient is 2 when $x \approx 0.375$.

Exercise 9D

For questions 1 and 2, use the method demonstrated in Worked Example 9.16 to find the derivative at the given point.

1 a If $y = 4x^2$, find $\frac{dy}{dx}$ when $x = 2$.
  b If $y = x^3 - 7x$, find $\frac{dy}{dx}$ when $x = 3$.

2 a If $A = 5 - 2h^{-1}$ find $\frac{dA}{dh}$ when $h = 4$.
  b If $\phi = \theta^3 - \theta^{-2}$, find $\frac{d\phi}{d\theta}$ when $\theta = 0.5$.

For questions 3 and 4, use the method demonstrated in Worked Example 9.16 to find the gradient of the tangent to the curve at the given point.

3 a The curve $y = x^4$ at the point where $x = 2$.
  b The curve $y = 3x + x^2$ at the point where $x = -5$.

4 a The curve $y = x^2 + \frac{1}{x}$ at the point where $x = \frac{3}{2}$.
  b The curve $y = x^2 - \frac{4}{x}$ at the point where $x = 0.2$.

For questions 5 and 6, find and evaluate an appropriate derivative.

5 a How quickly does $f = 4T^3$ change as T changes when $T = 3$?
  b How quickly does $g = 2y^3$ change as y changes when $y = 1$?

6 a What is the rate of increase of W with respect to q when q is -3 if $W = -4q^2$?
  b What is the rate of change of M with respect to c when $c = 4$ if $M = \frac{3}{c} + 5$?

For questions 7 to 9, use the method demonstrated in Worked Example 9.17 to find the value(s) of x at which the curve has the given gradient.

7 a gradient 6 on the graph of $y = x^4 + 2x$
  b gradient 36 on the graph of $y = 3x^3$

8 a gradient 2.25 on the graph of $y = x - 5x^{-1}$
  b gradient -27 on the graph of $y = 5x + \frac{8}{x}$

9 a gradient -4 on the graph of $y = x^3 + 6x^2 + 5x$
  b gradient -1.5 on the graph of $y = x^3 - 3x + 2$.

9D Tangents and normals at a given point and their equations

For questions 10 to 12, use the method demonstrated in Worked Example 9.18 to find the equation of the tangent of the given curve at the given point in the form $ax + by = c$.

10   a $y = x^2 + 3$ at $x = 2$ b $y = x^2 + x$ at $x = 0$

11   a $y = 2x^3 - 4x^2 + 7$ at $x = 1$ b $y = x^3 + x^2 - 8x + 1$ at $x = -1$

12   a $y = \dfrac{3}{x}$ at $x = 3$ b $y = 1 - \dfrac{2}{x}$ at $x = 1$

For questions 13 to 16, use the method demonstrated in Worked Example 9.20 to find the equation of the normal of the given curve at the given point in the form $y = mx + c$.

13   a $y = x + 7$ at $x = 4$ b $y = 3 - 0.5x$ at $x = 6$

14   a $y = x^3 + x$ at $x = 0$ b $y = 2x^3 - 5x + 1$ at $x = -1$

15   a $y = x^3 - 3x^2 + 2$ at $x = 1$ b $y = 2x^4 - x^3 + 3x^2 + x$ at $x = -1$

16   a $y = \dfrac{2}{x}$ at $x = 2$ b $y = 1 + \dfrac{4}{x}$ at $x = -2$

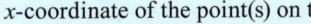

For questions 17 to 20, use the method demonstrated in Worked Example 9.21 (that is, use technology) to find the equation of the tangent of the given curve at the given point in the form $y = mx + c$.

17   a $y = \sqrt{x}$ at $x = 4$ b $y = \dfrac{1}{\sqrt{x}}$ at $x = 4$

18   a $y = \dfrac{1}{x+1}$ at $x = 1$ b $y = \dfrac{1}{(x-1)^2}$ at $x = 5$

19   a $y = \ln x$ at $x = 1$ b $y = x \ln x$ at $x = 1$

20   a $y = 2^x$ at $x = 2$ b $y = 3^x$ at $x = 0$

For questions 21 to 24, use technology to find the equation of the normal of the given curve at the given point.

21   a $y = \sqrt{x+1}$ at $x = 3$ b $y = \sqrt{2x+1}$ at $x = 0$

22   a $y = 2\sqrt{x} - \dfrac{2}{\sqrt{x}}$ at $x = 4$ b $y = \dfrac{x^2+4}{\sqrt{x}}$ at $x = 4$

23   a $y = \ln 3x$ at $x = 1$ b $y = 4 + 2\ln 3x$ at $x = 1$

24   a $y = xe^x$ at $x = 0$ b $y = (x-2)e^x$ at $x = 0$

For questions 25 to 27, use the method demonstrated in Worked Example 9.22 (that is, use technology) to find the x-coordinate of the point(s) on the curve with the given gradient.

25   a   gradient 4 on $y = \sqrt{x+1}$ b   gradient 2 on $\sqrt{1-x^2}$

26   a   gradient 0.5 on $y = \ln x$ b   gradient 2 on $y = e^x$

27   a   gradient -1 on $y = \dfrac{1}{x+2}$ b   gradient 0 on $y = x^2 e^x$

28   A graph is drawn of $y = x^4 - x$.

    a   Find an expression for $\dfrac{dy}{dx}$.

    b   What is the gradient of the tangent of the graph at $x = 0$?

    c   Hence, find the equation of the normal to the curve at $x = 0$.

29   A graph is drawn of $y = f(x)$ with $f(x) = x^3 + \dfrac{1}{x}$.

    a   Find $f'(x)$.

    b   What is the gradient of the tangent of the graph at $x = 1$?

    c   Hence, show that the tangent to the curve at $x = 1$ passes through the origin.

30 Use technology to find the equation of the tangent to the curve $y = x\sqrt{x+1}$ at $x = 3$.

31 Use technology to find the equation of the normal to the curve $y = \dfrac{1}{x+4}$ at $x = -2$.

32 Find the equation of the normal to the curve $y = x^2 e^x$ at $x = 0$.

33 Find the coordinates of the point at which the normal to the curve $y = x^2$ at $x = 1$ meets the curve again.

34 Find the coordinates of the point at which the normal to the curve $y = \dfrac{1}{x}$ at $x = 2$ meets the curve again.

35 Find the equation of the tangent to the curve $y = x^2$ which is parallel to $y + 2x = 10$.

36 Find the equation of the normal to the curve $y = x^2 + 2x$ which is parallel to $y = \dfrac{x}{4}$.

37 Find the equations of the normals to the curve $y = x^3$ which are perpendicular to the line $y = 3x + 5$.

38 Find the coordinates of the points at which the tangent to the curve $y = x^3 - x$ is parallel to $y = 11x$.

39 Find all points on the curve $y = x^2 + 4x + 1$ where the gradient of the tangent equals the y-coordinate.

40 The tangent at point P on the curve $y = \dfrac{1}{x}$ passes through $(4, 0)$. Find the coordinates of P.

41 The tangent at point P on the curve $y = \dfrac{4}{x}$ passes through $(1, 3)$. Find the possible x-coordinates of P.

42 Find the coordinates of the points on the curve $y = x^2$ for which the tangent passes through the point $(2, 3)$.

43 A tangent is drawn on the graph $y = \dfrac{1}{x}$ at the point where $x = a$ ($a > 0$). The tangent intersects the y-axis at P and the x-axis at Q. If O is the origin, show that the area of the triangle OPQ is independent of a.

44 The point P on the curve $y = \dfrac{a}{x}$ has x-coordinate 2. The normal to the curve at P is parallel to $x - 5y + 10 = 0$. Find the constant a.

45 Show that the tangent to the curve $y = x^3 + x + 1$ at the point with x-coordinate k meets the curve again at a point with x-coordinate $-2k$.

TOOLKIT: Modelling

Draw a circle centred on the origin. Measure the gradient of the tangent at several points, including an estimate of the error. How easy is it to manually find the gradient?

See if you can find a relationship between the x-coordinate of the point, the y-coordinate and the gradient of the tangent at that point. Does the radius of the circle matter? See if you can prove your conjecture.

Mixed Practice

Checklist

- You should be able to estimate the value of a limit from a table or graph.

- You should have an informal understanding of the gradient of a curve as a limit:
  - ☐ The gradient of a curve at a point is the gradient of the tangent to the curve at that point.
  - ☐ The gradient of a tangent at a point is the limit of the gradient of smaller and smaller chords from that point.

- You should know about different notation for the derivative.

- You should be able to identify intervals on which functions are increasing or decreasing:
  - ☐ If when x increases $f(x)$ increases (so that $f'(x) > 0$), the function is increasing.
  - ☐ If when x increases $f(x)$ decreases (so that $f'(x) < 0$), the function is decreasing.

- You should be able to interpret graphically $f'(x) > 0$, $f'(x) = 0$, $f'(x) < 0$.

- You should be able to differentiate functions of the form $f(x) = ax^n + bx^m + ...$
  - ☐ If $f(x) = ax^n + bx^m + ...$, where a and b are real constants and $n, m \in \mathbb{Z}$, then $f'(x) = anx^{n-1} + bmx^{m-1} + ...$

- You should be able to find the equations of the tangent and normal to a function at a given point:
  - ☐ The equation of the tangent to the curve $y = f(x)$ at the point where $y = a$ is given by $y - y_1 = m(x - x_1)$, where
    - — $m = f'(a)$
    - — $x_1 = a$
    - — $y_1 = f(a)$

- The equation of the normal to the curve $y = f(x)$ at the point where $x = a$ is given by $y - y_1 = m(x - x_1)$, where:
  - ☐ $m = -\dfrac{1}{f'(a)}$
  - ☐ $x_1 = a$
  - ☐ $y_1 = f(a)$

■ Mixed Practice

1 A curve has equation $y = 4x^2 - x$.
  - **a** Find $\dfrac{dy}{dx}$.
  - **b** Find the coordinates of the point on the curve where the gradient equals 15.

2 **a** Use technology to sketch the curve $y = f(x)$ where $f(x) = x^2 - x$.
  - **b** State the interval in which the curve is increasing.
  - **c** Sketch the graph of $y = f'(x)$.

3 The function f is given by $f(x) = 2x^3 + 5x^2 + 4x + 3$.
  - **a** Find $f'(x)$.
  - **b** Calculate the value of $f'(-1)$.
  - **c** Find the equation of the tangent to the curve $y = f(x)$ at the point $(-1, 2)$.

4 Use technology to suggest the limit of $\dfrac{\ln(1+x) - x}{x^2}$ as x gets very small.

5 Find the equation of the tangent to the curve $y = x^3 - 4$ at the point where $y = 23$.

6 Water is being poured into a water tank. The volume of the water in the tank, measured in m^3, at time t minutes is given by $V = 50 + 12t + 5t^2$.
  - **a** Find $\dfrac{dV}{dt}$.
  - **b** Find the values of V and $\dfrac{dV}{dt}$ after 6 minutes. What do these values represent?
  - **c** Is the volume of water in the tank increasing faster after 6 minutes or after 10 minutes?

7 The accuracy of an x-ray (*A*) depends on the exposure time (*t*) according to
$A = t(2 - t)$ for $0 < t < 2$.
  a Find an expression for the rate of change of accuracy with respect to time.
  b At $t = 0.5$, find:
   i the accuracy of the x-ray
   ii the rate at which the accuracy is increasing with respect to time.
  c Find the interval in which *A* is an increasing function.

8 Consider the function $f(x) = 0.5x^2 - \frac{8}{x}$, $x \neq 0$.
  a Find $f(-2)$.
  b Find $f'(x)$.
  c Find the gradient of the graph of f at $x = -2$.

Let *T* be the tangent to the graph of f at $x = -2$.
  d Write down the equation of *T*.
  e Sketch the graph of f for $-5 \leq x \leq 5$ and $-20 \leq y \leq 20$.
  f Draw *T* on your sketch.

The tangent, *T*, intersects the graph of f at a second point, *P*.
  g Use your graphic display calculator to find the coordinates of *P*.

Mathematical Studies SL May 2015 P2 TZ2

9 A curve has equation $y = 2x^3 - 8x + 3$. Find the coordinates of two points on the curve where the gradient equals -2.

10 A curve has equation $y = x^3 - 6x^2$.
  a Find the coordinates of the two points on the curve where the gradient is zero.
  b Find the equation of the straight line which passes through these two points.

11 A curve has equation $y = 3x^2 + 6x$.
  a Find the equations of the tangents at the two points where $y = 0$.
  b Find the coordinates of the point where those two tangents intersect.

12 The gradient of the curve $y = 2x^2 + c$ at the point $(p, 5)$ equals -8. Find the values of *p* and *c*.

13 A company's monthly profit, $P, varies according to the equation $P(t) = 15t^2 - t^3$, where *t* is the number of months since the foundation of the company.
  a Find $\frac{dP}{dt}$.
  b Find the value of $\frac{dP}{dt}$ when $t = 6$ and when $t = 12$.
  c Interpret the values found in part **b**.

14 The diagram shows the graphs of the functions $f(x) = 2x + 1$ and $g(x) = x^2 + 3x - 4$.

  a **i** Find $f'(x)$.
     ii Find $g'(x)$.
  b Calculate the value of *x* for which the gradient of the two graphs is the same.
  c On a copy of the diagram above, sketch the tangent to $y = g(x)$ for this value of *x*, clearly showing the property in part **b**.

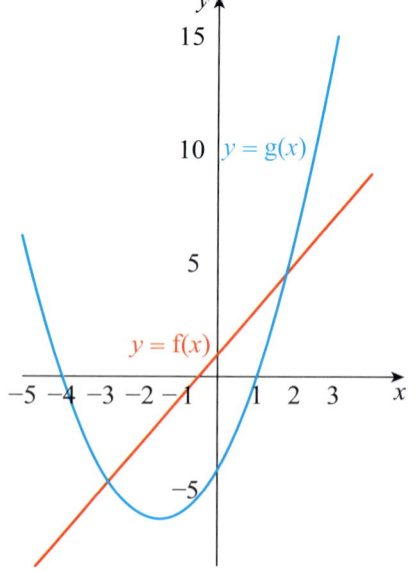

15 Consider the graph on the right:

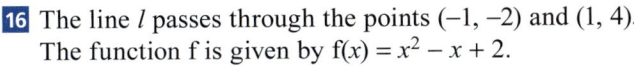

  a If the graph represents $y = f(x)$:

  i use inequalities to describe the interval in which $f(x)$ is decreasing

  ii sketch $y = f'(x)$.

  b If the graph represents $y = f'(x)$:

  i use inequalities to describe the interval in which $f(x)$ is decreasing

  ii sketch a possible graph of $y = f(x)$.

16 The line l passes through the points $(-1, -2)$ and $(1, 4)$. The function f is given by $f(x) = x^2 - x + 2$.

  a Find the gradient of l.

  b Differentiate $f(x)$ with respect to x.

  c Find the coordinates of the point where the tangent to $y = f(x)$ is parallel to the line l.

  d Find the coordinates of the point where the tangent to $y = f(x)$ is perpendicular to the line l.

  e Find the equation of the tangent to $y = f(x)$ at the point $(3, 8)$.

  f Find the coordinates of the vertex of $y = f(x)$ and state the gradient of the curve at this point.

17 Given that $f(x) = x^2 + x - 5$:

  a write down $f'(x)$

  b find the values of x for which $f'(x) = f(x)$.

18 The gradient of the curve $y = ax^2 + bx$ at the point $(2, -2)$ is 3. Find the values of a and b.

19 The gradient of the normal to the curve with equation $y = ax^2 + bx$ at the point $(1, 5)$ is $\frac{1}{3}$. Find the values of a and b.

20 For the curve with equation $y = 5x^2 - 4$, find the coordinates of the point where the tangent at $x = 1$ intersects the tangent at the point $x = 2$.

21 Let $f(x) = \dfrac{\ln(4x)}{x}$, for $0 < x \leq 5$.

Points $P(0.25, 0)$ and Q are on the curve of f. The tangent to the curve of f at P is perpendicular to the tangent at Q. Find the coordinates of Q.

Mathematics SL May 2015 Paper 2 TZ1

10 Core: Integration

ESSENTIAL UNDERSTANDINGS

- Integration describes the accumulation of limiting areas. Understanding these accumulations allows us to model, interpret and analyse real-world problems and situations.

In this chapter you will learn…

- how to integrate functions of the form $f(x) = ax^n + bx^m$, where $n, m \in \mathbb{Z}$, $n, m \neq -1$
- how to use your GDC to find the value of definite integrals
- how to find areas between a curve $y = f(x)$ and the x-axis, where $f(x) > 0$
- how to use integration with a boundary condition to determine the constant term.

CONCEPTS

The following key concepts will be addressed in this chapter:
- Areas under curves can be **approximated** by the sum of areas of rectangles which may be calculated even more accurately using integration.

PRIOR KNOWLEDGE

Before starting this chapter, you should already be able to complete the following:

1 Write $\dfrac{(x-2)^2}{x^4}$ in the form $ax^n + bx^m + cx^k$.

2 Differentiate the function $f(x) = 4x^{-3} + 2x - 5$.

■ Figure 10.1 What is being accumulated in each of these photographs?

Core: Integration

In many physical situations it is easier to model or measure the rate of change of a quantity than the quantity itself. For example, if you know the forces acting on an object that cause it to move, then you can find an expression for its acceleration, which is the rate of change of velocity. You cannot directly find an expression for the velocity itself, however, so to find that you now need to be able to undo the process of differentiation to go from rate of change of velocity to velocity.

Starter Activity

Look at the images in Figure 10.1. Discuss how, in physical situations such as these, rate of change is related to accumulation of a quantity.

Now look at this problem:

Estimate the shaded area between the graph and the x-axis.

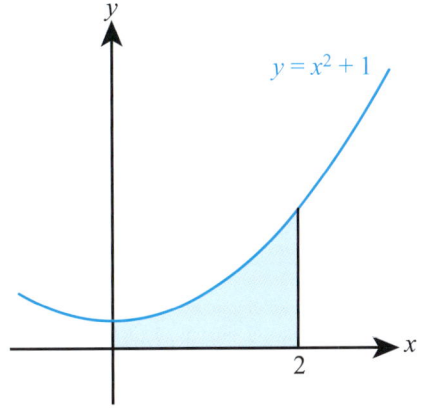

> **LEARNER PROFILE – Thinkers**
> Have you ever tried making an 'essay plan' for a problem? Sometimes it is good to map out a route through a problem without getting bogged down in the algebraic and arithmetic details.

10A Anti-differentiation

■ Introduction to integration as anti-differentiation of functions of the form $f(x) = ax^n + bx^m + ...$

If you are given the gradient function f(x), you can find the function f(x) by 'undoing' the differentiation. This process of reversing differentiation is known as **integration**.

However, it isn't possible to fully specify the original function f(x) without further information. For example, if you were given $f'(x) = 2x$ you would know that f(x) must contain x^2 since this differentiates to give $2x$, but you wouldn't know whether you had $f(x) = x^2 + 3$, $f(x) = x^2 - \frac{2}{5}$, or in fact $f(x) = x^2 + c$ for any constant c.

So, we say that if $f'(x) = 2x$ then $f(x) = x^2 + c$, or using the integration symbol,

$$\int 2x \, dx = x^2 + c$$

The dx indicates that the variable you are integrating with respect to is x, just as in $\frac{dy}{dx}$ it indicates that the variable you are differentiating with respect to is x.

Reversing the basic rule for differentiation gives a basic rule for integration:

$$\int x^n \, dx = \frac{1}{n+1} x^{n+1} + c$$

where $n \in \mathbb{Z}$, $n \neq -1$ and $c \in \mathbb{R}$.

> **Tip**
> Note the condition $n \neq -1$, which ensures that you are not dividing by zero.

WORKED EXAMPLE 10.1

If $f'(x) = x^{-4}$ find an expression for f(x).

Integrate to find f(x) .... $f(x) = \int x^{-4} \, dx$

Use $\int x^n \, dx = \frac{1}{n+1} x^{n+1} + c$ .... $= \frac{1}{-4+1} x^{-4+1} + c$

$= -\frac{1}{3} x^{-3} + c$

As with differentiation, the basic rule is unaffected by multiplication by a constant, and it can be extended to deal with sums.

KEY POINT 10.1

$$\int (ax^n + bx^m + ...) \, dx = \frac{a}{n+1} x^{n+1} + \frac{b}{m+1} x^{m+1} + ... + c$$

where a, b and c are real constants and $n, m \in \mathbb{Z}$, $n, m \neq -1$.

10A Anti-differentiation

WORKED EXAMPLE 10.2

If $\dfrac{dy}{dx} = 4x^3 + 10x^4 + 3$ find an expression for y.

Integrate to find y

$$y = \int (4x^3 + 10x^4 + 3)dx$$

Integrate term by term. You can think of 3 as $3x^0$, although after a while you will probably just be able to write the integral of a k as kx without going through this process every time

$$= \dfrac{4}{3+1}x^{3+1} + \dfrac{10}{4+1}x^{4+1} + \dfrac{3}{0+1}x^{0+1} + c$$

$$= x^4 + 2x^5 + 3x + c$$

Again, as with differentiation, you don't have a rule for integrating products or quotients of functions. So, before you can integrate these you will need to convert them into sums of terms of the form ax^n.

> The integral sign was first use by the German mathematician Gottfried Leibniz in 1675. It is an old letter S, which originates from one of its interpretations as a sum of infinitely small rectangles.

WORKED EXAMPLE 10.3

Find $\displaystyle\int \dfrac{2x-5}{x^3} dx$.

Use the laws of exponents to rewrite as a sum of functions of the form ax^n

Then integrate

$$\int \dfrac{2x-5}{x^3} dx = \int \left(\dfrac{2x}{x^3} - \dfrac{5}{x^3}\right) dx$$

$$= \dfrac{2}{-1}x^{-1} - \dfrac{5}{-2}x^{-2} + c$$

$$= \int (2x^{-2} - 5x^{-3})$$

$$= -2x^{-1} + \dfrac{5}{2}x^{-2} + c$$

Be the Examiner 10.1

Find $\displaystyle\int 2x(x^2 + 4) dx$.

Which is the correct solution? Identify the errors made in the incorrect solutions.

Solution 1	Solution 2	Solution 3
$\displaystyle\int 2x(x^2+4)dx = x^2\left(\dfrac{x^3}{3}+4x\right)+c$ $= \dfrac{x^5}{3}+4x^3+c$	$\displaystyle\int 2x(x^2+4)dx = \int(2x^3+8x)dx$ $= \dfrac{1}{2}x^4+4x^2+c$	$\displaystyle\int 2x(x^2+4)dx = 2x\int(x^2+4)dx$ $= 2x\left(\dfrac{x^3}{3}+4x\right)+c$ $= \dfrac{2}{3}x^4+8x^2+c$

Anti-differentiation with a boundary condition

You can determine the constant when you integrate if, alongside the gradient function f'(x), you are also given the value of a function at a particular value of *x*. This extra information is known as a boundary condition.

WORKED EXAMPLE 10.4

If $\frac{dy}{dx} = x^2 + 1$ and $y = 5$ when $x = 3$, find an expression for *y* in terms of *x*.

Integrate to find *y*
$$y = \int (x^2 + 1) dx$$
$$= \frac{1}{3}x^3 + x + c$$

Substitute in the boundary condition $x = 3$, $y = 5$ in order to find *c*

When $x = 3$, $y = 5$:
$$5 = \frac{1}{3}(3)^3 + 3 + c$$
$$5 = 12 + c$$
$$c = -7$$

Make sure you state the final answer

So, $y = \frac{1}{3}x^3 + x - 7$

Exercise 10A

For questions 1 to 3, use the method demonstrated in Worked Example 10.1 to find an expression for f(x).

1. a $f'(x) = x^3$
   b $f'(x) = x^5$
2. a $f'(x) = x^{-2}$
   b $f'(x) = x^{-3}$
3. a $f'(x) = 0$
   b $f'(x) = 1$

For questions 4 to 14, use the method demonstrated in Worked Example 10.2 to find an expression for *y*.

4. a $\frac{dy}{dx} = 3x^2$
   b $\frac{dy}{dx} = -5x^4$
5. a $\frac{dy}{dx} = -7x^3$
   b $\frac{dy}{dx} = 3x^7$
6. a $\frac{dy}{dx} = \frac{3}{2}x^5$
   b $\frac{dy}{dx} = -\frac{5}{3}x^9$
7. a $\frac{dy}{dx} = -3x^{-4}$
   b $\frac{dy}{dx} = 5x^{-6}$
8. a $\frac{dy}{dx} = 6x^{-5}$
   b $\frac{dy}{dx} = -4x^{-3}$
9. a $\frac{dy}{dx} = -\frac{2}{5}x^{-2}$
   b $\frac{dy}{dx} = \frac{7}{4}x^{-8}$
10. a $\frac{dy}{dx} = 3x^2 - 4x + 5$
    b $\frac{dy}{dx} = 7x^4 + 6x^2 - 2$
11. a $\frac{dy}{dx} = x - 4x^5$
    b $\frac{dy}{dx} = 6x^3 - 5x^7$
12. a $\frac{dy}{dx} = \frac{3}{4}x^2 + \frac{7}{3}x$
    b $\frac{dy}{dx} = \frac{4}{5} - \frac{2}{3}x^4$
13. a $\frac{dy}{dx} = 2x^{-5} + 3x$
    b $\frac{dy}{dx} = 5 - 9x^{-4}$
14. a $\frac{dy}{dx} = \frac{5}{2}y^{-2} - \frac{10}{7}y^{-6}$
    b $\frac{dy}{dx} = \frac{4}{3}y^{-3} - \frac{2}{5}y^{-7}$

For questions 15 to 17, find the given integrals by first expanding the brackets.

15 a $\int x^2(x+5)\,dx$
16 a $\int (x+2)(x-3)\,dx$
17 a $\int (3x-x^{-1})^2\,dx$

b $\int 3x(2-x)\,dx$
b $\int (4-x)(x-1)\,dx$
b $\int (2x+x^{-3})^2\,dx$

For questions 18 and 19, use the method demonstrated in Worked Example 10.3 to find the given integrals by first writing the expression as a sum of terms of the form ax^n.

18 a $\int \dfrac{3x^2-2x}{x}\,dx$
b $\int \dfrac{5x^3-3x^2}{x^2}\,dx$

19 a $\int \dfrac{4x-7}{2x^4}\,dx$
b $\int \dfrac{2x^3-3x^2}{4x^5}\,dx$

For questions 20 to 22, use the method demonstrated in Worked Example 10.4 to find an expression for y in terms of x.

20 a $\dfrac{dy}{dx}=3x^2$ and $y=6$ when $x=0$
b $\dfrac{dy}{dx}=5x^4$ and $y=5$ when $x=0$

21 a $y'=x^3-12x$ and $y=-10$ when $x=2$
b $y'=3-10x^4$ and $y=5$ when $x=1$

22 a $y'=\dfrac{6}{x^3}+4x$ and $y=16$ when $x=3$
b $y'=9x^2-\dfrac{2}{x^2}$ and $y=-9$ when $x=-1$

23 Find $\int \dfrac{4}{3t^2}-\dfrac{2}{t^5}\,dt$.

24 A curve has the gradient given by $\dfrac{dy}{dx}=3x^2-4$ and the curve passes through the point $(1, 4)$. Find the expression for y in terms of x.

25 Find the equation of the curve with gradient $\dfrac{dy}{dx}=\dfrac{4}{x^2}-3x^2$ which passes through the point $(2, 0)$.

26 Find $\int (3x-2)(x^2+1)\,dx$.

27 Find $\int z^2\left(z+\dfrac{1}{z}\right)dz$.

28 Find $\int \dfrac{x^5-2x}{3x^3}\,dx$.

29 The rate of growth of the mass (m kg) of a weed is modelled by
$\dfrac{dm}{dt}=kt+0.1$ where t is measured in weeks.
When $t=2$ the rate of growth is 0.5 kg per week.

a Find the value of k.

b If the weed initially has negligible mass, find the mass after 5 weeks.

30 A bath is filling from a hot water cylinder. At time t minutes the bath is filling at a rate given by $\dfrac{80}{t^2}$ litres per minute. When $t=1$ the bath is empty.

a How much water is in the bath when $t=2$?

b At what time will the bath hold 60 litres?

c The capacity of the bath is 100 litres. Using technology, determine whether the bath will ever overflow if this process is left indefinitely.

10B Definite integration and the area under a curve

■ Definite integrals using technology

So far you have been carrying out a process called **indefinite integration** – it is indefinite because you have the unknown constant each time.

A similar process, called **definite integration**, gives a numerical answer – no *x*s and no unknown constant.

You only need to be able to evaluate definite integrals on your GDC, but to do so you need to know the upper and lower limit of the integral: these are written at the top and bottom respectively of the integral sign.

> **WORKED EXAMPLE 10.5**
>
> Evaluate $\int_{1}^{4} x^3 \, dx$
>
> Use your GDC to evaluate the definite integral: $\int_{1}^{4} x^3 \, dx = 63.75$
>
>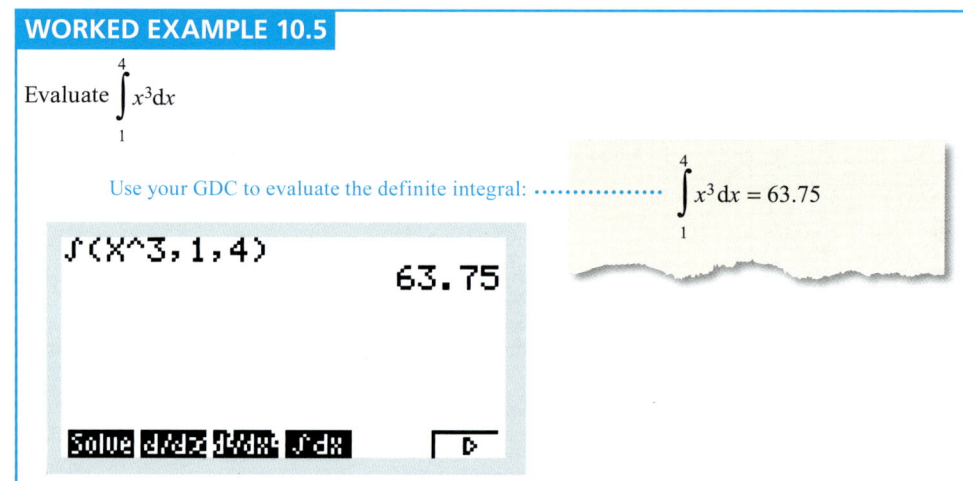

Tip

In the expression $\int_{1}^{4} x^3 \, dx$, there is nothing special about *x*. It could be replaced with *y* or *t* or any other letter and the answer would be the same. However, your calculator might need you to turn any such expressions into one involving *x*.

■ Areas between a curve $y = f(x)$ and the *x*-axis where $f(x) > 0$

In the same way that the value of the derivative at a point has a geometrical meaning (it is the gradient of the tangent to the curve at that point) so does the value of a definite integral.

Perhaps surprisingly, the value of a definite integral is the area between a curve and the *x*-axis between the limits of integration.

10B Definite integration and the area under a curve

Tip

In a question where you are asked to find the area under a curve, it is important that you write down the correct definite integral before finding the value with your GDC.

KEY POINT 10.2

- $A = \int_a^b y \, dx$

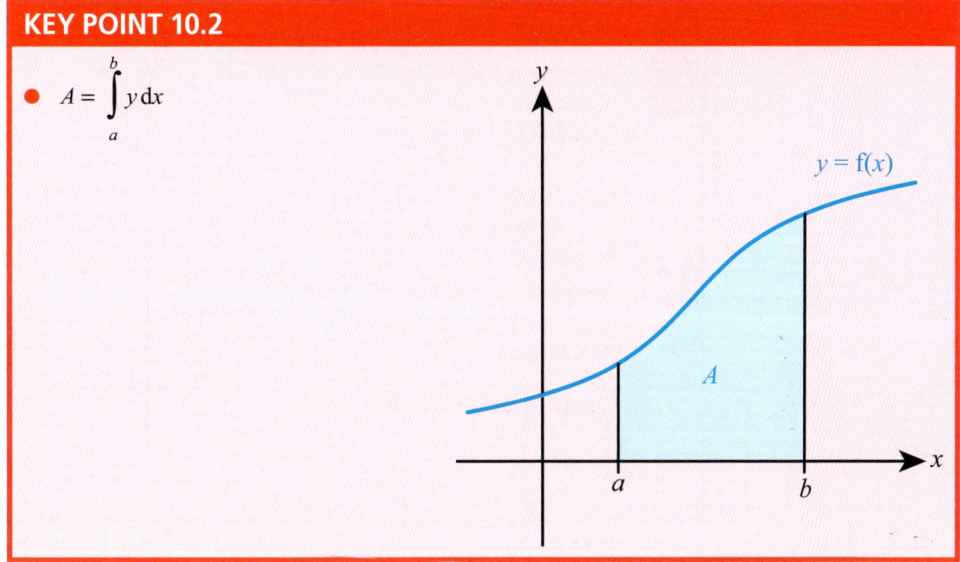

WORKED EXAMPLE 10.6

The diagram on the right shows the curve $y = x^2 - x^3$.

Find the shaded area.

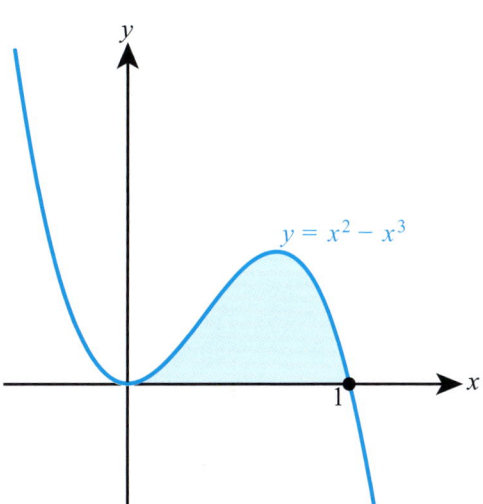

The area is the definite integral between the limits of 0 and 1. Write down the definite integral to be found ········ Area = $\int_0^1 (x^2 - x^3) \, dx$

Use your GDC to evaluate the definite integral ········ = 0.0833

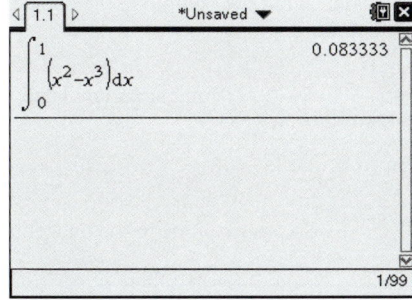

 **TOOLKIT: Problem Solving**

Find an expression for the sum of the areas of the three rectangles bounded by the curve $y = x^2$ between $x = 1$ and $x = 2$.

Find an expression in terms of n for n rectangles of equal width between $x = 1$ and $x = 2$.

Use your GDC to find the limit of your expression as n gets very large.

Compare this to the value of
$$\int_1^2 x^2 \, dx.$$

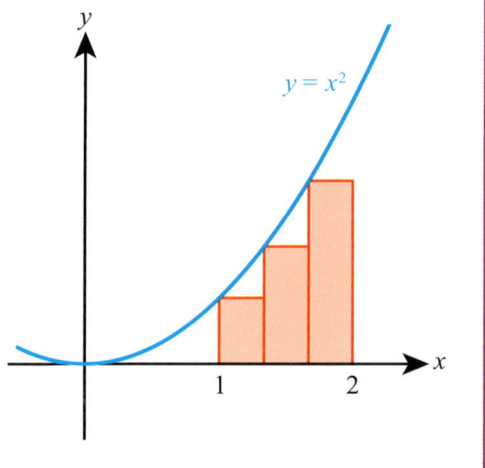

CONCEPTS – APPROXIMATION

In the problem-solving task above, the area under the curve is not the area of the rectangle – it is only an **approximation**. However, this approximation gets better and better the more rectangles you have. How many would you need to have a 'good enough' approximation for a real-world situation?

It gets so good, that we use this approximation to define the area under the curve. The idea of 'perfect approximations' might make your head spin, but it was put on a firm foundation by the work of eminent mathematicians such as Augustin Cauchy and it forms the basis for modern calculus.

The problem-solving activity above suggests an important further interpretation of integration as the amount of material accumulated if you know the rate of accumulation. We can write this as shown below.

Tip

You can think of integration as adding up lots of little bits of f(t).

KEY POINT 10.3

- $f(a) = f(b) + \int_b^a f'(t) \, dt$

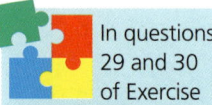 In questions 29 and 30 of Exercise 10A you were asked to do a similar task without using definite integration. You could now do these on a calculator.

WORKED EXAMPLE 10.7

Water is dripping into a bucket at a rate of t^2 cm^3 per minute, after a time of t minutes. The volume in the bucket after 1 minute is 5 cm^3. Find the volume in the bucket after 4 minutes.

The volume accumulated is the integral of the rate between 1 and 4 minutes. We use these as the limits on the integral. ···· Volume accumulated $= \int_1^4 t^2 \, dt$

Use your GDC to evaluate this ···· $= 21$ cm^3

We need to add the volume accumulated to the volume already in the bucket ···· So total volume in the bucket is $21 + 5 = 26$ cm^3

Exercise 10B

For questions 1 and 2, use the method demonstrated in Worked Example 10.5 to evaluate the given definite integrals with your GDC.

1. **a** $\displaystyle\int_{-1}^{3} 2x^4 \, dx$ **b** $\displaystyle\int_{1}^{4} 3x^{-2} \, dx$

2. **a** $\displaystyle\int_{1}^{2} (2 - 7t^{-3}) \, dt$ **b** $\displaystyle\int_{2}^{5} (t - 4t^2) \, dt$

For questions 3 and 4, use the method demonstrated in Worked Example 10.6 to find the shaded area.

3. **a**   **b**

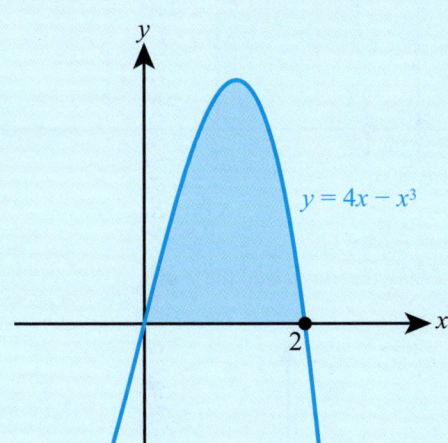
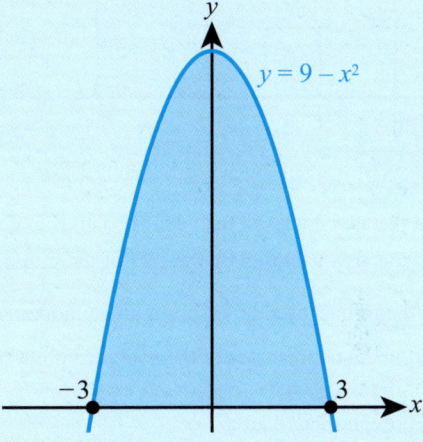

4. **a**   **b**

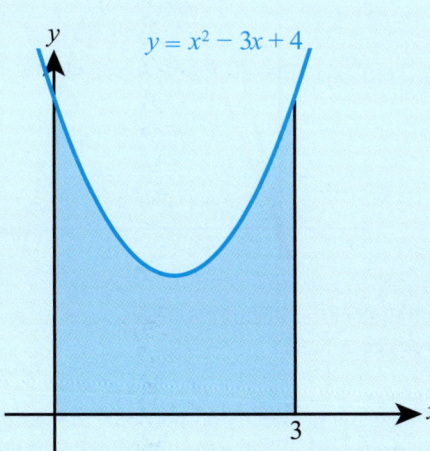
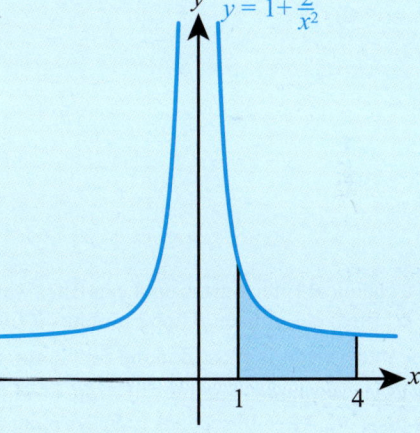

5. Evaluate $\displaystyle\int_{2}^{5} 2x + \frac{1}{x^2} \, dx$.

6. Evaluate $\displaystyle\int_{1}^{2} (x - 1)^3 \, dx$.

7 Find the shaded area.

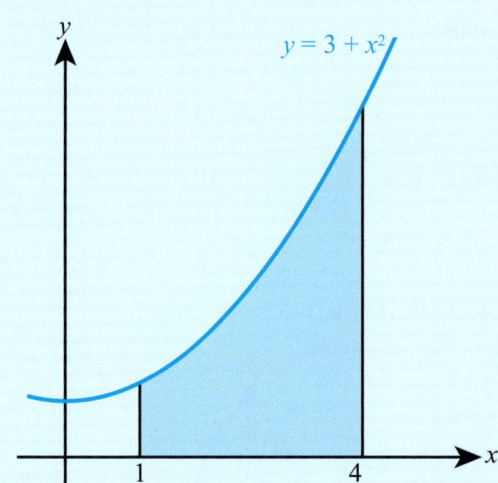

8 Find the shaded area.

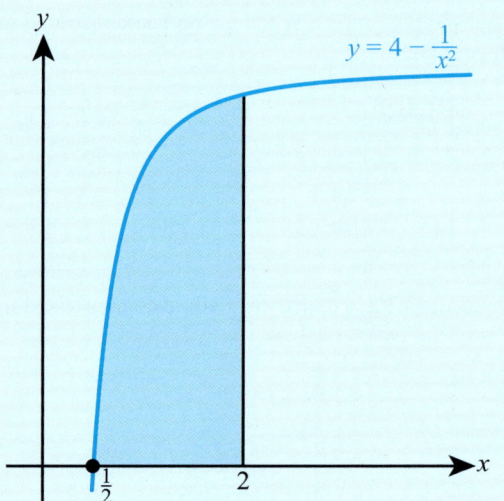

9 Find the area enclosed between the x-axis and the graph of $y = -x^2 + 8x - 12$.

10 Find the area enclosed by the part of the graph of $y = 9 - x^2$ for $x > 0$, the x-axis and the y-axis.

11 The diagram shows the graphs with equations $y = 5x - x^2$ and $y = \dfrac{5x}{2}$.

  a Find the coordinates of the non-zero point of intersection of the two graphs.

  b Find the shaded area.

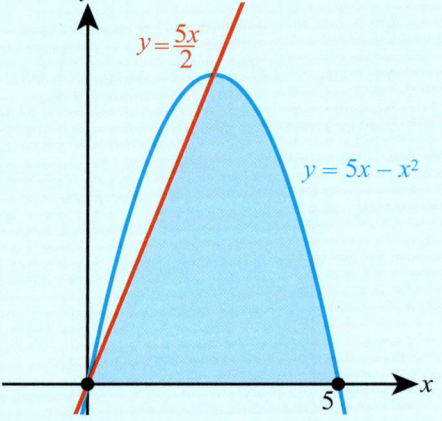

12 As a chemical filter warms up it can filter water more quickly. At a time t after it starts, it filters water at a rate of $5 + kt^2$ litres per minute. After 2 minutes it filters at a rate of 9 litres per minute.

How much water is filtered in the first 2 minutes?

13 For the first minute of activity, the rate at which a paint sprayer emits paint is directly proportional to the time it has been on. After 10 seconds it sprays paint at a rate of 20 grams per second. What is the total amount of paint it sprays in the first minute of activity?

14 A model suggests that sand falls through a timer at a rate of $\dfrac{100}{t^3}$ g where t is the time in seconds after it was started. When $t = 1$ there are $10\,\text{g}$ of sand in the base of the timer.

  a Find the amount of sand in the base 5 seconds after it was started.

  b Use technology to estimate the amount of sand that will eventually fall into the base.

  c Provide one criticism of this model.

15 The function f satisfies $f'(x) = x^2$ and $f(0) = 4$. Find $\displaystyle\int_0^3 f(x)\,dx$.

16 Given that $\displaystyle\int f(x)\,dx = 4\left(x^3 + \dfrac{1}{x^2} + c\right)$, find an expression for f(x).

17 Find the area enclosed by the curve $y = x^2$, the y-axis and the line $y = 4$.

18 A one-to-one function $f(x)$ satisfies $f(0) = 0$ and $f(x) \geq 0$. If $\int_0^a f(x)dx = A$ for $a > 0$, find an expression for $\int_0^{f(a)} f^{-1}(x)dx$.

Checklist

- You should be able to integrate functions of the form $f(x) = ax^n + bx^m$, where $n, m \in \mathbb{Z}$, $n, m \neq -1$:
  - $\int (ax^n + bx^m + \ldots)dx = \dfrac{a}{n+1}x^{n+1} + \dfrac{b}{m+1}x^{m+1} + \ldots + c$

  where a, b and c are constants and $n, m \in \mathbb{Z}$, $n, m \neq -1$.

- You should be able to use integration with a boundary condition to determine the constant term.

- You should be able to use your GDC to find the value of definite integrals.

- You should be able to find areas between a curve $y = f(x)$ and the x-axis, where $f(x) > 0$:
  - $A = \int_a^b y\,dx$

  where the area is bounded by the lines $x = a$ and $x = b$

- You should be able to use integration to find the amount of material accumulated:
  - $f(b) = f(a) + \int_a^b f'(t)dt$

Mixed Practice

1 Find $\int x^3 - \dfrac{3}{x^2} dx$.

2 Find $\int 4x^2 - 3x + 5 dx$.

3 Evaluate $\int_1^5 \dfrac{2}{x^4} dx$.

4 A curve has gradient given by $\dfrac{dy}{dx} = 3x^2 - 8x$ and passes through the point (1, 3). Find the equation of the curve.

5 Find the value of a such that $\int_2^a 2 - \dfrac{8}{x^2} dx = 9$.

6 The graph of $y = 9x - x^2 - 8$ is shown in the diagram.

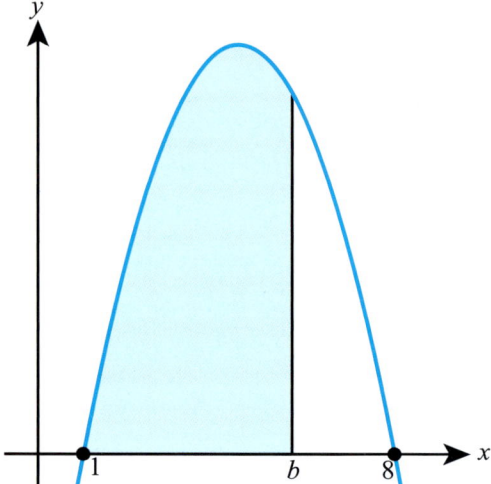

Given that the shaded area equals 42.7, find the value of b correct to one decimal place.

7 The diagram shows the graph of $y = -x^3 + 9x^2 - 24x + 20$.

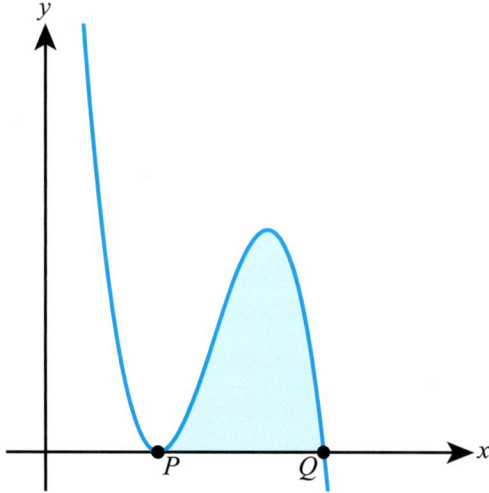

a Find the coordinates of P and Q.
b Find the shaded area.

8 The diagram shows the curve with equation $y = 0.2x^2$ and the tangent to the curve at the point where $x = 4$.

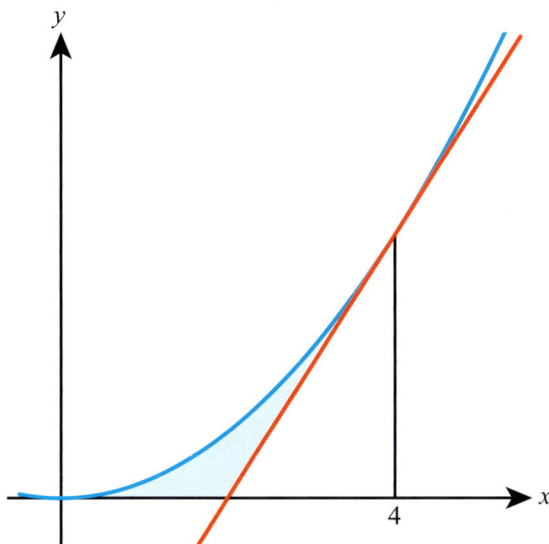

a Find the equation of the tangent.
b Show that the tangent crosses the x-axis at the point $(2, 0)$.
c Find the shaded area.

9 The diagram shows the graph of $y = \dfrac{1}{x^2}$ and the normal at the point where $x = 1$.

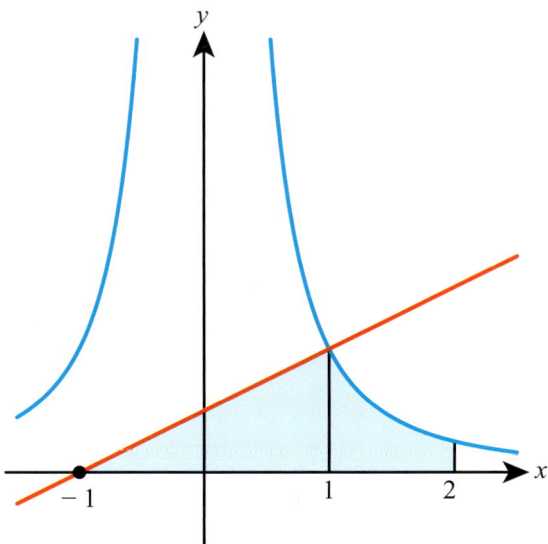

a Show that the normal crosses the x-axis at $(-1, 0)$.
b Find the shaded area.

10 The gradient of the normal to the curve $y = f(x)$ at any point equals x^2. If $f(1) = 2$, find $f(2)$.

11 The rate at which water is accumulated in a rainwater measuring device in a storm is given by $20t \, \text{cm}^3/\text{minute}$.

If the container is initially empty, find the volume of water in the container after 10 minutes.

12 A puppy at age 6 months has a mass of 2.3 kg. Its growth after this point is modelled by

$\dfrac{A}{20} + c$ kg/month where A is the puppy's age in months.

When the puppy is 10 months old it is growing at a rate of 1.5 kg per month. Use this model to estimate the mass of the puppy when it is 18 months old.

13 The function $f(x)$ is such that $f'(x) = 3x^2 + k$ where k is a constant. If $f(1) = 13$ and $f(2) = 24$, find $f(3)$.

14 A nutritionist designs an experiment to find the energy contained in a nut. They burn the nut and model the energy emitted as $\dfrac{k}{t^2}$ calories for $t > 1$ where t is in seconds.

They place a beaker of water above the burning nut and measure the energy absorbed by the water using a thermometer.
   a State one assumption that is being made in using this experimental setup to measure the energy content of the nut.
   b When $t = 1$ the water has absorbed 10 calories of energy. When $t = 2$ the water has absorbed 85 calories of energy. Use technology to estimate the total energy emitted by the nut if it is allowed to burn indefinitely.

15 The gradient at every point on a curve is proportional to the square of the x-coordinate at that point. Find the equation of the curve, given that it passes through the points $(0, 3)$ and $\left(1, \dfrac{14}{3}\right)$.

16 The function f satisfies $f'(x) = 3x - x^2$ and $f(4) = 0$. Find $\displaystyle\int_0^4 f(x)\,dx$.

17 Given that $\displaystyle\int f(x)\,dx = 3x^2 - \dfrac{2}{x} + c$, find an expression for $f(x)$.

Mixed Practice

18 The diagram shows a part of the curve with equation $y = x^2$. Point A has coordinates $(0, 9)$.

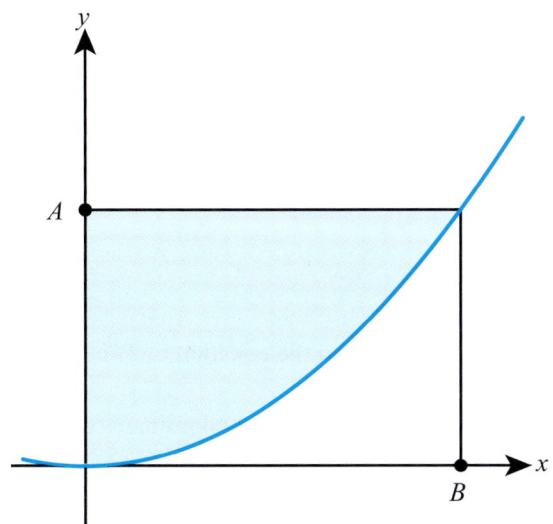

a Find the coordinates of point B.
b Find the shaded area.

Core SL content: Review Exercise

1 An arithmetic sequence has first term 7 and third term 15.
   a Find the common difference of the series.
   b Find the 20th term.
   c Find the sum of the first 20 terms.

2 Line L has equation $2x - 3y = 12$.
   a Find the gradient of L.
   b The point $P(9, k)$ lies on L. Find the value of k.
   c Line M is perpendicular to L and passes through the point P. Find the equation of M in the form $ax + by = c$.

3 a Sketch the graph of $y = (x + 2)e^{-x}$, labelling all the axis intercepts and the coordinates of the maximum point.
   b State the equation of the horizontal asymptote.
   c Write down the range of the function $f(x) = (x + 2)e^{-x}$, $x \in \mathbb{R}$.

4 A function is defined by $f(x) = \sqrt{x - a}$. The graph of $y = f(x)$ is shown in the diagram.

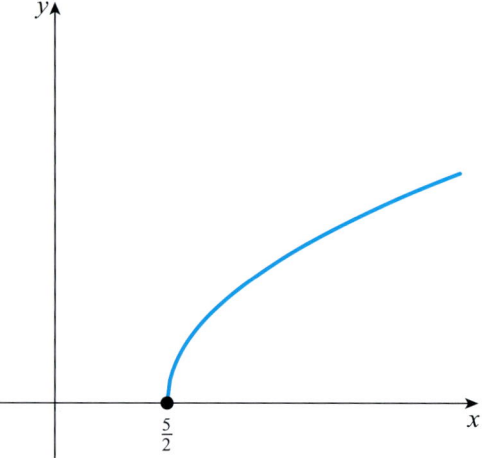

   a Write down the value of a and state the domain of f.
   b Write down the range of f.
   c Find $f^{-1}(4)$.

5 In triangle ABC, $AB = 12$ cm, $BC = 16$ cm and $AC = 19$ cm.
   a Find the size of the angle ACB.
   b Calculate the area of the triangle.

6 A random variable X follows a normal distribution with mean 26 and variance 20.25.
   a Find the standard deviation of X.
   b Find $P(21.0 < x < 25.3)$.
   c Find the value of a such that $P(X > a) = 0.315$.

7 The curve in the diagram has equation $y = (4 - x^2)e^{-x}$.

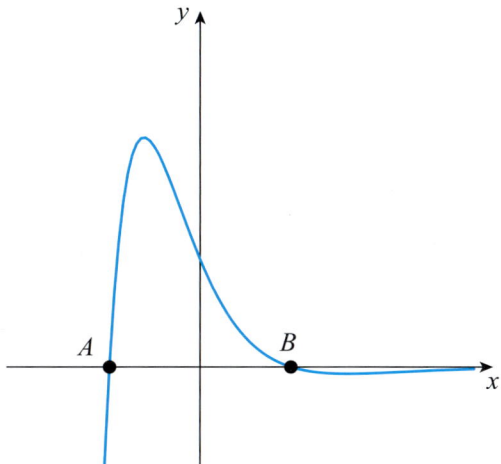

   a Find the coordinates of the points A and B.
   b Find the shaded area.

8 The IB grades attained by a group of students are listed as follows.
   6 4 5 3 7 3 5 4 2 5
   a Find the median grade.
   b Calculate the interquartile range.
   c Find the probability that a student chosen at random from the group scored at least a grade 4.

Mathematical Studies SL May 2015 Paper 1 TZ1 Q2

9 *ABCDV* is a solid glass pyramid. The base of the pyramid is a square of side 3.2 cm. The vertical height is 2.8 cm. The vertex *V* is directly above the centre *O* of the base.

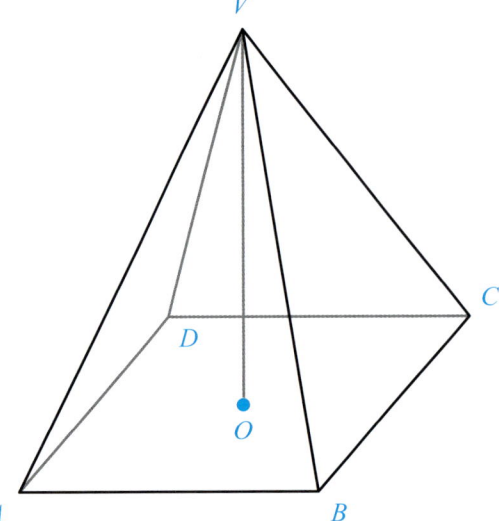

  a Calculate the volume of the pyramid.
  b The glass weighs 9.3 grams per cm³. Calculate the weight of the pyramid.
  c Show that the length of the sloping edge *VC* of the pyramid is 3.6 cm.
  d Calculate the angle at the vertex $B\hat{V}C$.
  e Calculate the total surface area of the pyramid.

  Mathematical Studies SL November 2007 Paper 2 Q2

10 a Sketch the graph of the function $f(x) = \dfrac{2x+3}{x+4}$, for $-10 \leq x \leq 10$, indicating clearly the axis intercepts and any asymptotes.
  b Write down the equation of the vertical asymptote.
  c On the same diagram sketch the graph of $g(x) = x + 0.5$.
  d Using your graphical display calculator write down the coordinates of one of the points of intersection on the graphs of f and g, giving your answer correct to five decimal places.
  e Write down the gradient of the line $g(x) = x + 0.5$.
  f The line *L* passes through the point with coordinates (−2, −3) and is perpendicular to the line $g(x)$. Find the equation of *L*.

  Mathematical Studies May 2008 Paper 2 TZ1 Q1

11 Find the equation of the normal to the curve $y = x + 3x^2$ at the point where $x = -2$. Give your answer in the form $ax + by + c = 0$.

12 a Find the value of x given that $\log_{10} x = 3$.
  b Find $\log_{10} 0.01$.

13 a Expand and simplify $(2x)^3(x - 3x^{-5})$.
  b Differentiate $y = (2x)^3(x - 3x^{-5})$.

14 A geometric series has first term 18 and common ratio r.
  a Find an expression for the 4th term of the series.
  b Write down the formula for the sum of the first 15 terms of this series.
  c Given that the sum of the first 15 terms of the series is 26.28 find the value of r.

15 Theo is looking to invest £200 for 18 months. He needs his investment to be worth £211 by the end of the 18 months. What interest rate does he need? Give your answer as a percentage correct to one decimal place.

16 The distribution of a discrete random variable X is given by $P(X = x) = \dfrac{k}{x^2}$ for $X = 1, 2, 3, 4$.
Find $E(X)$.

17 A box holds 240 eggs. The probability that an egg is brown is 0.05.
  a Find the expected number of brown eggs in the box.
  b Find the probability that there are 15 brown eggs in the box.
  c Find the probability that there are at least 10 brown eggs in the box.

Mathematics SL May 2011 Paper 2 TZ1 Q5

18 The vertices of quadrilateral $ABCD$ as shown in the diagram are $A(3, 1)$, $B(0, 2)$, $C(-2, 1)$ and $D(-1, -1)$.

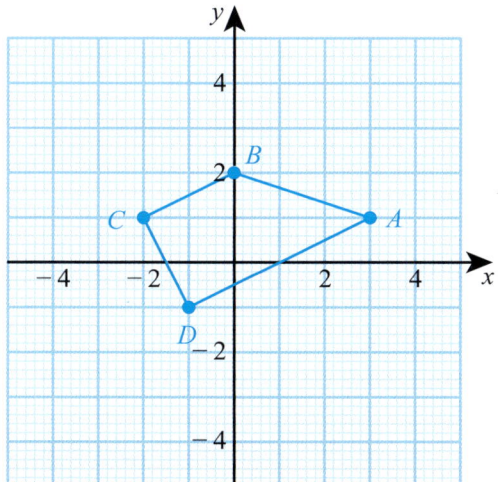

  a Calculate the gradient of line CD.
  b Show that line AD is perpendicular to line CD.
  c Find the equation of line CD. Give your answer in the form $ax + by = c$ where $a, b, c \in \mathbb{Z}$.
  Lines AB and CD intersect at point E. The equation of line AB is $x + 3y = 6$.
  d Find the coordinates of E.
  e Find the distance between A and D.
  The distance between D and E is $\sqrt{20}$.
  f Find the area of triangle ADE.

Mathematical Studies May 2009 Paper 2 TZ1 Q3

19 a Jenny has a circular cylinder with a lid. The cylinder has height 39 cm and diameter 65 mm.
  i Calculate the volume of the cylinder **in cm³**. Give your answer correct to **two** decimal places.

  The cylinder is used for storing tennis balls.

  Each ball has a **radius** of 3.25 cm.

  ii Calculate how many balls Jenny can fit in the cylinder if it is filled to the top.

  iii I Jenny fills the cylinder with the number of balls found in part **ii** and puts the lid on. Calculate the volume of air inside the cylinder in the spaces between the tennis balls.

  II Convert your answer to **iii I** into cubic metres.

b An old tower (BT) leans at 10° away from the vertical (represented by line TG).

The base of the tower is at B so that $M\hat{B}T = 100°$.

Leonardo stands at L on flat ground 120 m away from B in the direction of the lean.

He measures the angle between the ground and the top of the tower T to be $B\hat{L}T = 26.5°$.

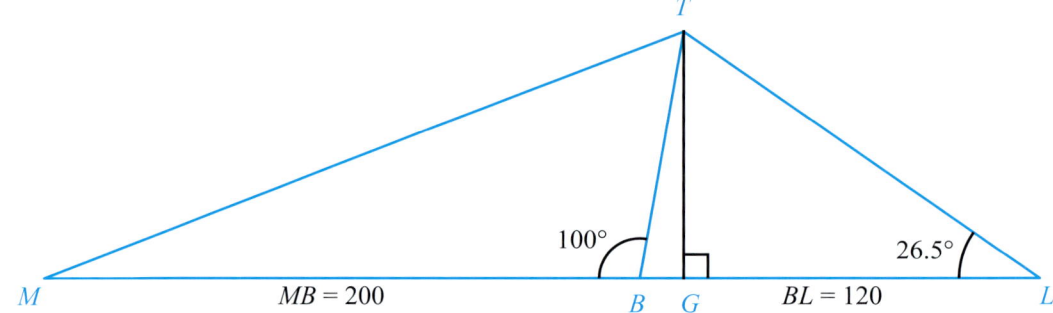

  i I Find the value of angle $B\hat{T}L$.

  II Use triangle BTL to calculate the sloping distance BT from the base B to the top, T of the tower.

  ii Calculate the vertical height TG of the top of the tower.

  iii Leonardo now walks to point M, a distance 200 m from B on the opposite side of the tower. Calculate the distance from M to the top of the tower at T.

  Mathematical Studies SL May 2007 Paper 2 Q2

20 The following table shows the number of bicycles, x, produced daily by a factory and their total production cost, y, in US dollars (USD). The table shows data recorded over seven days.

	Day 1	Day 2	Day 3	Day 4	Day 5	Day 6	Day 7
Number of bicycles, x	12	15	14	17	20	18	21
Production cost, y	3900	4600	4100	5300	6000	5400	6000

a i Write down the Pearson's product–moment correlation coefficient, r, for these data.
  ii Hence comment on the result.
b Write down the equation of the regression line y on x for these data, in the form $y = ax + b$.
c Estimate the total cost, to the nearest USD, of producing 13 bicycles on a particular day.

All the bicycles that are produced are sold. The bicycles are sold for 304 USD each.

d Explain why the factory does not make a profit when producing 13 bicycles on a particular day.
e i Write down an expression for the total selling price of x bicycles.
  ii Write down an expression for the profit the factory makes when producing x bicycles on a particular day.
  iii Find the least number of bicycles that the factory should produce, on a particular day, in order to make a profit.

Mathematical Studies May 2015 Paper 2 TZ2 Q6

21 It is known that, among all college students, the time taken to complete a test paper is normally distributed with mean 52 minutes and standard deviation 7 minutes.
  a Find the probability that a randomly chosen student completes the test in less than 45 minutes.
  b In a group of 20 randomly chosen college students, find the probability that:
    i exactly one completes the test in less than 45 minutes
    ii more than three complete the test in less than 45 minutes.

22 In a class of 26 students, 15 study French, 14 study biology and 8 study history.

Of those students, 7 study both French and biology, 4 study French and history, and 3 study biology and history.

  a Using a Venn diagram, or otherwise, find how many students study all three subjects.
  b Find the probability that a randomly selected student studies French only.
  c Given that a student studies French, what is the probability that they do not study biology?
  d Two students are selected at random. What is the probability that at least one of them studies history?

11 Applications and interpretation: Number and finance

ESSENTIAL UNDERSTANDINGS

- Number and algebra allow us to represent patterns, show equivalences and make generalizations which enable us to model real-world situations.
- Algebra is an abstraction of numerical concepts and employs variables to solve mathematical problems.

In this chapter you will learn…
- how to round numbers to a given number of decimal places or significant figures
- how to find the range of possible values for a number that has been rounded
- how to calculate percentage errors when approximating a value
- how to check whether the result of a calculation is reasonable
- how to use the TVM solver on your GDC to solve problems in financial mathematics involving savings accounts and loan repayments.

CONCEPTS

The following key concepts will be addressed in this chapter:
- Mathematical financial **models** such as compounded growth allow computation, evaluation and interpretation of debt and investment both **approximately** and accurately.
- **Approximation** of numbers adds uncertainty or inaccuracy to calculations, leading to potential errors, but can be useful when handling extremely large or small **quantities**.

LEARNER PROFILE – Balanced

Is being a good mathematician the same as being good at doing mathematics exams? Always remember that in the real world you are rarely asked to do things in silence, alone and without any reference materials. Although examinations are important, they need to be put into perspective and should not be seen as the only purpose of studying mathematics.

■ Figure 11.1 How accurate is accurate enough?

Applications and interpretation: Number and finance

> **PRIOR KNOWLEDGE**
>
> Before starting this chapter, you should already be able to complete the following:
>
> 1 Round these numbers to the given level of accuracy:
>
>   **a** 1268.7 to the nearest integer
>
>   **b** 1364.8 to the nearest 10
>
> 2 The population of bacteria increases by 15% every hour. Initially there were 1200 bacteria.
>
>   **a** Use technology to find the size of the population after 12 hours.
>
>   **b** After how many whole hours will the population first exceed 10 000?

Calculations in real-life applications of mathematics are hardly ever exact, and you need to be able to pick the most appropriate degree of accuracy for each situation. When calculations are made with rounded numbers, the final uncertainty can be a lot greater than the uncertainty in the starting values. Understanding how measurement error is compounded by calculation has a number of important applications, including in the design of efficient computer algorithms.

This chapter also explores some further financial applications of mathematics. Most of us will have a savings account or take out a loan, such as a mortgage to buy a house, at some point in our lives. Banks and other organizations will often offer several deals for you to choose from. Financial mathematics helps you understand how interest payments are calculated and enables you to make sensible decisions.

Starter Activity

Look at the images in Figure 11.1. In small groups discuss what level of accuracy would be appropriate for each measurement.

Now look at this problem:

Mario borrows $10 000 to buy a new car. His bank charges an annual interest rate of 6%, compounded annually. Mario pays back $1500 at the end of each year, starting from the end of the first year. His payment is made after the interest for that year has been added.

Estimate how much Mario will pay back in total, and what percentage of this amount will be spent on paying the interest. Then use a spreadsheet to check your answer.

11A Approximation

■ Decimal places and significant figures

An answer to a calculation will often have many (or possibly infinitely many) digits. You will usually want to round your answer to a certain number of decimal places or significant figures.

Decimal places are the digits after the decimal point. For example, 4.25 is given to two decimal places, 142.3 to one decimal place, and 0.00405 to five decimal places.

Significant figures are the digits after any leading zeros (in decimals) or before any trailing zeroes (in very large numbers). For example, 342000, 0.00503 and 0.0310 are all given to three significant figures. It is only the zeros at the start of the number that do not count as significant. Note that in a whole number, the zeros at the end may or may not be significant; for example, 1203 rounded to two significant figures is 1200, and rounded to three significant figures is also 1200.

The rules for rounding are the same for both decimal places and significant figures: you look at the next digit and round up if it is 5 or more.

WORKED EXAMPLE 11.1

Evaluate $\ln\left(\dfrac{\pi}{3}\right)$ giving your answer to:

a two decimal places

b three significant figures.

Evaluate $\ln\left(\dfrac{\pi}{3}\right)$ on your calculator and write down more significant figures than you need for the final answer ...... $\ln\left(\dfrac{\pi}{3}\right) = 0.04611759...$

The first three decimal places are 0.046
Since $6 \geqslant 5$, round the 4 up ...... **a** $= 0.05$ (2 d.p.)

The first four significant figures are 0.04611
Since $1 < 5$, keep the previous 1 unchanged ...... **b** $= 0.0461$ (3 s.f.)

When rounding an answer to a calculation, it is important to choose an appropriate level of accuracy. A common rule is to choose the lowest level of accuracy among the numbers used in the calculation. This is because your final answer should not be given to a higher level of accuracy than the information used to calculate it.

WORKED EXAMPLE 11.2

If the width of a rectangle is 3.2 to 2 significant figures and the length is 5.18 to two decimal places, find the area of the rectangle, giving your answer to an appropriate level of accuracy.

The appropriate level of accuracy will be the lowest level of accuracy to which any of the information is given ...... 3.2 is given to 2 s.f. and 5.18 is given to 3 s.f., so give the answer to 2 s.f.

Calculate the area using the given information ...... $3.2 \times 5.18 = 16.576$

The first three significant figures in 16.576 are 16.5
Since $5 \geqslant 5$, round the 6 up ...... $\therefore$ area $= 17$ (2 s.f.)

TOK Links

What is −0.5 rounded to the nearest whole number? How much of maths is community knowledge?

11A Approximation

Tip
Sometimes people get confused because the upper bound would itself get rounded up to a different number. However, it is just a boundary and anything less (for example 32.4 or 32.49) would still get rounded down to 32.

■ Upper and lower bounds of rounded numbers

If you are given a number that has been rounded (for example, a measurement or an answer to a calculation) you may want to know what the original number could have been. The value range that the original number could have been are described with the **upper** and **lower bounds**. The lower bound is the smallest value that rounds to the given number. The upper bound is the smallest number above the given number that does not round to it.

For example, if you are told that a length is 32 cm to the nearest centimetre, you know that the lower bound for the length is 31.5 cm because anything just below 31.5 would round down to 31. The upper bound is 32.5 because anything on or just above this would round up to 33.

The easiest way to identify upper and lower bounds is to use a number line.

> **WORKED EXAMPLE 11.3**
>
> If $x = 4.1$ to one decimal place, write down an inequality for the possible values x can take.
>
> To one decimal place, the numbers either side of 4.1 are 4.0 and 4.2
> Represent this on a number line
>
> Look for the values halfway between these numbers
> 4.05 rounds up to 4.1 so it is included ⋯⋯⋯ $4.05 \leq x < 4.15$
> 4.15 rounds up to 4.2 so is not included

You can find upper and lower bounds for the result of a calculation by considering the bounds on the individual values used in the calculation. When the calculation only involves addition and multiplication this works as you would expect.

> **KEY POINT 11.1**
>
> - upper bound for $x + y$ = (upper bound for x) + (upper bound for y)
> - lower bound for $x + y$ = (lower bound for x) + (lower bound for y)
>
> Analogous rules hold for multiplication.

> **WORKED EXAMPLE 11.4**
>
> The length of a rectangle is 11 to the nearest whole number. The width is 5.2 to one decimal place. Find the upper and lower bound for the area of the rectangle.

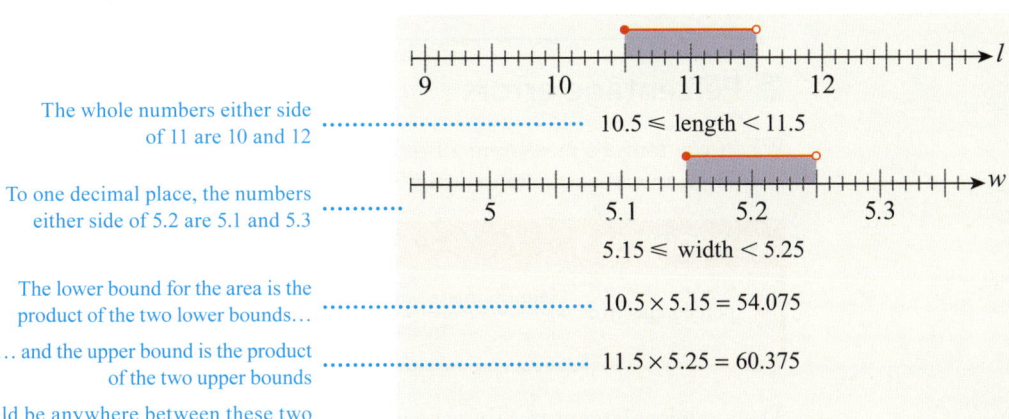

The whole numbers either side of 11 are 10 and 12 ⋯⋯⋯ $10.5 \leq \text{length} < 11.5$

To one decimal place, the numbers either side of 5.2 are 5.1 and 5.3 ⋯⋯⋯ $5.15 \leq \text{width} < 5.25$

The lower bound for the area is the product of the two lower bounds… ⋯⋯⋯ $10.5 \times 5.15 = 54.075$

…and the upper bound is the product of the two upper bounds ⋯⋯⋯ $11.5 \times 5.25 = 60.375$

The area could be anywhere between these two values. Remember that 10.5 and 5.15 were included, so 54.075 is a possible value, while 60.375 is not ⋯⋯⋯ $\therefore 54.075 \leq \text{area} < 60.375$

Links to: Sciences

The method used in Worked Example 11.4 is appropriate for dealing with systematic measurement error. However, in more advanced error analysis in the sciences, the term 'error' does not normally refer to the largest and smallest possible values, but a range of one standard deviation reflecting the fact that there is a random component to the error. There are rules for combining errors from different variables which are derived using calculus.

When a calculation involves subtraction and division, you need to be more careful when finding upper and lower bounds. This is because subtracting a smaller number gives a larger answer (and similarly for division).

KEY POINT 11.2

- upper bound for $x - y$ = (upper bound for x) – (lower bound for y)
- lower bound for $x - y$ = (lower bound for x) – (upper bound for y)

Analogous rules hold for division.

WORKED EXAMPLE 11.5

If $a = 10$, $b = 4$ and $c = 2$, each given to the nearest whole number, find the upper bound for $\dfrac{a}{b-c}$.

First write down the upper and lower bounds for a, b and c
$$9.5 \leqslant a < 10.5$$
$$3.5 \leqslant b < 4.5$$
$$1.5 \leqslant c < 2.5$$

The upper bound for $\dfrac{a}{b-c}$ is the upper bound for a divided by the *lower* bound for $(b-c)$

$$\text{upper}\left(\dfrac{a}{b-c}\right) = \dfrac{\text{upper}(a)}{\text{lower}(b-c)}$$

The lower bound for $b - c$ will be the lower bound for b minus the *upper* bound for c

$$= \dfrac{\text{upper}(a)}{\text{lower}(b) - \text{upper}(c)}$$

$$= \dfrac{10.5}{3.5 - 2.5}$$

$$= 10.5$$

■ Percentage errors

When you round a measurement or an answer to a calculation, it is important to know how large an error you are making. This is usually measured as a percentage.

KEY POINT 11.3

- percentage error $= \left|\dfrac{\text{approximate value} - \text{true value}}{\text{true value}}\right| \times 100$

The modulus sign ensures that the answer is always positive, regardless of whether the approximate value is larger or smaller than the true value.

11A Approximation

> **WORKED EXAMPLE 11.6**
>
> Find the percentage error in estimating π as $\frac{22}{7}$.
>
> Use the formula for the percentage error .... % error $= \left|\dfrac{\text{approximate} - \text{true}}{\text{true}}\right| \times 100$
>
> The 'approximate' value is $\frac{22}{7}$
>
> For the 'true' value use the π button on your calculator .... $= \left|\dfrac{\frac{22}{7} - \pi}{\pi}\right| \times 100$
>
> Give your answer to 3 s.f. .... $\approx 0.0402\%$

If the numbers used in the calculation have been rounded, you do not know the 'true' value, but can find upper and lower bounds for it.

> **WORKED EXAMPLE 11.7**
>
> What is the maximum percentage error in the area of a circle if the radius measured is 2.5 cm to one decimal place?
>
> Find the upper and lower bounds for the radius... .... $2.45 \leq r < 2.55$
>
> ...and for the true value of the area .... $A = \pi r^2$ ∴ $6.0025\pi \leq A < 6.5025\pi$
>
> The approximate value of the area is
>
> $2.5^2 \pi = 6.25\pi$
>
> So
>
> Use % error
> $= \left|\dfrac{\text{approximate} - \text{true}}{\text{true}}\right| \times 100$ .... % error $= \left|\dfrac{6.25\pi - \text{true}}{\text{true}}\right| \times 100$
>
> Check the value of the percentage error at each end of the interval for the true value .... When true area $= 6.0025\pi$:
>
> $\left|\dfrac{6.25\pi - 6.0025\pi}{6.0025\pi}\right| \times 100 = 4.12\%$
>
> When true area $= 6.5025\pi$:
>
> $\left|\dfrac{6.25\pi - 6.5025\pi}{6.5025\pi}\right| \times 100 = 3.88\%$
>
> The maximum percentage error is 4.12%.

■ Estimation

In the Mathematics: applications and interpretation SL course you will always have access to a calculator. However, working out a rough estimate for an answer can help you identify whether you have made a mistake. You should always check whether your answer is sensible in the context of the question, and whether it is given to an appropriate degree of accuracy.

WORKED EXAMPLE 11.8

Joanna measures the angle of elevation of a tree as 33° at a distance 12 m from the tree. She calculated the height as:

$h = 12 \tan 33° = -903.75618$ m

a   Give two reasons why you know that the final answer is incorrect.

b   State one further criticism of the final answer.

Look for things such as negative lengths, or numbers that are unreasonably large

a   Height cannot be negative.
    The height of a tree cannot be over 900 m.

Consider the level of accuracy of the answer – it should be the lowest of the accuracies of the given information (in this case 2 s.f.)

b   It is given to too many significant figures.

The reason Joanna got the answer wrong is that she had her calculator in radians mode. Radians are an alternative unit of angle to degrees which you will learn more about if you are studying Mathematics: applications and interpretation HL course. Having your calculator in the wrong mode is a common cause of such errors.

TOOLKIT: Modelling

One vital skill in modelling is the ability to validate your answers against your real-world experience. To do this you need to be able to estimate real-world quantities. This is often called a Fermi estimate after Enrico Fermi, who made a remarkably accurate estimation of the power of an atomic bomb by measuring the distances pieces of paper were blown by the blast in a viewing station some miles away from the explosion.

■ **Figure 11.2** Enrico Fermi

Here is a typical Fermi estimate to estimate the number of piano tuners in Chicago, USA:

- There are roughly 10 million people in Chicago.
- About 1 in 50 people have a piano that they have tuned.
- Pianos are tuned roughly once each year.
- Each piano takes 2 hours to tune.
- A piano tuner works eight hours a day, five days a week, fifty weeks a year.

Therefore, there needs to be:

$$\frac{10\,000\,000 \times 2}{50 \times 8 \times 5 \times 50} \approx 200 \text{ piano tuners}$$

(There are actually about 290 piano tuners in Chicago).

Try the following Fermi estimates:

- Could the entire population of the world fit in the area of the island of Ireland standing with enough room to stretch out their arms?
- Estimate the volume of gasoline used since cars have been invented. How does it compare to the volume of the Atlantic Ocean?
- How long would it take to count to a million?

11A Approximation

- Could you fit one billion dollars in $1 bills into your classroom?
- How much money is spent in your school canteen in a year?
- How long would it take to drink all the water in a swimming pool?
- What weight of trash is thrown out by your family each year? What volume of waste is created by your country each year?

CONCEPTS – APPROXIMATION

You only need 39 decimal places of π to calculate the perimeter of the observable universe to within the radius of one hydrogen atom. So why do people sometimes memorise π to hundreds of digits? In many applied areas, increasing accuracy is only important if it can improve our understanding of the real world and often this is not appropriate. Consider the following joke:

Museum curator: This dinosaur is 65 000 003 years old.

Child: Wow – that's amazing! How do you know that so accurately?

Museum curator: Well, three years ago a palaeontologist visited the museum and told me it was 65 million years old!

You should always think about the accuracy of your answers. Do not give figures which are wildly over-precise, as these are either not useful or, even worse, convey an unjustifiable level of precision.

Exercise 11A

For questions 1 and 2, use the method demonstrated in Worked Example 11.1 to round these numbers to one decimal place.

1. a 32.761 54
   b 654.038 21
2. a 0.249 04
   b 0.051 67

For questions 3 and 4, round the numbers from questions 1 and 2 to two decimal places.

3. a 32.761 54
   b 654.038 21
4. a 0.249 04
   b 0.051 67

For questions 5 and 6, round the numbers to three decimal places.

5. a 32.761 54
   b 654.038 21
6. a 0.249 04
   b 0.051 67

For questions 7 and 8, round the numbers to one significant figure.

7. a 32.761 54
   b 654.038 21
8. a 0.249 04
   b 0.051 67

For questions 9 and 10, round the numbers to two significant figures.

9. a 32.761 54
   b 654.038 21
10. a 0.249 04
    b 0.051 67

For questions 11 and 12, round the numbers to three significant figures.

11. a 32.761 54
    b 654.038 21
12. a 0.249 04
    b 0.051 67

For questions 13 and 14, use the method demonstrated in Worked Example 11.2 to give the stated quantities to an appropriate degree of accuracy.

13. a pq if $p = 12.4$ and $q = 8.7$
    b ab if $a = 5.50$ and $b = 2.23$
14. a $\dfrac{x}{y}$ if $x = 3.4$ and $y = 21$
    b $\dfrac{c}{d}$ if $c = 7$ and $d = 2.49$

For questions 15 to 18, use the method demonstrated in Worked Example 11.3 to write an inequality for the possible values x can take.

15   a $x = 12.3$ to 1 d.p.
     b $x = 0.8$ to 1 d.p.

16   a $x = 0.49$ to 2 d.p.
     b $x = 5.73$ to 2 d.p.

17   a $x = 60$ to 1 s.f.
     b $x = 0.3$ to 1 s.f.

18   a $x = 470$ to 2 s.f.
     b $x = 24$ to 2 s.f.

For questions 19 and 20, use the method demonstrated in Worked Example 11.4 to find the upper and lower bounds of the given quantity.

19   a $a + b$ where $a = 8$ to 1 s.f. and $b = 4.18$ to 2 d.p.
     b $c + d$ where $c = 3.4$ to 2 s.f. and $d = 0.23$ to 2 s.f.

20   a pq where $p = 20$ to 1 s.f. and $q = 0.1$ to 1 d.p.
     b rs where $r = 0.6$ to 1 d.p. and $s = 5.0$ to 1 d.p.

For questions 21 and 22, use the method demonstrated in Worked Example 11.5 to find the upper and lower bounds of the given quantity.

21   a $a - b$ where $a = 8$ to 1 s.f. and $b = 4.18$ to 2 d.p.
     b $c - d$ where $c = 3.4$ to 2 s.f. and $d = 0.23$ to 2 s.f.

22   a $\dfrac{p}{q}$ where $p = 20$ to 1 s.f. and $q = 0.1$ to 1 d.p.
     b $\dfrac{r}{s}$ where $r = 0.6$ to 1 d.p. and $s = 5.0$ to 1 d.p.

For questions 23 and 24, use the method demonstrated in Worked Example 11.6 to find the percentage error in each case.

23   a   Estimating $\sqrt{2}$ as 1.41
     b   Estimating $\sqrt{3}$ as 1.73

24   a   Estimating e as $\dfrac{19}{7}$
     b   Estimating ln 2 as $\dfrac{9}{13}$

For questions 25 and 26, use the method demonstrated in Worked Example 11.7 to find the maximum percentage error in each case.

25   a   The volume of a cube if the side length is 25 cm to the nearest whole number.
     b   The volume of a sphere if the radius is 0.42 m to 2 d.p.

26   a   The area of rectangle if the side lengths are 2.3 m and 1.8 m both to 2 s.f.
     b   The volume of a cylinder if the radius is 3.5 cm and the height 12.2 cm both to 1 d.p.

27   A cuboid has length 2.4 cm, width 5.1 cm and height 3.8 cm.
     a   Find the exact volume of the cuboid in cm³.
     b   Round your answer to part **a** to:
         i   two decimal places
         ii   two significant figures.

28   You are given $r = \ln\left(\dfrac{s^3}{t}\right) - \sqrt{t}$, $s = 8.72$ and $t = 5.3$.
     a   Find the value of r to an appropriate degree of accuracy.
     John estimates the value of r by first rounding both s and t to one significant figure.
     b   Calculate John's estimate of r to an appropriate degree of accuracy.

29   The exact value of x is 128.37.
     a   Write the value of x correct to
         i   one decimal place   ii   two significant figures.
     b   Find the percentage error in rounding x to two significant figures.

30   The volume of fizzy drink in five cans of a particular brand was measured and found to be:
     328 ml, 330.1 ml, 329.7 ml, 330.4 ml, 329.5 ml
     a   Find the exact mean volume of this sample.
     b   The manufacturer claims that the cans each contain 330 ml. Find the percentage error based on your answer to part **a**.

31   A shelf is 85 cm long to the nearest centimetre.
     Find the upper and lower bounds of the length.

32 The mass of beans in a can is stated as 130 g to three significant figures. What is the minimum mass of beans in a multipack of six cans of beans?

33 A right-angled triangle ABC has $AB = 4.7$ cm to one decimal place and $BC = 12.3$ to one decimal place as shown.

a Write an inequality for the possible length of AC.
b Find the maximum percentage error in any stated length of AC within the range found in part a.

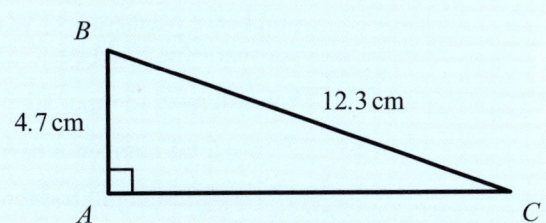

34 Voltage (V), current (I) and resistance (R) in an electrical circuit are related by the formula $V = IR$.

Eva measures the voltage to be 240 V to two significant figures and the resistance to be 18 Ω to two significant figures.

Find the upper and lower bounds of the current.

35 Given that $a = \dfrac{b}{c-d}$ and that $b = 8$, $c = 5$ and $d = 2$, all to the nearest whole number, find the upper and lower bounds of a.

36 The value p is given by the inequality $7 \leqslant p < 9$. The value q is given by the inequality $1 \leqslant q < 5$.

Find the maximum percentage error if the central values are used to calculate:

a $p + q$ b $p - q$

37 Jamie uses a model to predict the proportion of people in her class who will go to a party (P) based on the time in hours she spends talking about it (t).

The model predicts that:

$P = 0.3t(4 - t)$

a Evaluate the predicted proportion if t is

i 1 ii 2 iii 4.5.

b Which of the predictions made in part a are obviously not applicable? Justify your answer.

38 It is known that $a = 1.45$ to two decimal places.

a Write down an inequality to describe the possible values of a.
b Write down an inequality to describe the possible values of 10^a to two decimal places.
c Hence give 10^a to a suitable degree of accuracy, justifying your answer.

39 A pallet can have mass as given by the inequality $1.0 \leqslant m < 1.8$ tonnes. A truck can carry at most 12 tonnes.

A trucking company must transport 52 pallets. How many trucks must be used to guarantee transporting all 52 pallets? State one assumption your calculation requires.

40 A rectangle has width 10 cm to one significant figure and length 30 cm to one significant figure.

Find the upper and lower bounds of the area of the rectangle.

11B Further financial mathematics

Amortization and annuities using technology

When you take out a loan you are expected to pay it back with **interest**. The amount borrowed is called the **principal**. The interest is added at regular intervals and is calculated as a percentage of the amount currently owed. By the end of the loan period you need to pay back both the principal and any interest that has accumulated.

Amortization means that each of your payments is split between paying off the interest and paying off the principal. This way you pay off the loan gradually, making equal payments at regular intervals. The size of each payment depends on the amount borrowed (the principal), the interest rate and how often the interest is compounded, the length of the loan and the frequency of payments.

You can perform calculations related to amortized loans using the TVM Solver on your GDC. In all the examples in this section, the interest rate given will be annual, and the number of payments per year will be the same as the number of compounding periods per year (this could be, for example, annually, monthly or daily).

You will need to input all except one of the following pieces of information:

n = total number of payments

I% = annual interest rate

PV = the principal (positive for loan, negative for investment)

PMT = size of one payment

FV = final value – in this context, amount still owed; set to zero when the loan is repaid

P/Y = the number of payments made per year

C/Y = the number of compounding periods – this should be same as P/Y

WORKED EXAMPLE 11.9

Caleb takes out a mortgage of $100 000 which he needs to pay off over 25 years. If the annual interest rate is 5% (compounded monthly), what will be his monthly payment?

The payments are made monthly over 25 years .... $n = 25 \times 12 = 300$

The payments and the compounding are monthly, so there are 12 per year

FV = 0 as the loan is repaid at the end
Select PMT as that is what you want to find

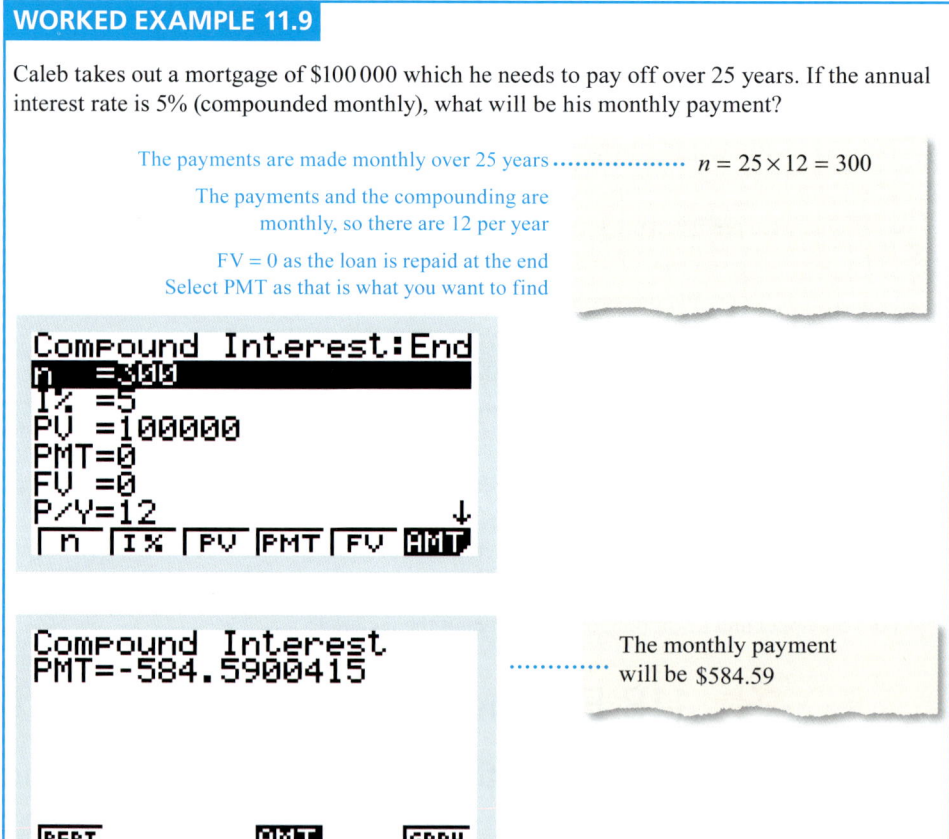

The monthly payment will be $584.59

TOOLKIT: Problem Solving

$100 is invested in an account at 7% annual interest, paid annually. Use a spreadsheet to compare the effect of the bank:
a using the exact value of the investment
b using the value in the account to the nearest cent
c rounding down to complete cents.

WORKED EXAMPLE 11.10

Ana can afford to pay £500 each month for a loan. If the period of the loan is amortized over 10 years at a 6% annual interest rate (compounded monthly), find the maximum amount that Ana can afford to borrow.

There are 12 payments a year for 10 years ········· $n = 10 \times 12 = 120$

FV is also 0 as the loan is repaid at the end
Select PV, as that is the amount required

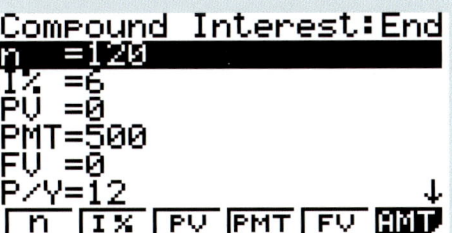

The maximum amount she can borrow is £45 036.72

You are the Researcher

The formula for the payment each month, A, for a loan of P at a rate of $r\%$ each period over n periods is

$$A = \frac{Pr\left(1+\frac{r}{100}\right)^n}{\left(1+\frac{r}{100}\right)^n - 1}$$

You might like to research how this formula is derived.

An **annuity** is a series of payments made at regular intervals, for example regular payments into (or withdrawals from) a savings account or monthly mortgage payments.

WORKED EXAMPLE 11.11

Wasim invests $50 000 at a fixed annual rate of 4% (compounded annually). What annuity can he draw if he expects to draw it annually for 30 years?

The payments are now taken once a year ········ $n = 30$
The interest is also compounded annually, so P/Y = C/Y = 1 $I\% = 4$
Set FV to 0 as Wasim wants to withdraw all the ···· $PV = -50\,000$
money by the end of the 30 years
You want to find PMT

```
Compound Interest
PMT=2891.504957
```

····· Using the GDC, he can draw $2891.50 per year.

[REPT]   [AMT]   [GRPH]

TOOLKIT: Modelling

A bank offers a mortgage with an initial fee of £1000 and an interest rate of 2.5% or no initial fee and an annual interest rate of 3%. Charlotte has the option of paying the initial fee and getting the lower interest rate, or investing the initial fee of £1000 in a bank account offering 2% annual interest. Either way, she wants a ten-year mortgage. Use technology to investigate which choice Charlotte should make. What are the modelling assumptions you are making?

Exercise 11B

For questions 1 and 2, use the method demonstrated in Worked Example 11.9 to find the required payments to repay a loan of $150 000 over 20 years.

1   a   Interest rate of 4% compounded annually; repayments annually.
    b   Interest rate of 3% compounded monthly; repayments monthly.
2   a   Interest rate of 2.5% compounded monthly; repayments annually.
    b   Interest rate of 3.5% compounded quarterly; repayments annually.

For questions 3 and 4, use the method demonstrated in Worked Example 11.10 to find the maximum amount that could be borrowed if the maximum monthly repayment that can be afforded is $750 and the loan is amortized over 15 years.

3   a   Interest rate of 5% compounded monthly; repayments monthly.
    b   Interest rate of 6% compounded annually; repayments annually.
4   a   Interest rate of 5.5% compounded quarterly; repayments annually.
    b   Interest rate of 4.5% compounded quarterly; repayments monthly.

For questions 5 and 6, use the method demonstrated in Worked Example 11.11 to find the annuity in the given period that can be drawn over 25 years on an investment of €100 000.

5   a   Interest rate of 3% compounded annually; payments annually.
    b   Interest rate of 2.5% compounded quarterly; payments quarterly.
6   a   Interest rate of 2% compounded monthly; payments annually.
    b   Interest rate of 2% compounded quarterly; payments monthly.

7 Ann has $60 000 in an account paying 4% interest. She wants to withdraw a regular annuity each year for 20 years. How much can she withdraw each year?

8 Patrick has $150 000 in an account earning 3% annual interest, paid annually. He wants to be able to withdraw $15 000 at the end of each year.

  a For how many years will he be able to do this?

  b How much will he be able to withdraw at the end of the year after he has withdrawn his final $15 000?

9 At the beginning of each year Nadine puts $1000 into a savings account which pays 2.5% annual interest, compounded annually. How much interest will have been paid in total at the end of the tenth year when she does this? Give your answer to the nearest cent.

10 Anwar takes out a loan of €1500 at an annual interest rate of 10%, compounded monthly.

  He has two different options:

  Option A: Take a payment holiday for the first year, then pay back in 48 monthly instalments.

  Option B: Pay back in 60 monthly instalments.

  a Find the size of the monthly instalments in each case.

  b How much more interest is paid using option A compared to option B?

  c Which option would you recommend? Justify your answer.

11 Martina takes out a loan of $10 000. Her monthly payments are $400. In the first two years the interest rate is 5%. After that the interest rate is 8%. Both interest rates are compounded monthly.

  a How many months will it take for Martina to pay off the loan?

  b What will her final payment be?

12 Sofia wishes to take out a mortgage of €125 000. She has two options:

  Mortgage A: 3% per annum interest rate compounded monthly. Term of mortgage 25 years.

  Mortgage B: 3.5% per annum interest rate compounded annually. Term of mortgage 30 years.

  Both plans require monthly repayments.

  Fully justifying your answer, find which mortgage Sofia should take if she wants to:

  a minimize her monthly repayments

  b minimize the total amount repaid over the lifetime of the mortgage.

13 Jon needs a loan of $15 000 to buy a new car.

  The terms of the loan are as follows:
  - Must be repaid in full in 5 years.
  - 5% interest compounded monthly for the first two years.
  - 6.5% interest compounded monthly for the remaining 3 years.

  Jon can afford a maximum monthly repayment of $300.

  Fully justifying your answer, explain whether or not he can afford to take the loan.

Checklist

- You should be able to round a number to a given degree of decimal places or significant figures, or to the nearest integer, nearest ten, etc.

- The level of accuracy of your final answer should not be greater than the level of accuracy of the given information.

- The percentage error made when approximating a value is given by

$$\text{percentage error} = \left|\frac{\text{approximate value} - \text{true value}}{\text{true value}}\right| \times 100$$

- When a number has been rounded, the largest and smallest values that the original number could have been are called the upper and lower bounds.

- When rounded numbers are used in a calculation,
  - For addition:
    — upper bound for $x + y$ = (upper bound for x) + (upper bound for y)
    — lower bound for $x + y$ = (lower bound for x) + (lower bound for y)
    — There are analogous rules for multiplication.
  - For subtraction:
    — upper bound for $x - y$ = (upper bound for x) − (lower bound for y)
    — lower bound for $x - y$ = (lower bound for x) − (upper bound for y)
    — There are analogous rules for division.

- You should be able to use estimation to judge whether an answer to a calculation is reasonable, and whether it is given to an appropriate degree of accuracy.

- You should be able to use the TVM solver on your GDC to solve problems in financial mathematics involving savings accounts and loan repayments.
  - In particular, you should be able to find how long it takes to repay a loan, or find the monthly payments in order to repay (amortize) a loan in a given time.

Mixed Practice

1 Given that $x = 12.475$, express the value of x to:
  a one decimal place
  b two significant figures.

2 Find the maximum percentage error in the area of a square if the side length is measured to be $0.06\,\text{m}$ to two decimal places.

3 Find the percentage error when $\sqrt{2}$ is approximated to one decimal place.

4 Amit invests £60 000 at a fixed annual rate of 3% (compounded annually). His investment will pay an annuity. What is the maximum annual annuity he can draw if he wants to ensure there is at least £10 000 remaining after 25 years of drawing the annuity?

5 José stands 1.38 kilometres from a vertical cliff.
  a Express this distance in metres.

José estimates the angle between the horizontal and the top of the cliff as 28.3° and uses it to find the height of the cliff.
  b Find the height of the cliff according to José's calculation. **Express your answer in metres, to the nearest whole metre**.
  c The actual height of the cliff is 718 metres. Calculate the percentage error made by José when calculating the height of the cliff.

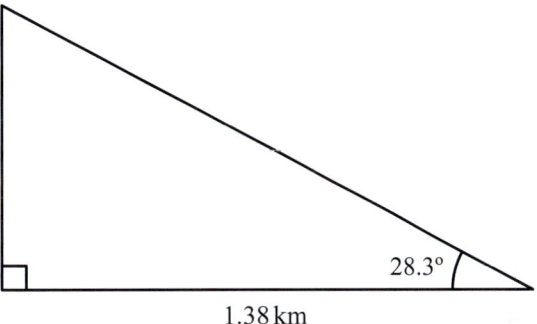

Mathematical Studies SL May 2010 TZ2 Paper 1 Q1

6 Given $p = x - \dfrac{\sqrt{y}}{z}$, $x = 1.775$, $y = 1.44$ and $z = 48$,
  a calculate the value of p.

Barry **first** writes x, y and z correct to one significant figure and **then** uses these values to estimate the value of p.
  b i Write down x, y and z each correct to one significant figure.
  ii Write down Barry's estimate of the value of p.
  c Calculate the percentage error in Barry's estimate of the value of p.

Mathematical Studies SL May 2011 TZ2 Paper 1 Q1

7 $T = \dfrac{(\tan(2z)+1)(2\cos(z)-1)}{y^2 - x^2}$, where $x = 9$, $y = 41$ and $z = 30°$.
  a Calculate the **exact** value of T.
  b Give your answer to T correct to
   i two significant figures;
   ii three decimal places.

Pyotr estimates the value of T to be 0.002.
  c Calculate the percentage error in Pyotr's estimate.

Mathematical Studies SL May 2015 TZ1 Paper 1 Q1

8 On the right is an extract from a questionnaire response used to survey people attending a doctor's surgery. Use estimation to determine which of the answers should be discarded.

Name:	John Sullie
Age:	27/11/1979
Height in metres:	5.11
Weight in kg:	75
Distance travelled in km:	1.4
Method of travel:	Car

9 Sam measures the circumference of a circle to be 15.1 cm to one decimal place and the diameter of the circle to be 4.8 cm to one decimal place. She uses these measurements to estimate π.
  a Find, to four decimal places, the upper and lower bounds of her estimate.
  b Find the percentage error if she uses her central values to estimate π.
  c Find the largest possible percentage error if she quotes a range as her estimate.

10 A ball is thrown vertically downwards with speed $u = 2.1 \, \text{m s}^{-1}$ to two significant figures. Its speed t seconds later is measured to be $v = 16 \, \text{m s}^{-1}$ to two significant figures. Its acceleration is known to be $a = 9.81 \, \text{m s}^{-2}$ to three significant figures.

Using the formula $v = u + at$, find the upper and lower bounds of t to four significant figures.

11 A rectangular wall has dimensions 3.4 m by 5.2 m, each to two significant figures. A can of paint says it contains 5.00 litres to two decimal places. 1 metre squared of wall requires 2.4 litres to two significant figures.

Joseph wants to buy enough paint to guarantee covering the wall. How many cans should he buy?

12 The value p is given by the inequality $8 \leqslant p < 10$. The value q is given by the inequality $3 \leqslant q < 7$.

Find the maximum percentage error if the central values are used to calculate:
  a $p + q$
  b $p - q$

13 Yousef uses a model to predict the probability of a popcorn kernel popping (p) based on the temperature (°C). The model claims that:

$$p = e^{\frac{T}{100}} - 1.5$$

  a Evaluate the predicted probability if T is:
    i 0
    ii 50
    iii 100.
  b Which of the predictions made in **a** are obviously not applicable? Justify your answer.

14 Lianne invests $1000 at 5% interest for 10 years, compounded monthly.
  a She estimates that she will get $50 interest each year, so a total of $500 in interest. What is the percentage error in her estimate?
  b How much could Lianne withdraw each month if she wants to have $500 remaining at the end of the ten-year period?

15 Kanmin takes out a loan of ¥24 000. The interest rate for the first year is 2%, compounded monthly. The subsequent annual interest rate is 8%, compounded monthly. If Kanmin makes repayments of ¥300 each month find:
  a the number of months required
  b the final month's payment.

16 Orlaigh wants to borrow £200 000. She has two options:

Option A: Pay 5% interest rate for 25 years, compounded quarterly.
Option B: Pay 4% interest rate for 30 years, compounded monthly.

Orlaigh will make the minimum required repayment in each compounding period.

Which option should Orlaigh choose if she wishes to minimize the total amount repaid? Justify your answer.

Mixed Practice   **295**

17 Chad takes out a loan of $100 000 over 20 years. The interest rate is 4.8% compounded monthly.
  a How much will Chad pay each month?
  b If Chad overpays by $100 each month, how much interest will he save in total? Give your answer to the nearest $100.

18 Quinn takes out a loan of €6000 over 10 years at an annual rate of 6%, compounded monthly.
  a What is the monthly repayment?
  b What percentage of repayments are on interest?
  c What percentage of the interest paid is paid in the first five years?

19 If $y = -5$ and $x = 20$, both given to one significant figure, find the maximum value of $(x - y)(x + y)$, giving your answer to two decimal places.

20 It is known that $a = 2.6$ to one decimal place.
  a Write down an inequality to describe the possible values of a.
  b Write down an inequality to describe the possible values of 10^a to one decimal place.
  c Hence give 10^a to a suitable degree of accuracy, justifying your answer.

12 Applications and interpretation: Solving equations with technology

ESSENTIAL UNDERSTANDINGS

- Algebra is an abstraction of numerical concepts and employs variables to solve mathematical problems.

In this chapter you will learn…
- how to use technology to solve systems of equations
- how to set up a system of equations to model a given situation
- how to use technology to solve polynomial equations
- how to rearrange an equation into polynomial form.

CONCEPTS

The following key concepts will be addressed in this chapter:
- **Quantities** and values can be used to describe key features and behaviours of functions and **models**, including quadratic functions.

PRIOR KNOWLEDGE

Before starting this chapter, you should already be able to complete the following:

1 Solve the system of equations.
$$\begin{cases} 3x - 2y = 17 \\ x + 4y = 5 \end{cases}$$

2 Write the following equations in the form $ax^2 + bx + c = 0$:
   a $(3x - 1)(x + 4) = 7$
   b $3x - \dfrac{4}{x} = 1$

■ Figure 12.1 How do we describe locations?

Applications and interpretation: Solving equations with technology

Systems of equations, where there is more than one unknown quantity, are used to describe systems of several interlinked variables. They also arise when fitting models to data and in the design of complex systems such as neural networks. Although there are analytical ways of solving such systems, in practice they are almost always solved using technology.

Polynomial equations can be used to model a wide variety of situations, from describing the state of a gas and studying vibration frequencies to modelling variations in population size. They are also essential in the design of encryption algorithms and in computer graphics. In pure mathematics, polynomials are important because they can be used to approximate more complicated functions. Analytical methods exist for polynomial equations involving terms up to x^4, although beyond quadratic equations they are impractically complicated. Other polynomial equations have to be solved using technology.

🌐 It is often claimed that the first computer was invented by Charles Babbage and the first computer program written by Ada Lovelace, both British Victorian scientists. However, there are also arguments that their work was just a natural evolution of the ancient Chinese abacus.

Starter Activity

Look at Figure 12.1. In small groups, discuss how many numbers (or other pieces of information) are required to precisely describe a position in each of these situations.

Now look at this problem:

a Find at least four pairs of numbers, x and y, that satisfy the equation $2x + y = 10$. Find at least four pairs of numbers, x and y, which satisfy the equation $x + 3y = 20$. How many pairs can you find which satisfy both equations?

b Find at least four sets of numbers, x, y and z, which satisfy the equation $x + y + z = 6$. Can you find at least two different sets of numbers that satisfy both $x + y + z = 6$ and $2x + y + z = 8$? How many sets of numbers can you find that satisfy these three equations: $x + y + z = 6$, $2x + y + z = 8$ and $3x + y + z = 10$?

LEARNER PROFILE – Principled

One of the main drivers of the development of the computer was the code breaking by various countries in the Second World War. Even now, quantum computing is often being driven by the need for international espionage. Is mathematics devoid of ethical responsibility for its applications?

12A Systems of linear equations

You can solve systems of two or three linear equations using your GDC.

WORKED EXAMPLE 12.1

Solve the following simultaneous equations.

$$\begin{cases} 2x + y + 5z - 1 = 0 \\ x + 2z = 2y + 5 \\ 2x + z + 2 = 0 \end{cases}$$

Write the equations in standard form, with all the variables on the left and the constants on the right

$$\begin{cases} 2x + y + 5z = 1 \\ x - 2y + 2z = 5 \\ 2x + z = -2 \end{cases}$$

Use the simultaneous equation solver on your GDC, making sure that you enter each term in the correct place:

$$x = -\frac{31}{19}, \; y = -\frac{39}{19}, \; z = \frac{24}{19}$$

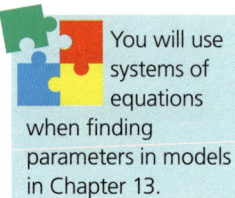

You will use systems of equations when finding parameters in models in Chapter 13.

 If you are studying the Mathematics: Applications and interpretation HL course, you will also learn how to solve systems of equations using matrices.

You may need to write a system of equations from given information. It is important to remember that the number of equations you need is the same as the number of unknowns you are trying to find. If you do not have enough equations, read the question carefully to ensure you have used all the given information!

WORKED EXAMPLE 12.2

Three friends go shopping at the same market. Xinyi buys four apples, four bananas and seven carrots and pays £6.08. Yanni buys one apple, four bananas and three carrots and pays £3.77. Zara buys two bananas and eight carrots and pays £3.22. How much does one banana cost?

Define the variables, in this case the cost of each item

Let
x = the cost of one apple
y = the cost of one banana
z = the cost of one carrot

Write the information as a system of three equations

Then
$$\begin{cases} 4x + 4y + 7z = 6.08 \\ x + 4y + 3z = 3.77 \\ 2y + 8z = 3.22 \end{cases}$$

Solve using GDC. The cost of one banana is y

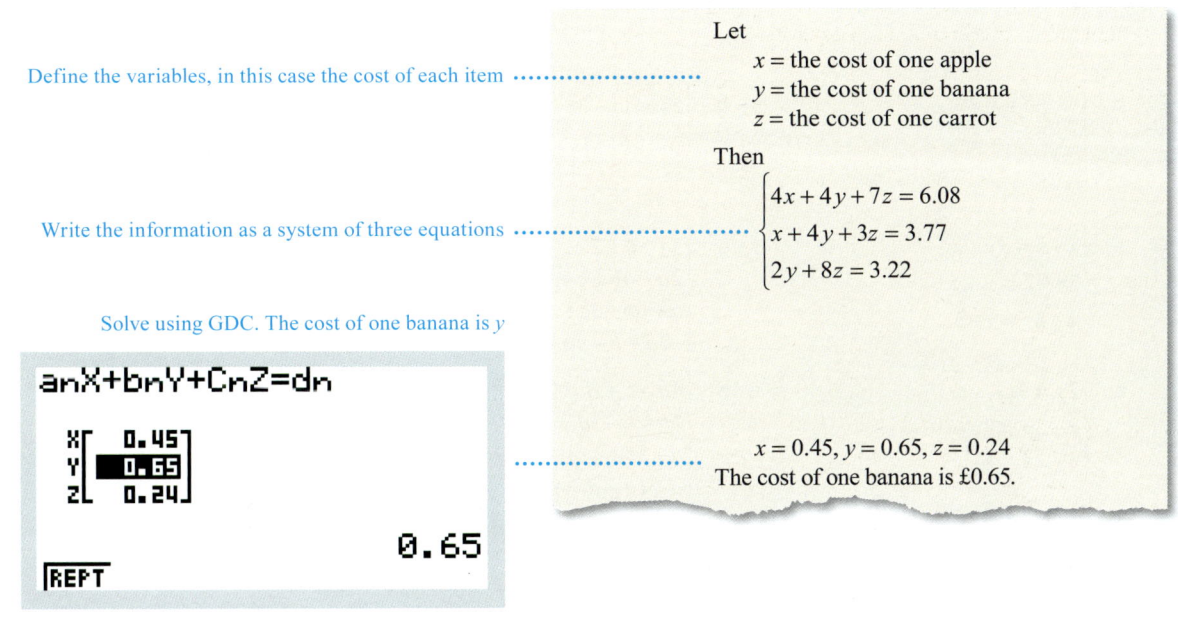

$x = 0.45, y = 0.65, z = 0.24$
The cost of one banana is £0.65.

CONCEPTS – QUANTITY

One of the most important parts of modelling is deciding upon and clearly defining the **quantities** involved. For example, in Worked Example 12.2, students who do not explicitly write their definitions down often confuse the cost of an apple with the number of apples bought.

However, in real-world situations, there is also a danger in trying to quantify variables which are too complex to capture in a single value. For example, happiness, intelligence, health are all quantities which people try to quantify, but you should be wary of how valid these measurements are.

TOOLKIT: Problem Solving

Sometimes it is possible to solve equations in two unknowns if you know that they are positive integers. Look at the problem below, for example.

Adult tickets to a theme park cost €23 and child tickets cost €19. Angelina spends €233 on tickets. How many adult tickets does she buy?

Exercise 12A

For questions 1 to 6, use the method demonstrated in Worked Example 12.1 to solve the simultaneous equations.

1   a $\begin{cases} 2x - 3y = 16 \\ 3x + 4y = 7 \end{cases}$

    b $\begin{cases} 4x + 7y = -2 \\ 5x + 3y = 4 \end{cases}$

2   a $\begin{cases} x = 6y + 10 \\ 4y = 7x - 5 \end{cases}$

    b $\begin{cases} 5y = 11 - 3x \\ y - 16 = 4x \end{cases}$

3   a $\begin{cases} 3x - y + z = 2 \\ 5x + 2y - 3z = 0 \\ x + y + 2z = 3 \end{cases}$

    b $\begin{cases} 2x - 5y + 4z = 10 \\ x + 3y - 2z = 5 \\ 4x + 2y - z = -4 \end{cases}$

4   a $\begin{cases} x + 5 = 3y - z \\ 3x + 7 = y + z \\ x + 3y = 1 - 5z \end{cases}$

    b $\begin{cases} 2y = 3x - z \\ x + 3y = 2z - 3 \\ 4z - 5 = 2x - y \end{cases}$

5   a $\begin{cases} 3a + b + 2c + 4d = -13 \\ 2a - 4b + c + 5d = -14 \\ a + 6b + 3c + d = -5 \\ -2a + b - c + 3d = 16 \end{cases}$

    b $\begin{cases} 4a - b + d = 10 \\ 2b - c - 3d = -12 \\ a + 3b + 3c + 2d = 25 \\ a + c - 4d = 8 \end{cases}$

6   a $\begin{cases} 2b + d = 3c + 8 \\ b + 4c = a - 1 \\ 2a + c = 20 - 5d \\ 6b + 2d = 3a + c - 6 \end{cases}$

    b $\begin{cases} 4c + 2d = 24 + 2b - a \\ 3b + 2 = c - 2a \\ a - 5b + 3d = 2c + 13 \\ 4a + 6b = 6 - d \end{cases}$

7   There are 60 children on a school trip. There are twice as many girls as boys.
    a Write this information as two simultaneous equations.
    b How many girls are there?

8   Matthew buys 3 widgets and 7 gizmos. He spends $14.10. Mark buys 5 widgets and 4 gizmos. He spends $12.
    a Write this information as two simultaneous equations.
    b What is the total cost of a widget and a gizmo?

9   An arithmetic sequence has first term a and common difference d. The sixth term of the sequence is 16 and the sum of the first 10 terms is 169.
Find a and d.

10   An arithmetic sequence has first term a and common difference d. The sum of the first 5 terms of this sequence is 19 and the sum of the first 11 terms is 48.4.
Find a and d.

11   Three groups of friends go to the cinema to see a film.
The first group buy 7 tickets for the film, 6 of them buy drinks and 4 of them buy popcorn. In total they spend $129.
The second group buy 6 tickets for the film, 4 of them buy drinks and 3 of them buy popcorn. In total they spend $102.
The third group buy 5 tickets for the film, 2 of them buy drinks and 2 of them buy popcorn. In total they spend $75.
Find the cost of a ticket, the cost of a drink and the cost of popcorn.

12   Jack wishes to invest $10 000 in three different accounts. He wants to make the best possible return on his investment at the end of the year.
Account A pays 2% interest and has no limit on how much can be invested, account B pays 3% interest and account C pays 4% interest. Both accounts B and C have limits on the amount that can be invested.
He ends up investing $300 more in account A than account B. At the end of the year he has earned $286.
How much did he invest in each of the accounts?

13 Marta bakes three types of cake: chocolate cakes, fruit cakes and sponge cakes. The recipe for each cake includes flour and sugar and she uses 3.3 kg of flour and 3.1 kg of sugar in total.

Each chocolate cake uses 225 g of flour and 300 g of sugar, each fruit cake uses 320 g of flour and 250 g of sugar and each sponge cake uses 400 g of flour and 325 g of sugar.

Marta sells the chocolate cakes for £12, the fruit cakes for £10 and the sponge cakes for £8 and makes a total of £114.

How many of each cake does she make?

14 Three cows and eleven sheep are worth £136 more than 32 pigs. One cow is worth seven sheep or eight pigs. How much, in total, is a cow, a sheep and a pig?

15 A survey of the colour of the cars of members of a company is summarized in the pie chart.

It is known that:
- The sector representing grey cars has an angle of 192°.
- There are two more blue cars than red cars.
- The total of red and blue cars is two less than the number of grey cars.

What is the probability that a randomly chosen car is red?

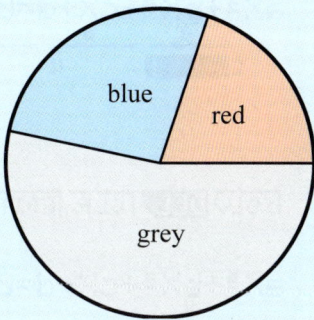

16 The sum of the digits of a three-digit number is 16. The third digit is twice the difference between the first and second digits. When the three digits are reversed the number decreases by 297.

Find the three-digit number.

17 Solve:
$x^2 + y^2 + z^2 = 26$
$x^2 + 2y^2 + 3z^2 = 67$
$x^2 - y^2 + z^2 = 8$

18 The mean of three numbers is twice the median. The range is five times the median. The difference between the two smallest numbers is 1. Find the largest number.

12B Polynomial equations

A **polynomial** is an expression featuring only non-negative powers of x, such as $3x^2 - 4x + 1$ or $-12x^3 + x + 3$. The **zeros of a polynomial** are the values of x for which the polynomial expression equals zero. For example, the zeros of the polynomial $x^3 + 4x^2 - 5$ are the solutions or the **roots of the equation** $x^3 + 4x^2 - 5 = 0$.

A polynomial equation can have several roots. In fact, if the highest power of x in the polynomial is x^n, there can be up to n roots. Many calculators have a separate equation solver for polynomials.

WORKED EXAMPLE 12.3

Find the roots of the equation $x^3 + 4x^2 = 5$.

In order to use the polynomial equation solver on your GDC, the equation needs to be written in the standard form, with the right-hand side being zero .... $x^3 + 4x^2 - 5 = 0$

Make sure that the coefficients are entered in the correct place

From GDC,
$x = 1, -1.38$ or -3.62

Sometimes your GDC will give you solutions containing the letter i. You should ignore those solutions for the moment, as in this course you only need to find real roots of equations.

The numbers involving i are called complex numbers. You will learn about them if you are studying Mathematics: applications and interpretation HL course.

WORKED EXAMPLE 12.4

Find the real root of the equation $3x^3 - x + 9 = 0$.

Use the polynomial equation solver on your GDC .... The only real root is $x = -1.52$

12B Polynomial equations

Sometimes an equation is not obviously polynomial, but it can be rearranged into that form.

> **WORKED EXAMPLE 12.5**
>
> Given that x satisfies the equation $2x - \dfrac{1}{x} = 4$:
>
> a show that $2x^2 - 4x - 1 = 0$
>
> b hence find the possible values of x.
>
> Multiply both sides by x to eliminate fractions .... **a** $\quad 2x^2 - 1 = 4x$
> $\quad 2x^2 - 4x - 1 = 0$
>
> Use the polynomial equation solver .... **b** From GDC,
> $\quad x = 2.22$ or -0.225

Exercise 12B

For questions 1 to 6, use the method demonstrated in Worked Example 12.3 to find the roots of the following equations.

1. a $\;x^2 - 5x + 2 = 0$
   b $\;2x^2 + x - 7 = 0$
2. a $\;4x^2 - 12x + 9 = 0$
   b $\;9x^2 - 6x + 1 = 0$
3. a $\;2x^3 - x^2 - 8x + 6 = 0$
   b $\;x^3 - 4x + 2 = 0$
4. a $\;9x^3 - 15x^2 - 8x + 16 = 0$
   b $\;2x^3 + 3x^2 - 12x - 20 = 0$
5. a $\;x^4 + 4x^3 - 5x^2 - 7x + 3 = 0$
   b $\;3x^4 - 9x^3 + 6x - 1 = 0$
6. a $\;x^4 - 7x^3 + 12x^2 + 9x - 27 = 0$
   b $\;4x^4 + 4x^3 - 23x^2 - 12x + 36 = 0$

For questions 7 to 9, use the method demonstrated in Worked Example 12.4 to find the real roots of the following equations.

7. a $\;x^2 - 3x + 8 = 0$
   b $\;3x^2 - 4x + 2$
8. a $\;3x^3 + 2x^2 - x + 4 = 0$
   b $\;x^3 - 2x^2 + 3x - 5 = 0$
9. a $\;x^4 + 5x^3 - 2x + 3 = 0$
   b $\;2x^4 - 3x^3 - x^2 + 4 = 0$

10. The base of a triangle is 2 cm larger than its perpendicular height. The area is 40 cm². Find the height.

11. A geometric sequence has first term 6. The sum of the first three terms is 58.5.
    Find the possible values of the common ratio.

12. A geometric sequence has first term 3.4. The sum of the first four terms is 51.
    Find the common ratio.

13. a Show that the equation
    $$2x - 1 = \dfrac{5}{x+3}$$
    can be written as $2x^2 + 5x - 8 = 0$.
    b Hence find the roots of the equation
    $$2x - 1 = \dfrac{5}{x+3}.$$

14. a Show that the equation $x^2 = 3 + x^{-1}$ can be written in the form $ax^3 + bx^2 + cx + d = 0$, where a, b, c and d are whole numbers.
    b Hence solve the equation $x^2 = 3 + x^{-1}$.

15. A right-angled triangle has sides of length x, $2x - 1$ and $x + 5$ as shown:
    a Show that $x^2 - 7x - 12 = 0$
    b Find both solutions for the equation in part **a**.
    c Hence state the value of x for the triangle, explaining your answer.

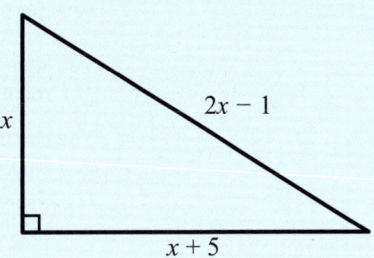

16 A path is to be laid round the back and one side of a house as shown. It is formed from two rectangles. All lengths are in metres.

The total area of the path is $19.8\,\text{m}^2$.

a Form a quadratic equation of the form $ax^2 + bx + c = 0$.

b Solve this equation to find the value of x.

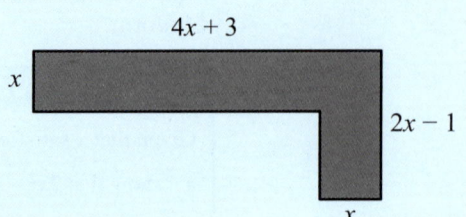

17 A piece of cardboard 50 cm long and 40 cm wide has squares of side x cm cut out from each corner.

The cardboard is folded up to make an open box of volume $6000\,\text{cm}^3$.

Find the possible values of x.

18 A closed cylindrical tin can has radius r and height h. The volume of the can is $600\,\text{cm}^3$ and the surface area is $450\,\text{cm}^2$.

a Show that $h = \dfrac{225 - \pi r^2}{\pi r}$

b Hence show that $\pi r^3 - 225r + 600 = 0$

c Hence find the possible radius of the can.

19 The numerical values of the volume of a cube and its surface area are added together. The answer is 160. Find the length of one edge of the cube.

20 An arithmetic sequence has first term 1 and common difference 6. The sum of the first n terms is 1160. Find the value of n.

21 The difference between two consecutive perfect cube numbers is 61. Find the possible numbers.

22 a Find an expression for the number of diagonals of a regular n-sided polygon.

b Hence find the number of sides of a polygon with 35 diagonals.

TOK Links

The four-colour problem is about how to colour maps on a flat surface. To prove it, mathematicians broke it down into a finite number of cases to consider, but there were far too many for it to be humanly possible, so they used a computer to check them all. No human has ever checked all the cases, but they have checked the computer program which was used. Is this a valid way of proving something?

You are the Researcher

Interestingly, the four-colour problem on a plane is incredibly hard to solve, but there is an analogous problem on a donut, called the seven-colour problem, which is much easier!

Checklist

- You should be able to use technology to solve systems of linear equations.
  - You need to have the same number of equations as unknowns.

- You should be able to use technology to find the roots of polynomial equations.
  - A polynomial is a function where x appears with positive integer powers.
  - The zeros of a polynomial f(x) are the roots (solutions) of the equation f(x) = 0.
  - A polynomial equation where the highest power of x is n may have up to n real roots.

Mixed Practice

1 The base of a triangle is 6 cm larger than its perpendicular height. The area is 20 cm². Find the height.

2 a Show that the equation
$(2x + 5)(x - 4) = 10$
can be written in the form
$ax^2 + bx + c = 0$
where a, b, and c are whole numbers to be found.
b Hence solve the equation
$(2x + 5)(x - 4) = 10$

3 a Show that the equation
$x^2 - 7 = \dfrac{2}{x + 4}$
can be written in the form
$ax^3 + bx^2 + cx + d = 0$
where a, b, c and d are whole numbers to be found.
b Hence solve the equation
$x^2 - 7 = \dfrac{2}{x + 4}$

4 Two numbers differ by 1. The product of the numbers is 10. The smaller number is x.
a Write this information as a quadratic equation in the form $ax^2 + bx + c = 0$
b Hence find the possible values of x to three significant figures.

5 A square-based cuboid has one side 2 cm longer than the other two. The volume of the cuboid is 10 cm³. Find the length of the shortest side of the cuboid to three significant figures.

6 A group of insects contains only flies and spiders. Flies have six legs and spiders have eight legs, but both insects have one head. There are 142 legs and 20 heads in total.
a Write down simultaneous equations to describe this situation.
b How many spiders are there?

7 Jack and Jill buy 'pails of water' and 'vinegar and brown paper'.
Jack buys 7 pails and 5 vinegar and brown papers. He spends $70.
Jill buys 5 pails and 7 vinegar and brown papers. She spends $74.
How much more expensive is vinegar and brown paper than a pail of water?

8 Daniel and Alessia are buying books. Science books cost £8 and art books cost £12. Daniel spends £168 on books. Alessia buys twice as many art books than Daniel but half as many science books than Daniel. She spends £264. How many books do they buy in total?

9 10 000 people attended a sports match. Let x be the number of adults attending the sports match and y be the number of children attending the sports match.
a Write down an equation in x and y.
The cost of an adult ticket was 12 Australian dollars (AUD). The cost of a child ticket was 5 Australian dollars (AUD).
b Find the total cost for a family of 2 adults and 3 children.
The total cost of tickets sold for the sports match was 108 800 AUD.
c Write down a second equation in x and y.
d Write down the value of x and the value of y.

Mathematical Studies SL May 2011 TZ2 Paper 1 Q7

10 In an arithmetic sequence, $S_{40} = 1900$ and $u_{40} = 106$. Find the value of u_1 and of d.

Mathematics SL November 2009 Paper 2 Q1

11 The length of a square garden is $(x+1)$ m. In one of the corners a square 1 m in length is used only for grass. The rest of the garden is only for planting roses and is shaded in the diagram opposite.
The area of the shaded region is A.
  a Write down an expression for A in terms of x.
  b Find the value of x given that $A = 109.25\,\text{m}^2$.
  c The owner of the garden puts a fence around the shaded region. Find the length of this fence.

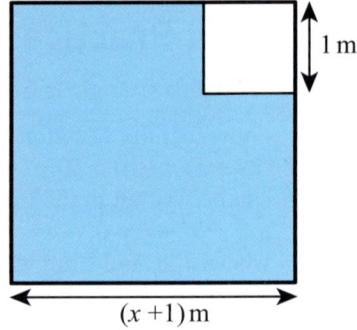

12 The product of three consecutive integers is 504. Find the mean of the numbers.

13 A geometric sequence has first term 8 and the sum of the first three terms is 14. Find the possible values of the common ratio.

14 A geometric sequence has first term 2 and the sum of the first four terms is -40. Find the common ratio.

15 The triangle ABC has $AB = x$, $BC = x + 2$, $AC = 2x$ and $B\hat{A}C = 60°$ as shown opposite.
  a Show that $x^2 - 2x - 2 = 0$.
  b Hence find the value of x.

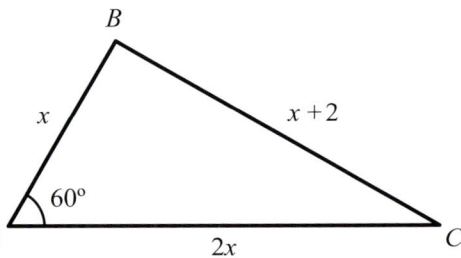

16 The first term of an arithmetic sequence is 100 and the common difference is -2. The sum of the first n terms is 2440. Find the possible values of n.

17 A tank in the shape of a cuboid has the dimensions shown in the diagram.
  a Write down an expression for the volume of the tank, V, in terms of x, expanding and simplifying your answer.
The volume of the tank is $31.5\,\text{m}^3$.
  b Find the dimensions of the tank.

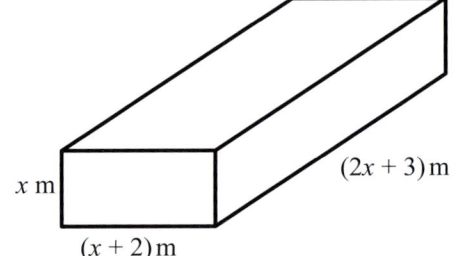

18 Clara organizes cans in triangular piles, where each row has one less can than the row below. For example, the pile of 15 cans shown has 5 cans in the bottom row and 4 cans in the row above it.
  a A pile has 20 cans in the bottom row. Show that the pile contains 210 cans.
  b There are 3240 cans in a pile. How many cans are in the bottom row?
  c i There are S cans and they are organized in a triangular pile with n cans in the bottom row. Show that $n^2 + n - 2S = 0$.
   ii Clara has 2100 cans. Explain why she cannot organize them in a triangular pile.

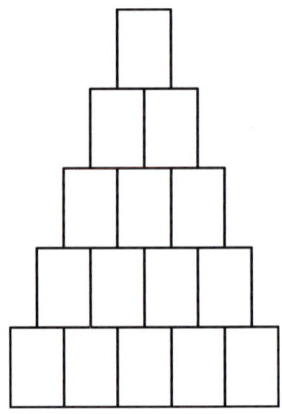

Mathematics SL November 2006 Paper 2 Q3

Mixed Practice

19 In a class of 30, everybody learns either French or Spanish or both. The number of people who learn Spanish is one more than the number who learn French. The number who learn both is 12 less than the number who only learn one language. A student is picked at random. What is the probability that they learn both French and Spanish?

20 The curve $y = ax^2 + bx$ passes through the point (1, 7) with gradient 12. Find the value of y when $x = 2$.

21 The numerical values of the volume of a cube and its surface area are added together. The answer is 100. Find the length of one edge of the cube, giving your answer to three significant figures.

22 Two numbers differ by one. Their cubes differ by two. Find the possible values of the smaller number.

23 Solve the following simulataneous equations:
$x^3 + y^3 + z^3 = 8$
$x^3 + 2y^3 + 3z^3 = 23$
$x^3 - y^3 + 2z^3 = 18$

24 n people meet and everybody shakes everyone else's hand. There are a total of 190 handshakes. Find the value of n.

13 Applications and interpretation: Mathematical models

ESSENTIAL UNDERSTANDINGS

- Models are depictions of real-life events using expressions, equations or graphs, while a function is defined as a relation or expression involving one or more variables.
- Creating different representations of functions to model the relationships between variables, visually and symbolically as graphs, equations and/or tables represents different ways to communicate mathematical ideas.

In this chapter you will learn…
- how to recognise the shape and main features of a variety of graphs: linear, quadratic, cubic, reciprocal, exponential and trigonometric
- how to choose an appropriate equation to model a given graph or a real-life situation
- how to use given information to determine the parameters in the model, with the aid of technology
- how to evaluate the appropriateness of a model and suggest modifications and improvements.

CONCEPTS

The following key concepts will be addressed in this chapter:
- Different **representations** of functions, symbolically and visually as graphs, equations and tables provide different ways to communicate mathematical **relationships**.
- The parameters in a function or equation may correspond to notable geometrical features of a graph and can represent physical **quantities** in **spatial** dimensions.
- **Changing** the parameters of a trigonometric function **changes** the position, orientation and shape of the corresponding graph.
- Different **representations** facilitate **modelling** and interpretation of physical, social, economic and mathematical phenomena, which support solving real-life problems.
- Technology plays a key role in allowing humans to **represent** the real world as a **model** and to **quantify** the appropriateness of the **model**.

PRIOR KNOWLEDGE

Before starting this chapter, you should already be able to complete the following:

1 Find the equation of the straight line passing through the points (1, 3) and (6, 10), giving your answer in the form $y = mx + c$.

2 Use technology to find the roots of the equation $3x^3 - 4x + 1 = 0$

■ Figure 13.1 How do we decide which variables to factor into a model?

Applications and interpretation: Mathematical models

LEARNER PROFILE –
Reflective
Is being really good at arithmetic the same as being really good at mathematics?

3 Use technology to solve the system of equations.
$$\begin{cases} 4x + z = 5 \\ 3x + y + z = 8 \\ 7x + 2y = 1 \end{cases}$$
4 Use graphs to solve the equation $2 \times 3^{-x} = 0.4$

A mathematical model is a representation of a real-life situation using equations or graphs. Representing a situation as a mathematical model enables us to analyze it mathematically and make predictions which can then be tested, so that the model can either be accepted or improved.

In order to be able to suggest a suitable model for a given situation, you need to recognize some common models and their graphs. Once you have decided on the type of model that is most appropriate for your situation, you need to use information about particular values of the variables to determine the parameters of the model: these are the numbers that appear in the equation.

In creating a mathematical model, we often need to identify the most common features of the situation, include those, and ignore others. It is important to realize that, however many variables we include in a model, it will always be a simplification of the real-life situation. You need to consider how accurate the predictions from the model are; in some situations, the model will work well for a certain range of values but not for others.

For example, when modelling the flight of an object we often ignore air resistance. That kind of model may give good predictions for certain types of objects, but depending on the shape, size and speed of the object, the effect of air resistance may become significant. We then need to modify our model to take this into account.

Starter Activity

Look at Figure 13.1. In small groups, discuss which variables would need to be considered when modelling the motion of each object. Pick the three most important variables you would include in the model for each situation.

Now look at this problem:

For each situation, think about how you would expect the given quantity to vary with time, and sketch a graph to illustrate this. You do not need to include any numerical detail, just a general shape of the graph.

1 The population of the Earth over the next 1000 years.
2 The height of a ball thrown vertically upwards.
3 The distance travelled by an object moving at a constant speed.
4 The temperature of a cup of coffee left on the kitchen table.
5 The cost of a phone call.

13A Linear models

A **linear model** is a model that can be represented by a straight-line graph. You already know from Chapter 4 that you need two pieces of information to uniquely define a straight line: either coordinates of two points, or the gradient and one point.

> **KEY POINT 13.1**
>
> - The gradient of the line connecting the points (x_1, y_1) and (x_2, y_2) is $m = \dfrac{y_2 - y_1}{x_2 - x_1}$.
> - The equation of the straight line with gradient m passing through the point (x_1, y_1) is $y - y_1 = m(x - x_1)$.
> - The equation of the straight line with gradient m and y-intercept $(0, c)$ is $y = mx + c$.

> **WORKED EXAMPLE 13.1**
>
> Find a linear model that goes through the points (1, 3) and (5, 11), giving your answer in the form $y = mx + c$.
>
> | | |
> |---|---|
> | Find the gradient of the line using $m = \dfrac{y_2 - y_1}{x_2 - x_1}$ | $m = \dfrac{11 - 3}{5 - 1} = 2$ |
> | Use the equation of a straight line in the form $y - y_1 = m(x - x_1)$ | $y - 3 = 2(x - 1)$ |
> | Expand the brackets and rearrange into the required form | $y - 3 = 2x - 2$
 $y = 2x + 1$ |

> **CONCEPTS – REPRESENTATION**
>
> What is the difference between **representing** a function using a straight-line model and as an arithmetic sequence?

■ Piecewise linear models

Sometimes a different straight-line model applies in different parts of the domain, producing a **piecewise linear model**.

> **WORKED EXAMPLE 13.2**
>
> A mobile phone contract has a monthly charge of $50 plus $1 for every 10 minutes of call time for the first 100 minutes, then $1 for every 20 minutes of call time thereafter.
>
> **a** Form a piecewise linear model for the monthly cost (C) in terms of the call time (t minutes).
>
> **b** Find the total monthly cost in a month when the total call time was 350 minutes.

13A Linear models

First find the points where the equation changes. The initial charge is $50	**a** When $t = 0$, $C = 50$
The charge for 100 minutes is $50 + 10 \times 1$	When $t = 100$, $C = 60$
Pick any point with $t > 100$, for example $t = 200$. The cost is broken down into two parts: 10 lots of 10 minutes plus 5 lots of 20 minutes	When $t = 200$, $C = 50 + (10 \times 1) + (5 \times 1) = 65$
Now find the equations of straight lines connecting those points. First find the gradient, then use $y - y_1 = m(x - x_1)$, but here, x is t and y is C	From $(0, 50)$ to $(100, 60)$: $$m = \frac{60 - 50}{100 - 0} = 0.1$$ $$C - 50 = 0.1(t - 0)$$ $$C = 0.1t + 50$$ From $(100, 60)$ to $(200, 65)$: $$m = \frac{65 - 60}{200 - 100} = 0.05$$ $$C - 60 = 0.05(t - 100)$$ $$C = 0.05t + 55$$
Piecewise linear models are usually written as separate equations, indicating the domain on which each applies	$$\therefore C = \begin{cases} 0.1t + 50 & \text{for } 0 \leq t \leq 100 \\ 0.05t + 55 & \text{for } t > 100 \end{cases}$$
350 minutes falls into the $t > 100$ part of the domain, so use the second equation	**b** When $t = 350$, $C = 0.05 \times 350 + 55 = 72.5$ The cost for 350 minutes is $72.50

TOOLKIT: Modelling

A pool ball is hit directly towards the side of the pool table at a speed of $0.5\,\text{m s}^{-1}$. It starts 1 metre away from the side. It bounces back off the side at the same speed.

a Sketch a graph of distance from the side against time.

b Find T, the time taken to return to a position 1 metre from the side.

c Each of the following factors have been ignored in your model. Determine whether including these factors would increase, decrease or have no effect on T. If there is an effect, try to suggest if the effect would be significant, moderate or negligible.

  i Friction
  ii The loss of energy when the ball bounces
  iii The country in which the pool ball was hit
  iv The temperature in the room
  v The fact that the pool ball has a diameter of 5 cm
  vi Air resistance
  vii The spin on the pool ball
  viii The ball is not hit exactly at right angles to the table side
  ix The colour of the pool ball
  x The table is not perfectly level, but sloping with an initial angle of depression of 1°

Exercise 13A

For questions 1 to 6, use the method demonstrated in Worked Example 13.1 to find a linear model that passes through the given points.

1. **a** (4, 2) and (10, 14)
   b (2, 5) and (8, 11)

2. **a** (1, 12) and (4, 3)
   b (10, 3) and (2, 15)

3. **a** (−3, −1) and (7, 4)
   b (−2, −24) and (3, −4)

4. **a** (1, 7) and (6, −5)
   b (−1, 6) and (3, −2)

5. **a** (2.5, 1) and (4, 8.5)
   b (0.9, 3.1) and (5.4, 13.6)

6. **a** (3.8, 12.5) and (7.6, −2.7)
   b (−5.8, −2.5) and (−0.3, −4.7)

7. The mileage, M thousand miles, of a second-hand car is modelled by the function $M = 8.2y + 29.4$ where y is the number of years after it was purchased.
   a Interpret in this context the meaning of
      i 8.2
      ii 29.4
   b Find the number of miles the car is predicted to have travelled after 5 years.
   c In which year will the mileage exceed 100 000 miles?

8. A taxi firm charges an initial fee of $2.50 and then a further $1.25 per kilometre.
   a Write down a linear function to model the cost, C ($), of a journey of d kilometres.
   b How much would a journey of 3.2 km cost?
   Ryan needs to get home from a friend's house which is 6.5 km away. He has $10 in his wallet.
   c Explain whether or not he can afford the taxi fare home.

9. The percentage of charge in a mobile phone battery, C%, after t hours of use can be modelled by the function $C = mt + k$. Initially, the battery is fully charged. The battery life is 12.5 hours.
   a i State the value of k
      ii Calculate the value of m
   b Determine a suitable domain for the function.

10. It is thought that there is a linear relationship between the cost of 3-bedroom houses, C (thousands of £), and the distance from the centre of London, d (miles).
    Suki has found that a 3-bedroom house 0.8 miles from the centre of London costs £630 000 and a similar house 3 miles from the centre of London costs £520 000.
    a Find a linear model that relates C to d.
    b Use your model to predict the cost of a 3-bedroom house 4.8 miles from the centre of London.

11. The weight of a baby can be modelled by the function $w = an + b$ where w is the mass in kilograms and n is the age of the baby in weeks.
    a If a baby weighs 4.08 kg aged 2 weeks and weighs 4.98 kg aged 6 weeks, find the values of a and b.
    b State, in this context, what
       i a represents
       ii b represents.

12. An electricity company charges a monthly fee of £10 plus £0.12 per kilowatt hour (kWh) of electricity used for the first 250 kWh and £0.15 per kWh for any further usage.
    a Form a piecewise linear model for the monthly cost (£C) in terms of the electricity usage (e kWh).
    b Find the total monthly cost in a month when the total electricity usage was 300 kWh.

13 Brian is moving house and needs a removal company to move his possessions. He is considering two options:

Hare Removals charge a flat fee of £80 plus £7.50 per mile.

Tortoise Removals charge a flat fee of £125 plus £5 per mile.

a Write down functions to model the cost, C (£) to move possessions m (miles) for

   i Hare removals

   ii Tortoise removals.

b How far does Brian's new house need to be from his old house for Tortoise Removals to be the cheaper option?

14 In Economics a demand function gives the number of items, D, that consumers are willing to buy at price p ($).

For every extra dollar a particular item costs, demand falls by 15. Demand falls to zero when the price reaches $75.

a Find the demand function $D(p)$.

A supply function gives the number of items, S, that the producer is willing to provide at price p ($).

For every extra dollar the item costs, supply increases by 12. The producer is not prepared to produce the item if the price falls below $30.

b Find the supply function $S(p)$.

The market is said to be at equilibrium when demand is equal to supply.

c Find the equilibrium price of an item.

15 A swimming pool is 25 m long. Its depth, d metres, is given by

$$d = \begin{cases} 0.04x + 1.5 & \text{for } 0 \leq x \leq 20 \\ ax + b & \text{for } 20 < x \leq 25 \end{cases}$$

where x (metres) is the horizontal distance from the shallow end of the pool.

The depth at the deep end of the pool is 3.2 m.

a State the minimum depth of the pool.

b Assuming that there is no step in the pool, find the value of:

   i a

   ii b

c Find the depth of the pool 15 m from the shallow end.

d Find the distance from the shallow end where the depth is 2.5 m.

16 The rate of tax paid depends on income according to the following rules:

On the first £12 500 of income there is no tax to pay.

On the next £37 500 the tax rate is 20%

On the next £100 000 the tax rate is 40%

On any further income the tax rate is 45%.

a Form a piecewise linear function for the amount of tax T (thousand £s) to be paid on an income of I (thousand £s).

b Calculate the tax payable on

   i £35 000

   ii £70 000

c Find the minimum income required to have to pay tax of £50 000.

13B Quadratic models

A **quadratic model** has the form $y = ax^2 + bx + c$. The graphs of these models have a **turning point**, so this type of model is frequently used when the situation being modelled has a maximum or a minimum value.

TOOLKIT: Modelling

For each of the following situations, decide whether a function with the same shape as graph Ⓐ, graph Ⓑ or neither would form a suitable model.

Situation	y-variable	x-variable
Throwing a ball from one person to another	Height of ball	Horizontal distance from thrower
Throwing a ball from one person to another	Height of ball	Time since ball thrown
Making widgets in a factory	Cost per widget	Number of widgets
A satellite dish facing directly upwards	Height of surface of satellite dish	Distance from centre of dish
Profit made by a company	Profit made	Amount spent on advertising
Population of rabbits on an island	Growth rate of rabbits	Number of rabbits
Modelling a railway tunnel	Height of tunnel	Distance from edge of tunnel

Can you suggest any other situations where a quadratic model might be appropriate?

You learned how to solve systems of three equations in Section 12A.

You need three pieces of information to determine the parameters a, b and c. Using the coordinates of three points, you can form a system of three equations.

13B Quadratic models

WORKED EXAMPLE 13.3

Find a quadratic model that goes through the points (1, 9), (2, 18) and (3, 31).

Write a general equation for a quadratic model ··· Let $y = ax^2 + bx + c$

Substitute in the given coordinates to get three equations for a, b, and c
$x = 1, y = 9$: $a + b + c = 9$
$x = 2, y = 18$: $4a + 2b + c = 18$
$x = 2, y = 31$: $9a + 3b + c = 31$

Solve the system of three equations using your GDC:

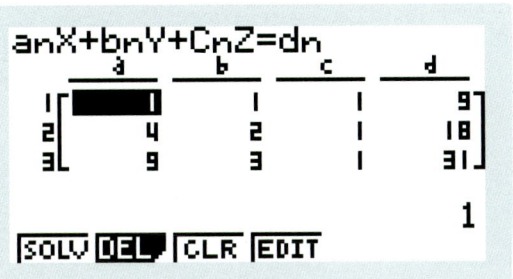

From GDC, $a = 2$, $b = 3$, $c = 4$
Hence, $y = 2x^2 + 3x + 4$

The graph of a quadratic model is called a **parabola**. There are several features of a parabola that you need to know.

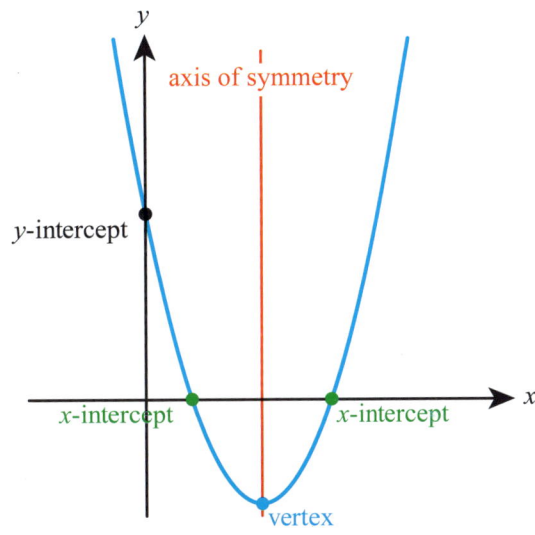

KEY POINT 13.2

For a parabola with equation $y = ax^2 + bx + c$

- the **axis of symmetry** has equation $x = -\dfrac{b}{2a}$
- the turning point of the parabola is called the **vertex of the graph** and has x-coordinate $-\dfrac{b}{2a}$
- the y-intercept is $(0, c)$
- the x-intercepts are the roots of the equation $ax^2 + bx + c = 0$
- the vertex is halfway between the x-intercepts.

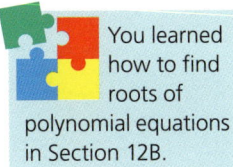

You learned how to find roots of polynomial equations in Section 12B.

WORKED EXAMPLE 13.4

A quadratic model has the following properties:
- the axis of symmetry of $y = f(x)$ is at $x = 4$
- the y-intercept is at $(0, 12)$
- the curve passes through $(1, 5)$.

a Find the roots of $f(x) = 0$.

b Find the minimum value the model takes.

Write down the general form for a quadratic model, then use the given information to find the coefficients ...... Let $y = ax^2 + bx + c$

The axis of symmetry is $x = -\dfrac{b}{2a}$ ...... $-\dfrac{b}{2a} = 4$ ①

The y-intercept is c ...... $c = 12$ ②

When $x = 1, y = 5$ ...... $a(1^2) + b(1) + c = 5$

Use the fact that $c = 12$ ...... $a + b = -7$ ③

You have two equations ① and ③ for a and b

① : $-b = 8a$
$8a + b = 0$

Write them in the standard simultaneous equations form and solve using your GDC
$$\begin{cases} a + b = -7 \\ 8a + b = 0 \end{cases}$$

From GDC, $a = 1, b = -8$

Write down the model ...... So, $y = x^2 - 8x + 12$

Use GDC to find the roots of the quadratic equation ...... **a** When $x^2 - 8x + 12 = 0$:
$x = 2$ or 6

You can find the minimum value by using the graph on the GDC, or use the fact that the vertex lies on the line of symmetry ...... **b** Minimum value is when $x = 4$
$y = 4^2 - 8(4) + 12 = -4$

CONCEPTS – SPACE

Sketch the following curves using technology:

$y = x^2$

$y = x^4$

$y = 2x + 2^{-x} - 2$

Can you describe the relative **spatial** features and differences between these graphs? How might you decide which of them to use to model any particular situation?

Exercise 13B

For questions 1 to 4, use the method demonstrated in Worked Example 13.3 to find a quadratic model that passes through the given points.

1. a (1, 3), (2, 11) and (3, 21)
   b (1, 4), (3, 14) and (4, 25)
2. a (−2, 12), (−1, 1) and (1, −3)
   b (−3, −19), (−1, −3) and (2, 6)
3. a (0.2, 6.12), (1.8, 9.32) and (3.3, 3.02)
   b (0.6, 0.38), (5.8, 1.42) and (7, 5.5)
4. a (−1, −7.6), (−3.5, −4.35) and (−4.7, 1.65)
   b (−10, −5.9), (−6, 5.3) and (2, −1.1)

For questions 5 and 6, use the method demonstrated in Worked Example 13.4 to find a quadratic model with the given properties.

5. a ■ Zeros at $x = -1$ and $x = 3$
      ■ y-intercept at $y = -12$
   b ■ Zeros at $x = -2$ and $x = 5$
      ■ y-intercept at $y = 10$
6. a ■ Vertex at $x = 3$
      ■ y-intercept at $(0, -8)$
      ■ Passes through $(1, 2)$
   b ■ Line of symmetry $x = -2$
      ■ y-intercept at $(0, -5)$
      ■ Passes through $(-2, -17)$

7. The axis of symmetry of the graph of a quadratic function has equation $x = -\frac{3}{2}$.
   The graph intersects the x-axis at the point $P(2, 0)$ and at the point Q.
   a State the coordinates of Q.
   The graph of the function has equation $y = x^2 + bx + c$.
   b Find the values of b and c.
   c Sketch the graph of the function, labelling all axis intercepts.

8. The quadratic function $f(x) = ax^2 + bx + c$ has zeros at $x = -5.5$ and $x = -0.5$.
   The graph of $y = f(x)$ has a vertex at the point P.
   a State the x value of P.
   b Given that the y value of P is 12.5, write down three equations in a, b and c.
   c Find the values of a, b and c.
   d Sketch the graph, labelling all axis intercepts.

9. The owner of a business believes the weekly profits, P (£), of the business can be modelled by the function $P = ax^2 + bx + c$ where x is the number of items produced.
   She has the following data:

x	50	200	400
P (£)	1350	5100	3100

   a Write down three equations in P and x.
   b Hence determine the values of a, b and c.
   c Interpret the value of c in this context.
   d i Find how many items should be produced to maximize the profit.
      ii Calculate the maximum profit.

10. The top surface of a road is built in the shape of a parabola to allow water to drain away on either side. The road is 7.5 m wide and at a distance of 1 m from each edge the road needs to be 5 cm high to allow drains to be fitted.
    Find the maximum height of the road.

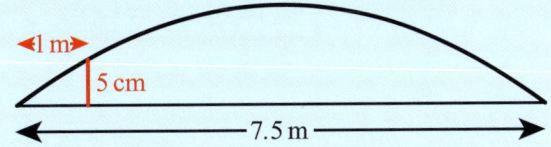

11 A ball is thrown vertically upwards from a ledge 1.45 m above the ground. The height, h metres, of the ball above the ground at t seconds can be modelled by the equation $h = at^2 + bt + c$.

  a State the value of c.

The ball reaches its highest point above the ground when $t = 1.4$ seconds and hits the ground after a further 1.5 seconds.

  b i Write down two equations in a and b.

  ii Hence find the values of a and b.

  c Find the times when the ball is 5 m above the ground.

12 The surface area of algae on a pond is modelled as a circle of radius R. The value of R is modelled by a linear function. Initially the surface area is 2 m². After 4 days the surface area is 3 m².

  a Use the model to estimate the surface area after 7 days.

  b State one assumption of this model.

  c State one limitation of the model.

13C Exponential models

An **exponential model** has the variable in the exponent (power). It can represent **exponential growth** or **exponential decay**.

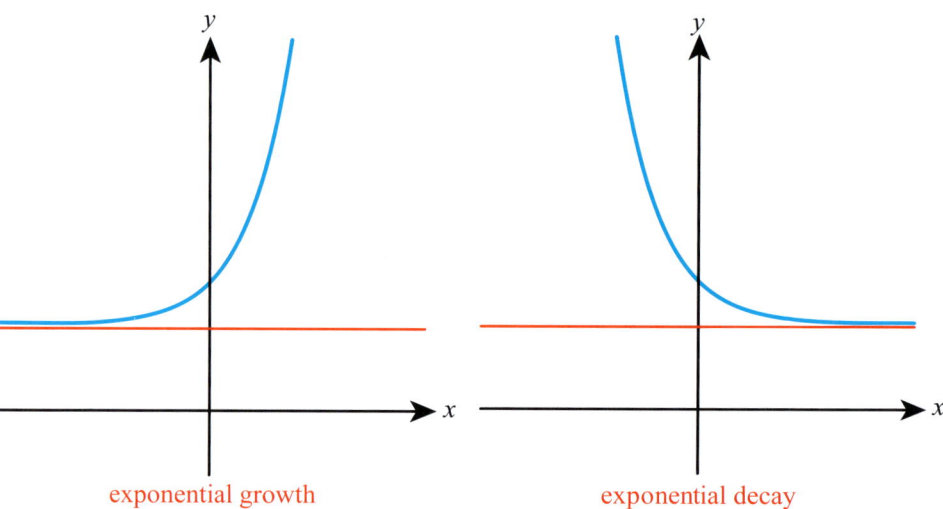

exponential growth   exponential decay

All exponential models have a horizontal asymptote which the curve approaches on one side.

KEY POINT 13.3

An exponential model of the form $y = ka^{rx} + c$
- represents exponential growth when $r > 0$ and exponential decay when $r < 0$
- has horizontal asymptote $y = c$
- has y-intercept $k + c$

13C Exponential models

WORKED EXAMPLE 13.5

Sketch the function $y = 10 \times 3^{-x} + 2$, labelling the axis intercept and the asymptote.

Use your GDC to sketch the graph and find the y-intercept. Notice that the y-intercept is when $x = 0$:
$y = 10 \times 3^0 + 2 = 12$ .... (0, 12)

The asymptote is not shown on the GDC display. You should know that it is $y = 2$ .... $y = 2$

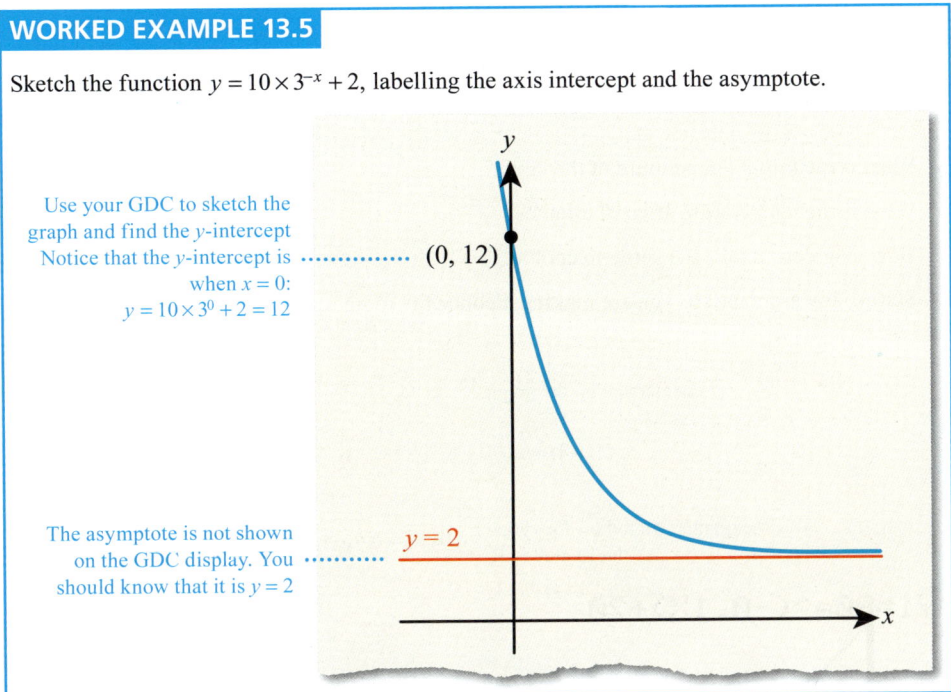

You can use information about x and y values to find the parameters of the model. This often involves simultaneous equations.

WORKED EXAMPLE 13.6

An exponential model has the form $y = k \times 2^x + c$. It is known that it passes through (1, 4) and (2, 10). Find the y value that the model predicts when $x = 4$.

Substitute the given values into the equation to form two equations for k and c
$$k \times 2^1 + c = 4$$
$$k \times 2^2 + c = 10$$

Simplify and write in the standard simultaneous equations form
$$\begin{cases} 2k + c = 4 \\ 4k + c = 10 \end{cases}$$

Solve (using your GDC if needed) .... From GDC: $k = 3$, $c = -2$

Substitute back into the model and use the given value of x .... The model is $y = 3 \times 2^x - 2$
When $x = 4$: $y = 3 \times 2^4 - 2 = 46$

Exponential growth and decay models can be used to describe how quantities such as population size, amount of radioactive substance or temperature of a liquid vary with time. There is some common terminology you should understand in these contexts.

KEY POINT 13.4

- The **initial value** is when $t = 0$. It corresponds to the y-intercept of the graph.
- **Long term behaviour** refers to the value of y when x is large. It corresponds to the horizontal asymptote of the graph.

WORKED EXAMPLE 13.7

The temperature of a kettle ($T\,°C$) at a time t minutes after it has boiled is modelled by

$T = 80e^{-0.1t} + 20$

a What is the initial temperature of the kettle?

b What is the temperature after 10 minutes?

c How long does it take the kettle to cool to $30\,°C$?

d Stating any assumptions you are making, deduce the room temperature where the kettle is kept.

Find the value of T when $y = 0$, or the y-intercept on the graph

a When $t = 0$,
$T = 80e^0 + 20 = 100\,°C$

Use $t = 0$

b When $t = 10$,
$T = 80e^{-0.1 \times 10} + 20 \approx 49.4\,°C$

Use the graph to find the x for which the y value is 30

c Using GDC, when $T = 30$,
$t = 20.8$

It takes 20.8 minutes.

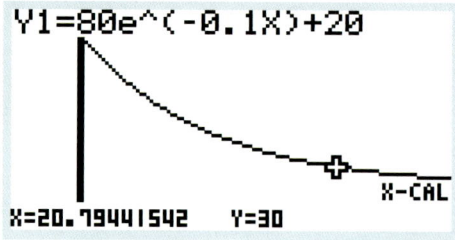

Assuming this model continues to hold, in the long term the temperature of the kettle will approach the room temperature. The long-term value of T is given by the horizontal asymptote

d Assuming that the model can be applied until the kettle cools fully, the room temperature is $20\,°C$.

TOOLKIT: Modelling

Judging a model is not just about whether it fits observations well. Many curves look fairly similar (e.g. $y = x^2$ and $y = x^4$), so a good model should be judged on more than just whether it has roughly the right shape. A lot of models can be justified mathematically. One of the most common starting places for creating a model is a differential equation which describes the rate of increase of a quantity. It turns out that we can often describe these rates far more easily than the underlying quantities themselves, for example, we might say that the birth rate of a population is proportional to the size of the population. These differential equations often lead to functions including exponential terms, which justifies exponential models being so widely used.

An important feature of exponential models is that a constant increase or decrease in x results in y increasing by a constant factor. For example, if $y = 5 \times 3^x$:

x	0	1	2
y	5	15	45

×3   ×3

13C Exponential models

> **Proof 13.1**
>
> If $y = ka^x$, prove that y increases by a factor of a when x increases by 1.
>
> Consider two x values, x_1 and $x_2 = x_1 + 1$, and their corresponding y values .... When $x = x_1$:
> $$y_1 = ka^{x_1}$$
> When $x = x_2 = x_1 + 1$:
> $$y_2 = ka^{x_1+1}$$
>
> Use laws of exponents to express y_2 in terms of y_1 .... $= ka^{x_1} a^1$
> $= ay_1$
>
> Hence, when 1 is added to x, y is multiplied by a.

From Chapter 2, you know that the nth term of a geometric series is $u_1 r^{n-1}$. This looks very much like the exponential functions you have been studying in this chapter and they have lots of similar properties. The big difference is that n can only take positive integer values, while the exponential functions you have been using take all real values.

Exercise 13C

For questions 1 to 4, use the method demonstrated in Worked Example 13.5 to sketch the given exponential functions, labelling any axis intercepts and asymptotes.

1 a $y = 5 \times 2^x + 3$
   b $y = 4e^x - 1$
2 a $y = 3e^{-x} - 2$
   b $y = 7 \times 4^{-x} + 1$
3 a $y = 2 \times 0.8^x + 4$
   b $y = 0.4^x - 2.5$
4 a $y = 2 - e^x$
   b $y = 5 - 3 \times 2^{-x}$

For questions 5 to 8, use the method demonstrated in Worked Example 13.6 to find an exponential model of the stated form that passes through the given points.

5 a $y = k \times 3^x + c$ passing through (1, 5) and (2, 17)
   b $y = k \times 4^x + c$ passing through (−1, 1.75) and (2, 17.5)
6 a $y = k \times 5^{-x} + c$ passing through (1, 8) and (3, 6.08)
   b $y = k \times 2^{-x} + c$ passing through (−1, 7) and (1, −2)
7 a $y = k \times 1.2^x + c$ passing through (1, 4.2) and (2, 4.44)
   b $y = k \times 0.5^x + c$ passing through (−2, 18) and (3, 2.5)
8 a $y = ke^{-x} + c$ passing through (0, 3) and $\left(\ln \frac{1}{3}, 13\right)$
   b $y = ke^x + c$ passing through (0, 5) and (ln 2, 6)

9 The number of people in the world with a mobile phone, N billion, is modelled by
$$N = 4.68 \times 1.02^t$$
where t is the number of years after 2019.

   a In 2019, how many people have a mobile phone?
   b What is the yearly percentage increase in the number of people with a phone predicted by this model?
   c How many people does this model predict will have a mobile phone in 2025?

10 The amount of medication, M mg l^{-1}, in a patient's bloodstream decreases over time, t hours, according to the model $M = 1.5 \times 0.82^t$, $t \geq 0$
   a Write down the amount of medicine in the bloodstream at $t = 0$.
   b Calculate the amount of medicine in the bloodstream after 5 hours.
   c Find to the nearest minute, how long it takes for the concentration to reach 0.25 mg l^{-1}.

11 The number of bacteria in a population, N thousand, is modelled by the function $N = ke^{rt}$ where t is the time in minutes since the bacteria were first observed. Initially there were 2500 bacteria and after 5 minutes there were 7000.
   a Find the values of k and r.
   b According to this model, how many bacteria will there be after 30 minutes?

12 The mass of a substance in a chemical reaction, m grams, is modelled by $m = ka^{-t}$ where t is the time in seconds after the reaction has started. After 3 seconds there is 29.6 g remaining, and after 10 seconds there is 11.1 g remaining.
   a Find the values of k and a.
   b How long does it take for the mass to fall to 1 g?

13 The value of a new car is $8500. One year later the value is $6800.
   a Assuming the value continues to decrease by the same percentage each year, express the value, V, after time t years in the form $V = ka^t$.
   b Find the value of the car after 10 years predicted by this model.

   A company guarantees to pay $500 for any car, no matter how old it is. A refined model is
   $V = cb^t + d$
   c State the value of c and d.

14 The radioactive decay of sodium-24 can be modelled by $m = ke^{-ct}$ where m is the mass in grams and t is the time in hours. The sodium takes 15 hours to decay to half of its original mass.
   a Find the value of c.
   b Find how long it takes for the sodium to decay to 1% of its original mass.

15 The speed, v metres per second, of a parachutist t seconds after jumping from an aeroplane is modelled by the function $v = 50(1 - e^{-0.1t})$.
   What speed does the model predict the parachutist will eventually reach?

16 A bowl of soup is served 65 °C above room temperature 20 °C. Every 5 minutes, the difference between the temperature of the soup and room temperature decreases by 25%.
   The temperature of the soup, T °C, can be modelled by $T = ka^t + c$ where t is the time in minutes since the soup was served.
   a Find the temperature of the soup after 8 minutes.
   b If the soup were put into a thermos flask instead of a bowl, state how this would affect the value of
      i k
      ii a

13D Direct and inverse variation and cubic models

■ Direct proportion

Two quantities are directly proportional when one is a constant multiple of the other. We can also say that y varies directly with x^n. These graphs all show **direct proportion**:

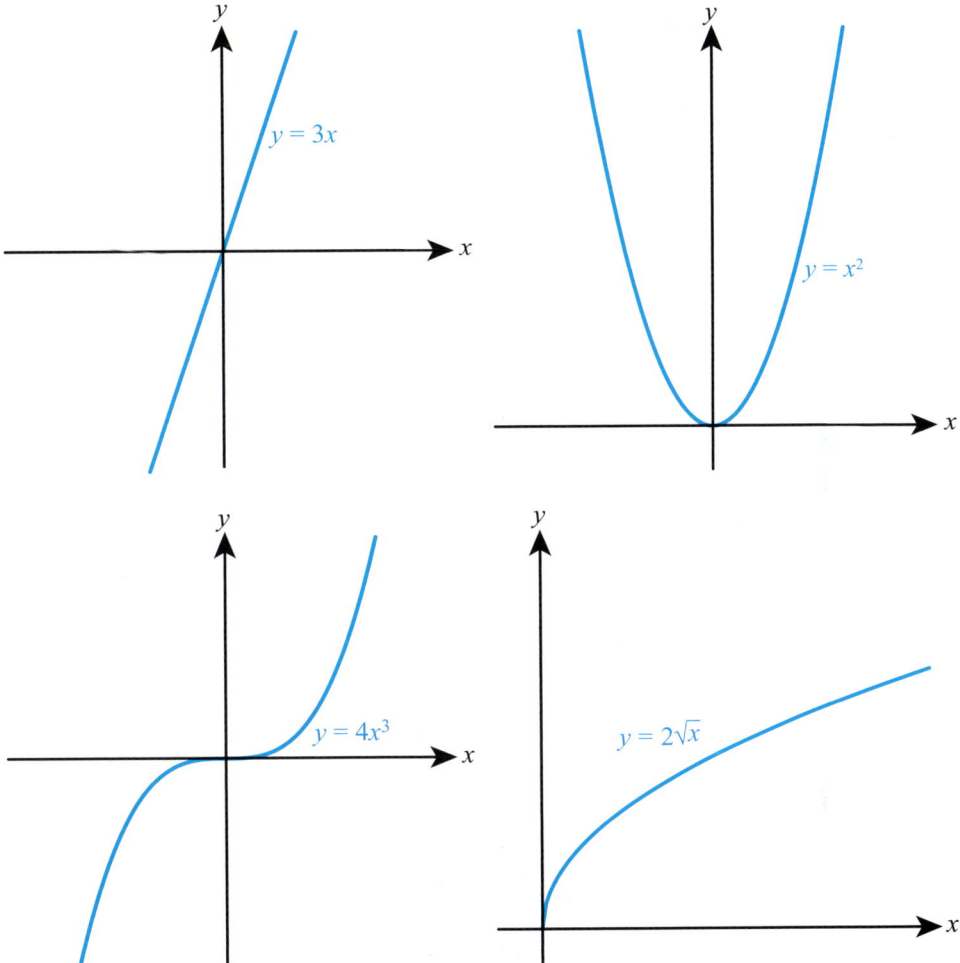

> **KEY POINT 13.5**
>
> If y is directly proportional to x^n then
> - $y = kx^n$
> - when $x = 0$, $y = 0$

You need one pair of (x, y) values to find k.

WORKED EXAMPLE 13.8

y is directly proportional to x^3. When $x = 1$, $y = 4$. Find the value of y when $x = 3$.

Write down the general formula for direct proportion ⋯ $y = kx^3$

Substitute the given values to find k ⋯ $x = 1, y = 4$:
$4 = k \times 1^3$
$k = 4$

Now use $x = 3$ ⋯ When $x = 3$, $y = 4 \times 3^3 = 108$

■ Inverse proportion

Inverse proportion (or inverse variation) is a relationship of the form $y = \dfrac{k}{x^n}$. The graphs below all show inverse proportion:

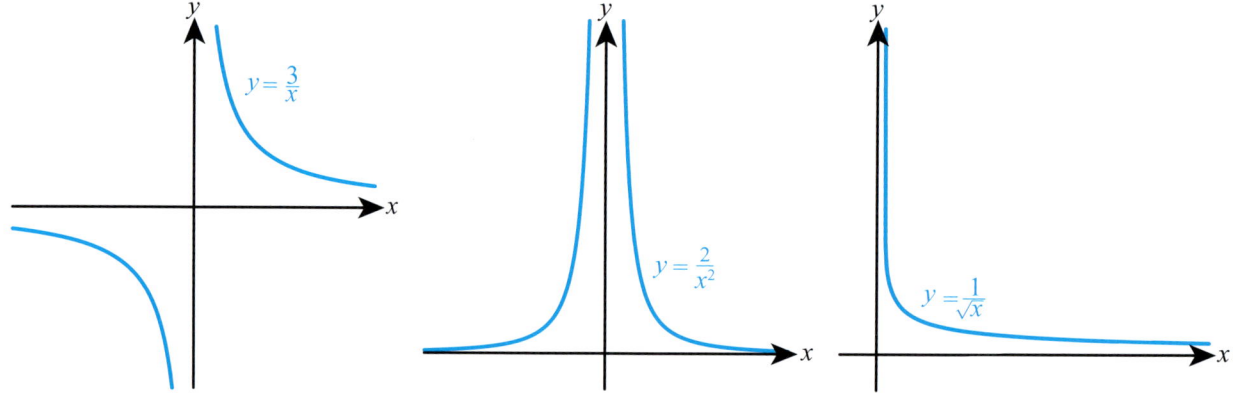

KEY POINT 13.6

If y is inversely proportional to x^n then
- $y = \dfrac{k}{x^n}$
- the y-axis is an asymptote of the graph of y against x

WORKED EXAMPLE 13.9

y is inversely proportional to x^2. When $y = 1$, $y = 4$. Find the positive value of x when $y = 16$.

Write down the general formula for inverse proportion ⋯ $y = \dfrac{k}{x^2}$

Substitute the given values to find k ⋯ $x = 1, y = 4$:
$4 = \dfrac{k}{1^2}$
$k = 4$

13D Direct and inverse variation and cubic models

	When $y = 16$:
Now use $y = 16$	$\dfrac{4}{x^2} = 16$
	$x^2 = \dfrac{4}{16} = \dfrac{1}{4}$
We need the positive value of x	$x = \dfrac{1}{2}$

■ Cubic models

A **cubic model** has the form $y = ax^3 + bx^2 + cx + d$. Graphs of cubic functions can have zero or two turning points.

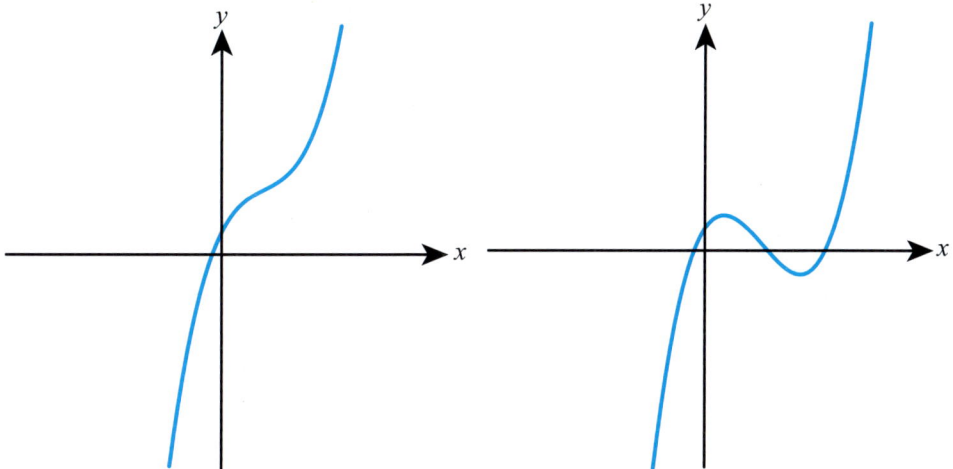

As with quadratic models, you can use given x and y values to find the coefficients.

WORKED EXAMPLE 13.10

$f(x)$ is modelled by a cubic function. It is known that $f(0) = -5$, $f(1) = -2$, $f(2) = 7$ and $f(3) = 34$. Find $f(4)$.

Write a general equation for a cubic model	Let $f(x) = ax^3 + bx^2 + cx + d$
Substitute given values to get four equations	$\begin{cases} d = -5 \\ a + b + c + d = -2 \\ 8a + 4b + 2c + d = 7 \\ 27a + 9b + 3c + d = 34 \end{cases}$
Solve using your GDC	From GDC, $a = 2, b = -3, c = 4, d = -5$
Now evaluate f when $x = 4$	$f(4) = 2(4^3) - 3(4^2) + 4(4) - 5 = 91$

Tip

Notice that d is the y-intercept, which is the value of $f(0)$; therefore, you can immediately say that $d = -5$. This is useful if your GDC can only solve a system of three equations.

TOK Links

A function is known to have f(0) = 0, f(1) = 1, f(2) = 2.

This could be modelled by the function $f(x) = x$, or $f(x) = x^3 - 3x^2 + 3x$, or $f(x) = 2x^3 - 6x^2 + 5x$. In the absence of any other information, it is difficult to know which one to choose, but there is an idea from philosophy called Occam's razor which says that, generally, the simpler solution is more likely to be correct. Is this a valid way of choosing between models in mathematics? How about in science?

Exercise 13D

For questions 1 and 2, use the method demonstrated in Worked Example 13.8 to find the required direct proportion relationships.

1. a y is directly proportional to x^2; $y = 36$ when $x = 3$
   b y is directly proportional to x^4; $y = 8$ when $x = 2$

2. a y is directly proportional to $\sqrt[3]{x}$; $y = 6$ when $x = 8$
   b y is directly proportional to $\sqrt{x}$; $y = 6$ when $x = 16$

For questions 3 and 4, use the method demonstrated in Worked Example 13.9 to find the required inverse proportion relationships.

3. a y is inversely proportional to x; $y = 10$ when $x = 0.5$
   b y is inversely proportional to x^3; $y = 0.25$ when $x = 2$

4. a y is inversely proportional to $\sqrt{x}$; $y = 2.5$ when $x = 36$
   b y is inversely proportional to $\sqrt[4]{x}$; $y = 4$ when $x = 81$

For questions 5 to 8, use the method demonstrated in Worked Example 13.10 to find a cubic model that passes through the given points.

5. a (0, 4), (1, 2), (2, 10) and (3, 34)
   b (0, 2), (1, 0), (2, 8) and (3, 44)

6. a (−2, 0), (−1, −1), (0, −6) and (2, 20)
   b (−2, −2), (0, 2), (1, −2) and (2, −18)

7. a (0, 1), (0.5, 2.25), (1.5, 7.75) and (2.5, 7.25)
   b (0, 10), (1, 7.8), (2, 2.6) and (6, 11.8)

8. a (−1, −5.3), (−0.5, −4.35), (0, −2.5) and (2, −4.1)
   b (−2, −10.4), (−1, −0.1), (0, −3.8) and (2, 18.8)

9. The force applied to a spring, F Newtons, is directly proportional to the extension, x cm, of the spring.

   When the force applied is 12 N, the extension is 5 cm.

   a Find the relationship between F and x.
   b Hence find:
   i   the force required to produce an extension of 4 cm
   ii the extension when a 20 N force is applied.

10. The time taken, T seconds, for a pendulum to swing from one side to the other and back again (its period) is directly proportional to the square root of the length, l cm, of the pendulum.

    When the length is 25 cm, the period is 1 second.

    a Find the relationship between T and l.
    b Hence find:
    i   the period of the swing of pendulum of length 60 cm
    ii the length of a pendulum that has a period of 0.8 seconds.

11. The power, P, generated by a wind turbine is directly proportional to the cube of the wind speed, s.

    If the power generated is 2.05×10^6 Watts when the wind speed is $10 \,\text{m s}^{-1}$, find

    a the power generated when the wind speed is $14 \,\text{m s}^{-1}$.
    b the minimum wind speed needed to generate a power of 1×10^7 Watts.

13D Direct and inverse variation and cubic models

12 Boyle's Law states that at constant temperature, the pressure, P, of a gas is inversely proportional to its volume, V. When the volume of a particular gas is $20\,\text{cm}^3$, the pressure is $45\,000\,\text{Pa}$.
  a Find the relationship between P and V for this gas.
  b Hence find the value of:
   i P when $V = 125\,\text{cm}^3$
   ii V when $P = 80\,000\,\text{Pa}$.

13 Daily sales of sweatshirts, J, in a particular department store are thought to be inversely proportional to the average daytime temperature T (°C). When the temperature is $8\,°\text{C}$, 55 sweatshirts are sold.
Find how many sweatshirts this model predicts will be sold when the temperature is $15\,°\text{C}$.

14 A cubic model, $Q = ax^3 + bx^2 + cx + d$, is proposed to fit the following data:

x	−2	−1	1	2
Q	−1	−8	2	−5

  a Write down four equations in a, b, c and d.
  b Hence find the values of a, b, c and d.

15 The distance from the Sun, in Astronomical Units, and the orbital period, in Earth years, of four planets in the solar system is given below.

Planet	Distance from Sun (AU)	Orbital period (years)
Venus	0.723	0.615
Mars	1.52	1.88
Saturn	9.54	29.5
Neptune	30.1	165

  a From these data, form a cubic model to predict the orbital period, T, from a planet's distance from the Sun, d. Give the coefficients to 3 significant figures.
  b i Jupiter is $5.20\,\text{AU}$ from the Sun. What does the model predict is its orbital period?
    ii The true value is 11.9 years. Calculate the percentage error in the model's prediction.
  c According to this model, what is the maximum possible orbital period of a planet in the solar system?

16 x is inversely proportional to the square root of y and y is inversely proportional to the cube of z.
Find the relationship between x and z.

17 The force of attraction between two magnets is inversely proportional to the square of the distance between them.
Find the percentage decrease in the force of attraction when the distance between two magnets is increased by 10%.

18 $f(x)$ is modelled by a cubic function.
It is known that:
$f(1) = −2$
$f(2) = −4$
$f(3) = −2$
$f(4) = 10$
Find $f(5)$.

19 The time taken to build a housing estate is directly proportional to the number of houses on the estate and inversely proportional to the number of builders. If it takes 7 builders 7 days to build 7 houses, how long would it take 2 builders to build 2 houses?

20 Find all cubic models which satisfy:
$f(0) = 0$
$f(1) = 1$
$f(2) = 2$

13E Sinusoidal models

There are many quantities that vary periodically with time, meaning that they repeat at regular intervals. Examples include the height of a tide or the position of an object oscillating on a spring. Such quantities can often be described by **sinusoidal models**, which are functions of the form $y = a\sin(bx) + d$ or $y = a\cos(bx) + d$. The graphs of these functions oscillate about the **principal axis**. The maximum vertical distance from the principal axis is called the **amplitude**, and the horizontal distance between consecutive maximum (or minimum) points is the **period** of the graph.

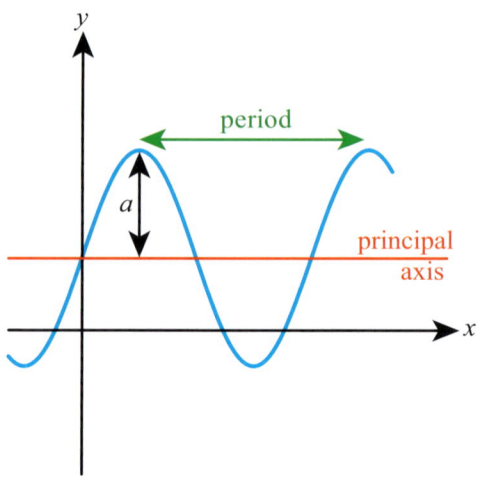

KEY POINT 13.7

For a sinusoidal model of the form $y = a\sin(bx) + d$ or $y = a\cos(bx) + d$
- the amplitude is a
- the period is $\dfrac{360}{b}$
- the principal axis is $y = d$
- the largest value of y is $d + a$ and the smallest value is $d - a$.

The sine and the cosine graphs have the same shape but a different starting point. You need to be able to distinguish between them.

KEY POINT 13.8

- The y-intercept of a sine graph is on the principal axis.
- The y-intercept of a cosine graph is at the maximum or minimum point.

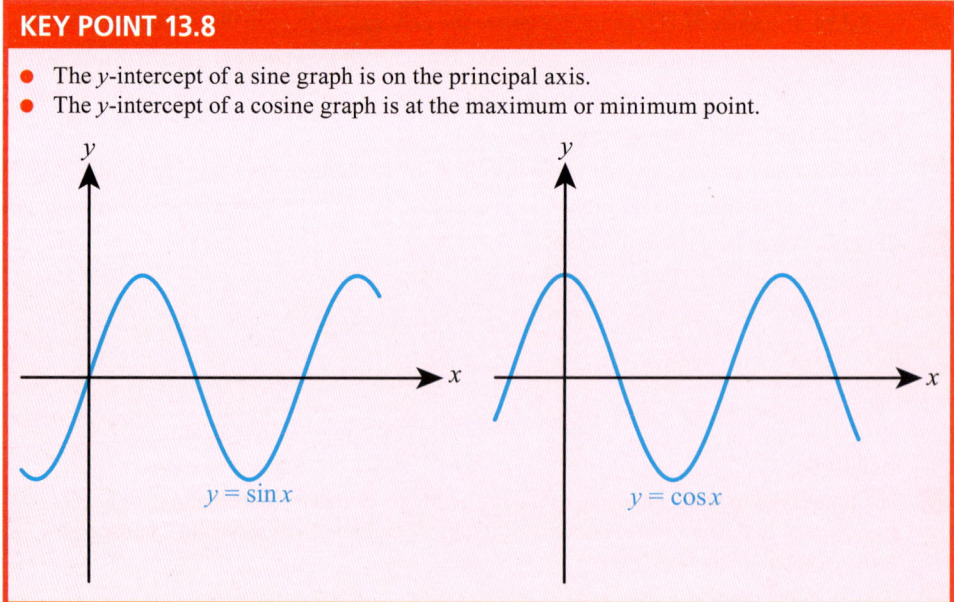

13E Sinusoidal models

WORKED EXAMPLE 13.11

A model has the form $f(x) = 2\cos(180x) + 1$

a Find the amplitude of this model.

b Find the period of this model.

c Find the equation of the principal axis of this model.

d What does the model predict is the value of $f(3)$?

e What does the model predict is the largest value of $f(x)$?

Compare the given equation to the standard sinusoidal model $y = a\cos(bx) + d$. The amplitude is a (the coefficient of cos)	**a** The amplitude is 2
The period is $\dfrac{360}{b}$	**b** period $= \dfrac{360}{180} = 2$
The principal axis is $y = d$	**c** The principal axis is $y = 1$
Use $x = 3$	**d** $f(3) = 2\cos(180 \times 3) + 1 = -1$
The largest value is $d + a$	**e** The largest value is $1 + 2 = 3$

You can use the properties from Key Point 13.7 to find the parameters given some information about a model.

WORKED EXAMPLE 13.12

The depth of water in a harbour, in metres, is modelled using a function of the form $h(t) = a\sin(bt) + d$ where t is the time, in hours, after midnight. It is believed that the depth varies between 3 m and 5 m with a period of 12 hours. Find the depth at 3 pm.

The largest value of h is $d + a$ and the smallest value is $d - a$	$\begin{cases} d - a = 3 \\ d + a = 5 \end{cases}$
Solve the system of equations (using your GDC)	$a = 1, d = 4$
The period is $\dfrac{360}{b}$	$\dfrac{360}{b} = 12$ $b = 30$
Use the model with $t = 15$ (3 pm is 15 hours after midnight)	At 3 pm: $h(3) = 1 \times \sin(30 \times 3) + 4$ $= 5\,\text{m}$

Exercise 13E

For questions 1 and 2, use the method demonstrated in Worked Example 13.11 to find the following for the given sinusoidal model:

    i amplitude   ii period   iii principal axis
    iv maximum value   v minimum value   vi y-intercept.

1. a $y = 3\sin(4x) + 2$
   b $y = 2\sin\left(\frac{x}{2}\right) - 5$

2. a $y = 0.5\cos\left(\frac{x}{3}\right) - 2$
   b $y = 5\cos(90x) + 3$

3. The diagram opposite shows part of the graph of the function $y = p\sin qx + r$.

   Find the value of:
   a p
   b q
   c r.

4. The diagram opposite shows part of the graph of the function $y = p\cos qx + r$.

   Find the value of:
   a p
   b q
   c r.

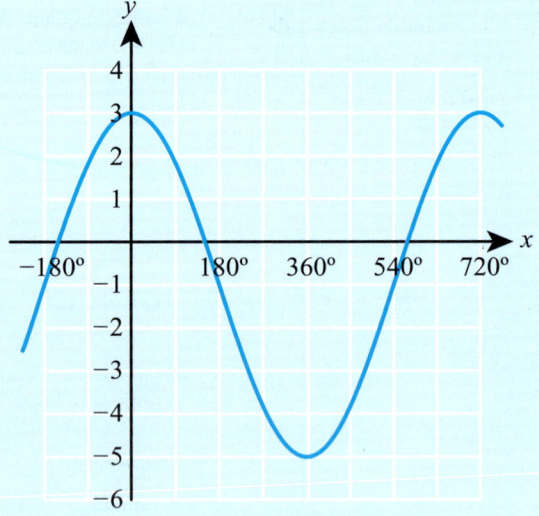

5. The depth, d, of water out at sea at a distance x from a fixed point on a given day is modelled by the function $d = 1.8\sin(72x) + 50$ where d and x are both measured in metres.
   a State the depth of the sea in completely calm water.
   b State the amplitude of the waves in this model.
   c Find the distance between two consecutive peaks of a wave.

13E Sinusoidal models

6 A swing at a playground is released from its maximum height above the ground. The height, h metres, of the swing above the ground at time t seconds after release can be modelled by the equation $h = 0.58\cos(bt) + 1.1$.

  a Find

    i the maximum height of the swing above the ground

    ii the minimum height of the swing above the ground.

  The swing takes 0.6 seconds to move from its highest point to its lowest point.

  b Find the value of b.

  c How high is the swing above the ground after 4 seconds?

7 An elastic string is suspended from a fixed point 2 m above the ground. A small weight is attached to the other end.

  The weight is pulled down and released so that it moves up and down. The height, h, of the weight above the ground at time t is given by $h = 140 - 12\cos(450t)$ where h is in centimetres and t in seconds.

  a Find the maximum height of the weight above the ground.

  b Find how long it takes for the weight to first reach its maximum height.

  c State the height of the weight above the ground before it was pulled down and released.

8 A cyclist cycles round a circular track, starting at a point due east of the centre. The track has a diameter of 80 m and the cyclist takes 20 s to complete each lap.

  At any given time t seconds, the cyclist is d metres north of the East–West diameter of the circuit. d is modelled by a function of the form $d = p\sin(qt) + r$.

  a Find the values of p, q and r.

  b How long after starting is the cyclist first 10 m south of the East–West diameter?

9 The temperature in a 24-hour period is modelled using a function of the form $T = a\sin(bt) + d$ where t is the time in hours after 09:00.

  The temperature varies between 6 °C and 15 °C. The minimum temperature is achieved once in the 24 hour period at 03:00.

  Find the temperature predicted by this model at 12:00.

10 The water level in a tidal lake can be modelled by the function $h = a\cos(bt) + d$ where h is the depth of the water in metres and t is the time in hours after high tide at 06:00.

  At high tide the depth of water is 5.5 m and at low tide the depth is 2 m. The time between consecutive high tides is 12 hours.

  Find the depth of water in the lake at 14:30.

13F Modelling skills

When trying to find a suitable function to model a given situation, you first need to decide what type of model to use, and then find the parameters from the information you have about specific values. The easiest way to decide on the type of model is to plot the data on a graph.

WORKED EXAMPLE 13.13

5 data points found when trying to model f(x) are shown opposite:

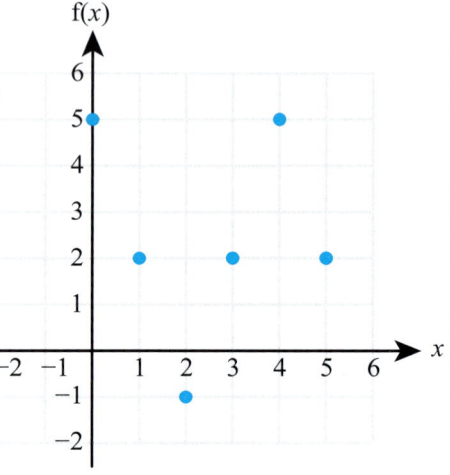

a Choose one of the following functions to model, justifying your answer.

  A $f(x) = ax + b$
  B $f(x) = ax^2 + bx + c$
  C $f(x) = a \sin bx + d$
  D $f(x) = a \cos bx + d$
  E $f(x) = ke^{rx} + c$

b Use your model to predict the value of f(0.5).

c The value of f(0.5) was found to be 3.5. Based on this, suggest an improved model.

The points form an oscillating pattern, suggesting a sinusoidal model
It starts from the maximum point, so it is cos rather than sin

a The graph shows an oscillating quantity, with the maximum value when $x = 0$.
   So the best model is D.

To find a and d, look at maximum and minimum values

b max: $d + a = 5$
   min: $d - a = -1$
   So $d = 2$, $a = 3$

To find b look at the period, which is the distance between the two maximum points

period = 4 : $\frac{360}{b} = 4$

$b = 90$

The model is: $f(x) = 3\cos(90x) + 2$

Use x = 0.5

It predicts $f(0.5) = 3\cos(45) + 2 = 4.12$

The point (0.5, 3.5) lies on the straight line segment from (0, 5) to (1, 2), suggesting that the given points might be connected by straight lines

c A piecewise linear model, with straight line segments between maximum and minimum points.

13F Modelling skills

> **CONCEPTS – REPRESENTATION**
>
> The data show in Worked Example 13.13 could be **represented** by many other functions – for example a negative cubic function.
>
> When choosing which type of model to use, it is helpful if you know something about the situation you are trying to model. In Worked Example 13.13 you may have a strong reason to believe that the sinusoidal model is a better representation of the situation because you know it should have periodic behaviour. In that case you may decide to stick with your original model and treat the information that $f(0.5) = 3.5$ as experimental error. In real-world situations you often have to consider the reliability of your data when constructing a model.

WORKED EXAMPLE 13.14

A fitness coach wants to open a new gym. He expects to start with 50 members and the membership to double every year.

a Construct a model for the number of members of the gym after n years.

b If the total population of the town is 100 000, suggest the largest feasible domain for your model.

c Use your model to predict the gym membership after 6 months.

d After 6 months, he actually has 72 members. Comment on this in light of your answer to part **c**.

e Suggest two reasons why this model is unrealistic.

Since the number increases by a constant factor of 2 every year, this is an exponential model: $M = a \times 2^n$

a Let M be the number of members after n years.
Then $M = a \times 2^n$

Use $n = 1$ to find the value of a

After 1 year there are 100 members:
$100 = a \times 2^1$
$a = 50$
Hence,
$M = 50 \times 2^n$

The number of members cannot exceed 100 000
Use a graph on your GDC to find the value of n for which $M = 100\,000$

b $50 \times 2^n = 100\,000$
$n = 10.97$
The largest feasible domain is 11 years.

Use $n = 0.5$
Note that, although originally we constructed the model as a sequence, we want it to apply for fractional values of n as well

c $M = 50 \times 2^{0.5} = 70.7$

The model predicts 71 members after six months.

This is close to the value predicted by the model

d The model seems to give a good prediction.

The model predicts indefinite growth, but in reality the membership is limited by the size of the town's population

e The model predicts that eventually the whole town will join the gym.

The value of n varies continuously; but there may be different joining rates at different times of the year

The model does not take into account different joining rates at different times of the year.

> Remember that the domain of a function is the set of all possible values of the input variable – see Chapter 3.

TOK Links

Just because a model is a good fit to known data, is it still valid to use it to predict future data?

WORKED EXAMPLE 13.15

A company predicts that the number of books it will sell each week (n), if the price is P, will be $1000 - kP$.

a The company knows that if the price is $50 it will sell 500 books each week. Determine the value of k in the model.

b Use the model to predict the number of books that will be sold each week if the price is $150. Comment on your answer.

c Determine a suitable domain for the model.

d Suggest an improved model, justifying your answer. You do not need to find an equation for your model.

Substitute the given values into the model

a $P = 50$, $n = 500$:
$$1000 - k \times 50 = 500$$
$$50k = 500$$
$$k = 10$$

Use $P = 150$

b $P = 150$:
$$n = 1000 - 10 \times 150 = -500$$

The model predicts negative sales, which is not possible

According to the model, they will not sell any books at this price.

This model is only valid if it predicts a non-negative value of n

c Need $n \geq 0$:
$$1000 - 10P \geq 0$$
$$10P \leq 1000$$
$$P \leq 100$$

The price itself also needs to be positive

For this model, $0 < P \leq 100$.

You need a function that decreases but never reaches zero, so an exponential model seems suitable

d An exponentially decreasing model, because it never gives negative n.

There is more than one possible answer in part **d**. For example, you could use a piecewise linear model,

$$n = \begin{cases} 1000 - 10P & \text{for } 0 < P \leq 100 \\ 0 & \text{for } P > 100 \end{cases}$$

You need more information to select which of the two models is better.

13F Modelling skills

> **You are the Researcher**
>
> Governments use models to try to predict how their populations will grow and their economies will develop. Try researching the factors used in your country's model. Do you think that the factors will be mainly things your government can control, or international factors?

Exercise 13F

For questions 1 to 4, use the method demonstrated in Worked Example 13.13 to decide which of the following models is appropriate for the given data. In each case, find the values of the constants.

A $f(x) = ax + b$
B $f(x) = ax^2 + bx + c$
C $f(x) = ax^3 + bx^2 + cx + d$
D $f(x) = a \sin bx + d$
E $f(x) = a \cos bx + d$
F $f(x) = k \times 3^{rx} + c$

1 a 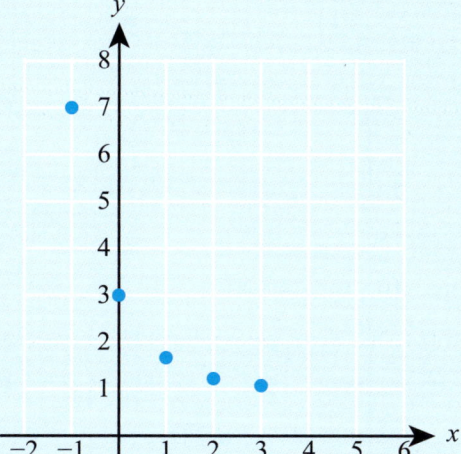   **b**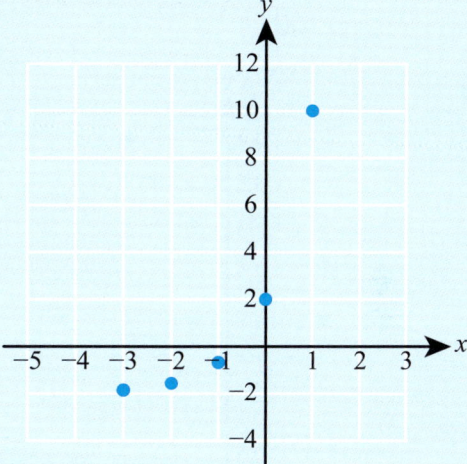

2 a 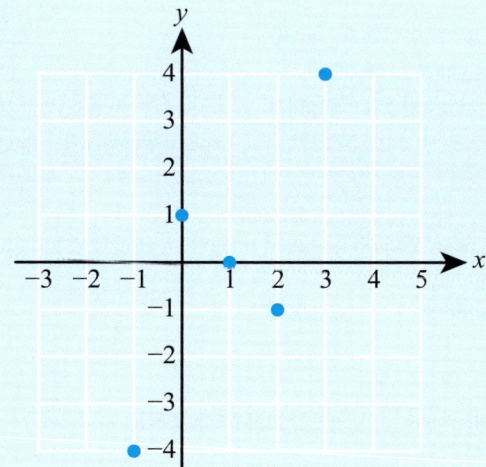   **b**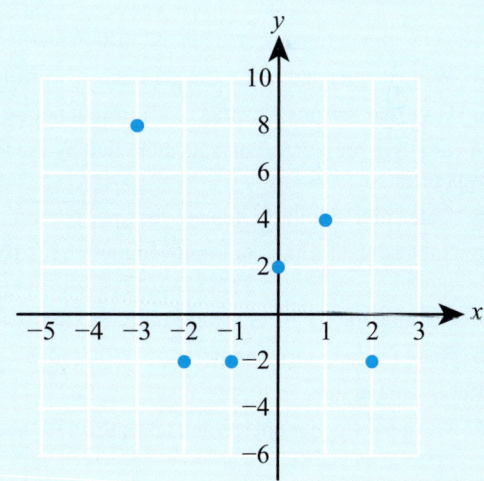

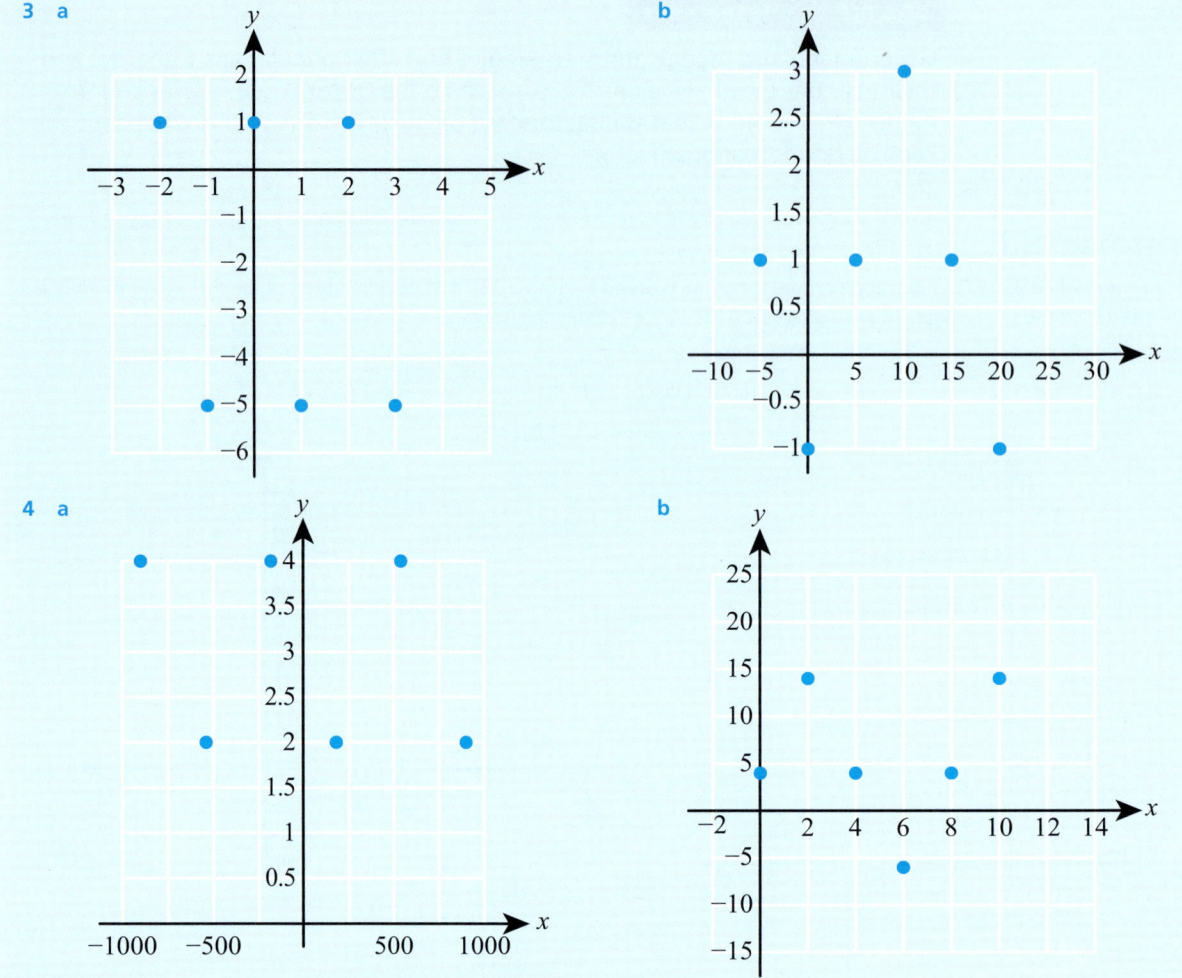

5 The UK population is currently 67 million and is predicted to grow by 1% per year.

 a Write down a model for the population, P million, at time n years from now.

 b Give two reasons why this model might not be suitable for predicting the population of the UK in the year 2100.

6 A car's tyre pressure, P bars, is modelled by the function $P = 2.2 - 0.04t$ where t is the time in weeks after the tyre was inflated.

 a Determine a suitable domain for the function.

 b Suggest the form of an improved model, justifying your answer. You do not need to find an equation for your model.

7 The temperature $T\,°C$ of water t minutes after boiling is modelled by $T = 100 \times 0.87^t$.

 a State the long-term temperature of the water.

 Room temperature is $20\,°C$.

 b Form a refined model which takes this into account, but which has the same initial temperature and rate of cooling as before.

Checklist

> **You are the Researcher**
>
> Modelling is a great topic for your exploration. Try to start from a question and think about how it might be modelled mathematically (your teacher might be able to point you towards a variety of techniques in addition to the ones mentioned in this chapter). Then collect real world data to either estimate parameters of your model or test your model and reflect on how accurate your model predictions were and perhaps improve your model. Some examples of possible questions might include:
>
> - How long does it take a bottle to empty?
> - How does changing the price of a drink change how many are sold?
> - How does changing your initial speed change your time in a race?

Checklist

- You should know that a linear model can be represented by a straight line graph.
    - The gradient of the line connecting the points (x_1, y_1) and (x_2, y_2) is $m = \dfrac{y_2 - y_1}{x_2 - x_1}$ and its equation is $y_1 = m(x - x_1)$.

- You should know that a piecewise linear model uses different straight line segments on different parts of the domain.

- You should know that a quadratic model has the form $y = ax^2 + bx + c$ and its graph is a parabola
    - the axis of symmetry has equation $x = -\dfrac{b}{2a}$
    - the turning point of the parabola is called the vertex and has x-coordinate $-\dfrac{b}{2a}$
    - the y-intercept is $(0, c)$
    - the x-intercepts are the roots of the equation $ax^2 + bx + c = 0$
    - the vertex is halfway between the x-intercepts.

- You should know that an exponential model has the form $y = ka^{rx} + c$
    - it represents exponential growth when $r > 0$ and exponential decay when $r < 0$
    - the horizontal asymptote is $y = c$
    - the y-intercept is $k + c$
    - in the case $c = 0$, when x increases by one unit, y increases/decreases by a factor of a^r.

- You should know that if y is directly proportional to x^n, then $y = kx^n$. When $x = 0$, $y = 0$.

- You should know that if y is inversely proportional to x^n, then $y = \dfrac{k}{x^n}$. The y-axis is the vertical asymptote of the graph.

- You should know that a cubic model has the form $y = ax^3 + bx^2 + cx + d$. Graphs of cubic functions can have zero or two turning points.

- You should know that a sinusoidal model has the form $y = a\sin(bx) + d$ or $y = a\cos(bx) + d$
    - the amplitude is a
    - the period is $\dfrac{360}{b}$
    - the principal axis is $y = d$
    - the largest value of y is $d + a$ and the smallest value is $d - a$
    - the y-intercept of a sin graph is on the principal axis, while for a cos graph it is at a maximum or minimum point.

- You should be able to select an appropriate model based on the shape of a graph, or on a description of a real-life situation.

- You should be able to find the parameters of the model from given information. This often involves solving a system of equations.

- You should be able to use models to make predictions, and to evaluate and suggest improvements to models.

■ Mixed Practice

1 The percentage of charge in a laptop battery, $C\%$, after t minutes of charging can be modelled by the function $C = 1.2t + 16$.
 a What is the percentage of charge in the battery when charging begins?
 b What percentage is the battery at after half an hour?
 c How long does it take to fully charge the battery?

2 The graph of the quadratic function $f(x) = ax^2 + bx + c$ intersects the y-axis at the point $(0, 3)$ and has its vertex at the point $(6, 15)$.
 a Write down the value of c.
 b By using the coordinates of the vertex, or otherwise, write down two equations in a and b.
 c Find the values of a and b.

3 The function $f(x) = p \times 0.4^x + q$ is such that $f(0) = 5$ and $f(1) = 4.1$.
 a Find the values of p and q.
 b State the equation of the horizontal asymptote of the graph of $y = f(x)$.

4 Charles' Law states that at constant pressure the volume of a gas, $V\,\text{cm}^3$, is proportional to its absolute temperature, T kelvins. When the temperature of a gas is 350 kelvins its volume is $140\,\text{cm}^3$.
 a Find the relationship between volume and absolute temperature for the gas.
 b Given that 0 kelvin = $-273\,°C$, find the volume of the gas at room temperature of $20\,°C$.

5 The number of apartments in a housing development has been increasing by a constant amount every year. At the end of the first year the number of apartments was 150, and at the end of the sixth year the number of apartments was 600. The number of apartments, y, can be determined by the equation $y = mt + n$, where t is the time, in years.
 a Find the value of m.
 b State what m represents **in this context**.
 c Find the value of n.
 d State what n represents **in this context**.

Mathematical Studies SL November 2014 Paper 1 Q7

6 Consider the function $f(x) = px^3 + qx^2 + rx$. Part of the graph of f is shown opposite.

The graph passes through the origin O and the points $A(-2, -8)$, $B(1, -2)$ and $C(2, 0)$.
 a Find three linear equations in p, q and r.
 b Hence find the value of p, of q and of r.

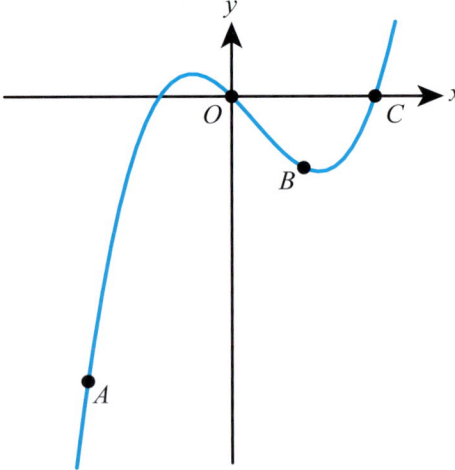

Mathematics SL November 2010 Paper 2 Q5

Mixed Practice

7 The diagram shows the graph of a cosine function, $g(x) = a \cos bx + c$ for $-360° \leq x \leq 360°$, and a sine function, $f(x)$.
  a Write down
   i the amplitude of $f(x)$
   ii the period of $f(x)$.
  b Write down the value of
   i a
   ii b
   iii c.
  c Write down the number of solutions of $f(x) = g(x)$ in the domain $-180° \leq x \leq 360°$.

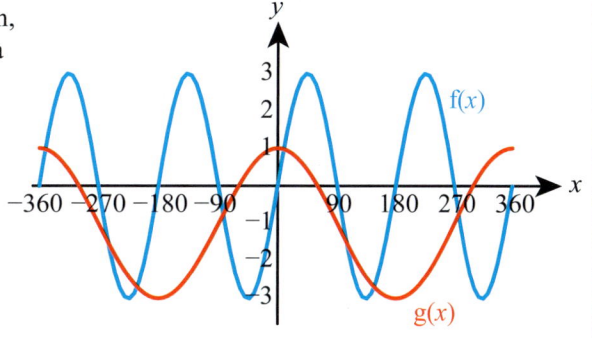

Mathematical Studies SL November 2013 Paper 1 Q14

8 The graph shows the cost, in dollars, of posting letters with different weights.
  a Write down the cost of posting a letter weighing 60 g.
  b Write down the cost of posting a letter weighing 250 g.

Kathy pays 2.50 dollars to post a letter.
  c Write down the range for the weight, w, of the letter.

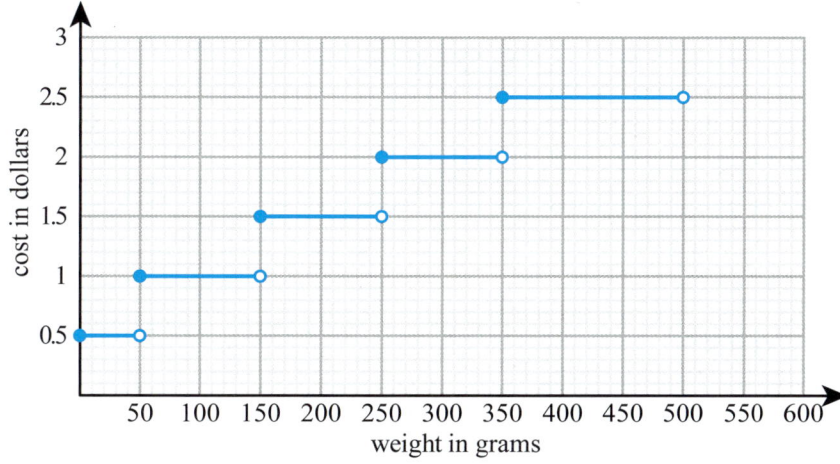

Mathematical Studies SL November 2005 Paper 1 Q9

9 The function $f(x) = a^x + b$ is defined by the mapping diagram opposite.
  a Find the values of a and b.
  b Write down the image of 2 under the function f.
  c Find the value of c.

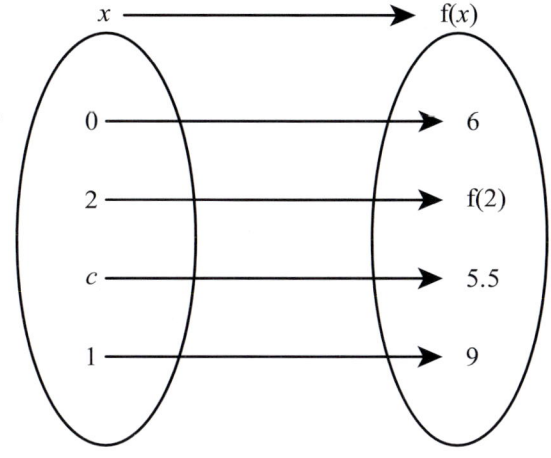

Mathematical Studies SL November 2008 Paper 1 Q14

10 The gravitational force on an object is inversely proportional to the square of the distance of the object from the centre of the Earth. A satellite is launched into orbit 1500 km above the Earth's surface.

Given that the radius of the Earth is 6000 km, find the percentage decrease in the gravitational force on the satellite when launched into orbit.

11 George leaves a cup of hot coffee to cool and measures its temperature every minute. His results are shown in the table below.

Time, t (minutes)	0	1	2	3	4	5	6
Temperature, y (°C)	94	54	34	24	k	16.5	15.25

a Write down the decrease in the temperature of the coffee
  i during the first minute (between $t = 0$ and $t = 1$)
  ii during the second minute
  iii during the third minute.
b Assuming the pattern in the answers to part **a** continues, show that $k = 19$.
c Use the **seven** results in the table to draw a graph that shows how the temperature of the coffee changes during the first six minutes.
Use a scale of 2 cm to represent 1 minute on the horizontal axis and 1 cm to represent 10 °C on the vertical axis.

The function that models the change in temperature of the coffee is $y = p(2^{-t}) + q$.
d i Use the values $t = 0$ and $y = 94$ to form an equation in p and q.
  ii Use the values $t = 1$ and $y = 54$ to form a second equation in p and q.
e Solve the equations found in part **d** to find the value of p and the value of q.

The graph of this function has a horizontal asymptote.
f Write down the equation of this asymptote.

George decides to model the change in temperature of the coffee with a linear function using correlation and linear regression.
g Use the **seven** results in the table to write down
  i the correlation coefficient
  ii the equation of the regression line y on t.
h Use the equation of the regression line to estimate the temperature of the coffee at $t = 3$.
i Find the percentage error in this estimate of the temperature of the coffee at $t = 3$.

Mathematical Studies SL May 2013 TZ1 Paper 2 Q3

12 The amount of electrical charge, C, stored in a mobile phone battery is modelled by $C(t) = 2.5 - 2^{-t}$, where t, in hours, is the time for which the battery is being charged.
a Write down the amount of electrical charge in the battery at $t = 0$.

The line L is the horizontal asymptote to the graph.
b Write down the equation of L.

To download a game to the mobile phone, an electrical charge of 2.4 units is needed.
c Find the time taken to reach this charge.
Give your answer correct to the nearest minute.

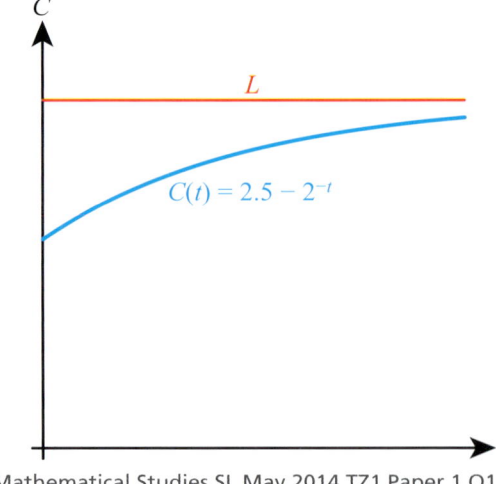

Mathematical Studies SL May 2014 TZ1 Paper 1 Q11

Mixed Practice

13 The diagram opposite represents a big wheel, with a diameter of 100 metres.

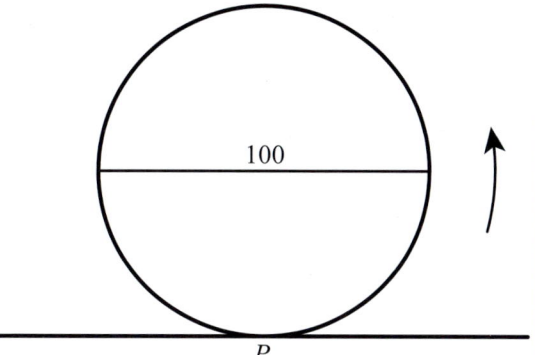

Let P be a point on the wheel. The wheel starts with P at the lowest point, at ground level. The wheel rotates at a constant rate, in an anticlockwise (counterclockwise) direction. One revolution takes 20 minutes.

a Write down the height of P above ground level after
  i 10 minutes
  ii 15 minutes.

Let h(t) metres be the height of P above ground level after t minutes. Some values of h(t) are given in the table below.

t	h(t)
0	0.0
1	2.4
2	9.5
3	20.6
4	34.5
5	50.0

b i Show that h(8) = 90.5.
  ii Find h(21).
c **Sketch** the graph of h, for $0 \leq t \leq 40$.
d Given that h can be expressed in the form $h(t) = a \cos bt + c$, find a, b and c.

14 Applications and interpretation: Geometry

ESSENTIAL UNDERSTANDINGS

- Geometry allows us to quantify the physical world, enhancing our spatial awareness in two and three dimensions.
- This topic provides us with the tools for analysis, measurement and transformation of quantities, movements and relationships.

In this chapter you will learn…
- how to find the length of an arc and the area of a sector of a circle
- how to find the equation of the perpendicular bisector of a line segment
- how to construct, interpret and use Voronoi diagrams.

CONCEPTS

The following key concepts will be addressed in this chapter:
- The Voronoi diagram allows us to navigate, path-find or establish an optimum position in two-dimensional **space**.

PRIOR KNOWLEDGE

Before starting this chapter, you should already be able to complete the following:

1 A line segment connects the points $A(-2, 4)$ and $B(1, -6)$. Find
   a the length of AB,
   b the coordinates of the midpoint of AB.

2 Line L_1 has equation $5x - 2y = 30$. Line L_2 is perpendicular to L_1 and passes through the point $(1, 5)$. Find the equation of L_2, giving your answer in the form $ax + by = c$ where a, b and c are integers.

■ **Figure 14.1** How do we decide which school, hospital or supermarket to go to, or which team to support?

Applications and interpretation: Geometry

Many real-life situations require an understanding of spatial relationships. Urban or ecological planning, disease management and delivery scheduling all utilize distance calculations to make decisions. What area should each hospital cover? How can we use information from weather stations to estimate temperatures at different sites? Where should we build a new feeding station in a forest so it is as far away as possible from any existing ones? All of these questions can be answered with the aid of Voronoi diagrams.

Starter Activity

Look at the images in Figure 14.1. In small groups, discuss how you could find your nearest example of each of these types of places. Do you attend your local school, or support your local team? What other factors have to be taken into consideration?

Now look at this problem:

If you wanted to open a new ice cream shop, what factors would you consider when deciding on the best location for it?

LEARNER PROFILE – Risk-takers
Do you ever start a mathematics problem without knowing whether you will be able to finish it? Are you happy to try new methods even if you are not sure you have fully mastered them? Good mathematical problem solvers often take a reasoned attitude towards risk and know that even if they fail in solving one problem, they can learn from it and do better with the next one!

14A Arcs and sectors

■ Length of an arc

The arc AB **subtends** an angle θ at the centre of the circle opposite.

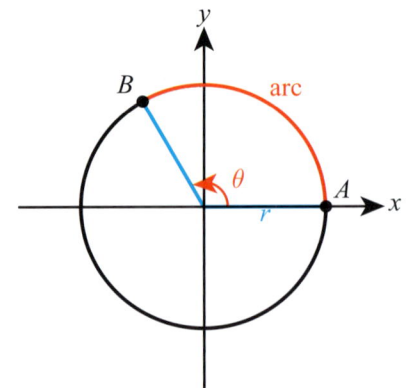

Since the ratio of the length of an arc, s, to the circumference will be the same as the ratio of θ to 360°, this gives us:

$$\frac{s}{2\pi r} = \frac{\theta}{360}$$

Rearranging gives the formula for arc length (when θ is measured in degrees):

KEY POINT 14.1

The length of an arc is

$$s = \frac{\theta}{360} \times 2\pi r$$

where r is the radius of the circle and θ is the angle subtended at the centre.

WORKED EXAMPLE 14.1

Find the length of the arc AB in the circle shown.

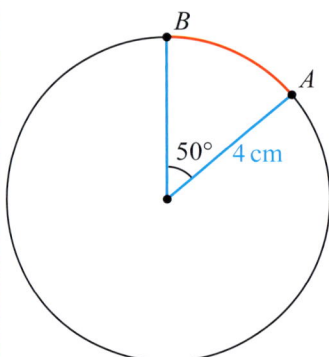

Use $s = \frac{\theta}{360} \times 2\pi r$ ⋯⋯⋯⋯⋯⋯⋯⋯

$s = \frac{\theta}{360} \times 2\pi r$

$= \frac{50}{360} \times 2\pi \times 4$

$= 3.49 \, \text{cm}$

Area of a sector

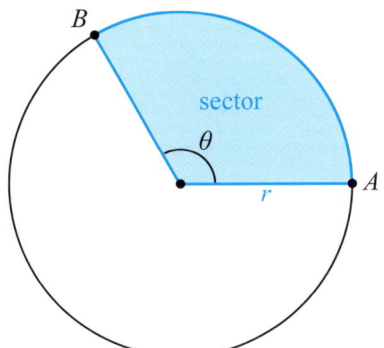

By a very similar argument to that for the length of an arc, we can obtain a formula for the area of a sector:

Since the ratio of the area of a sector, A, to the area of the circle will be the same as the ratio of θ to 360°, this gives us:

$$\frac{A}{\pi r^2} = \frac{\theta}{360}$$

Rearranging gives the formula for the area of a sector (when θ is measured in degrees):

KEY POINT 14.2

The area of a sector is

$$A = \frac{\theta}{360} \times \pi r^2$$

where r is the radius of the circle and θ is the angle subtended at the centre.

WORKED EXAMPLE 14.2

Find the area of the sector AOB in the circle shown.

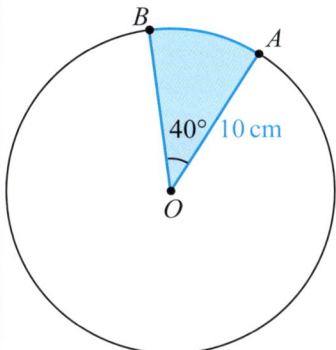

Use $A = \frac{\theta}{360} \times \pi r^2$

$A = \frac{\theta}{360} \times \pi r^2$

$A = \frac{40}{360} \times 10^2 \pi$

$= 34.9 \, \text{cm}^2$

> You met the sine rule, cosine rule, and area of a triangle formula in Section 5B.

Tip

If you are using $\sin \theta$, $\cos \theta$ or $\tan \theta$ with θ in degrees, make sure your calculator is set to degree mode.

 If you are studying the Mathematics: applications and interpretation HL course you will meet another unit for measuring angles, called radians. When using this unit, the formulae for the arc length and area of a sector are slightly different.

You will often need to combine the results for arc length and area of sector with the results you know for triangles.

WORKED EXAMPLE 14.3

The diagram shows a sector of a circle with radius 8 cm and angle 40°.

For the shaded region, find:

a the perimeter

b the area.

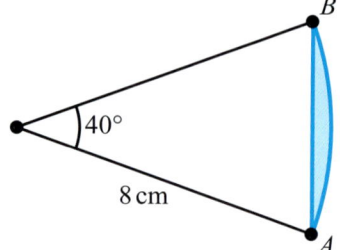

Use $s = \dfrac{\theta}{360} \times 2\pi r$ to find the arc length	a $s = \dfrac{40}{360} \times 2\pi \times 8 = 5.585$
Use the cosine rule to find the length of the chord AB. Make sure your calculator is in degree mode	By cosine rule, $AB^2 = 8^2 + 8^2 - 2 \times 8 \times 8 \cos 40°$ $= 29.946$ So, $AB = 5.472$
The perimeter of the shaded region is the length of the arc plus the length of the chord	Hence, $p = 5.585 + 5.472$ $= 11.1 \text{ cm}$
Use $A = \dfrac{\theta}{360} \times \pi r^2$ to find the area of the sector	b Area of sector $= \dfrac{40}{360} \times 8^2 \pi = 22.34$
Use $A = \dfrac{1}{2} ab \sin C$ with $a = b = r$ and $C = \theta$ to find the area of the triangle	Area of triangle $= \dfrac{1}{2} \times 8^2 \sin 40° = 20.57$
The area of the shaded region is the area of the sector minus the area of the triangle	$A = 22.34 - 20.57$ $= 1.77 \text{ cm}^2$

14A Arcs and sectors

Exercise 14A

For questions 1 to 3, use the methods demonstrated in Worked Examples 14.1 and 14.2 to find the length of the arc AB that subtends the given angle, and the area of the corresponding sector.

1 a 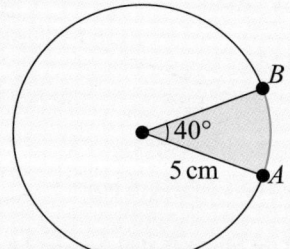   b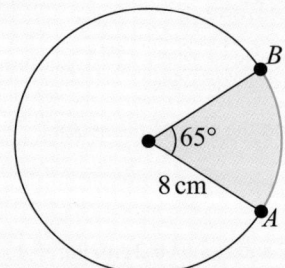

2 a 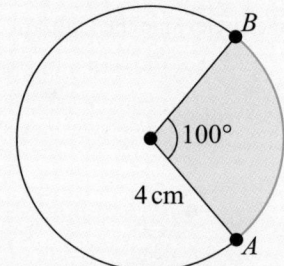   b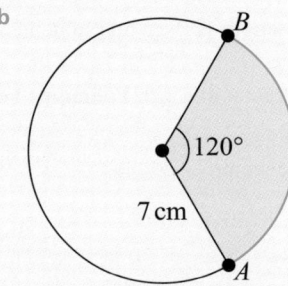

3 a 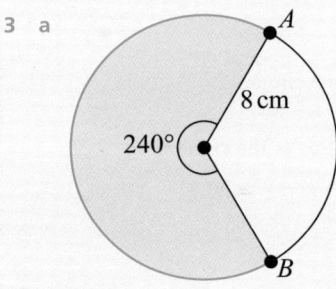   b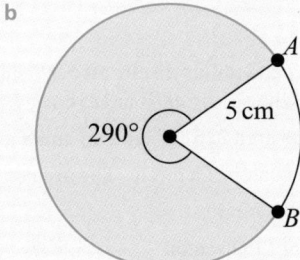

For questions 4 to 6, use the methods demonstrated in Worked Example 14.3 to find the area and the perimeter of the shaded region.

4 a 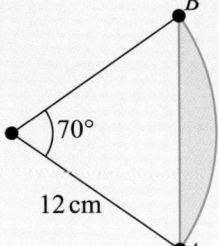   b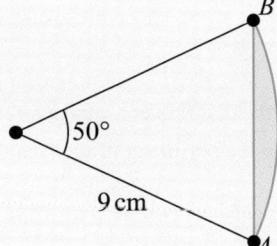

5 a 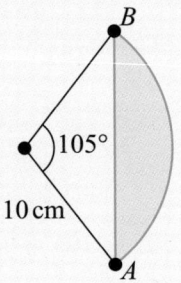   b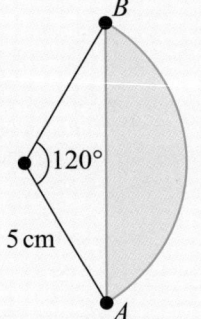

6 a 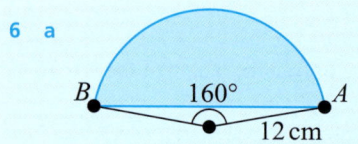   b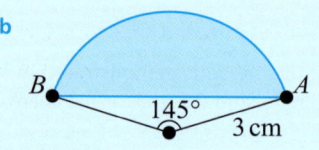

7 The diagram shows a sector *AOB* of a circle of radius 8 cm. The angle of the centre of the sector is 35°.

Find the perimeter and the area of the sector.

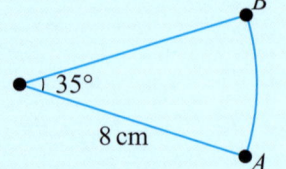

8 A circle has centre *O* and radius 6.2 cm. Point *A* and *B* lie on the circumference of the circle so that the arc *AB* subtends an angle of 140° at the centre of the circle. Find the perimeter and the area of the sector *AOB*.

9 An arc of a circle has length 12.3 cm and subtends an angle of 70° at the centre of the circle. Find the radius of the circle.

10 The diagram shows a sector of a circle of radius 5 cm. The length of the arc *AB* is 7 cm.

   a Find the value of θ.
   b Find the area of the sector.

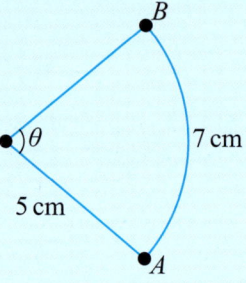

11 A circle with centre *O* has radius 23 cm. Arc *AB* subtends angle θ at the centre of the circle. Given that the area of the sector *AOB* is 185 cm², find the value of θ.

12 A sector of a circle has area 326 cm² and an angle at the centre of 155°. Find the radius of the circle.

13 The diagram shows a sector of a circle. The area of the sector is 87.3 cm².

   a Find the radius of the circle.
   b Find the perimeter of the sector.

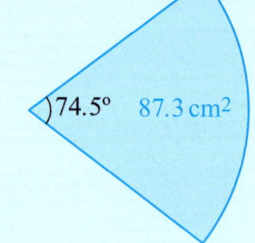

14 The figure shown in the diagram consists of a rectangle and a sector of a circle. Calculate the area and the perimeter of the figure.

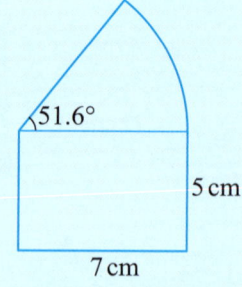

15 The diagram shows a sector of a circle. The perimeter of the sector is 26 cm. Find the radius of the circle.

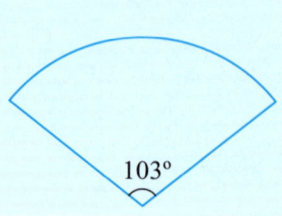

16 A sector of a circle has area 18 cm² and perimeter 30 cm. Find the possible values of the radius of the circle.

17 A circle has centre O and radius 8 cm. Chord PQ subtends angle 50° at the centre. Find the difference between the length of the arc PQ and the length of the chord PQ.

18 The arc AB of a circle of radius 12 cm subtends an angle of 40° at the centre.
Find the perimeter and the area of the shaded region.

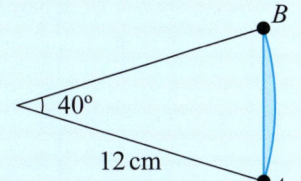

19 The diagram shows two circular sectors with angle 60° at the centre. The radius of the smaller circle is 15 cm and the radius of the larger circle is x cm larger.
Find the value of x so that the area of the shaded region is 16.5π cm².

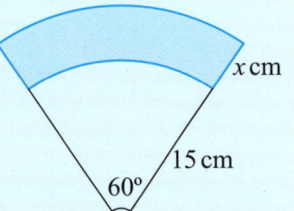

20 A piece of paper has a shape of a circular sector with radius r cm and angle θ. The paper is rolled into a cone with height 22 cm and base radius 8 cm.
Find the values of r and θ.

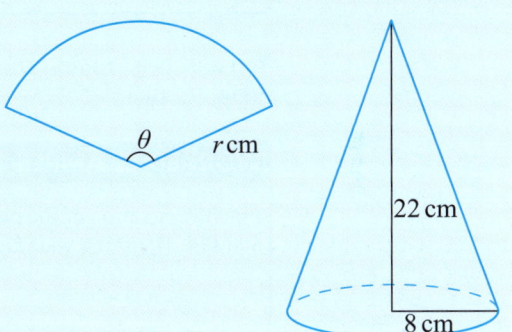

21 Two identical circles each have radius 8 cm. They overlap in such a way that the centre of each circle lies on the circumference of the other, as shown in the diagram.
Find the perimeter and the area of the shaded region.

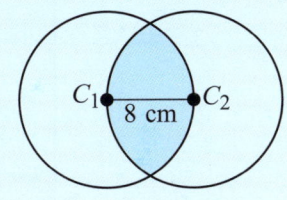

14B Voronoi diagrams

■ Perpendicular bisectors

The **perpendicular bisector** of the line segment connecting points A and B is the line which is perpendicular to AB and passes through its midpoint.

All the points on the perpendicular bisector are the same distance from A as from B.

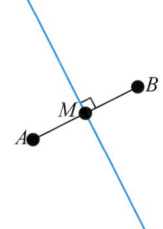

You learned about gradients of perpendicular lines in Chapter 4.

KEY POINT 14.3

For a line segment AB with gradient m and midpoint M, the perpendicular bisector that passes through M has gradient $-\dfrac{1}{m}$.

WORKED EXAMPLE 14.4

The line segment AB has equation $3x + 5y = 30$ and midpoint $(2, 7)$. Find the equation of the perpendicular bisector of AB. Give your answer in the form $px + qy = r$.

Find the gradient of AB: rearrange the equation into the form $y = mx + c$

$$3x + 5y = 30$$
$$5y = -3x + 30$$
$$y = -\frac{3}{5}x + 6$$

The gradient of AB is $-\frac{3}{5}$

The gradient of a perpendicular line is $-\frac{1}{m}$ .... The gradient of the perpendicular bisector is $\frac{5}{3}$

Use $y - y_1 = m(x - x_1)$ for the equation of a straight line, with $(x_1, y_1) = (2, 7)$ .... The equation is: $y - 7 = \frac{5}{3}(x - 2)$

Rearrange the equation into the required form; remove the fraction first

$$3y - 21 = 5x - 10$$
$$5x - 3y = -11$$

You may need to find the gradient of the line segment and the coordinates of the midpoint first.

WORKED EXAMPLE 14.5

Find the equation of the perpendicular bisector of the line segment connecting points $A(-2, 5)$ and $B(4, 1)$, giving your answer in the form $y = mx + c$.

The midpoint is given by $\left(\frac{x_1 + x_2}{2}, \frac{y_1 + y_2}{2}\right)$ .... Midpoint: $\left(\frac{-2 + 4}{2}, \frac{5 + 1}{2}\right) = (1, 3)$

The gradient of AB is given by $\frac{y_2 - y_1}{x_2 - x_1}$ .... Gradient of AB: $m = \frac{1 - 5}{4 + 2} = -\frac{2}{3}$

The perpendicular gradient is $-\frac{1}{m}$ .... Gradient of perpendicular bisector: $-\frac{1}{m} = \frac{3}{2}$

Use $y - y_1 = m(x - x_1)$ for the equation of a straight line .... The equation is: $y - 3 = \frac{3}{2}(x - 1)$

$$y - 3 = \frac{3}{2}x - \frac{3}{2}$$
$$y = \frac{3}{2}x + \frac{3}{2}$$
$$2y = 3x + 3$$

■ Introducing Voronoi diagrams

A **Voronoi diagram** is like a map with borders drawn between points on it, indicating the areas that are closest to each point. So, for example, in a town with several schools, a Voronoi diagram could be used to determine the geographical catchment area for each of the schools.

14B Voronoi diagrams

> **KEY POINT 14.4**
>
> In a Voronoi diagram:
> - the initial points are called **sites**
> - the region containing the points which are closer to a given site than any other site is called the site's **cell**
> - the boundary lines of the cells are called **edges**.

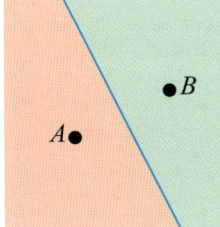

Consider a simple situation with just two sites, A and B. The Voronoi diagram will be made up of two cells: the points in the orange cell are closer to A, and the points in the green cell are closer to B.

The boundary line between the two cells consist of points which are equidistant to the two sites. This line is therefore the perpendicular bisector of AB.

If instead there are three sites, we need to draw all three perpendicular bisectors, of AB, AC and BC. They divide the plane into three cells. The points in the orange cell, for example, are closer to A than to either B or C.

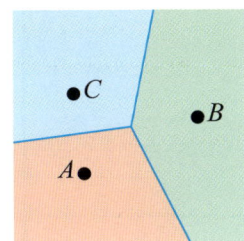

Notice that the three perpendicular bisectors meet at a single point, called a **vertex** of the Voronoi diagram.

> **KEY POINT 14.5**
>
> - The edges of a Voronoi diagram lie along the perpendicular bisectors of pairs of sites.
> - The points where the edges meet are called vertices. Each vertex has at least three edges meeting there.

With four points there are more options. The perpendicular bisectors still form the edges, but each vertex can have three or more edges meeting there.

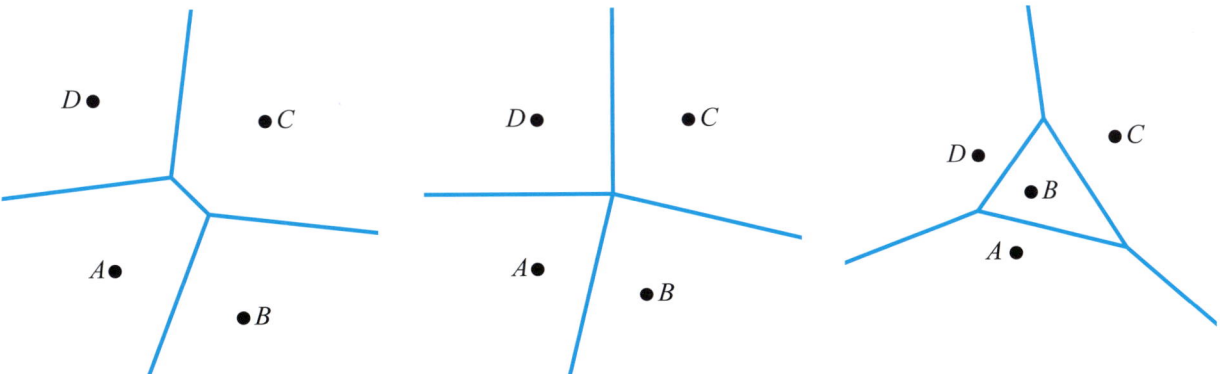

WORKED EXAMPLE 14.6

The following graph shows 5 sites:

$A(1, 1)$ $B(1, 3)$ $C(4, 5)$ $D(6, 0)$ $E(3, 2)$.

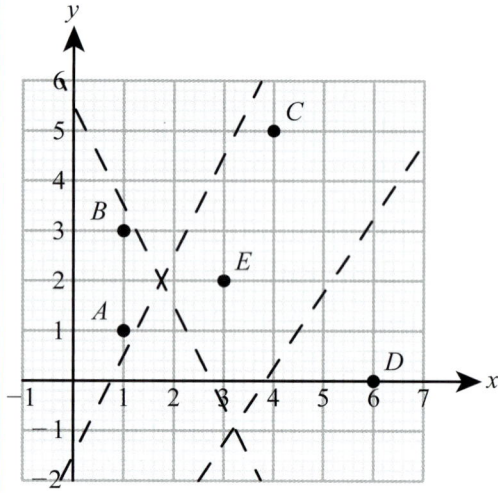

The dashed lines show three perpendicular bisectors which form edges of the cell containing site E.

a Find the equation of the line which would form the final edge of the cell containing E.

b Add this line to a sketch of the diagram, and shade the cell containing E.

c What can you deduce about the point $(3, 0)$ by the fact that it is in the cell containing E?

The three dashed lines are perpendicular bisectors of EA, EB and ED. The remaining edge is formed by the perpendicular bisector of EC

a Perpendicular bisector of EC:

midpoint: $\left(\dfrac{3+4}{2}, \dfrac{2+5}{2}\right) = (3.5, 3.5)$

gradient of $EC = \dfrac{5-2}{4-3} = 3$

perpendicular gradient $= -\dfrac{1}{3}$

Find the equation of the line through $(3.5, 3.5)$ with gradient $-\dfrac{1}{3}$

$y - 3.5 = -\dfrac{1}{3}(x - 3.5)$

$y = -\dfrac{1}{3}x + \dfrac{14}{3}$

b

Add a line through $(3.5, 3.5)$ perpendicular to EC

The cell containing E is the quadrilateral formed by the four perpendicular bisectors

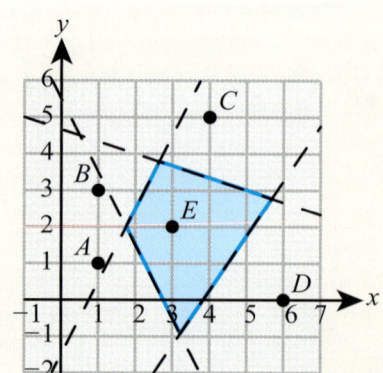

c E is the closest site to the point $(3, 0)$.

The complete Voronoi diagram for Worked Example 14.6 is given in the diagram:

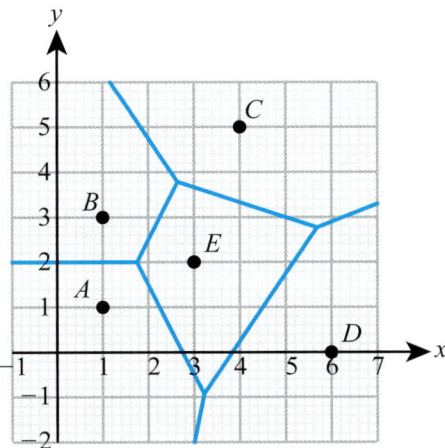

 **TOOLKIT: Problem Solving**

Investigate the position of the vertex of a three-site Voronoi diagram in these three situations.

a The three points form an acute triangle.
b The three points form a right-angled triangle.
c The three points form an obtuse triangle.

The vertex in this situation is called the orthocentre of the triangle and it has many interesting properties which you might like to research.

■ Addition of a site to an existing Voronoi diagram

Constructing a Voronoi diagram can be complicated because you need to decide which part of each perpendicular bisector forms an edge. The diagram below shows all ten perpendicular bisectors for five sites, with the actual Voronoi diagram in blue.

Tip

This diagram looks horrible. This is why you should always construct a Voronoi diagram using the incremental algorithm given below instead.

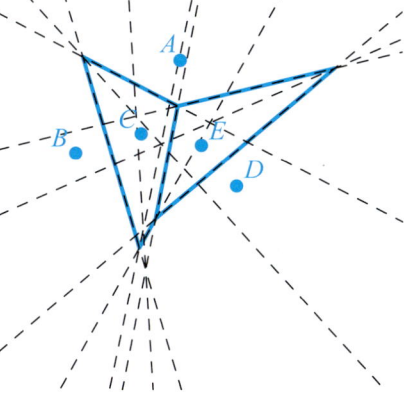

The **incremental algorithm** builds the Voronoi diagram one site at a time. When you add a new site, you draw perpendicular bisectors between it and all the existing sites. You then travel along the new perpendicular bisectors, from one existing edge to the next, until you have completed the cell containing the new site.

Tip

When generating the new cell for an added site, you might hit the boundary of the whole region. If the new cell is not complete, return to the starting point and travel in the opposite direction to complete the edges within the region.

You are the Researcher

Voronoi diagrams are related to another type of diagram called a Delaunay triangulation. Together, they form a pair called a dual in which vertices and faces are interchanged. This has some lovely mathematical properties and real-world applications that you might like to investigate.

WORKED EXAMPLE 14.7

The diagram below shows (in solid green lines) a Voronoi diagram for the sites A, B, C, D and E. An additional site F is added, and the perpendicular bisectors of line segments connecting F to other relevant sites are shown as dashed lines. On a sketch, show the new Voronoi diagram for the sites A, B, C, D, E and F.

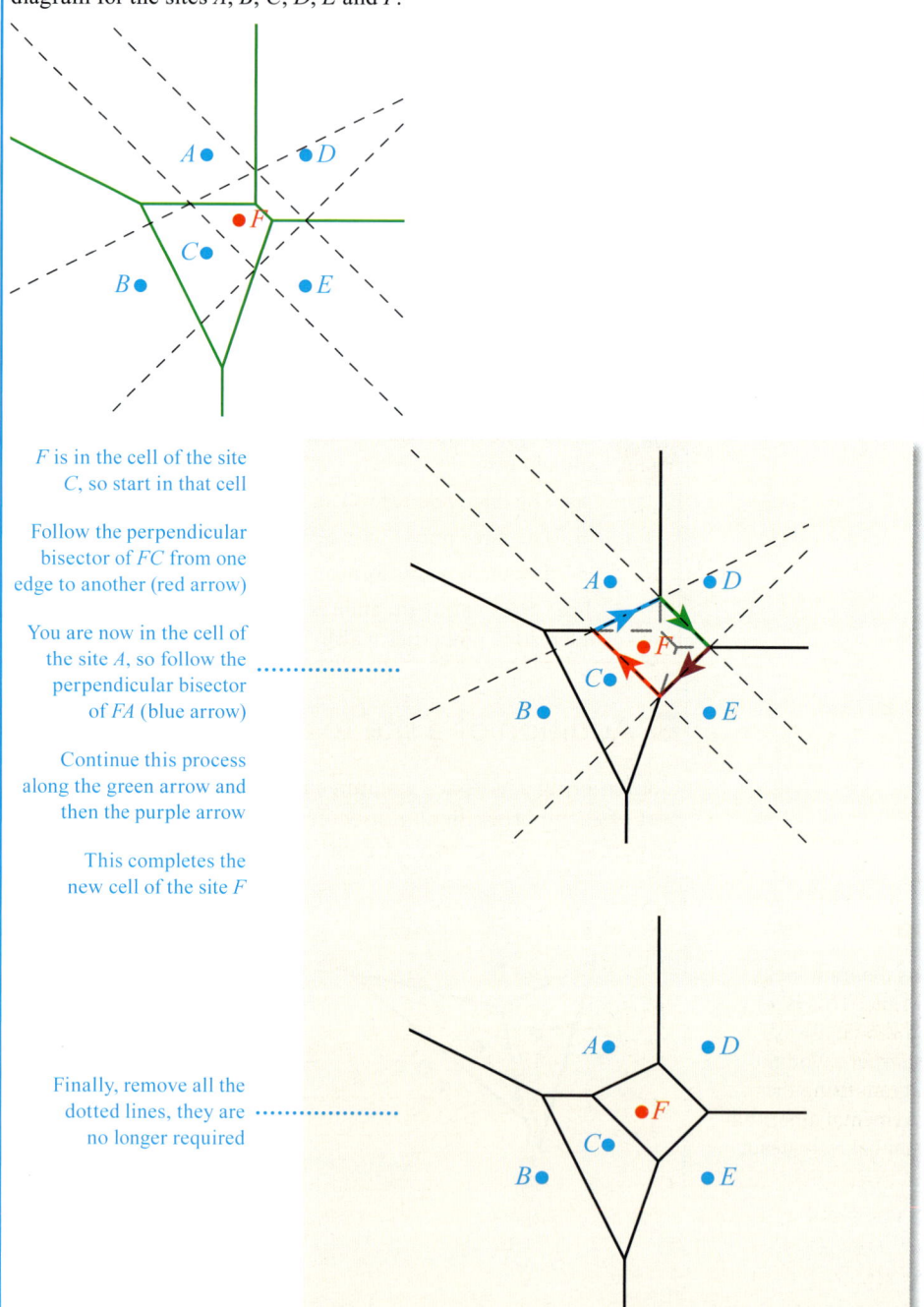

F is in the cell of the site C, so start in that cell

Follow the perpendicular bisector of FC from one edge to another (red arrow)

You are now in the cell of the site A, so follow the perpendicular bisector of FA (blue arrow)

Continue this process along the green arrow and then the purple arrow

This completes the new cell of the site F

Finally, remove all the dotted lines, they are no longer required

Nearest neighbour interpolation

Suppose f is a function which takes different values at different points in the plane, and that you know the values of f at all the sites of a Voronoi diagram. For example, the sites may represent weather stations and the function f could be the rainfall recorded on a particular day. How would you estimate the value of f at some other point in the plane?

There are many ways to answer this question. The simplest method considers which cell the point is in.

> **KEY POINT 14.6**
>
> Nearest neighbour interpolation:
>
> Given the values of a function at the sites of a Voronoi diagram, the value of the function at any other point is estimated to be the same as the value at the closest site.
>
> This means that the estimates for all the points in one cell will be equal.

WORKED EXAMPLE 14.8

The function f has known values at the following sites.

Site	f	Site	f
A	16	F	4
B	15	G	20
C	9	H	12
D	8	I	11
E	6	J	15

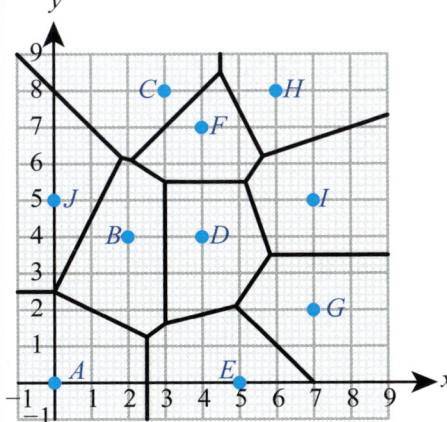

Use nearest neighbour interpolation to estimate the value of the function at the point (6, 3).

This will be the same as the value of the function at the nearest site

The point (6, 3) is in the cell containing site G, so the value of the function is 20.

You are the Researcher

Other methods of estimation that you might be interested in researching include natural neighbour interpolation.

The toxic waste dump problem

Another common application of Voronoi diagrams is to find the point which is as far as possible from any of the sites. This is the **toxic waste dump problem**, also known as 'the largest circle problem'.

This question only makes sense if the choice of points is confined to a finite region of the plane (otherwise you can keep moving away from any site indefinitely). It turns out that the required point is either on the boundary of the region, or at one of the vertices of the Voronoi diagram. In this course, the solution will always be one of the vertices.

> **KEY POINT 14.7**
>
> To solve the toxic waste dump problem:
> - For each vertex, calculate its distance from the sites in neighbouring cells.
> - Select the vertex that gives the greatest distance.

Tip

Remember that a vertex is at the same distance from all the neighbouring sites. This means that you only need to calculate one distance for each vertex.

The diagram below shows a Voronoi diagram with four sites, A, B, C and D. The circles are drawn around each vertex so that there are no sites within any of the circles. The orange circle is the largest of the three, so vertex V is further away from any of the sites than the other two vertices.

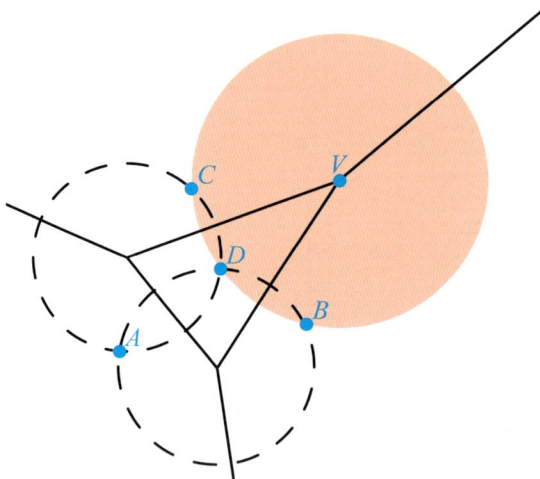

14B Voronoi diagrams

WORKED EXAMPLE 14.9

An island is modelled by the polygon $ABCD$. Towns are located at points A, B, C, D, E and F. A toxic waste dump must be located somewhere on the island, but the mayor wants it to be as far from any town as possible. Find the coordinates of the required location.

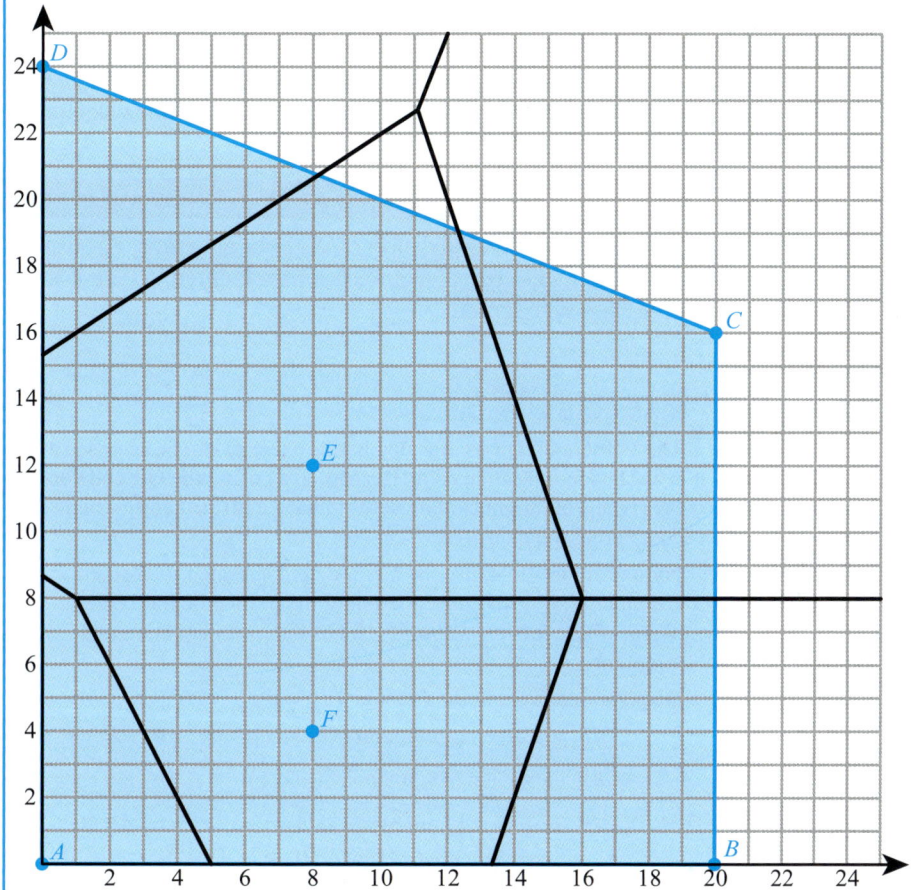

The possible points are the vertices of the Voronoi diagram which are on the island	Vertices: (1, 8) and (16, 8)
Find the distance of each vertex to one of the neighbouring sites	(1, 8) to A(0, 0): $\sqrt{1^2 + 8^2} = \sqrt{65}$ (16, 8) to B(20, 0): $\sqrt{4^2 + 8^2} = \sqrt{80}$
The required vertex is the one with the greatest distance	The required location for the toxic waste dump is at (16, 8).

CONCEPTS – SPACE

What are the limitations when you use a two-dimensional Voronoi diagram to model real **space**? What assumptions do you have to make?

Links to: Computer Science

If you are studying Computer Science, there is an efficient algorithm for generating Voronoi diagrams that you might be interested in called Fortune's algorithm.

You are the Researcher

Try to find a local application of Voronoi diagrams. For example, look at how service activities run by your school are located, to see if there are any geographical areas less well served.

Exercise 14B

The Voronoi diagrams included in this exercise can be accessed at www.hoddereducation.com/IBextras or by using the QR code on the right. You might find it useful to have a printed copy when completing the questions.

For questions 1 to 4, use the method demonstrated in Worked Example 14.4 to find the equation of the perpendicular bisector of AB, where the line segment AB has the given equation and midpoint.

Give your answers in the form $ax + by = c$.

1 a $AB: y = 2x + 11$; midpoint $(-4, 3)$ b $AB: y = -3x + 7$; midpoint $(6, -11)$

2 a $AB: y = -\frac{4}{3}x - 10$; midpoint $(-12, 6)$ b $AB: y = \frac{2}{5}x + 1$; midpoint $(10, 5)$

3 a $AB: 4x + 5y = 13$; midpoint $(2, 1)$ b $AB: 3x - 4y = 9$; midpoint $(-1, -3)$

4 a $AB: 6x - 24y = -37$; midpoint $\left(\frac{1}{2}, \frac{5}{3}\right)$ b $AB: 4x + 12y = 33$; midpoint $\left(-\frac{9}{4}, \frac{7}{2}\right)$

For questions 5 to 8, use the method demonstrated in Worked Example 14.5 to find the equation of the perpendicular bisector of the line segment connecting the given points.

Give your answers in the form $y = mx + c$.

5 a $(5, 2)$ and $(13, 6)$ b $(3, 1)$ and $(21, 7)$

6 a $(-4, 5)$ and $(5, -1)$ b $(-2, -3)$ and $(2, 7)$

7 a $\left(\frac{17}{3}, \frac{1}{4}\right)$ and $\left(\frac{7}{3}, \frac{21}{4}\right)$ b $\left(\frac{3}{2}, \frac{4}{5}\right)$ and $\left(\frac{9}{2}, \frac{34}{5}\right)$

8 a $\left(-\frac{23}{2}, -\frac{31}{6}\right)$ and $\left(-\frac{7}{2}, \frac{17}{6}\right)$ b $\left(\frac{11}{2}, -\frac{13}{3}\right)$ and $\left(-\frac{5}{2}, -\frac{1}{3}\right)$

9 For each of the following Voronoi diagrams, give all the examples of the:

   i sites   ii vertices

   iii finite edges   iv finite cells.

a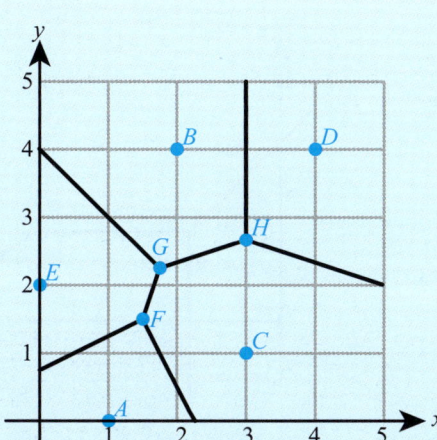

b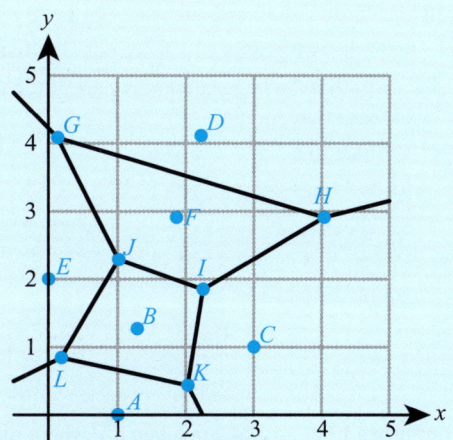

10 For each of the following Voronoi diagrams find the site closest to
  i (0, 1)   **ii** (3, 2)

a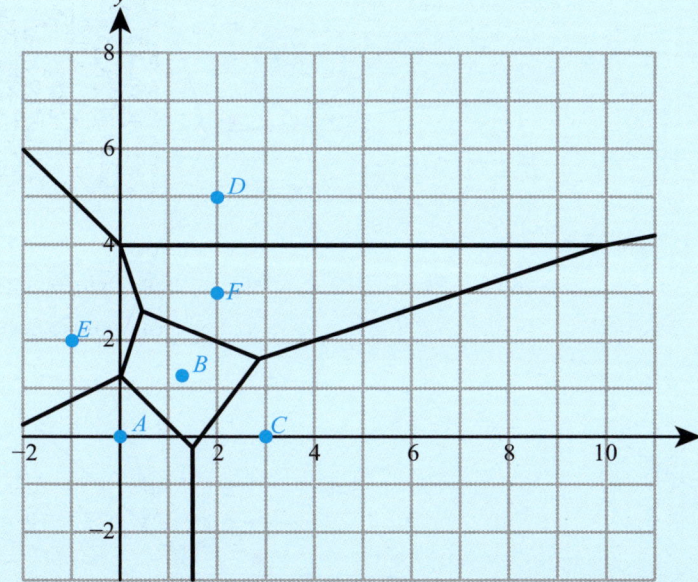

b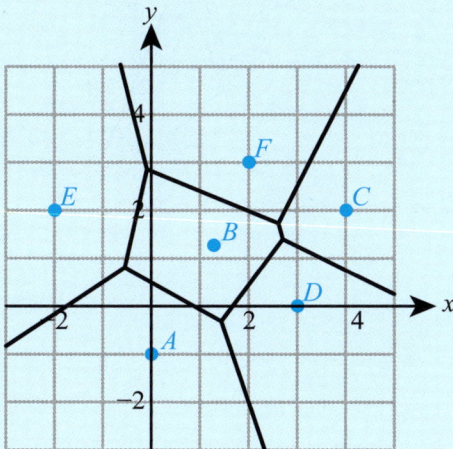

11 In the Voronoi diagram opposite, for the given line segment:
  i state the two points which are equally distant from that line
  ii verify that the line segment is part of the perpendicular bisector of these two points.
   a blue line
   b red line
   c green line

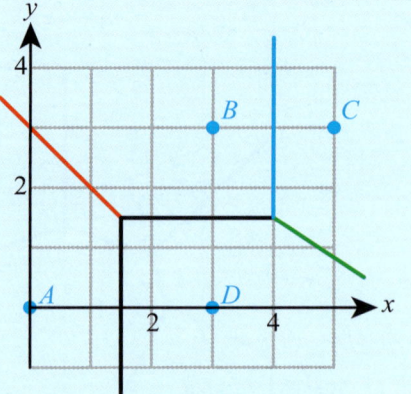

For questions 12 to 14, you are given a Voronoi diagram (black lines), a new site (red point), and the perpendicular bisectors (grey lines) between the new site and some of the existing sites. On a sketch of the diagram, use the method demonstrated in Worked Example 14.7 to add the new site to the Voronoi diagram.

12 a

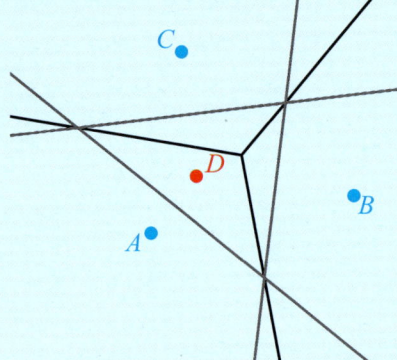

b

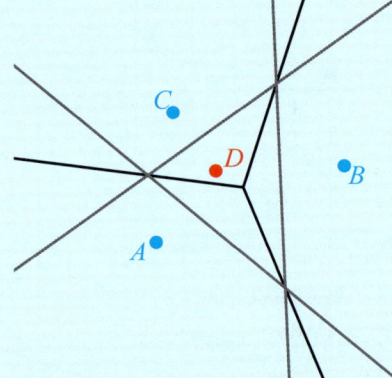

13 a

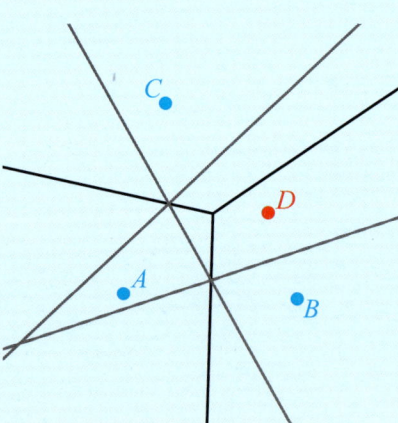

b

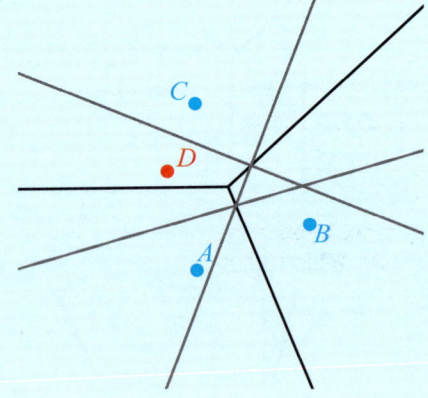

14 a 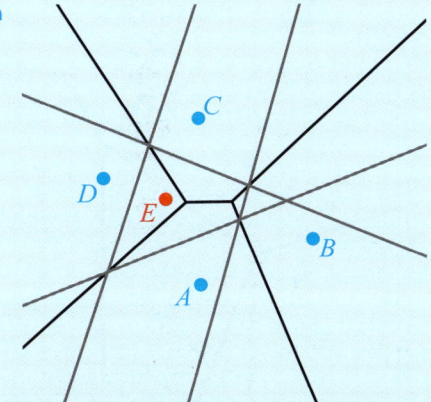 b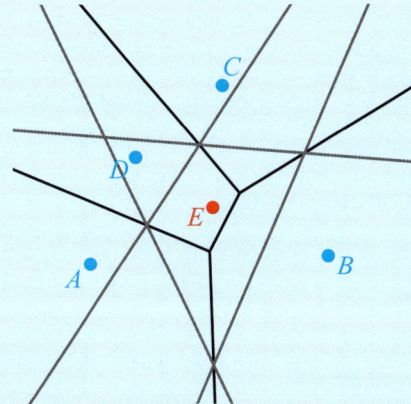

15 Two farmers want to split their land so that the boundary is an equal distance from their houses. On a map, their farmhouses are at coordinates (1, 2) and (3, 1). Find the equation of the boundary line.

16 Seven weather stations are located on an island. The diagram shows the Voronoi diagram with vertices at the weather stations.

  a One of the edges in the diagram is missing. Write down its equation.

The following measurements are taken at the weather stations:

Station	A	B	C	D	E	F	G
Temperature (°C)	18	22	20	18	19	23	20
Air pressure (mbar)	1025	1020	1015	1020	1020	1025	1025
Daily rainfall (mm)	38	42	27	35	31	29	41

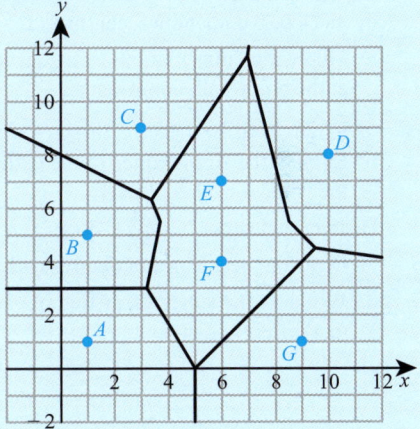

Use the data in the table to estimate:

  b the temperature at the location with coordinates (3, 5)
  c the air pressure at the location with coordinates (7, 0)
  d the rainfall at the location with coordinates (6, 9).

17 The sites of the Voronoi diagram are the locations of nine schools in a city. A child can choose a school if their house is located in that school's cell, including the edges.

  a Which schools can a child attend if their house is located at the point:
    i M
    ii P?
  b There is one location in the city which has the largest choice of schools. Which schools are they?

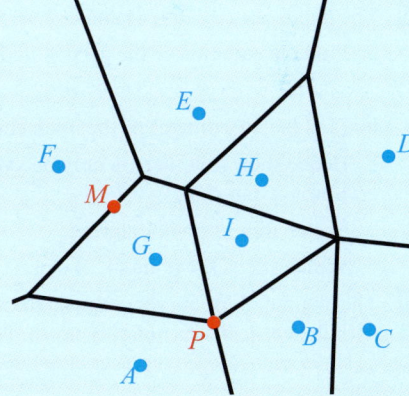

18 A botanist suspects that contaminated water is causing a certain disease in plants in a park. The park has four small ponds and she thinks that one of them is the source of the disease.
The ponds are shown as the sites of the Voronoi diagram opposite.

The botanist selects a sample of locations around the park and counts the number of affected plants at each location. The results are summarized in the table.

Coordinates	(1, 2)	(3, 5)	(6, 4)	(0, 9)	(11, 9)	(13, 2)	(9, 9)	(1, 8)
Number of plants	3	8	3	11	6	12	2	0

Which of the four ponds is most likely to be contaminated? Justify your answer.

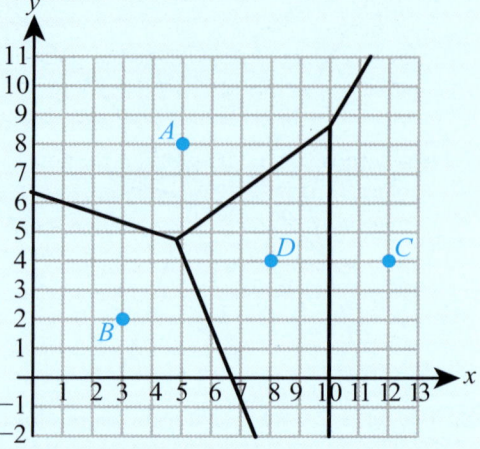

19 A national supermarket chain has five distribution centres, shown as the sites of the Voronoi diagram. Each store is served by its closest distribution centre.

  a Find which distribution centre serves the store located at the point with coordinates:
   i (4, 5)   **ii** (14, 1)   **iii** (9, 8)
  b Explain the significance of the cell containing site A in the context of this problem.

Distribution centre C closes down.

  c Draw the new Voronoi diagram with sites A, B, D and E.
  d Which of the stores identified in part **a** will need to switch to a different distribution centre? State the new distribution centre.

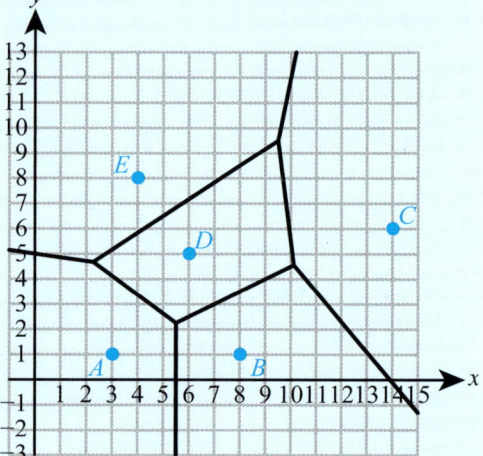

20 The points $A(0, 0)$, $B(4, 4)$ and $C(2, 4)$ are sites on a Voronoi diagram.
  a Find the equation of the perpendicular bisector of
   i AC **ii** BC **iii** AB.
  b Find the coordinates of the intersection of the perpendicular bisectors of AB and AC.
  c Hence show that there is only one vertex on the Voronoi diagram.
  d Sketch this Voronoi diagram.
  e Which site is the point (4, 2) closest to?

21 The Voronoi diagram represents a map of a woodland area, with the vertical axis pointing north. Points $A(3, 2)$, $B(7, 9)$, $C(11, 7)$ and $D(10, 3)$ represent the locations of four existing holiday cottages.

The equation of the perpendicular bisector of CD is $y = -\frac{1}{4}x + \frac{61}{8}$.

  a Show that the equation of the perpendicular bisector of BC is $y = 2x - 10$.
  b Obi wants to build a new holiday cottage which is an equal distance from B, C and D. Find the coordinates of the location of his cottage.

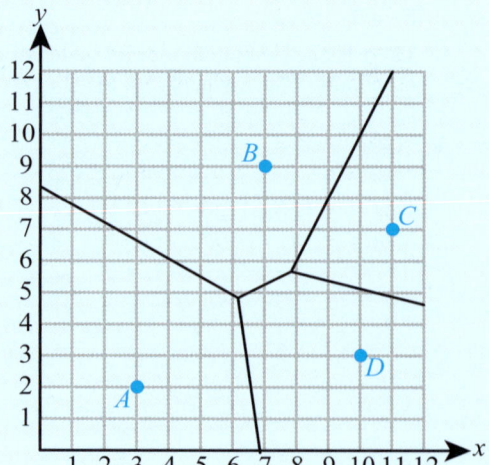

c Roshni wants her holiday cottage on a point with integer coordinates, closer to D than any other existing cottage (excluding Obi's new cottage), and as far north as possible. Which of the following points could Roshni choose:

   i (8, 8)   **ii** (7, 5)   **iii** (10, 3)?

A new cottage is actually built at the point $E(3, 6)$. The dotted lines are perpendicular bisectors between E and the other four sites.

d Copy and complete the new Voronoi diagram to include site E.

e The woodland is located on some hills. The altitude above sea level is known for the five cottages:

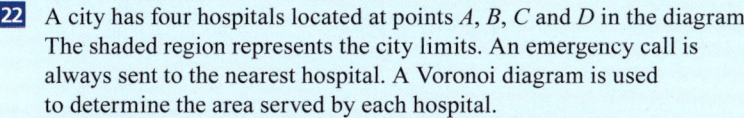

Cottage	A	B	C	D	E
Altitude (m)	348	412	265	317	382

Use nearest neighbour interpolation to estimate the altitude of the point (7, 7).

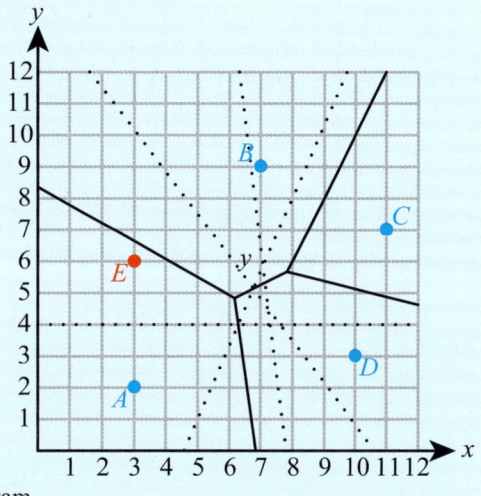

22 A city has four hospitals located at points A, B, C and D in the diagram. The shaded region represents the city limits. An emergency call is always sent to the nearest hospital. A Voronoi diagram is used to determine the area served by each hospital.

  a Find the equation of the line highlighted in red.

  b An accident takes place at the point with coordinates (8, 13). Which hospital should be called?

  c A new hospital is to be built, within the city, so that it is as far away as possible from any existing hospital. Find the coordinates of the location of the new hospital, showing your working clearly.

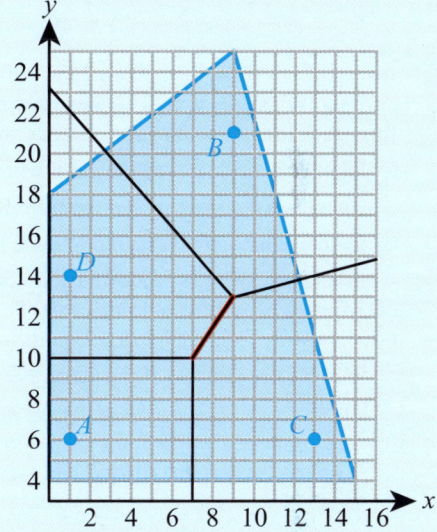

23 The Voronoi diagram shows locations of four restaurants within a shopping centre, which is represented by the shaded rectangle.

  a Customers tend to go the restaurant closest to them. A customer is located at the point with coordinates (7, 4). Which restaurant will they go to?

  b Write down the coordinates of the location which is equidistant from restaurants A, C and D.

  c One unit on the diagram represents 10 m. Find the area served by restaurant A.

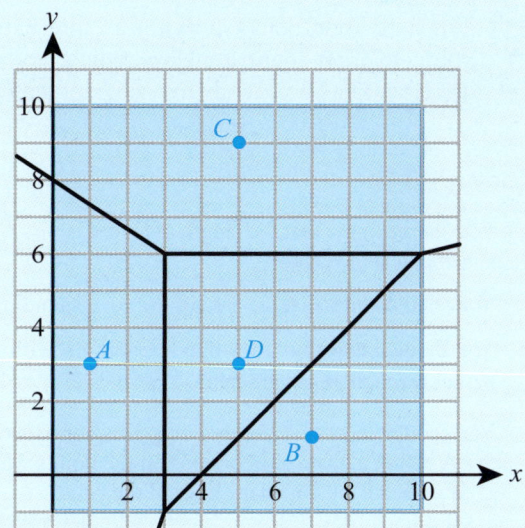

A new restaurant opens at the location $M(6,8)$. The dotted lines are the perpendicular bisectors of MB and MC.

d Find the equation of the perpendicular bisector of MD and add it to a copy of the diagram.

e Complete the new Voronoi diagram on all five vertices.

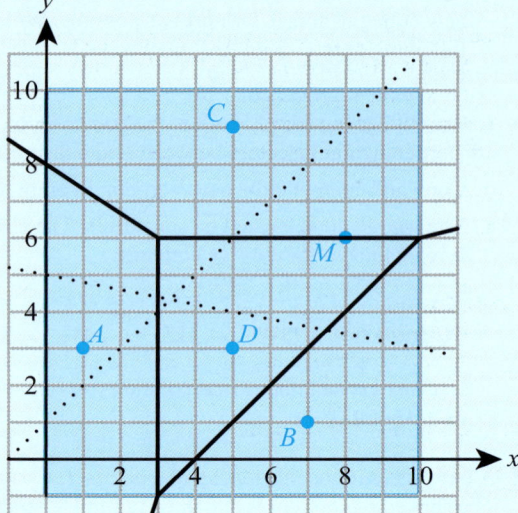

24 An island has three towns located at points with coordinates $(0, 0)$, $(30, 30)$ and $(50, 10)$, where the coordinates have units of kilometres. The government wants to locate a toxic waste dump at a position as far as possible away from each town. You may assume that this location is not on the edge of the island.

a Find the coordinates of the point where the toxic waste dump should be located.

b How far is the toxic waste dump from each city?

c State one modelling assumption you have made in your calculation.

25 An island has four towns located at points with coordinates $A(0, 0)$, $B(20, 0)$, $C(0, 30)$ and $D(60, 40)$ where the coordinates have units of kilometres. The government wants to locate a toxic waste dump at a position as far as possible away from each town. You may assume that this location is not on the edge of the island.

a Find the coordinates of the point where the toxic waste dump should be located, justifying your answer carefully.

b As a result of a petition, the government decides to take into account the relative size of the different towns. A census reveals that town B has roughly twice the population of each of the other towns. Without further calculation, suggest how this changes the position of the toxic waste dump relative to your answer above.

 The Eurovision Song Contest is an international musical competition in which each country awards points to other countries. How could a Voronoi diagram help to explain some of the scoring? Are there any situations in which being geographically close differs from being culturally close?

Checklist

- You should know that the length of an arc is $s = \frac{\theta}{360} \times 2\pi r$, where r is the radius of the circle and θ is the angle subtended at the centre.

- You should know that the area of the corresponding sector is $A = \frac{\theta}{360} \times \pi r^2$.

- You should know that, for a line segment AB with gradient m and midpoint M, the perpendicular bisector that passes through M has gradient $-\frac{1}{m}$.

- You should know that Voronoi diagrams consists of sites, cells, edges and vertices.
    - A cell or a given site contains all the points which are closer to that site than to any other.
    - Edges are the boundary lines of the cells.
        — Each edge lies along a perpendicular bisector of two of the sites.

- Vertices are the intersections of edges.
  — Each vertex has three or more edges meeting there.

■ You should be able to use the incremental algorithm to add an additional site (S) to an existing Voronoi diagram.
  □ Start with the site in the cell where S lies.
  □ Follow the perpendicular bisector between that site and S until you get to an edge.
  □ Continue along the perpendicular bisector between the next site and S, and so on.
  □ Stop when you either return to the starting point or hit the boundary.

■ You should know that in nearest neighbour interpolation, all points in a cell are allocated the same value (of some function) as the cell's site.

■ You should be able to solve the toxic waste dump (largest circle) problem, to find the point which is as far as possible from any of the sites. To solve the toxic waste dump problem:
  □ for each vertex, calculate its distance from the sites in neighbouring cells
  □ select the vertex that gives the greatest distance.

■ Mixed Practice

The Voronoi diagrams included in this exercise can be accessed at www.hoddereducation.com/IBextras or by using the QR code on the right. You might find it useful to have a printed copy when completing the questions.

1 A sector is formed from a circle of radius 10 cm. The sector subtends an angle of 32° at the centre of the circle.
  a Find the area of the sector.
  b Find the perimeter of the sector.

2 A pizza slice is a sector of radius 15 cm. It has an area of approximately 88 cm². How many equal slices was the full circular pizza cut into?

3 A 2 m long trough is modelled as a prism with the cross section opposite, formed from a circle of radius 0.5 m with a segment removed.

Find the volume of the trough.

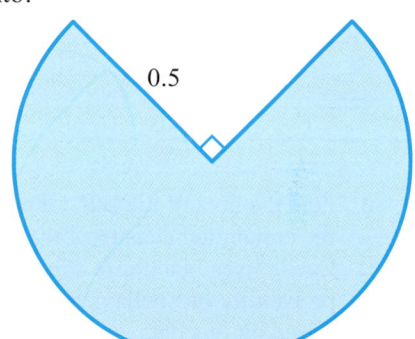

4 An international treaty splits a country into two. The border is determined to be a straight line at an equal distance from the two new capitals. The capitals have positions (330, 53) and (48, 15) on a map where the units are in kilometres.
  a Find the equation of the straight line along which the border will be drawn.
  b Find the closest distance from the capitals to the border line.

5 Park Rangers have installed feeding stations for animals at point A, B, C, D and E. On the map opposite, 1 cm represents 1 km.
  a Find, to the nearest 10 metres, the distance between the stations A and B.
  b Find the gradient of the line segment AB.
  c The dotted lines are a part of an incomplete Voronoi diagram. Find the equation of the line which needs to be added to complete the cell containing site A. Give your answer in the forma $ax + by = c$, where a, b, c are integers.
  d In the context of this question, explain the significance of the vertex marked P.

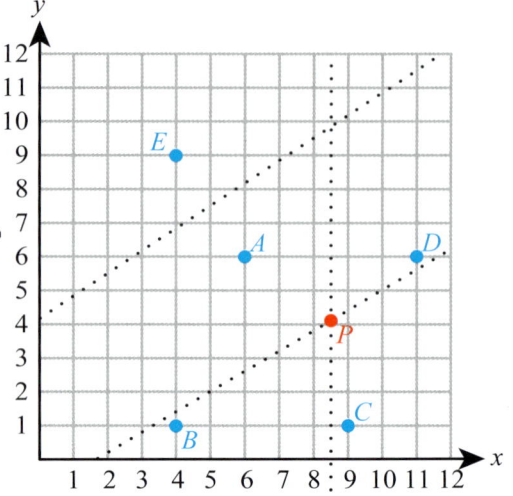

6 The diagram on the right shows the location of 7 water pumps on an island. These locations are used to form a Voronoi diagram. The units on the graph are km.

An epidemiologist uses the Voronoi diagram opposite to investigate the cause of a cholera outbreak on the island. He collects data for each cell:

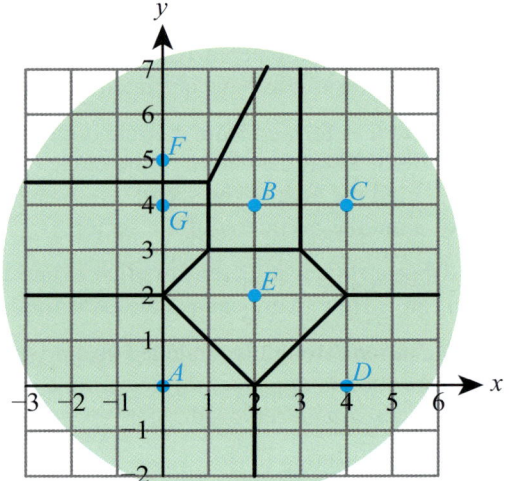

Pump Label	Population	Number of cholera cases
A	4807	318
B	1844	624
C	604	82
D	1253	125
E	2215	322
F	1803	421
G	3689	715

  a Find the area of the cell containing pump E. Hence find the population density of the cell containing E.
  b Loella lives in a house at coordinates (0, 3). Use nearest neighbour interpolation to estimate the probability of Loella getting cholera.
  c Based on this data, suggest which pump is the source of the cholera. Justify your answer.

🌐 The 1854 Broad Street cholera outbreak was studied by the physician John Snow. He used a method very similar to that in this question to split London into different cells based on proximity to different water pumps. He showed that there were distinctly different cholera rates in the different cells, and this provided the initial idea that cholera might be a water-borne disease.

7 An island has three towns located at points with coordinates (10, 10), (0, 30) and (20, 40) where the coordinates have units of kilometres. The government wants to locate a toxic waste dump at a position as far as possible away from each town. You may assume that this location is not on the edge of the island.
  a Find the coordinates of the point where the toxic waste dump should be located.
  b How far is the toxic waste dump from each city?

Mixed Practice

8 The Voronoi diagram with sites A, B, C and D is shown opposite. A new site, E, is to be added to the diagram. The perpendicular bisectors of EA, EB, EC and ED are shown as dotted lines.
  a Site A has coordinates $(2, 1)$ and site E has coordinates $(5, 3)$. Find the equation of the perpendicular bisector of EA, giving your answer in the form $ax + by = c$, where a, b and c are integers.
  b On a copy of the diagram, complete the new Voronoi diagram with all five sites.

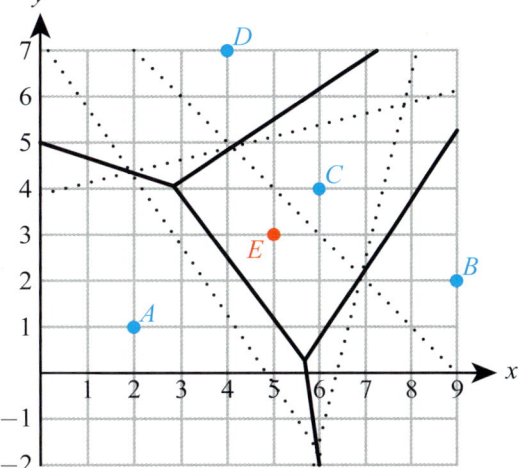

9 A town has five mobile phone masts located at $A(0, 0)$, $B(2, 2)$, $C(-2, 2)$, $D(-2, -2)$ and $E(2, -2)$, where all units are in kilometres. Mobile phones connect to the nearest mobile phone mast. The population density of the town is 3200 people per square kilometre. Estimate the number of mobile phones mast A must be able to connect with.

10 The points $A(0, 0)$, $B(6, 2)$ and $C(0, 2)$ represent the positions of supermarkets in a town with coordinates in kilometres.
  a Find the equation of the perpendicular bisector of
  i AC **ii** CB **iii** AB
  b Find the coordinates of the vertex on the Voronoi diagram with sites A, B and C.
  c Sketch the Voronoi diagram with sites A, B and C.
A company wants to open a new supermarket in the town as far as possible from any other supermarket.
  d Given that this location is not on the edge of the town, find the position where the new supermarket should open.
  e Sketch a new Voronoi diagram including this new point.
  f Supermarket C previously covered an area of $25\,\text{km}^2$. What percentage of this area has been taken over by the new supermarket?

11 An island has six weather stations, located at the sites A, B, C, D, E, F of the Voronoi diagram opposite. The edges of the diagram are shown as solid lines. The temperature recorded by the six stations on a particular morning is given in the table.

Station	A	B	C	D	E	F
Temperature (°C)	23	27	28	26	25	24

  a Estimate the temperature at the locations with the following coordinates:
  i $(2, 6)$ **ii** $(8, 3)$ **iii** $(6, 3)$

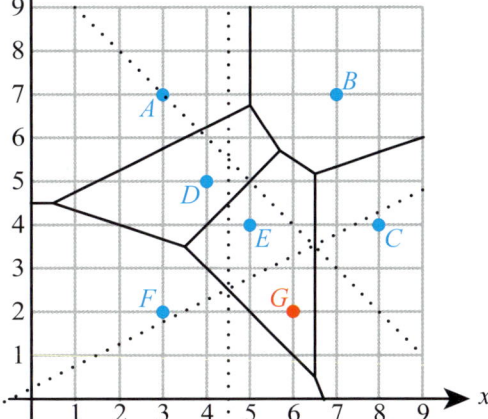

An amateur meteorologist is located at the point G (marked in red in the diagram opposite). He measures the temperature at G to be $24\,°C$.
  b A new weather station at site G is added as shown in the diagram. The relevant perpendicular bisectors are shown as dotted lines on the diagram. Draw the new Voronoi diagram representing all seven weather stations.
  c Which point from part **a** should have the temperature estimate adjusted? What is the new estimate?

12 A sector has area $3\pi\,\text{cm}^2$ and arc length $\pi\,\text{cm}$.
Find the radius of the sector.

13 A rectangle is drawn around a sector of a circle as shown.
If the angle of the sector is $57.3°$ and the area of the
sector is $7\,\text{cm}^2$, find the dimensions of the rectangle,
giving your answers to the nearest millimetre.

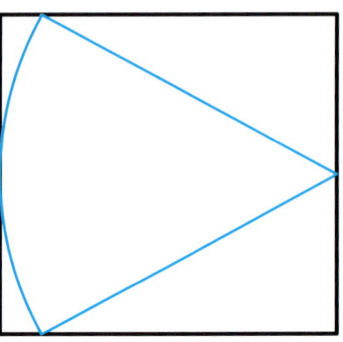

Mathematics HL May 2013 TZ1 Paper 2 Q5 Adapted

14 The diagram opposite shows two intersecting
circles of radii $4\,\text{cm}$ and $3\,\text{cm}$. The centre C of
the smaller circle lies on the circumference of the
bigger circle. O is the centre of the bigger circle
and the two circles intersect at points A and B.

Find:
a $B\hat{O}C$
b the area of the shaded region.

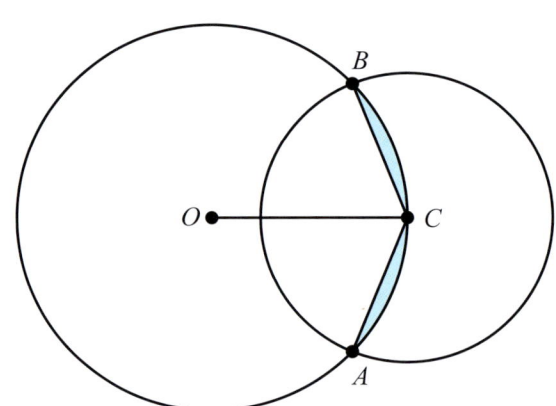

15 A circle has centre O and radius $12\,\text{cm}$. Chord AB
subtends angle θ at the centre. The area of the
shaded region is $12\,\text{cm}^2$.
  a Show that $0.4\pi\theta - 72\sin\theta = 12$.
  b Find the value of θ.
  c Find the perimeter of the shaded region.

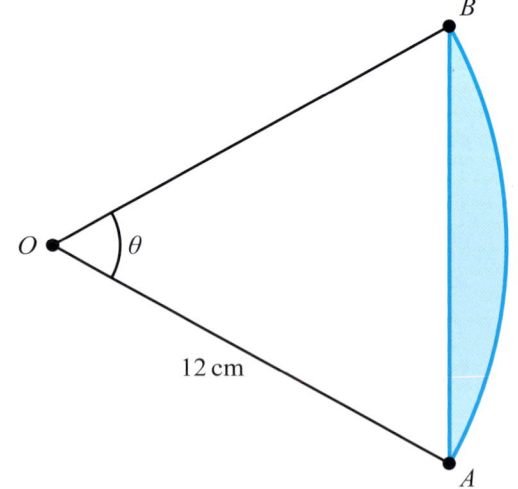

Mixed Practice

16 A bicycle chain is modelled by the arcs of 2 circles connected by 2 straight lines which are tangents to both circles.

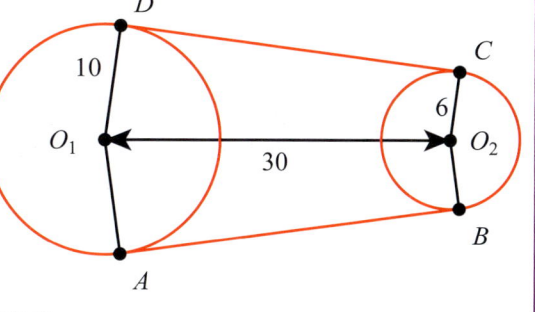

The radius of the larger circle is 10 cm and the radius of the smaller circle is 6 cm. The distance between the centre of the circles is 30 cm.
  a Find angle AO_1O_2.
  b Hence find the length of the bicycle chain, giving your answer to the nearest cm.

17 A lake is modelled as a circle, shown in blue in the diagram. The centre of the circle is the point M. Three cabins, A, B and C, are located on the circumference of the lake. The solid lines in the diagram are the edges of the Voronoi diagram with sites A, B and C.

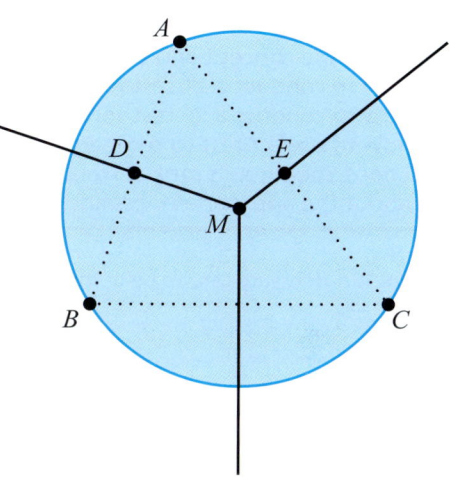

Each cabin has fishing rights for the region of the lake which is closer to it than to any other cabin.

Angle $B\hat{A}C$ is 50°.
  a Use the quadrilateral $ADME$ to find the size of the angle $D\hat{M}E$.
  b Hence determine the percentage of the lake where the fishing rights belong to cabin A.

The owners measured the temperature of the water outside their cabins on a particular morning:

Cabin	A	B	C
Temperature (°C)	17.3	18.2	16.8

  c A fishing boat is located halfway between B and M. Use nearest neighbour interpolation to estimate the temperature of the water near the fishing boat.

18 Four magnets are fixed in the base of a box such that their north poles are exposed to the interior. A magnet is allowed to dangle on the end of string into the box so that its north pole is in contact with the base of the box. It moves in such a way that it comes to rest as far as possible from the four fixed magnets. It is noted that this location is not on the edge of the box.

The coordinates of the fixed magnets are $A(1, 2)$, $B(0, 5)$, $C(3, 4)$, $D(4, 0)$. Find the final resting place of the free magnet. Justify your answer fully.

19 A park has four mobile phone masts located at $A(0, 1)$, $B(2, 3)$, $C(-2, 3)$ and $D(4, 1)$ where all units are in kilometres and the positive x-axis points east and the positive y-axis points north. Mobile phones connect to the nearest mobile phone mast.
  a Sketch the Voronoi diagram showing the zones covered by each mast.
  b Alex runs due east at constant speed. He notices that during his run, his mobile phone connects to each of the masts for the same time. How far does he run?

20 The rhombus $OPQR$ has vertex O at the origin and P at $(1, 0)$. Point R has coordinates (a, b) with $b \neq 0$.
  a Given that side OR has the same length as PQ find a condition on a and b.
  b Find the coordinates of point Q.
  c Find the perpendicular bisector of OQ and show that it passes through P and R.

15 Applications and interpretation: Hypothesis testing

ESSENTIAL UNDERSTANDINGS

- Statistics uses the theory of probability to estimate parameters, discover empirical laws, test hypotheses and predict the occurrence of events.
- Both statistics and probability provide important representations which enable us to make predictions, valid comparisons and informed decisions.
- The fields of statistics and probability have power and limitations and should be applied with care and critically questioned, in detail, to differentiate between the theoretical and the empirical/observed.

In this chapter you will learn…

- about the concept and underlying principles of a hypothesis test
- how to conduct a χ^2 test for goodness of fit
- how to conduct a χ^2 test for independence using contingency tables
- how to use a *t*-test to compare population means
- about Spearman's rank correlation coefficient for non-linear correlation
- about the appropriateness and limitations of Pearson's and Spearman's rank correlation coefficients.

■ **Figure 15.1** How sure can we be of the assumptions we make about populations based on data samples?

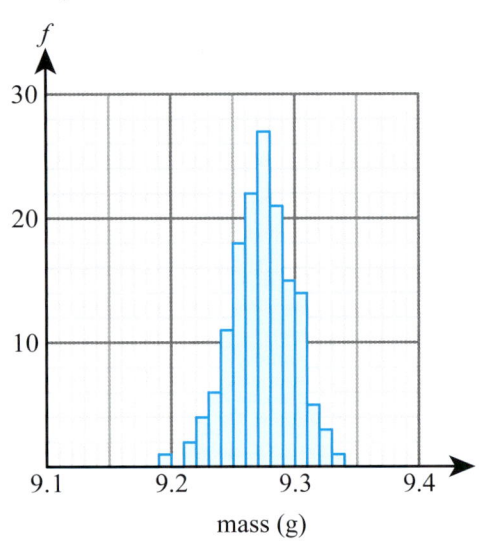
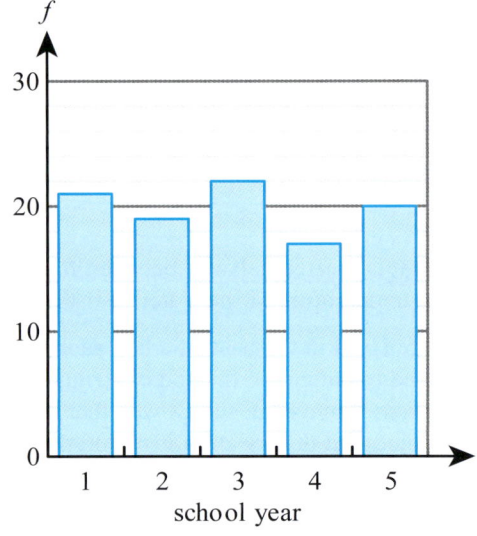
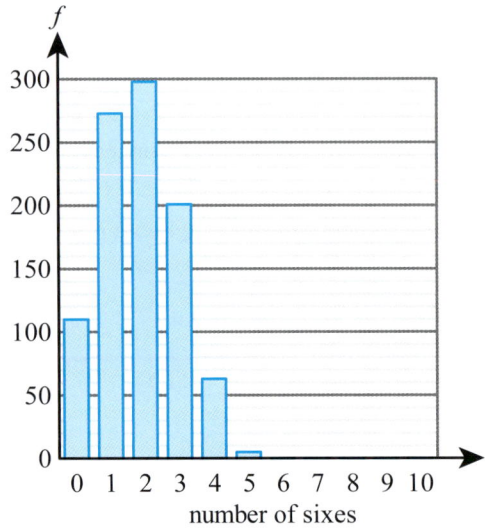

Applications and interpretation: Hypothesis testing

> **CONCEPTS**
>
> The following key concepts will be addressed in this chapter:
> - Organizing, **representing**, analysing and interpreting data, and utilizing different statistical tools facilitates prediction and drawing of conclusions.
> - Different statistical techniques require justification and the identification of their limitations and **validity**.
> - **Approximation** in data can approach the truth but may not always achieve it.
> - Correlation and regression are powerful tools for identifying **patterns**.

> **PRIOR KNOWLEDGE**
>
> Before starting this chapter, you should already be able to complete the following:
>
> **1** There are 263 students in a year group. The probability of a student being absent is 0.07. Find the expected number of absent students.
>
> **2** Given that $X \sim B(30, 0.23)$ find $P(12 \leq X \leq 18)$.
>
> **3** Given that $X \sim N(25, 4.8^2)$ find $P(X > 27)$.
>
> **4 a** Find the Pearson's product-moment correlation coefficient for this data:
>
> | 32 | 51 | 87 | 22 | 45 | 57 | 45 | 73 |
> |----|----|----|----|----|----|----|----|
> | 12 | 18 | 15 | 11 | 15 | 17 | 12 | 19 |
>
> **b** Interpret the value found in part **a**, given that the critical value of the correlation coefficient is 0.549.

In statistics, we use sample data to make inferences about a population. This always involves some degree of uncertainty, for example, the mean of a sample will not be, in general, exactly the same as the mean of the population. We can use probability to decide whether a value obtained from a sample is significantly different from what we expect.

Hypothesis testing is used to determine whether something about a population has changed, or whether two populations have significantly different characteristics. Different types of test are appropriate in different situations and it is important to understand the advantages and limitations of each test. Hypothesis tests are commonly used in a range of subjects, including Biology, Geography, Psychology and Exercise and Health Science.

> ## Starter Activity
>
> Each of the histograms in Figure 15.1 represent a sample from a different population. What can you suggest about the distribution of each population? How confident are you in your statement?
>
> **Now look at this problem:**
>
> The lengths of leaves of a certain plant are distributed normally with mean 18 cm and standard deviation 3 cm. Mo has measured the lengths of a sample of leaves. For each of these measurements decide whether it is possible, unusual or probably a measurement error.
>
> **a** 23 cm
> **b** 16 cm
> **c** 30 cm
>
> What if, instead, Mo's measurements are for a plant that has leaf lengths that are distributed normally with mean 18 cm and standard deviation 4 cm?

15A Chi-squared tests

■ Introduction to hypothesis testing

Consider the following questions:
- Has the mean temperature increased in the past fifty years?
- Does $r = 0.631$ suggest significant correlation?
- Is a normal distribution a good model for the data?

One common statistical technique for answering such questions is **hypothesis testing**. This procedure starts with a default position, called the **null hypothesis**, and determines whether sample data provides sufficient evidence against it, and in favour of an **alternative hypothesis**.

> **KEY POINT 15.1**
>
> The null hypothesis, denoted by H_0, is the default position, assumed to be true unless there is significant evidence against it.
>
> The alternative hypothesis, denoted by H_1, specifies how you think the position may have changed.

WORKED EXAMPLE 15.1

For each situation, write down the null and the alternative hypotheses for a hypothesis test.

a The mean January temperature in Dubai from 1968 to 1998 was 29 °C. In January 2018 it was 31 °C. Is there significant evidence that the mean temperature has increased?

b Hans collects data on the average house price and the distance from the nearest train station for eight villages. He calculates the value of the correlation coefficient to be −0.631. Does this provide significant evidence of negative correlation between the average house price and the distance from the nearest train station?

c Aranyaa knows that heights of trees in her local park follow a normal distribution with mean 12 m and standard deviation 2.5 m. She wants to test whether the heights of trees in a nearby forest follow the same distribution. She measures the heights of 50 randomly selected trees from the forest and obtains the following results:

Height (h m)	$h \leq 7$	$7 < h \leq 9.5$	$9.5 < h \leq 12$	$12 < h \leq 14.5$	$14.5 < h \leq 17$	$h > 17$
Frequency	3	7	20	15	4	1

Does this provide significant evidence that the heights of the forest trees do not follow the same normal distribution?

The default position is that the temperature has not increased
In this case it is easiest to write the hypotheses using equations and inequalities, defining the meaning of any letters you use

a $H_0: \mu = 29$, $H_1: \mu > 29$
where μ is the mean January temperature in Dubai, in °C

You can also write the hypotheses in words. In this case, the default position is that there is no correlation

b H_0: There is no correlation between the average house price and the distance from the train station.
H_1: There is a negative correlation between the average house price and the distance from the train station.

The default position is that the trees in the forest follow the same distribution as the trees in the park

Remember that $N(\mu, \sigma^2)$ stands for a normal distribution with mean μ and standard deviation σ

c H_0: The heights of the trees in the forest come from the distribution $N(12, 2.5^2)$
H_1: The heights of the trees in the forest do not come from the distribution $N(12, 2.5^2)$

In examples **a** and **b** on the previous page, the alternative hypotheses specified the anticipated direction of change: whether the temperature has *increased*, or whether the correlation is *negative*. In some situations, you may have a reason to believe that the mean value, for example, has changed, but you might not have any indication in which direction it has changed. In this case you need to use $\neq$, instead of $<$ or $>$, in the alternative hypothesis.

> **KEY POINT 15.2**
>
> In a **one-tailed test**, the alternative hypothesis specifies whether a population parameter has increased or decreased.
>
> In a **two-tailed test**, the alternative hypothesis just states that the parameter has changed.

> **WORKED EXAMPLE 15.2**
>
> For each situation, write down the null and alternative hypotheses, and state whether the test is one-tailed or two-tailed.
>
> a The proportion of students at the college taking Maths SL was 72%. Elsa wants to test whether this proportion has changed.
>
> b Asher has collected some data on the heights of trees and the lengths of their leaves and wants to test whether there is significant positive correlation.
>
> c Theo wants to find out whether there is any correlation between the number of hours of homework done by a student and their final examination grade.
>
> *The direction of change is not specified*
>
> **a** two-tailed test
> $H_0: p = 72$, $H_1: p \neq 72$
> where p is the percentage of students taking Maths SL
>
> **b** one-tailed test
>
> *Asher is only testing for positive correlation*
>
> H_0: There is no correlation between the height of a tree and the length of its leaves
>
> H_1: There is a positive correlation between the height of a tree and the length of its leaves
>
> **c** two-tailed test
>
> *Theo has not specified the type of correlation*
>
> H_0: There is no correlation between amount of homework and grade
>
> H_1: There is a correlation between amount of homework and grade

In a hypothesis test, you start with a default position (the null hypothesis) and ask whether there is significant evidence against it. Such evidence is found if, assuming the null hypothesis is true, it would be very unlikely to observe the sample data you collected. What 'very unlikely' means is a matter of judgement. It may be different in different contexts and is given by the **significance level** of the test.

When conducting a hypothesis test, you need to calculate the probability of obtaining the observed sample value, or a more extreme value, assuming the null hypothesis is true.

> **KEY POINT 15.3**
>
> The significance level of a hypothesis test specifies the probability that is sufficiently small to provide significant evidence against H_0.
>
> Assuming H_0 is correct, the probability of the observed sample value, or more extreme, is called the **p-value**.
>
> If the p-value is smaller than the significance level, there is sufficient evidence to reject H_0 in favour of H_1. Otherwise, there is insufficient evidence to reject H_0.

When writing the conclusion to a hypothesis test, it is important to state it in context, making it clear that it is not certain, but referring to the significance level.

> **WORKED EXAMPLE 15.3**
>
> State the conclusion of each hypothesis test.
>
> a Daniel wants to find out whether he spends more time on average playing computer games than his friends do. His hypotheses are $H_0: \mu = 1.2$, $H_1: \mu > 1.2$, where μ is the mean time (in hours) per day spent playing computer games. He uses a 5% significance level for his test. Daniel collects some data and calculates the p-value to be 0.065.
>
> b Alessia is investigating whether the times Daniel spends playing computer games each day come from a normal distribution with mean 1.2 hours and standard deviation 0.3 hours. She uses a 10% significance level for her test. Alessia collects some data and calculates the p-value to be 0.07.
>
> Compare the p-value to the significance level ....
>
> Write the conclusion in context, making it .... clear that it is uncertain
>
> **a** $0.065 > 0.05$, so there is insufficient evidence to reject H_0.
>
> There is insufficient evidence, at the 5% significance level, that the mean time Daniel spends playing computer games is more than 1.2 hours.
>
> Compare the p-value to the significance level ....
>
> Write the conclusion in context, making it .... clear that it is uncertain
>
> **b** $0.07 < 0.10$, so there is sufficient evidence to reject H_0.
>
> There is sufficient evidence, at the 10% significance level, that the times Daniel spends playing computer games do not come from the distribution $N(1.2, 0.3^2)$.

It is also possible to use critical values, rather than p-values, in establishing a conclusion of a hypothesis test. You met this approach in Chapter 6, Section D.

> # Be the Examiner 15.1
>
> Each of these conclusions has been incorrectly written. Explain why.
>
> | | |
> |---|---|
> | **A** | Accept H_0. There is insufficient evidence that the mean has changed. |
> | **B** | Do not reject H_0. The mean daily temperature is still 23 °C. |
> | **C** | Do not reject H_0. There is sufficient evidence that the mean height of 12-year-olds is 135 cm. |
> | **D** | Reject H_0. There is no correlation between height and weight of puppies. |
> | **E** | Reject H_0. The mean running time has decreased. |

The exact details of calculating the *p*-value depend on the type of test you are conducting. You will learn about several different types of test in the following sections.

TOOLKIT: Problem Solving

A hypothesis test rejects H_0 at the 5% significance level. What can you say about what the conclusion would be at the 10% significance level? What about the 2% significance level?

CONCEPTS – REPRESENTATION

The sample is just a snapshot of the whole population. If it has been collected in a sensible way, we hope it will be a **representative** sample. However, it is rarely of interest in its own right. Most mathematical statistics is about using information from a sample to infer properties of the population. Hypothesis testing is one of the most common ways of doing this.

■ χ^2 test for goodness of fit

The χ^2 test (pronounced chi-squared) is used to test whether sample data comes from a given distribution. The null hypothesis is that the data comes from the specified distribution. The **χ^2 statistic** measures how far the observed data values are from what would be expected if the null hypothesis were true. You can use your GDC to calculate the χ^2 statistic and the corresponding *p*-value from observed and expected frequencies. You can then compare the χ^2 statistic to the critical value, or the *p*-value to the significance level of the test.

KEY POINT 15.4

χ^2 test for goodness of fit:
- The data is recorded in a frequency or grouped frequency table. These are the **observed frequencies.**
- **Expected frequencies** are the frequencies for each group assuming the null hypothesis is true.
- The χ^2 statistic and the *p*-value can be obtained from a GDC.
- If the χ^2 statistic is larger than the critical value, or if the *p*-value is smaller than the significance level, there is sufficient evidence to reject H_0.

The critical value depends on the significance level and the number of groups in the table.

KEY POINT 15.5

The critical value for the χ^2 test depends on the significance level and the number of **degrees of freedom (*v*).**

For a frequency table with *n* groups, $v = n - 1$.

Although in the SL course you will only encounter situations where the number of degrees of freedom is $n - 1$, this number can in fact change depending on the number of parameters that have been estimated from the data. You will learn more about this if you are studying Mathematics: applications and interpretation HL.

WORKED EXAMPLE 15.4

A school offers a choice of three different languages. Igor knows that in his year, $\frac{1}{3}$ of the students selected Spanish, $\frac{1}{4}$ selected German and $\frac{5}{12}$ selected Russian. In order to investigate whether the choices in the year below follow the same distribution, he selected a sample of 60 students. Their choices were:

Language	Frequency
Spanish	26
German	16
Russian	18

Igor conducts a χ^2 test at the 5% significance level.

a State the null and alternative hypotheses.

b State the number of degrees of freedom.

c Fill out a table of observed and expected frequencies and calculate the χ^2 value and the p-value.

d The critical value for this test is 5.991. What should Igor conclude?

The null hypothesis is that the two distributions are the same

a H_0: The data comes from the same distribution as Igor's year.
H_1: The data does not come from the same distribution.

Use $v = n - 1$
The table has three groups

b $v = 3 - 1 = 2$

c

The expected frequencies are calculated assuming H_0 is true, so the proportions are the same as in Igor's year

Language	Observed frequency	Expected frequency
Spanish	26	$\frac{1}{3} \times 60 = 20$
German	16	$\frac{1}{4} \times 60 = 15$
Russian	18	$\frac{5}{12} \times 60 = 25$

From GDC,
$\chi^2 = 3.8267, p = 0.1476$

Use your GDC to find the critical value. You need to enter the observed and expected frequencies (into two different lists), and the number of degrees of freedom

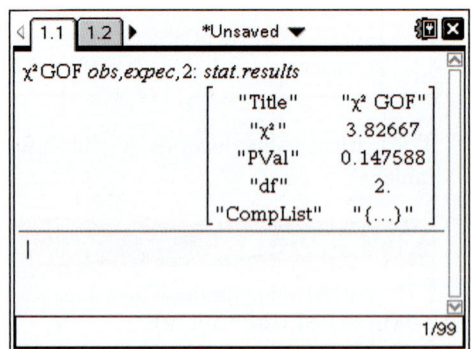

15A Chi-squared tests

Compare the χ^2 value to the critical value. A large χ^2 value provides evidence against H_0 ..... **d** $3.83 < 5.991$, so there is insufficient evidence to reject H_0.

Interpret the conclusion in context .... There is insufficient evidence, at the 5% significance level, that the choices in the year below come from a different distribution.

Sometimes you need to calculate probabilities before you can find expected frequencies. You do this by using the distribution specified in the null hypothesis.

WORKED EXAMPLE 15.5

The times taken by eight-year-old children to solve a puzzle can be modelled by a normal distribution with mean 12 minutes and standard deviation 2.5 minutes. The times taken to solve the same puzzle by a random sample of 50 ten-year-old children are as follows:

Time (min)	$t \leq 9$	$9 < t \leq 11$	$11 < t \leq 13$	$3 < t \leq 15$	$t > 15$
Frequency	10	11	20	5	4

Test, using a 10% significance level, whether the times of the ten-year-old children come from the same distribution.

State the hypotheses .... H_0: The times come from the distribution $N(12, 2.5^2)$
H_1: The times do not come from the distribution $N(12, 2.5^2)$

Calculate the expected frequencies. You need to find the probability for each group, using the distribution $N(12, 2.5^2)$, and then multiply the probabilities by 50

For example, the first probability is $P(X \leq 9)$ where $X \sim N(12, 2.5^2)$

Time	Observed frequency	Probability	Expected frequency
≤ 9	10	0.1151	5.76
9 – 11	11	0.2295	11.5
11 – 13	20	0.3108	15.5
13 – 15	5	0.2295	11.5
> 15	4	0.1151	5.76

Use your GDC to calculate the critical value and the p-value. You need the number of degrees of freedom

$v = 5 - 1 = 4$
From GDC:
$$\chi^2 = 8.62, p\text{-value} = 0.0713$$

You are not given the critical value, so compare the p-value to the significance level .... $0.0713 < 0.1$, so there is sufficient evidence to reject H_0.

Write the conclusion in context .... There is sufficient evidence, at the 10% significance level, that the times for the ten-year-old children do not come from the distribution $N(12, 2.5^2)$.

TOK Links

What type of error rate is acceptable to you in decisions you make? For example, when choosing which subjects to study, which people to be friends with, or which clothes to buy, what kind of mistakes do you make and how often do you make them? Are you ever completely certain that you are making the right decision? What would life be like if you could never make 'mistakes'?

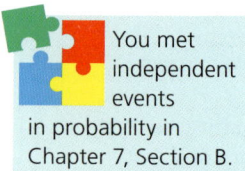 You met independent events in probability in Chapter 7, Section B.

■ χ^2 for independence: contingency tables

The χ^2 test can also be used to test whether two variables are independent. The principle is the same as for the goodness of fit test: the χ^2 statistic is a measure of difference between observed and expected frequencies. The null hypothesis is that the two variables are independent.

To carry out the test, the observed frequencies need to be recorded in a **contingency table**. Your GDC can then calculate the expected frequencies, the χ^2 value and the p-value.

 **TOOLKIT: Problem Solving**

You can use the fact that, under the null hypothesis, the two variables are independent to work out the expected frequencies:

Consider the contingency table:

	A	B
X	1	1
Y	2	4

a Show that the probability of A happening is $\frac{3}{8}$. Find the probability of X happening.

b Assuming that A and X occur independently, find the probability of A and X occurring.

c Given that the total sample size is fixed, find the expected frequency of $A \cap X$.

d Hence complete the expected frequency contingency table.

The number of degrees of freedom depends on the size of the contingency table.

KEY POINT 15.6

For a $m \times n$ contingency table the number of degrees of freedom is $v = (m - 1)(n - 1)$.

WORKED EXAMPLE 15.6

Julio investigates whether people's favourite sport depends on their age. He asks 30 adults and 50 children to select their favourite sport out of football, basketball and baseball. He records the results in the contingency table:

	Adults	Children
Football	8	23
Basketball	12	11
Baseball	10	16

Julio conducts a χ^2 test for independence.

a State the hypotheses for this test.

b Find the number of degrees of freedom.

c Find the expected frequencies and the χ^2 value.

d Julio looks up a list of critical values for his test and finds that the appropriate critical value is 9.21. What should Julio conclude from the test?

15A Chi-squared tests

The null hypothesis is that the two variables are independent	**a**	H_0: Age and favourite sport are independent. H_1: Age and favourite sport are not independent.
$v = (m-1)(n-1)$	**b**	$v = (3-1)(2-1) = 2$
Enter the observed frequency into a matrix	**c**	Expected frequencies:

	Adults	Children
Football	11.6	19.4
Basketball	8.63	14.4
Baseball	9.75	16.3

The GDC will record the expected frequencies in a different matrix

```
χ² Test
Observed:Mat A
Expected:Mat B
Save Res:None
Execute

[Mat][▶MAT]
```

```
χ² Test
  χ²=3.93190204
  p =0.14002265
  df=2

              [▶MAT]
```

From GDC, $\chi^2 = 3.93$

Compare the χ^2 value to the critical value. You need a large χ^2 value in order to reject H_0

d $3.93 < 9.21$, so insufficient evidence to reject H_0.

State the conclusion in context

There is insufficient evidence that age and favourite sport are not independent.

Tip

If a conclusion suggests things are not independent, then it implies a relationship and so they can also be called dependent.

> **You are the Researcher**
>
> There are some situations when the χ^2 test needs to be adapted. For example, the above procedure is not valid if any of the expected frequencies are smaller than five, or if the number of degrees of freedom is 1 (so for a 2 × 2 table). One possible alternative to the χ^2 test that you might like to investigate is Fisher's exact test.

Exercise 15A

For questions 1 to 3, use the method demonstrated in Worked Example 15.1 and 15.2. For each situation, write down suitable hypotheses and state whether the test is one-tailed or two-tailed.

1 **a** The mean temperature in Perth in 1980 was 23.4 °C. Sonya wants to test whether the average temperature in 2018 was higher.

  b Paulo's mean time for a 2 km run was 8.3 minutes. After following a new training regime, he wants to test whether his mean time has decreased.

2 **a** A bottle claims to contain 300 ml of water on average. Tamara wants to test whether the mean amount of water in a bottle is different.

  b The mean height of adult residents of Nairobi is 163 cm. Amandla wants to test whether the mean height of residents of Boston is different.

3 a Leila is investigating whether there is a negative correlation between maximum daily temperature and the number of people at a seaside café.

b Bashir is investigating whether there is any correlation between daily rainfall and minimum daily temperature.

For questions 4 to 6, use the method demonstrated in Worked Example 15.3 to state the conclusion of the hypothesis test.

4 a $H_0: \mu = 13.4$, $H_1: \mu > 13.4$, 5% significance level; p-value = 0.0638

b $H_0: \mu = 26$, $H_1: \mu > 26$, 5% significance level; p-value = 0.105

5 a $H_0: \mu = 2.6$, $H_1: \mu \neq 2.6$, 10% significance level; p-value = 0.0723

b $H_0: \mu = 8.5$, $H_1: \mu \neq 8.5$, 10% significance level; p-value = 0.103

6 a H_0: The data comes from the distribution $N(12, 3.4^2)$

H_1: The data does not come from the distribution $N(12, 3.4^2)$

5% significance level, p-value = 0.0492

b H_0: The data comes from the distribution $N(53, 8^2)$

H_1: The data does not come from the distribution $N(53, 8^2)$

5% significance level, p-value = 0.102

In questions 7 to 9, you are given expected probabilities and observed frequencies. The null hypothesis is that the observed data comes from the same distribution as the expected values. Use the method from Worked Example 15.4 to

i calculate the expected frequencies and the χ^2 value

ii state the number of degrees of freedom

iii find the p-value and state the conclusion of the test at 5% significance level.

7 a

Data value	Red	Blue	Yellow
Expected probability	$\frac{1}{4}$	$\frac{1}{2}$	$\frac{1}{4}$
Observed frequency	34	51	15

b

Data value	Apple	Orange	Banana
Expected probability	$\frac{1}{2}$	$\frac{3}{10}$	$\frac{1}{5}$
Observed frequency	45	33	22

8 a

Data value	1	2	3	4	5
Expected probability	0.2	0.2	0.2	0.2	0.2
Observed frequency	12	16	8	10	4

b

Data value	0	2	4	6
Expected probability	0.25	0.25	0.25	0.25
Observed frequency	6	18	13	23

9 a

Data value	French	Spanish	English	Mandarin
Expected probability	$\frac{1}{3}$	$\frac{1}{4}$	$\frac{1}{5}$	$\frac{13}{60}$
Observed frequency	52	51	30	67

b

Data value	Biology	Chemistry	Physics	Geology
Expected probability	$\frac{1}{3}$	$\frac{1}{6}$	$\frac{1}{5}$	$\frac{3}{10}$
Observed frequency	52	31	42	35

15A Chi-squared tests

In questions 10 to 12, you are given the null hypothesis, observed frequencies and the critical value for the goodness of fit test. Using the method demonstrated in Worked Example 15.5, conduct the test and state the conclusion.

10 a H_0: The data comes from the distribution B(4, 0.5)

Critical value = 9.49

Data value	0	1	2	3	4
Observed frequency	10	65	68	43	14

b H_0: The data comes from the distribution B(5, 0.6)

Critical value = 12.8

Data value	0	1	2	3	4	5
Observed frequency	7	28	103	188	132	42

11 a H_0: The data comes from the distribution B(10, 0.77)

Critical value = 9.24

Data value	≤ 5	6	7	8	9	10
Observed frequency	7	16	28	34	12	3

b H_0: The data comes from the distribution B(20, 0.08)

Critical value = 7.78

Data value	0	1	2	3	≥ 4
Observed frequency	19	35	22	15	9

12 a H_0: The data comes from the distribution $N(5.5, 2^2)$

Critical value = 7.78

Data value	≤ 2	2 to 4	4 to 6	6 to 8	> 8
Observed frequency	7	21	44	17	11

b H_0: The data comes from the distribution $N(36, 11^2)$

Critical value = 9.24

Data value	≤ 20	20–30	30–40	40–50	50–60	> 60
Observed frequency	25	92	140	112	29	2

In questions 13 to 15, you are given a contingency table of observed values of two variables and the critical value for a χ^2 test. Use the method demonstrated in Worked Example 15.6 to test whether the two variables are independent. State the χ^2 value and the conclusion of each test.

13 a

12	15	21
22	18	12

Critical value = 9.21

b

63	81	35
27	32	62

Critical value = 7.38

14 a

8	12	16
9	15	21
13	14	22

Critical value = 7.78

b

18	13	8
13	16	21
8	18	22

Critical value = 9.49

15 a

62	37	81
88	53	30
26	15	32
81	73	55

Critical value = 12.6

b

13	11	11
12	8	3
21	16	9
13	13	13

Critical value = 10.6

16 Ruby wants to test whether students' food preferences depend on their age. She conducts a survey in the school canteen, recording which option each student chooses.

	Veggie burger	Fish fingers	Peperoni pizza
Junior school	18	26	51
Senior school	53	38	47

Conduct a suitable χ^2 test at the 5% significance level, stating your hypotheses and conclusion clearly.

17 A zoologist investigates whether different types of insect are more common in different locations. She collects a random sample of insects from a meadow and a forest and counts the number of ants, bees and flies.

	Ants	Bees	Flies
Meadow	26	15	21
Forest	32	6	18

Use a χ^2 test with a 10% significance level to test whether the type of insect found depends on the location.

18 A theory predicts that three different types of flower should appear in the ratio 1 : 2 : 3. A sample of 60 flowers contains 14 flowers of type A, 18 flowers of type B and 28 flowers of type C. A χ^2 goodness of fit test is used to test the theory at the 10% significance level.

a Calculate the expected frequencies and state the number of degrees of freedom.

b Find the χ^2 value.

c The critical value is 4.605. State the conclusion of the test.

19 Zhao thinks that students at her large college are equally likely to study any of the four mathematics courses. She asks a random sample of 80 students and obtains the following results:

Course	Analysis and approaches HL	Analysis and approaches SL	Applications and interpretation HL	Applications and interpretation SL
Number of students	18	21	8	33

Zhao conducts a χ^2 test using a 5% significance level.

a State the null and alternative hypotheses.

b Write down the expected frequencies and the number of degrees of freedom.

c Find the *p*-value for the test and hence state the conclusion.

20 The table shows information about the mode of transport students use to get to school in four different cities. Use a χ^2 test to find out whether there is evidence, at the 5% significance level, that there is a relationship between the mode of transport and the city.

	Amsterdam	Athens	Houston	Johannesburg
Car	12	25	48	24
Bus	18	33	12	18
Bicycle	46	12	7	53
Walk	38	8	3	21

21 A six-sided dice is rolled 120 times, giving the following results:

Outcome	1	2	3	4	5	6
Frequency	26	12	16	28	14	24

Is there evidence, at the 2% significance level, that the dice is not fair?

22 Rajesh is practising tennis serves. He takes three serves at a time and records the number of successful serves. He believes that this number can be modelled by the binomial distribution B(3, 0.7).

Number of successful serves out of three	0	1	2	3
Frequency	7	28	95	70

  a State the hypotheses for a χ^2 goodness of fit test.
  b Find the expected frequencies and write down the number of degrees of freedom.
  c Calculate the χ^2 value.
  d The critical value for the test is 6.25. State the conclusion of the test.

23 Michelle tosses six coins simultaneously and records the number of tails. She repeats this 600 times. The results are shown in the table.

Number of tails	0	1	2	3	4	5	6
Frequency	9	62	120	178	152	67	12

  a State the distribution of the number of tails for an unbiased coin.
  b Hence work out the expected frequencies for Michelle's experiment.
  c Test at the 5% significance level whether there is evidence that the coins are biased.

24 Four friends are guessing answers to maths questions. They think that they each have the probability of 0.5 of guessing the correct answer to any question, independently of each other.

  a Let X be the number of correct answers to a single question. Accepting the assumptions above are correct, state the distribution of X.

In order to test whether their assumptions are correct, the friends guess answers to 100 questions and record the number of correct answers to each question.

Number of correct answers	0	1	2	3	4
Frequency	12	13	45	22	8

  b State appropriate hypotheses for a χ^2 goodness of fit test.
  c State the number of degrees of freedom.
  d Conduct the test at the 2% significance level and state your conclusion.

25 An athlete believes that her long jump distances follow a normal distribution with mean 5.8 m and standard deviation 0.8 m.

  a Assuming her belief is correct, copy the table below and fill in the missing probabilities:

Distance (m)	< 5	5 to 6	6 to 7	> 7
Probability	0.159			

In order to test her belief, she records her distances from a random sample of 100 jumps, obtaining the following results:

Distance (m)	< 5	5 to 6	6 to 7	>7
Frequency	17	42	38	3

  b State the number of degrees of freedom for a χ^2 goodness of fit test.
  c State suitable hypotheses.
  d Conduct the test at the 10% significance level.

26 A train company claims that times for a particular journey are distributed normally with mean 23 minutes and standard deviation 2.6 minutes. Sumaya takes this train to school and wants to test the company's claim. She decides to conduct a χ^2 test and records the durations of 50 randomly selected journeys:

Time (min)	< 21.5	21.5–22.5	22.5–23.5	23.5–24.5	> 24.5
Frequency	3	8	14	17	8

a Find the expected frequencies.
b Write down the number of degrees of freedom.
c Calculate the χ^2 value.
d The critical value for Sumaya's test is 9.49. State the conclusion in context.

27 A teacher suggests that exam grades at their college can be modelled by the following distribution:
$P(G = g) = \dfrac{g(11-g)}{140}$ for $g = 3, 4, 5, 6, 7$

A random sample of 40 students had the following grades:

Grade	3	4	5	6	7
Frequency	2	10	9	12	7

Test, using a 10% significance level, whether the teacher's model is appropriate for these data.

28 Katya wants to find out whether diet choices are dependent on age. She collects data from 200 pupils at her school and records it in the contingency table:

	Vegetarian	Vegan	Eats meat
11–13	12	32	21
14–15	22	16	30
16–19	26	18	23

a Conduct a χ^2 test for independence, using a 1% significance level. State your conclusion in context.

In order to get a better understanding of the data, Katya decides to split the 16–19 age group in two. The new data table is:

	Vegetarian	Vegan	Eats meat
11–13	12	32	21
14–15	22	16	30
16–17	13	12	10
18–19	13	6	13

b Repeat the test, stating the new conclusion clearly.

29 A teacher asked a group of 80 students for their preference out of three science subjects. The table shows the fraction of students in each group.

	Boys	Girls
Biology	$\dfrac{3}{20}$	$\dfrac{1}{5}$
Chemistry	$\dfrac{1}{10}$	$\dfrac{3}{20}$
Physics	$\dfrac{1}{4}$	$\dfrac{3}{20}$

a Test, using a 10% significance level, whether favourite science is independent of gender.
b The teacher decides to check her results by using a larger sample. She asks a group of 240 students and finds that the fractions in each group are the same. Repeat the test and state the new conclusion.

15B *t*-tests

Sometimes you know that a population can be modelled by a normal distribution, but not what its mean and variance are. The *t*-test can be used to test whether the mean has changed from a previously known value.

The null hypothesis specifies the old/assumed value of the mean. Given a sample, the **t-statistic** is calculated using its mean and standard deviation. The corresponding *p*-value tells you how likely this value of the *t*-statistic would be if the null hypothesis was true. As before, if the *p*-value is smaller than the significance level, there is sufficient evidence to reject the null hypothesis.

A *t*-test is not always appropriate. In particular, it relies on the knowledge (or an assumption) about the underlying distribution.

> **KEY POINT 15.7**
>
> A *t*-test is only valid if the distribution of the underlying population is normal.

> **WORKED EXAMPLE 15.7**
>
> Peter is investigating whether the average rainfall last year was higher than in the past. He knows that the average monthly rainfall in April over the past fifty years was 61 mm. He collected the rainfall data for April last year for eight different locations, obtaining the following results.
>
> | Location | 1 | 2 | 3 | 4 | 5 | 6 | 7 | 8 |
> |---|---|---|---|---|---|---|---|---|
> | Rainfall (mm) | 52 | 71 | 45 | 83 | 61 | 65 | 73 | 48 |
>
> Peter conducts a *t*-test using a 10% significance level.
>
> a State the null and alternative hypotheses.
>
> b Find the *t*-statistic and the corresponding *p*-value.
>
> c State the conclusion of the test and interpret it in context.
>
> d What assumption about the distribution of rainfall does Peter need to make?
>
> Peter is testing for an increase, so this is a one-tailed test ⋯⋯ **a** $H_0: \mu = 61$, $H_1: \mu > 61$
> where μ mm is the population mean rainfall in April
>
> This is a 1-sample *t*-test
> You need to enter the data into a list and set ⋯⋯ **b** From GDC, $t = 0.266$, $p = 0.399$
> the null and alternative hypotheses

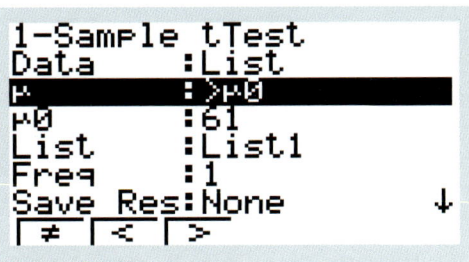

The calculator gives you the required values

```
1-Sample tTest
μ    >61
t    =0.26590801
p    =0.39898584
x̄    =62.25
xσn-1=13.2960789
n    =8
```

Compare the *p*-value to the significance level ⋯⋯ **c** 0.399 > 0.10, insufficient evidence to reject H_0.

Interpret the conclusion in context ⋯⋯ There is insufficient evidence, at the 10% significance level, that the mean April rainfall last year was higher than in the past.

The assumption refers to the underlying population distribution ⋯⋯ **d** The underlying population of rainfall is normally distributed.

Sometimes you may not be given all the sample data, but only the summary statistics: the sample mean and the estimate of the population variance. Notice that in a *t*-test, the null hypothesis specifies the value of the population mean, but not the variance. In fact, the *t*-test should only be used when the population variance is unknown and needs to be estimated from the sample.

WORKED EXAMPLE 15.8

In an athletics club, the mean 400 m times of all the members are known to be normally distributed with mean 67.3 seconds. The club employs a new coach. After a few months, they take a sample of the 400 m times of 20 randomly selected club members. The mean of the sample was 63.8 seconds and the estimate of the population variance was 7.84. Is there sufficient evidence, at the 5% significance level, that the mean 400 m time has changed?

State the hypotheses. We are testing for a change in mean, so this is a two-tailed test

$H_0: \mu = 67.3$, $H_1: \mu \neq 67.3$
where μ is the population mean 400 m time.

Using $\bar{x} = 63.8$, $\sigma = \sqrt{7.84} = 2.8$

You need to enter the sample mean ($\bar{x}$), the estimate of the standard deviation (usually denoted σ_{n-1}) and the sample size (n) ⋯⋯ From GDC, $t = -5.59$, $p = 2.17 \times 10^{-5}$

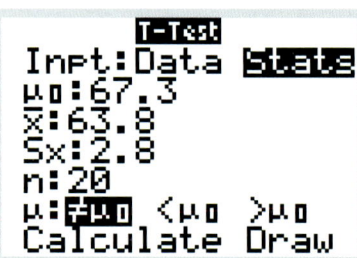

Compare the *p*-value to the significance level ⋯⋯ $2.17 \times 10^{-5} < 0.05$, sufficient evidence to reject H_0

State the conclusion in context ⋯⋯ There is sufficient evidence, at the 5% significance level, that the mean 400 m time has changed.

15B t-tests

Tip

In Worked Example 15.8, because the test was two-tailed, you can only conclude that the mean has changed, not that it has decreased.

CONCEPTS – APPROXIMATION AND VALIDITY

You might wonder about the lack of rigour when we throw around the assumption that populations are normally distributed. If the population is not normally distributed, does this **invalidate** the method? What if the population is only **approximately** normal? Statisticians have spent a lot of time worrying about this. A concept referred to as the 'robustness' of a test looks at how much error is introduced if the assumptions of a test are not perfectly met. The good news is that, based on computer simulations, it seems that the t-test is fairly robust – as long as the underlying distribution is vaguely symmetric and declining away from the centre (a much weaker assumption than normality), then the t-test still works reasonably well.

The t-test can also be used to test whether the means of two populations are different. You need to select a two-sample t-test and enter the two samples into two different lists. The two samples need not have the same size. In this course, it will always be assumed that the two populations have equal variance, which is estimated by combining the variances of the two samples. This is called the **pooled variance**.

KEY POINT 15.8

Assumptions for the **pooled sample t-test**:
- both populations are normally distributed
- the populations' variances are equal.

If you are given summary statistics rather than the sample data, you need to know the sample size, sample mean and the estimate of the population variance for each sample.

WORKED EXAMPLE 15.9

A teacher believes that eight-year-old children take less time to read a piece of text than seven-year-old children. He times a random sample of 20 eight-year-old children and 25 seven-year-old children. The eight-year-old children had sample mean time of 7.8 minutes and the estimate of the population standard deviation was 2.3 minutes. The seven-year-old children had sample mean time of 8.6 minutes and the estimate of the population standard deviation was 3.5 minutes.

Stating any necessary assumptions, use a suitable test with the 10% significance level to test the teacher's belief.

You need a two-sample t-test ····· Assumptions:
Both populations are normally distributed.
The two populations have equal variance.

State the hypotheses. The test is one-tailed. The null hypothesis is that the two population means are equal ····· $H_0: \mu_1 = \mu_2$, $H_1: \mu_1 < \mu_2$
where μ_1 is the population mean time for eight-year-olds, and μ_2 is the population mean time for seven-year-olds.

Remember to select the *pooled* two-sample t-test ····· From GDC, $t = -0.880$, $p = 0.192$

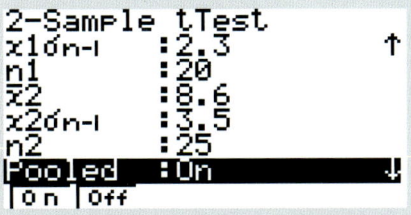

Compare the p-value to the significance level ····· $0.192 > 0.10$, insufficient evidence to reject H_0.

Interpret the conclusion in context ····· There is insufficient evidence, at the 10% significance level, that the mean time for eight-year-old children is less than the mean time for seven-year-old children.

> **You are the Researcher**
>
> The tests used in this section require several assumptions: that the population is normally distributed, that the population variance has been estimated from the sample and, in the case of the two-sample test, that the two populations have equal variance. There are several alternative tests that can be used if some of those assumptions do not hold. You might like to research Welch's test, the Wilcoxon test or the Mann–Whitney test. If you want to compare more than two groups there is a suite of tests called ANOVA.

Exercise 15B

In questions 1 to 3, you are given the null and alternative hypotheses, the significance level and a sample of data. Using the method demonstrated in Worked Example 15.7, perform a t-test to test whether there is sufficient evidence to reject the null hypothesis. Give the value of the t-statistic and the p-value.

1. **a** $H_0: \mu = 30$, $H_1: \mu > 30$, 5% significance level

23	29	32	27	37	38	28

   b $H_0: \mu = 50$, $H_1: \mu > 50$, 5% significance level

50	58	61	46	49	57	65

2. **a** $H_0: \mu = 220$, $H_1: \mu < 220$, 10% significance level

224	209	207	215	216	222

   b $H_0: \mu = 145$, $H_1: \mu < 145$, 10% significance level

141	143	139	144	142	150

3. **a** $H_0: \mu = 76.2$, $H_1: \mu \neq 76.2$, 2% significance level

69.4	78.3	71.0	75.5	77.4	72.9	68.8	73.3	70.5

   b $H_0: \mu = 14.7$, $H_1: \mu \neq 14.7$, 2% significance level

14.7	13.9	18.8	15.5	17.7	15.6	20.3	17.5	16.9

In questions 4 to 6, you are given the sample size (n), sample mean ($\bar{x}$) and the estimate of the population variance (σ_{n-1}^2). Using the method demonstrated in Worked Example 15.8, perform a t-test to test, at the given significance level, whether there is sufficient evidence to reject the null hypothesis.

4. **a** $H_0: \mu = 100$, $H_1: \mu > 100$, 10% significance level
   $n = 25$, $\bar{x} = 110$, $\sigma_{n-1}^2 = 25$

   b $H_0: \mu = -6$, $H_1: \mu > -6$, 10% significance level
   $n = 40$, $\bar{x} = -5$, $\sigma_{n-1}^2 = 10$

5. **a** $H_0: \mu = 0$, $H_1: \mu < 0$, 1% significance level
   $n = 60$, $\bar{x} = -0.6$, $\sigma_{n-1}^2 = 400$

   b $H_0: \mu = 50$, $H_1: \mu < 50$, 1% significance level
   $n = 15$, $\bar{x} = 48$, $\sigma_{n-1}^2 = 81$

6. **a** $H_0: \mu = 1$, $H_1: \mu \neq 1$, 5% significance level
   $n = 10$, $\bar{x} = 1.04$, $\sigma_{n-1}^2 = 0.25$

   b $H_0: \mu = 10$, $H_1: \mu \neq 10$, 5% significance level
   $n = 5$, $\bar{x} = 9.5$, $\sigma_{n-1}^2 = \frac{1}{9}$

In questions 7 to 9, you are given sample means and estimates of population variances for two samples. Using the method demonstrated in Worked Example 15.9, perform a *t*-test to find, at the given significance level, whether there is sufficient evidence to reject the null hypothesis.

7 a $H_0: \mu_1 = \mu_2$, $H_1: \mu_1 > \mu_2$, 5% significance level

	Sample 1	Sample 2
Mean	10	8
Variance	4	5
Size	5	6

b $H_0: \mu_1 = \mu_2$, $H_1: \mu_1 > \mu_2$, 2% significance level

	Sample 1	Sample 2
Mean	100	80
Variance	40	50
Size	50	60

8 a $H_0: \mu_1 = \mu_2$, $H_1: \mu_1 < \mu_2$, 10% significance level

	Sample 1	Sample 2
Mean	5.4	5.8
Variance	4.3	3.9
Size	48	35

b $H_0: \mu_1 = \mu_2$, $H_1: \mu_1 < \mu_2$, 10% significance level

	Sample 1	Sample 2
Mean	12	15
Variance	36	49
Size	50	50

9 a $H_0: \mu_1 = \mu_2$, $H_2: \mu_1 \neq \mu_2$, 2% significance level

	Sample 1	Sample 2
Mean	104	108
Variance	81	100
Size	100	100

b $H_0: \mu_1 = \mu_2$, $H_1: \mu_1 \neq \mu_2$, 5% significance level

	Sample 1	Sample 2
Mean	150	147
Variance	25	25
Size	10	10

10 Benga recorded the times taken to complete his homework over a whole academic year. He found that they follow a normal distribution with mean 34 minutes. The following year he changed classes and thinks that his new teacher sets more homework. A sample of times taken to complete the homework, recorded to the nearest minute, is as follows:

34, 27, 42, 45, 38, 26, 30, 42

Test whether there is sufficient evidence, at the 10% significance level, that the new teacher sets more homework.

11 A bottle states that it contains 300 ml of water. Olivia measures the amount of water in twenty randomly selected bottles, finding that the mean is 298 ml and the estimate of the population standard deviation is 6 ml. Test, using a 5% significance level, whether the mean amount of water in a bottle is different from 300 ml. You may assume that the amount of water in a bottle is normally distributed.

12 The heights of trees in a certain forest are known to be normally distributed with mean 23.6 m. Zhao measured the heights of ten trees in a different forest, recording the following results (in metres):

13.6, 22.5, 18.0, 21.3, 28.7, 31.6, 12.8, 22.5, 18.9, 23.1

Assuming that the heights of trees in this forest are also normally distributed, use a suitable test to determine whether there is evidence, at the 10% significance level, that the trees in this forest are shorter on average.

13 Yuka thinks that, in her town, the mean temperature is 14.3 °C. Hiroki thinks that the mean is higher. They collect 30 pieces of data and find that the sample mean is 17.3 °C and the estimate of the population standard deviation is 8.6 °C. Stating your hypotheses clearly, conduct a *t*-test using a 5% significance level, to determine whether the data support Hiroki's belief.

14 Hamid exercises regularly at his local gym and knows that, on average, he can run 10 km on a treadmill in 42 minutes. After an injury he has started exercising again and wants to check whether his times have returned to their previous values. A sample of eight times have the following values (to the nearest minute):

45, 42, 40, 44, 45, 39, 41, 42

a Test whether there is evidence, at a 10% significance level, that he is running slower than before.

b What assumption did you need to make about the running times in order to use your test?

15 A trainee pharmacist is supposed to be dispensing a particular medicine in 25 ml doses. The manager is concerned that they have not been dispensing the medicine correctly. She measures the amount in a random sample of 30 doses of medicine and finds that the mean amount is 23.8 ml and the unbiased estimate of the population variance is 3.8 ml^2.

   a State an assumption needed in order to use a *t*-test.

   b Test, using a 5% significance level, whether the pharmacist has been dispensing an incorrect amount of medicine.

16 Divya measures masses (in kg) of a random sample of dogs from two different breeds, with the following results:

   Breed A: 6.2, 5.1, 8.7, 12.1, 18.3, 5.4

   Breed B: 12.1, 7.5, 6.3, 4.8, 11.3, 8.3, 5.9, 7.0

   Is there evidence, at the 5% significance level, that the two breeds have different mean masses? You may assume that both populations are normally distributed and have equal variances.

17 A teacher records the length of time (in minutes) it takes two groups of pupils to complete a writing task.

	Sample size	Sample mean	Estimate of population variance
Group 1	23	18.3	12.3
Group 2	28	20.6	15.0

   The teacher uses a *t*-test to determine whether the first group completed the task significantly faster.

   a State two assumptions required to use a *t*-test.

   b State the appropriate hypotheses.

   c Conduct the test and state the conclusion at:

   i 5% significance

   ii 1% significance.

18 Sabrina is monitoring the share price of a particular company. She records the increase or decrease in price at the end of each day (so, for example, −1.23 means that the share price decreased by $1.23 since the previous day). The results for a period of ten days are:

   −1.23, 0.67, 1.45, −2.12, 0.30, 2.33, 1.05, −0.89, −0.22, 1.17

   Sabrina wants to use a *t*-test to determine whether there is an average upward trend in the share price.

   a State suitable hypothesis for the test.

   b What assumption do you need to make about the share prices?

   c Assuming this assumption holds, conduct the test at the 3% significance level.

19 A coffee shop has introduced two new drinks. Their advertising claims that MoccaLite contains less sugar than ChoccyFroth. The manager checks a sample of five cups of each drink and records the amount of sugar to the nearest gram:

   MoccaLite: 27, 22, 25, 27, 23

   ChoccyFroth: 28, 24, 27, 32, 22

   a In order to use a *t*-test, the manager needs to assume that the amounts of sugar in both types of drink can be modelled by a normal distribution. State one further assumption he needs to make.

   b Conduct a suitable test, using a 10% level of significance, to determine whether the data support the advertising claim.

20 In a trial of a drug designed to decrease blood pressure, a group of ten patients have their systolic blood pressure measured at the start and the end of the trial.

	A	B	C	D	E	F	G	H	I	J
Start	139	141	150	134	149	152	147	153	155	149
End	139	135	142	138	137	141	136	138	155	151

a Is there evidence, at the 5% significance level, that the mean blood pressure is lower at the end of the trial than before? You may assume that blood pressure before and after the trial are distributed normally with equal variances.

One doctor suggests calculating the change in blood pressure for each patient.

b Copy and complete the table:

	A	B	C	D	E	F	G	H	I	J
Change	0	−6								

c Suggest suitable null and alternative hypotheses to test whether the blood pressure has decreased on average.
d Conduct the new test at the 5% significance level.
e What assumption is needed for the second t-test to be valid?
f Why might the second test be preferred to the first test?

15C Spearman's rank correlation

TOK Links
You should know that correlation does not imply causation, but how can social scientists gain knowledge about causation?

In Chapter 6, Section D, you learned how to calculate Pearson's correlation coefficient and use it to determine whether there is a significant linear correlation between two variables. The scatter diagrams below show a clear relationship between two variables. In Diagram 1, as x increases so does y; in Diagram 2, as x increases y decreases. However, the Pearson's correlation coefficient between x and y is 0.843 for Diagram 1 and −0.761 for Diagram 2.

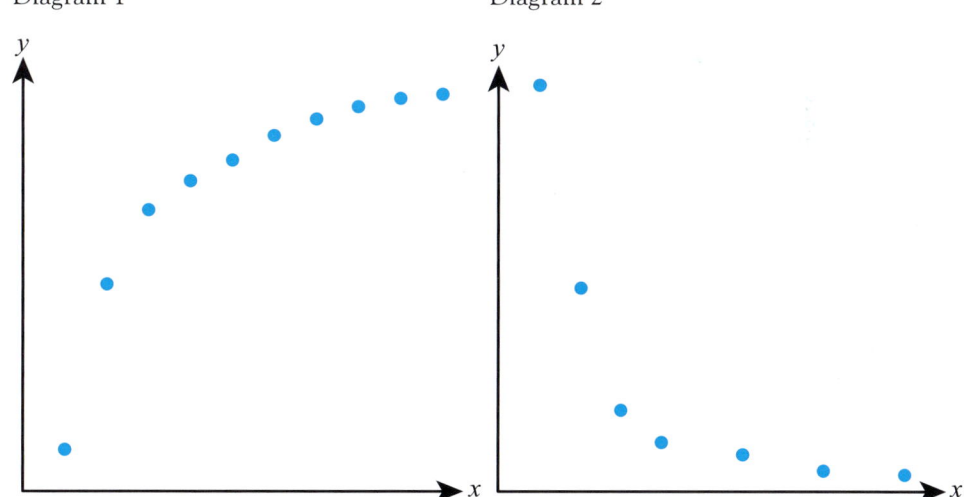

Diagram 1   Diagram 2

Spearman's rank correlation coefficient (denoted r_s) is a measure of non-linear correlation between two variables. It measures the extent to which the two variables increase or decrease together. Diagram 1 above has $r_s = 1$ and Diagram 2 has $r_s = -1$.

> **KEY POINT 15.9**
>
> To calculate Spearman's rank correlation coefficient:
> - rank both sets of data in order of magnitude
> - calculate the Pearson's correlation coefficient between the ranks.

In the two examples on the previous page, replacing actual data values by their ranks turns the graphs into a straight line.

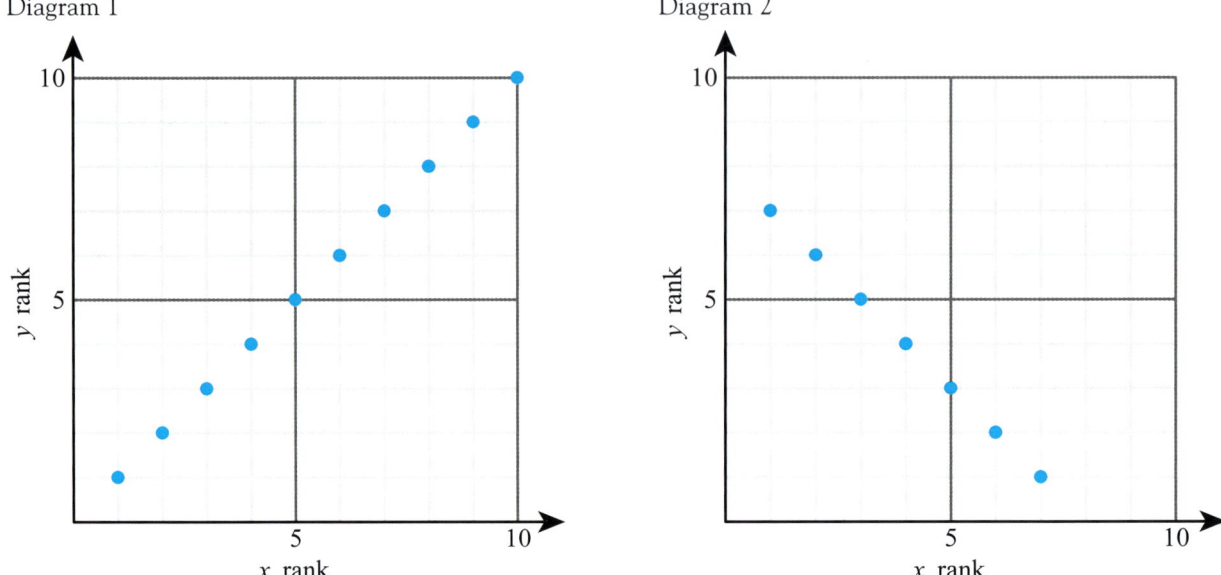

Both Spearman's and Pearson's correlation coefficients are appropriate in different situations and each has its limitations.

> **KEY POINT 15.10**
>
> - Pearson's correlation coefficient only detects a linear relationship between variables, while Spearman's rank can also detect a non-linear relationship.
> - Pearson's uses the actual data values, while Spearman's only uses their ranks.
> - Pearson's is more affected by outliers.

> **You are the Researcher**
>
> You might like to research other correlation coefficients, for example, Kendal's tau, which are appropriate in different situations.

As with Pearson's correlation coefficient, you can use a critical value to determine whether the correlation is significant. You can now express this in the language of hypothesis testing.

15C Spearman's rank correlation

WORKED EXAMPLE 15.10

Eight students took a Chemistry test and a History test. Their marks are given in the table.

Student	1	2	3	4	5	6	7	8
Chemistry mark	34	41	57	48	42	61	50	67
History mark	82	77	88	75	66	93	80	98

a Calculate the Spearman's rank correlation coefficient between the two sets of marks.

b The critical value of the correlation coefficient for the 5% significance level is 0.643. Stating your hypothesis clearly, test, at the 5% significance level, whether the data provides significant evidence of positive correlation between the Chemistry and History marks.

Rank each set of data from smallest to largest

a Ranks:

Student	1	2	3	4	5	6	7	8
Chemistry	1	2	6	4	3	7	5	8
History	5	3	6	2	1	7	4	8

Enter the ranks into two lists and calculate the correlation coefficient

From GDC, $r_s = 0.690$

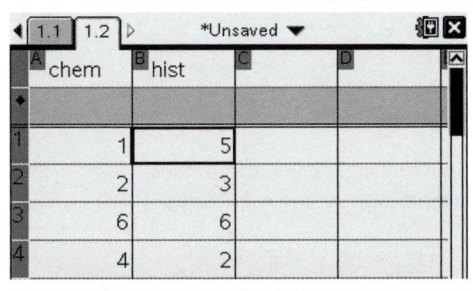

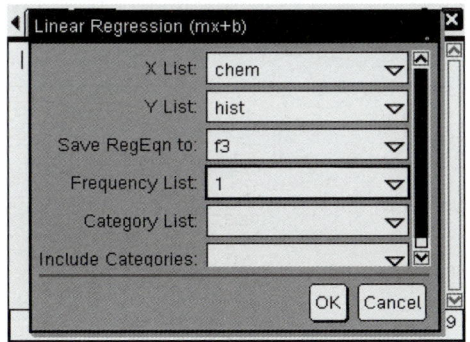

The null hypothesis is that there is no correlation

b H_0: There is no correlation between Chemistry and History marks.

The test is one-tailed, as we are only testing for positive correlation

H_1: There is positive correlation between Chemistry and History marks.

Compare r_s to the critical value. A large value of r_s is evidence of correlation

$r_s = 0.690 > 0.643$, sufficient evidence to reject H_0.

Interpret the conclusion in context

There is sufficient evidence, at the 5% significance level, that there is positive correlation between Chemistry and History marks.

If several data values are equal, you replace all of them by the average of their ranks.

WORKED EXAMPLE 15.11

Maryam works at a coffee shop. She wants to find out whether there is any correlation between the sugar content of a drink and the number of people who buy it. She records this information for a sample of ten drinks over a period of time.

Sugar content (g)	12	15	15	18	19	21	26	28	35	37
Sales	25	37	40	25	52	48	12	57	25	23

a Calculate the Spearman's rank correlation coefficient for this data.

b Test, using a 10% significance level, whether there is any correlation between the sugar content and the sales of a drink. The critical value for this test is 0.564.

Rank each set of data from smallest to largest

For the sugar content, the items ranked 2nd and 3rd are equal, so replace both ranks by 2.5

For the sales, the items ranked 3rd, 4th and 5th are equal, so replace each of those ranks by 4

a Ranks:

1	2.5	2.5	4	5	6	7	8	9	10
4	6	7	4	9	8	1	10	4	2

Use your GDC to calculate the correlation coefficient

From GDC, $r_s = -0.172$

State the hypotheses. The test is two-tailed, we do not specify whether the correlation is positive or negative

b H_0: There is no correlation between sugar content and sales
H_1: There is a correlation between sugar content and sales

Compare r_s to the critical value. Since r_s is negative, you actually need to look at its modulus

$|r_s| = 0.172 < 0.564$, insufficient evidence to reject H_0
There is insufficient evidence, at the 10% significance level, of a correlation between sugar content and sales.

You are the Researcher

A formula for the Spearman's rank correlation for n pairs of data is:

$$r_s = 1 - \frac{6 \sum d_i^2}{n(n^2 - 1)}$$

where d_i is the difference in the ranks of the corresponding data items.

You might like to find out how this formula is derived. When does it give the same answer as your method and when does it give a different answer? How different can the two answers be?

CONCEPT – PATTERNS

Is it more important to know that the relationship is linear or that it is increasing? Can you think of situations in which each **pattern** is more important to know about? Are there any other patterns in the data that you think would be useful to know?

Exercise 15C

For questions 1 to 4, use the method demonstrated in Worked Example 15.10 to calculate the Spearman's rank correlation coefficient between the two sets of data.

1 a

x	3	4	6	7	9	10	12	13
y	2	3	5	4	8	6	9	7

b

x	3	6	4	5	8	11	10	13
y	−3	0	1	−2	−1	2	4	3

2 a

x	5	7	11	8	6	10	12	14	17
y	6	7	9	3	5	4	0	1	2

b

x	14	13	15	19	21	22	23	24	25
y	14	12	10	9	7	4	3	1	0

3 a

x	1.3	3.5	6.8	4.7	5.1	3.6	10.8	7.1
y	16.3	11.8	15.5	13.7	21.8	18.6	22.1	13.5

b

x	6.3	3.5	2.8	5.7	2.6	9.8	8.7	9.0
y	2.0	4.7	4.1	2.7	3.8	0.7	1.8	2.1

4 a

x	12	9	10	14	21	18	11
y	−6	−3	−5	−1	−4	−8	−9

b

x	35	38	41	47	52	61	58
y	−4	−8	−2	−7	−5	−6	−11

For questions 5 to 8, use the method demonstrated in Worked Example 15.11 to calculate the Spearman's rank correlation coefficient between the two sets of data.

5 a

x	3	1	1	6	7	8	9
y	6	7	5	4	4	1	2

b

x	1	1	2	6	5	8	9
y	6	7	5	4	2	1	2

6 a

x	6	7	8	7	9	5	7
y	12	9	3	6	1	5	11

b

x	21	25	36	28	31	40	32
y	21	11	18	11	11	7	9

7 a

x	1	2	2	3	5	7	8	8
y	−3	0	1	−2	−1	2	4	3

b

x	1	1	2	3	5	6	6	8	9
y	6	7	5	3	8	2	0	−2	−1

8 a

x	4.5	6.1	6.1	4.5	8.1	7.3	4.5
y	6	5	8	5	8	7	8

b

x	12.8	8.1	12.8	8.7	9.1	6.2	7.0
y	3	10	7	10	10	5	5

9 Match each graph with the corresponding values of Spearman's rank (r_s) and Pearson's (r) correlation coefficient.

  a $r_s = r = 1$
  b $r_s = 1$ but $r = 0.7$
  c $r = 0.5$ and $r_s = 0.7$

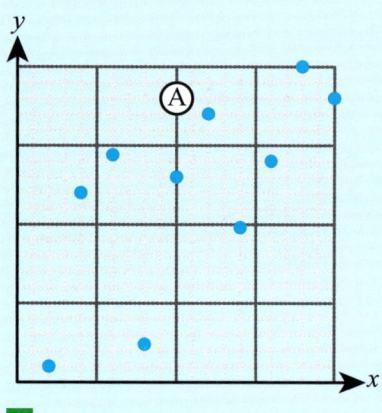
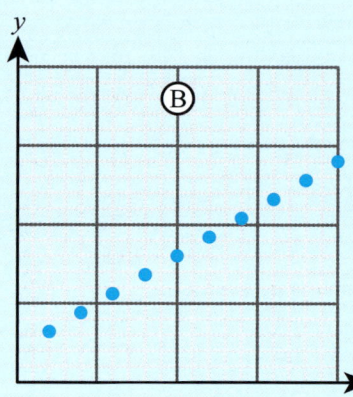
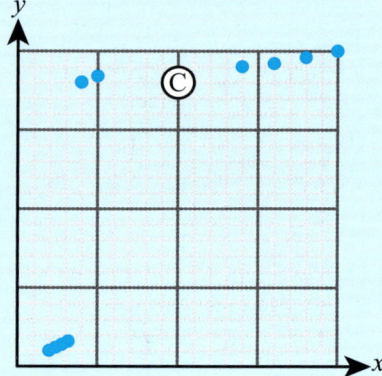

10 For the data in this table:

x	1.2	3.7	1.6	2.1	3.2	2.8
y	11	37	13	25	38	18

  a Calculate
    i   the Pearson's product-moment correlation coefficient
    ii the Spearman's rank correlation coefficient.
  b Describe the correlation between the two sets of data.

11 Elena collected some data on the number of people visiting a park in a day (y) and the maximum daily temperature (x °C) over a period of time. The results are shown in the table:

x	21	10	17	8	16	24	6	31
y	88	31	63	20	45	72	18	79

  a Calculate the Spearman's rank correlation coefficient for the data.
  b What does this value tell you about the relationship between the temperature and the number of people visiting the park?

12 The time taken to build a wardrobe (y minutes) is thought to depend on the number of people working on the task (x people).

x	1	2	3	3	5	6	7	10
y	96	52	35	30	25	20	15	12

  a Plot a scatter graph of y against x.
  b Use your scatter graph to explain why Spearman's rank is more appropriate than Pearson's product-moment correlation to investigate the association between x and y.
  c Calculate the Spearman's rank correlation coefficient and comment on your answer.

15C Spearman's rank correlation

13 The table shows the data for height and arm length for a sample of ten 15-year-olds.

Height (cm)	154	148	151	165	154	147	172	156	168	152
Arm length (cm)	65	63	58	71	59	65	75	62	61	61

Use Spearman's rank to test whether, at the 5% significance level, there is evidence of a positive correlation between height and arm length. State your hypotheses and conclusion clearly. The critical value for this test is 0.564.

14 Daniel records the amount of time he spends playing video games (x minutes) and the number of hours of sleep (y) he gets per night. Calculate the Spearman's rank correlation coefficient for the data and interpret the result in context.

x	25	35	30	75	50	0	45	40	20
y	8	9	8.5	9.5	9	8.5	9	10	10.5

15 A group of eight pupils take part in a 400 m race. Their ages, to the nearest half-year, and their times, to the nearest second, are recorded in this table:

Age	12	14	13.5	9	9.5	11	10	12.5
Time	88	79	87	126	102	98	121	82

a Calculate the Spearman's rank correlation coefficient for the data.

b Hassan claims that older pupils always run faster. Does the data support this claim?

16 Six students obtained the following marks out of 10 on a French test (x) and a Physics test (y).

x	3	7	2	5	4	9
y	5	10	6	8	7	9

a Calculate Spearman's rank correlation coefficient for the data. Give your answer to four significant figures.

b Test, using a 5% significance level, whether there is any correlation between the two sets of marks. The critical values are 0.829 for a one-tailed test and 0.886 for a two-tailed test.

17 The table shows the number of students taking Mathematics HL and History HL in seven different schools.

Mathematics HL	14	13	15	19	19	22	23
History HL	14	10	12	11	9	9	10

Use Spearman's rank to test, at the 10% significance level, whether there is a negative correlation between the number of students taking the two subjects. The critical value for this test is 0.571.

18 Five students were comparing their grades as awarded by two teachers on a scale from A to F:

Student	Alison	Bart	Chad	Dev	Ejam
Mr Wu	C	C	A	B	B
Miss Stevens	D	B	A	A	B

a Find the Spearman's rank correlation coefficient between the grades of Mr Wu and Miss Stevens.

The critical value at the 10% significance level is 0.9 for a two-tailed test and 0.7 for a one-tailed test.

b Determine if there is significant evidence of an association between Mr Wu's grades and Miss Stevens' grades.

c Determine if there is significant evidence that Mr Wu and Miss Stevens tend to agree.

19 Two judges ranked the eight competitors in a gymnastics competition. The competitors are labelled A to H and the two judges' ranks were as follows:

Rank	1	2	3	4	5	6	7	8
Judge 1	A	F	E	H	B	C	D	G
Judge 2	F	H	A	C	E	D	B	G

a Calculate the Spearman's rank correlation coefficient for the data.

b Test, using a 10% significance level, whether there is any agreement between the two judges. The critical value for this test is 0.524.

20 The volume of oxygen produced (V millilitres) when n cubes of liver are dropped into a fixed amount of hydrogen peroxide is shown opposite:

The Pearson's product-moment correlation coefficient (r) was found to be 0.760. The Spearman's rank correlation coefficient (r_s) was found to be 0.812. However, the plot marked in red was found to be misrecorded.

The new data is shown below, with the new position of the red data point.

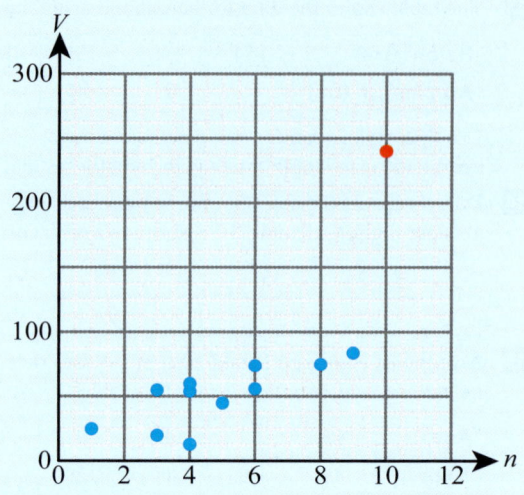

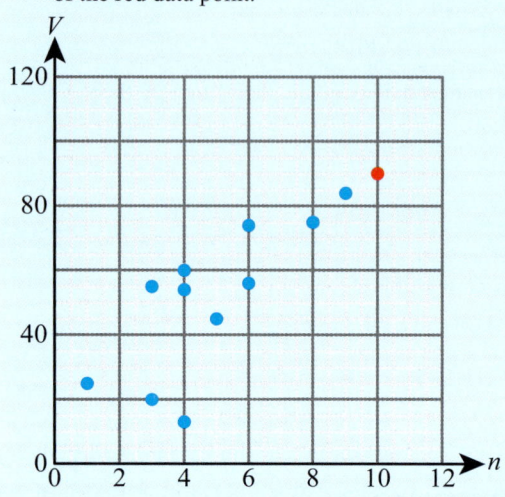

Without calculating the new value, state how the new value of

a r

b r_s

will compare with the value found originally.

c The units of V are converted into centilitres. 1 centilitre equals 10 millilitres. How does this affect the value of r_s?

Checklist

- You should know a hypothesis test can answer a question of the form 'Does the sample provide sufficient evidence that some aspect of the population distribution has changed?'

- You should know the null hypothesis (H_0) is the default position, assumed to be true unless there is significant evidence against it. The alternative hypothesis (H_1) specifies how you think the position may have changed.

- You should know that in a one-tailed test, the alternative hypothesis specifies whether a population parameter has increased or decreased. In a two-tailed test, the alternative hypothesis just states that it has changed.

- You should know that a test statistic is a value calculated from the sample and used to decide whether there is sufficient evidence against H_0. This is done by comparing it to a critical value. The exact conclusion depends on the type of test and on the alternative hypothesis.

- You should know that, assuming H_0 is correct, the probability of the observed value of the test statistic, or more extreme, is called the p-value.

- You should know the significance level of a hypothesis test specifies the probability that is sufficiently small to provide significant evidence against H_0.
    - If the p-value is smaller than the significance level, there is sufficient evidence to reject H_0 in favour in H_1.
    - Otherwise, there is insufficient evidence to reject H_0.

Mixed Practice

- You should know the χ^2 test for goodness of fit is used to test whether the data come from a prescribed distribution.
  - The null hypothesis is that the data does come from the prescribed distribution.
  - The test statistic, called the χ^2 value, is calculated from observed and expected frequencies in a (grouped) frequency table.
  - For a table with n groups, the number of degrees of freedom is $n - 1$.
  - If the χ^2 value is greater than the critical value, there is sufficient evidence to reject H_0.

- You should know the χ^2 test for independence uses data recorded in a contingency table to test whether two variables are independent.
  - The null hypothesis is that the two variables are independent.
  - For an $m \times n$ table, the number of degrees of freedom is $(m - 1)(n - 1)$.

- You should know *t*-tests can be used to test whether a population mean is different from an assumed value, or whether means of two populations are different.
  - A *t*-test is only valid if the underlying population distributions are normal.
  - When comparing two population means, it must also be assumed that the two populations have equal variance, which is estimated from the sample (pooled variance).
  - The null hypothesis is that the population means are equal.

- You should know Spearman's rank correlation coefficient (r_s) is a measure of (possibly non-linear) correlation.
  - r_s is the Pearson's product-moment correlation coefficient between the ranks of the data values.
  - If several data points have the same value, each is assigned the average of their ranks.
  - In a hypothesis test for correlation, the null hypothesis is that there is no correlation.
  - If $|r_s|$ is greater than the critical value, there is sufficient evidence to reject H_0.

- You should know Pearson's product-moment correlation coefficient only measures the strength of linear correlation. Spearman's rank coefficient also detects non-linear correlation and is less affected by outliers.

Mixed Practice

1 The following data shows the amount of oxygen produced (V ml) by a plant in an hour at different levels of carbon dioxide ($c\%$).

c	0.01	0.02	0.03	0.04	0.05	0.06	0.07
V	5.1	10.3	14.8	19.6	20.8	22.3	22.4

   a Find the Pearson's product-moment correlation coefficient of this data
   b Find the Spearman's rank correlation coefficient of this data.
   c What do your answers to **a** and **b** tell you about this data?

2 Juan plays a quiz game. The scores he achieves on the separate topics may be modelled by independent normal distributions.
   a On the topic of sport, the scores have the distribution $N(75, 12^2)$.
   Find the probability that Juan scores less than 57 points on the topic of sport.
   b Juan claims that he scores better in current affairs than in sport. He achieves the following scores on current affairs in 10 separate quizzes.
   91 84 75 92 88 71 83 90 85 78
   Perform a hypothesis test at the 5% significance level to decide whether there is evidence to support his claim.

   Mathematics HL November 2007 Paper 3 Q1, adapted (part b removed)

3 A six-sided dice is thrown 300 times and the outcomes recorded in the following table.

Score	1	2	3	4	5	6
Frequency	45	57	51	56	47	44

Perform a suitable test at the 5% level to determine if the dice is fair.

Mathematics HL November 2005 Paper 2 Q6 part (iv)

4 A six-sided dice is rolled 300 times and the following results are recorded:

Outcome	1	2	3	4	5	6
Frequency	42	38	55	61	46	58

A test is conducted to check whether the dice is biased.
a State suitable null and alternative hypotheses.
b Find the expected frequencies.
c State the number of degrees of freedom.
d State the conclusion of the test at the 10% level of significance and justify your answer.

5 Su-Yong is investigating the relationship between hours of sunshine per day and average daily temperature (in °C) in a number of different locations. She collects the following data:

Location	A	B	C	D	E	F	G
Hours of sunshine	7.3	8.7	8.1	5.9	4.3	7.2	6.5
Average temperature	11.4	18.2	12.7	9.8	10.0	17.5	11.2

a Calculate the Spearman's rank correlation coefficient for this data.
b Su-Yong thinks that there is a positive correlation between the hours of sunshine and temperature. Write down suitable hypotheses she could use to test this.
c The critical value for this test is 0.571. What should Su-Yong conclude?

6 A train company claims that times for a particular journey are normally distributed with mean 17.5 minutes. Lenka makes this journey to school every morning and thinks that they take longer on average. She records the times taken on eight randomly selected days, correct to the nearest half-minute:

17.5, 16.5, 19.0, 21.5, 16.0 20.0, 18.0, 19.5

Stating suitable hypotheses, conduct a *t*-test with the 10% level of significance to test Lenka's belief.

7 A fruit grower knows that the average yield of his apple trees is 16 bushels. He decides to stop using artificial fertilizers and wants to test whether this has led to decreased yield. The following year, the yields from 10 randomly selected trees had mean 14.7 bushels and the estimate of population standard deviation was 2.3 bushels. He conducts a *t*-test, using a 5% level of significance, to test whether there has been a decrease in average yield.
a State suitable hypotheses for this test.
b State the conclusion the fruit grower should draw.

8 A teacher works at two different schools and wants to find out whether the grades his students get depend on the school. She has the following data:

	3	4 or 5	6 or 7
School A	11	44	36
School B	16	83	40

The teachers decides to conduct a χ^2 test for independence, using the 5% level of significance.
a Write down suitable hypotheses for this test.
b Calculate the *p*-value and hence state the conclusion.

Mixed Practice

9 Francisco and his friends want to test whether performance in running 400 metres improves if they follow a particular training schedule. The competitors are tested before and after the training schedule.

The times taken to run 400 metres, in seconds, before and after training are shown in the following table.

Competitor	A	B	C	D	E
Time before training	75	74	60	69	69
Time after training	73	69	55	72	65

Apply an appropriate test at the 1% significance level to decide whether the training schedule improves competitors' times, stating clearly the null and alternative hypotheses. (It may be assumed that the distributions of the times before and after training are normal.)

Mathematics HL November 2013 Paper 3 Q4

10 A toy manufacturer makes a cubical dice with the numbers 1, 2, 3, 4, 5, 6 respectively marked on the six faces. The manufacturer claims that, when it is thrown, the probability distribution of the score X obtained is given by

$P(X = x) = \frac{x}{21}$ for $x = 1, 2, 3, 4, 5, 6$.

To check this claim, Pierre throws the dice 420 times with the following results.

x	Frequency
1	25
2	46
3	64
4	82
5	99
6	104

State suitable hypotheses and, using an appropriate test, determine whether or not the manufacturer's claim can be accepted at the 5% significance level.

Mathematics HL November 2006 Paper 3 Q3

11 A baker claims that his loaves of bread weigh 800g on average. A customer believes that the average weight is less than this. A random sample of ten loaves is weighed, with the following results (in grams):
803, 785, 780, 800, 801, 791, 783, 781, 807, 783
a Find the mean of the sample.
In spite of these results, the baker still insists that his claim is correct.
b Test this claim at the 10% significance level, stating your hypotheses and conclusion clearly. You may assume that the data comes from a normal distribution.

12 Hilary keeps two chickens. She wants to determine whether the eggs laid by the two chickens have the same average mass. She weighs the next ten eggs laid by each chicken.
a State a possible problem with this sampling technique.
The masses of the ten eggs (in grams) are:

| Chicken A | 53.1 | 52.7 | 56.3 | 51.2 | 53.7 | 50.9 | 51.2 | 55.1 | 52.8 | 52.0 |
| Chicken B | 56.3 | 54.1 | 52.7 | 50.2 | 51.4 | 55.0 | 50.9 | 51.3 | 54.9 | 56.2 |

b Write down suitable hypotheses.
c State two assumptions you need to make in order to use a t-test.
d Conduct the test at the 5% significance level and state your conclusion clearly.

13 The scores on a Maths test and a Geography test for a group of ten students are as follows.

Student	A	B	C	D	E	G	H	I	J	K
Maths	34	36	44	49	50	58	59	60	60	60
Geography	34	33	35	39	25	40	32	32	28	26

a Calculate the Spearman's rank correlation coefficient between the two sets of data.

A teacher wants to test whether there is any correlation between the two sets of scores.

b Write down suitable null and alternative hypotheses.

c Using the 5% level of significance, the critical value for a one-tailed test is 0.418 and for a two-tailed test is it 0.564. What should the teacher conclude? Justify your answer.

14 A manufacturer of batteries wants to test whether there is any difference in average lifetime of two types of battery. They test a sample of batteries, obtaining the following results (lifetime in hours).

	Sample size	Sample mean	Estimate of population variance
Type A	18	1286	110,403
Type B	11	1064	130,825

a Use a suitable test, with a 10% significance level, to determine whether there is any difference in mean lifetimes.

b State two assumptions you need to make in order to use your test.

15 Roy is practising archery. He thinks that he has the probability 0.45 of hitting the target.
To test this, he takes five shots at a time and records how many times he hit the target.
He repeats this 100 times, obtaining the following results.

Outcome	0	1	2	3	4 or 5
Observed	12	18	34	22	14

a Assuming Roy is correct, write down the distribution of the number of hits out of 5 shots.

b Hence find the expected frequencies.

c Write down the number of degrees of freedom for a χ^2 goodness of fit test.

d Test using the 5% level of significance whether Roy's belief is correct.

16 Ana tosses seven coins and counts the number of tails. She repeats the experiment 800 times and uses the results to test whether the coins are fair. The results are shown in the following table.

Number of tails	0	1	2	3	4	5	6	7
Frequency	12	34	151	218	223	126	32	4

a State suitable hypotheses for a χ^2 test.

b Calculate expected frequencies, giving your results to two decimal places.

c Find the *p*-value and the value of the χ^2 statistic.

d State the conclusion of the test at the 2% significance level.

17 Collectable cards come in packs of five. There are a number of special 'shiny cards' which are particularly sought after. The manufacturer claims that one-fifth of cards are 'shiny' and they are distributed at random among the packets. Anwar selected a random sample of 200 packs, and found the following results:

Number of shiny cards	0	1	2	3, 4 or 5
Frequency	70	80	40	10

a Use this information to estimate the average number of shiny cards in a pack. What does this tell you about the manufacturer's claim?

b Test the manufacturer's claim at the 5% significance level. State
   i the null and alternative hypotheses
   ii the value of the test statistic
   iii the *p*-value
   iv the conclusion of the test, in context.

Mixed Practice

18 Jamie wants to know if there is a tendency for people's 100 m times to decrease with age. He tracks his own best times each year:

Age	100 m time
10	16.2
11	14.8
12	13.9
13	14.1
14	14.4
15	13.6
16	13.4

a Assuming that the differences between times each year follow a normal distribution, is there evidence at the 5% significance level that the average change each year is negative?
b If the difference between times each year does not follow a normal distribution, the same question can be answered using Spearman's rank correlation coefficient.
   i State the null and alternative hypotheses.
   ii Find the value of Spearman's rank correlation coefficient.
   iii Given that the appropriate critical value at 5% significance is 0.714, what is the appropriate conclusion? Give your answer in context.

19 An examiner claims that test scores (out of 100) follow a normal distribution with mean 62 and variance 144. A teacher believes that the mean is lower. She looks at a sample of 80 scores and finds that the sample mean is 60.4 and the estimate of population variance is 144.2.
a Conduct a suitable test to show that, at the 5% level of significance, there is insufficient evidence for the teacher's belief.

A student reminds the teacher that, for her test to be valid, the population distribution of the scores must be normal. The 80 scores in the sample were distributed as follows:

Score (s)	$\leqslant 45$	$45 < s \leqslant 55$	$55 < s \leqslant 65$	$65 < s \leqslant 75$	$s > 75$
Frequency	11	23	22	18	6

b Show that, at the 5% level of significance, there is sufficient evidence that the scores do not come from the distribution N(62, 144).
c The examiner says that the above test shows that the scores are not distributed normally. The teacher says that there are other possible conclusions. State one possible alternative conclusion.
d State another reason, unrelated to the tests above, why a normal distribution may not be a suitable model for the scores.

20 The following table shows the results of a survey into student satisfaction in a school. The students were asked to give a satisfaction score on a scale from 1 to 6:

Score	1	2	3	4	5	6
Frequency	25	30	32	27	34	32

Jane wants to ask various questions about this data, using a 10% significance level. She remembers the following fact:

The critical values of Spearman's rank with this sample size are 0.829 and 0.657 for the one-tailed and two-tailed tests.

But she can't remember which critical value corresponds to which type of test.

Stating any necessary assumptions for each test, determine at the 10% significance level:
a if the mean score has changed from the previous year's average of 3.3
b if there is a tendency for higher scores to be more popular
c if the students were choosing numbers at random (i.e. each score is equally likely).

16 Applications and interpretation: Calculus

ESSENTIAL UNDERSTANDINGS

- Calculus describes rates of change between two variables and accumulation of limiting areas.
- Calculus helps us understand the behaviour of functions and allows us to interpret features of their graphs.

In this chapter you will learn…
- how to identify the points where the gradient of a curve is zero
- how to use technology to find the points with zero gradient
- how to find local maximum and minimum points on a curve
- how to solve optimization problems in context
- how to approximate areas using the trapezoidal rule.

CONCEPTS

The following key concepts will be addressed in this chapter:
- Numerical integration can be used to **approximate** areas in the physical world.
- Optimization of a function allows us to find the largest or smallest value that a function can take **in general** and can be applied to a specific set of conditions to solve problems.
- Stationary points identify maximum and minimum points to help solve optimization problems.
- The area under a function on a graph has a meaning and has applications in **space** and time.

■ **Figure 16.1** What factors affect the profits of a company, the yield of a harvest, our fitness and our happiness?

Applications and interpretation: Calculus

> **PRIOR KNOWLEDGE**
>
> Before starting this chapter, you should already be able to complete the following:
> 1. If $f(x) = x^2 + 2$, find $f'(x)$.
> 2. Find the gradient of the curve $y = x^3$ at $x = 1$.
> 3. Find the area enclosed by the graph of $y = x^3 + 2$, the x-axis and the lines $x = 1$ and $x = 2$.

You already know how to differentiate some simple functions, and how to use differentiation to find gradients and rates of change. In this chapter you will apply differentiation to find maximum and minimum values of functions and to find optimal solutions to practical problems.

You also know about the link between integration and finding areas. You will now learn about a method to estimate an area in situations where the required integral cannot be evaluated exactly.

> **Starter Activity**
>
> Look at the pictures in Figure 16.1. Think about what they represent and discuss the following question: What are the variables you might consider changing if you wanted to achieve an optimal result in each of these situations? Sketch graphs to show how the outcomes might vary depending on the different variables you have considered.
>
> **Now look at this problem:**
>
> The area of a rectangle is $60\,\text{cm}^2$. What is the smallest possible perimeter it can have?
>
> **LEARNER PROFILE – Communicators**
> Is mathematics more about getting the right answer or communicating that your answer must be right? How does your mathematical communication vary depending upon the person reading the solution?

TOK Links

Calculus is a good tool for optimizing things that can be measured, but can all objectives be measured. Can you measure health, intelligence or happiness?

16A Maximum and minimum points

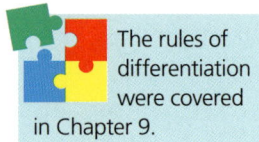
The rules of differentiation were covered in Chapter 9.

Values of x where the gradient of a curve is zero

Remember that differentiation gives you the gradient of a tangent to a curve. At the points where the gradient is zero, the tangent to the curve is horizontal.

WORKED EXAMPLE 16.1

The curve $y = x^2 - ax$ has a horizontal tangent at $x = 2$. Find the value of a.

Use differentiation to find the gradient of the tangent
$$\frac{dy}{dx} = 2x - a$$

You want the gradient at the point where $x = 2$
$$= 2(2) - a$$
$$= 4 - a$$

A horizontal line has gradient 0
$$4 - a = 0$$
$$a = 4$$

Solving $f'(x) = 0$

Given a function $f(x)$, you can use technology to produce the graph of the derivative function, $f'(x)$. This is a graph showing the value of the gradient at each point. You can then use this graph to find the point where the gradient is zero.

WORKED EXAMPLE 16.2

If $f(x) = x^3 - 6x^2 + 9x + 2$, solve $f'(x) = 0$.

Use your GDC to draw the graph of $f'(x)$

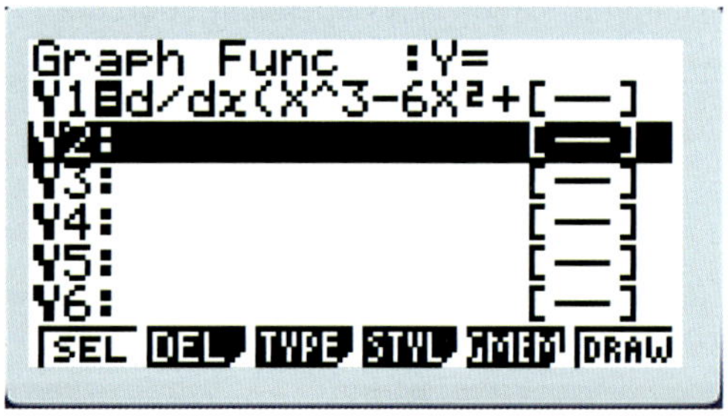

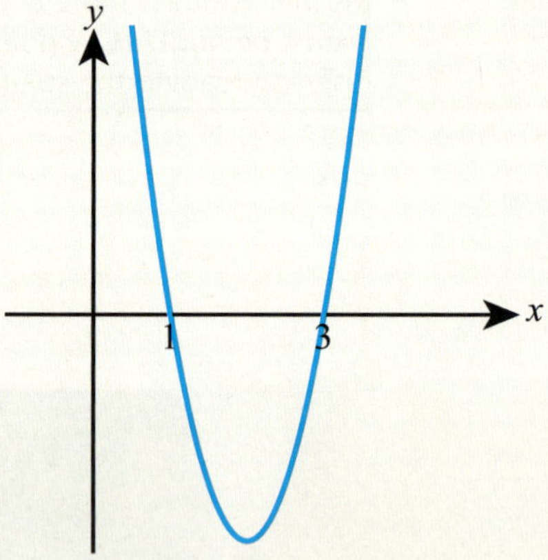

Sketch the graph as evidence of your working

Find the zeroes of this graph

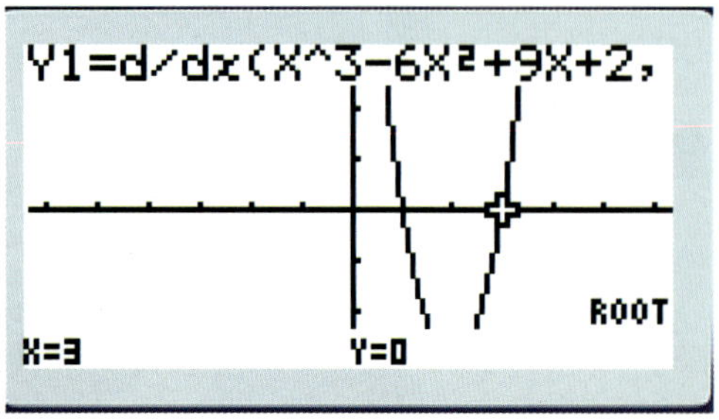

$x = 1$ or 3

Maximum and minimum points

Tip

You could also have done the question in Worked Example 16.2 by finding the expression for f′(x) which is $3x^2 - 12x + 9$ and solving $3x^2 - 12x + 9 = 0$ using the quadratic equation solver on your calculator.

In many real-life problems, you are interested in the maximum or minimum values that a certain function can take.

In the diagram below, points A and C are **local maximum points**, and B and D are **local minimum points**.

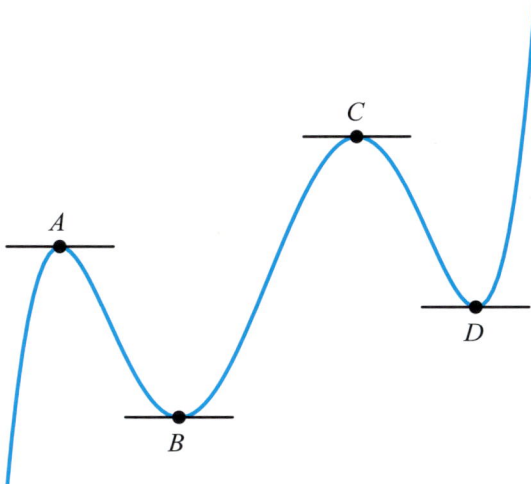

As the diagram suggests, the tangent to the graph at any local maximum or minimum point is horizontal.

CONCEPTS – GENERALIZATION

The term *local* means that these points have larger (or smaller) values of the function than other points near them. **In general**, there may be even larger (or smaller) values elsewhere on the graph. For example, A is a local maximum point even though C has a larger value. In any practical situation, you need to consider which of the points are relevant in the given context.

KEY POINT 16.1

If the graph of $y = f(x)$ has a local maximum or local minimum point at $x = a$ then $f'(a) = 0$.

Tip

The notation $n \in \mathbb{N}$ means that n is a natural number, that is, a positive integer or zero.

 **TOOLKIT: Problem Solving**

Just because $f'(a) = 0$, does not necessarily mean $y = f(x)$ has a local minimum or maximum at $x = a$. By investigating functions of the form $y = x^n$ where $n \in \mathbb{N}$ suggest some other possible shapes that might occur when $f'(x) = 0$.

 You can usually find local maximum and minimum points by using the graph on your calculator. You should be aware that the largest (or smallest) value of the function could be at one of the endpoints of the domain.

KEY POINT 16.2

The largest (or smallest) value of a function occurs either at one of the local maximum (or minimum) points or at one of the endpoints of the domain.

WORKED EXAMPLE 16.3

a Find the local maximum point on the curve $y = 2x^3 - x^2 + 1$ for $-1 < x < 1$.

b Find the maximum value on the curve in this interval.

Your calculator can find local maximum and minimum points. Make sure your window only shows $-1 < x < 1$ ········ **a**

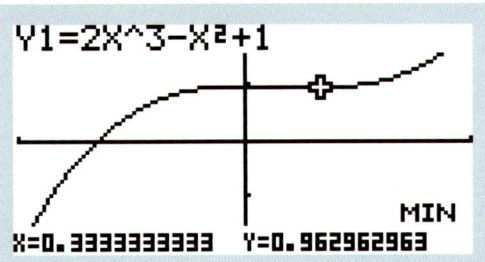

Sketch the graph as evidence of your method

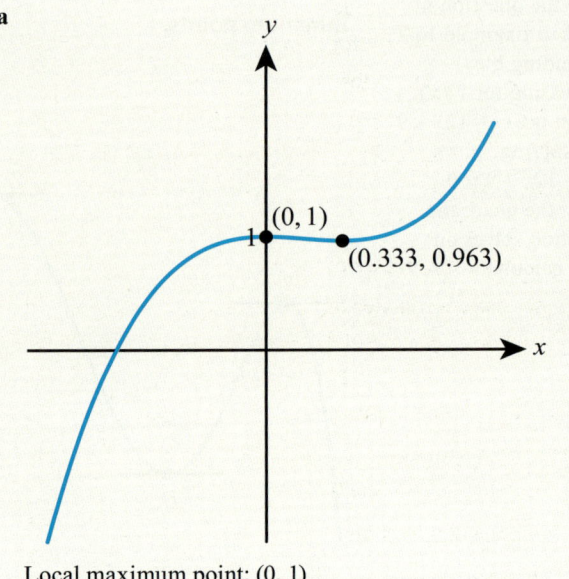

You should give both x and y coordinates ········ Local maximum point: $(0, 1)$
Local minimum point: $(0.333, 0.963)$

You can see from the graph that the largest value of y, for x between -1 and 1, is when $x = 1$ ········ **b** The largest value is $y = 2$ (when $x = 1$).

Tip

Maximum and minimum points may not show clearly on the graph, but the calculator can still find them. You can zoom in to see all the features clearly.

Exercise 16A

For questions 1 to 4, use the method demonstrated in Worked Example 16.1 to find the value of a such that the given curve has a horizontal tangent at the given point.

1 a $y = x^2 - ax$, $x = 4$
   b $y = x^2 - ax$, $x = 5$

2 a $y = x^3 - ax^2$, $x = -4$
   b $y = x^3 - ax^2$, $x = -3$

3 a $y = \dfrac{3}{x} + ax^2$, $x = 3$
   b $y = \dfrac{4}{x} + ax^2$, $x = 2$

4 a $y = \dfrac{a}{x^2} - 3x$, $x = 2$
   b $y = \dfrac{a}{x^2} - 4x$, $x = 2$

For questions 5 to 7, use the method demonstrated in Worked Example 16.2 to solve the equation $f'(x) = 0$.

5 a $f(x) = x^3 - 4x^2 + 5$
   b $f(x) = x^3 - 9x + 2$

6 a $f(x) = 3x^4 - 3x^2 + x + 1$
   b $f(x) = 2x^4 + x^2 - 3x + 4$

7 a $f(x) = 3x + \dfrac{2}{x^2}$
   b $f(x) = \dfrac{2}{x} + 5x$

In questions 8 to 11, use the method demonstrated in Worked Example 16.3 to find
  i   any local maximum and minimum points on the curve
  ii the maximum value on the curve.

8   a $y = 2x^3 - 2x + 1$ for $-1 \leqslant x \leqslant 1$
    b $y = x^3 - 4x^2 + 5$ for $-1 \leqslant x \leqslant 3$
9   a $y = 4x^3 + x^2 - 3$ for $-0.5 \leqslant x \leqslant 0.5$
    b $y = -x^3 + x^2 - 1$ for $-1 \leqslant x \leqslant 1$
10 a $y = x^4 - 5x^2 + 1$ for $-1 \leqslant x \leqslant 2$
    b $y = x^2 - x^4 + 1$ for $-1 \leqslant x \leqslant 1$
11 a $y = x^4 - x + 2$ for $-1 \leqslant x \leqslant 1$
    b $y = x^4 - x^3 + 2$ for $-1 \leqslant x \leqslant 1$
12 The curve $y = x^4 + bx^2 + c$ has a horizontal tangent at $(1, 2)$. Find the values of b and c.
13 The tangent to the curve $y = x^4 + bx + c$ at the point $(1, -2)$ is horizontal. Find the values of b and c.
14 Given that $f(x) = x^4 - 3x - 1$, solve the equation
    a $f(x) = 0$ b $f'(x) = 0$
15 Find the coordinates of the points on the graph of $y = 4x^3 - x^4$ where the tangent is horizontal.
16 Find the coordinates of the local maximum point on the graph of $y = \dfrac{2}{x^3} - 3x^2$.
17 A function is defined by $f(x) = 2x^3 + 5x^2 - 2x + 8$.
    a   Find the coordinates of the local minimum point on the curve with equation $y = f(x)$.
    b   Find the smallest value of the function on the domain $-4 \leqslant x \leqslant 1$.
18 Find the greatest value of the function $f(x) = 3x^2 + x^3 - 0.3x^5$ for $-3 \leqslant x \leqslant 3$.
19 Find the range of the function $f(x) = 3x^2 + \dfrac{11}{x}$ defined on the domain $1 \leqslant x \leqslant 3$.
20 Find the range of the function $f(x) = 3x^4 - 4x^3 - 12x^2 + 2$.
21 The graph of $y = ax^3 - bx^2 - 4x$ has a horizontal tangent at $(2, -12)$. Find the values of a and b.
22 The graph of $y = ax^2 - \dfrac{b}{x}$ has a horizontal tangent at $(-1, 9)$. Find the values of a and b.
23 The graph of $y = ax^5 + 4x^2 - bx$ has a local minimum point at $(1, -8)$. Find the coordinates of the local maximum point on the graph.

16B Optimization

WORKED EXAMPLE 16.4

The profit, P million dollars, that a company makes if it spends a fraction, f, of its income on advertising is modelled by
$$P = 2f^2(1 - f)$$
Find the fraction the company should spend if it is to maximize its profit.

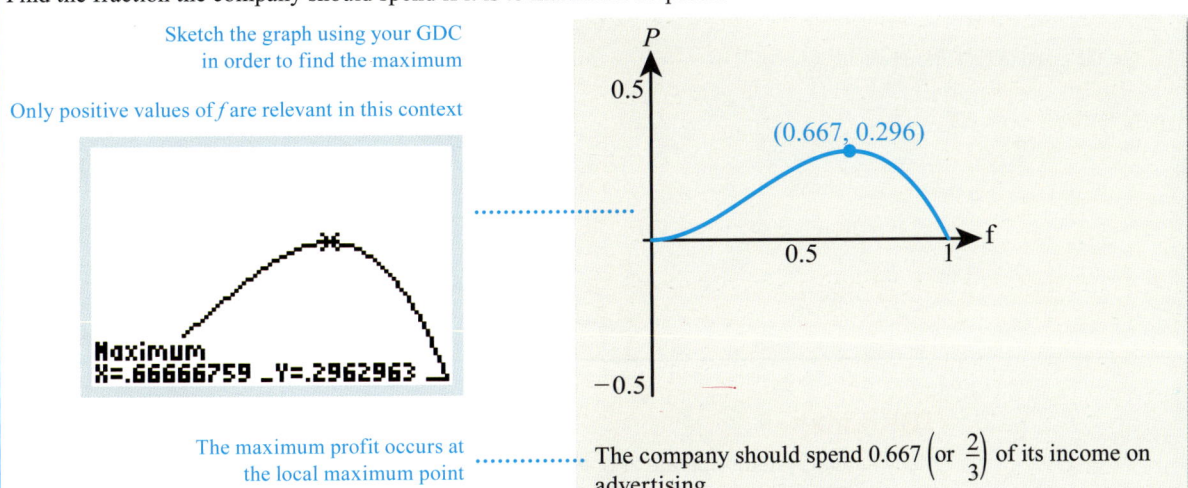

Sketch the graph using your GDC in order to find the maximum

Only positive values of f are relevant in this context

The maximum profit occurs at the local maximum point

The company should spend 0.667 $\left(\text{or } \dfrac{2}{3}\right)$ of its income on advertising.

Sometimes you need to choose appropriate variables and write an expression that you want to find the maximum or minimum value for. This may involve combining two equations into one.

WORKED EXAMPLE 16.5

Find the maximum volume of a square-based cuboid with surface area 600 cm².

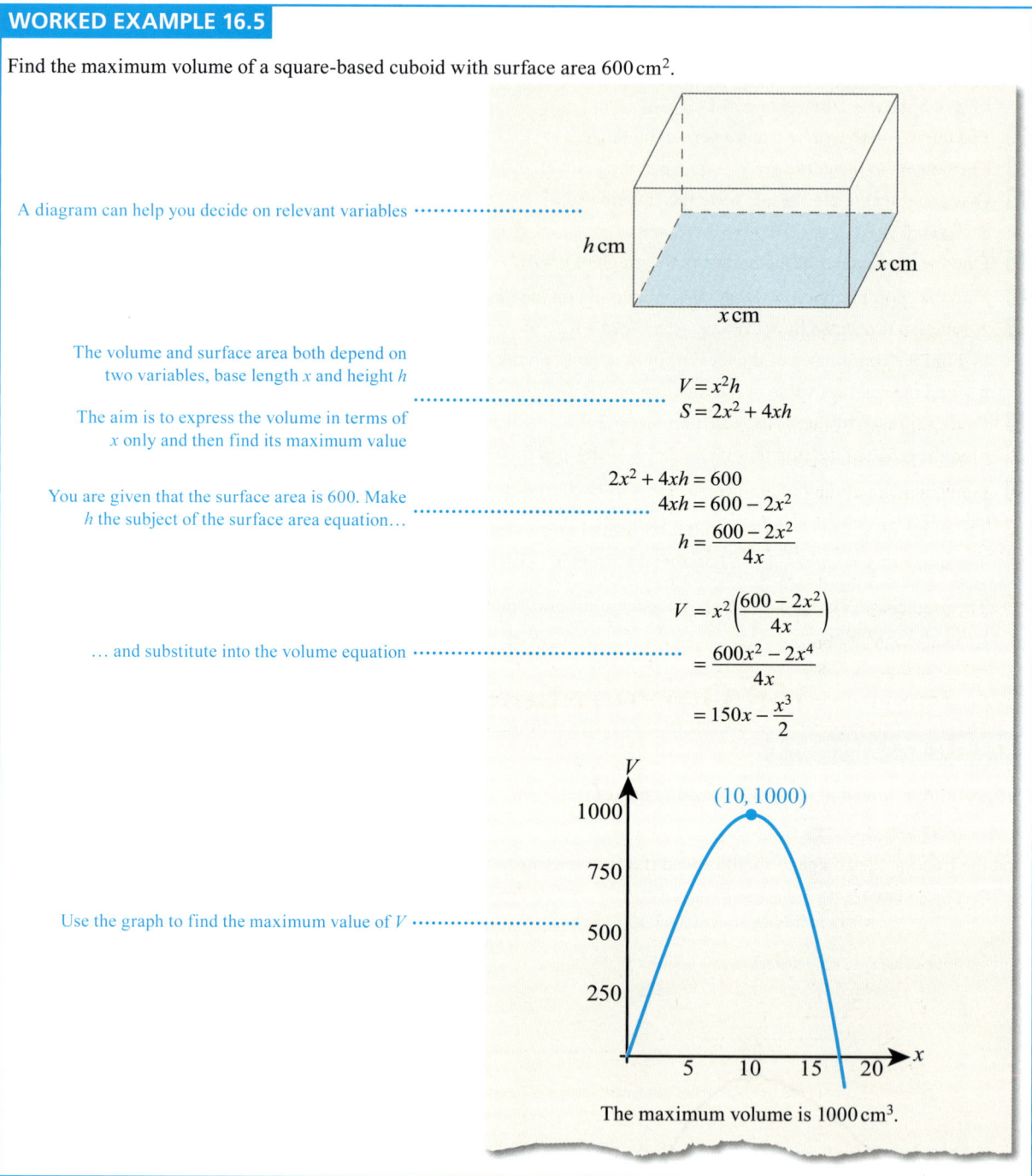

A diagram can help you decide on relevant variables

The volume and surface area both depend on two variables, base length x and height h

$$V = x^2 h$$
$$S = 2x^2 + 4xh$$

The aim is to express the volume in terms of x only and then find its maximum value

You are given that the surface area is 600. Make h the subject of the surface area equation…

$$2x^2 + 4xh = 600$$
$$4xh = 600 - 2x^2$$
$$h = \frac{600 - 2x^2}{4x}$$

… and substitute into the volume equation

$$V = x^2 \left(\frac{600 - 2x^2}{4x}\right)$$
$$= \frac{600x^2 - 2x^4}{4x}$$
$$= 150x - \frac{x^3}{2}$$

Use the graph to find the maximum value of V

The maximum volume is 1000 cm³.

TOK Links

Many natural systems operate at or very close to theoretical maximum values – for example, the sex ratio of fig wasps. Does this mean that fig wasps have an intuitive knowledge of calculus? Does knowing abstract theory always improve the practical outcome?

Exercise 16B

1. The profit, P million, made by a business which invests x million in advertising is given by
   $P = 10x^2 - 10x^4$ for $x \geq 0$
   Find the maximum profit the company can make based on this model.

2. A manufacturer produces smartphone covers. They know that if they sell n thousand covers, they will make a profit of P hundred, and they use the model $P = 20n - 3n^2 - n^5$. Find, to the nearest dollar, the maximum profit they can make according to this model.

3. The fuel consumption of a car, F litres per 100 km, varies with the speed, v km h^{-1}, according to the equation
   $F = (3 \times 10^{-6})v^3 - (1.2 \times 10^{-4})v^2 - 0.035v + 12$
   At what speed should the car be driven in order to minimize fuel consumption?

4. The rate of growth, R, of a population of bacteria, t hours after the start of an experiment, is modelled by
   $R = \dfrac{6}{t} - \dfrac{47}{t^4}$ for $t \geq 2$. Find the time when the population growth is the fastest.

5. A rectangle has width x cm and length $20 - x$ cm.
   a Find the perimeter of the rectangle.
   b Find the maximum possible area of the rectangle.

6. A rectangle has sides $3x$ cm and $\dfrac{14}{x}$ cm.
   a Find the area of the rectangle.
   b Find the smallest possible perimeter of the rectangle.

7. A cuboid is formed by a square base of side length x cm. The other side of the cuboid is of length $9 - x$ cm. Find the maximum possible volume of the cuboid.

8. A rectangle has area 36 cm^2. Let x cm be the length of one of the sides.
   a Express the perimeter of the rectangle in terms of x.
   b Hence find the smallest possible perimeter.

9. A farmer wants to build a rectangular enclosure, using the side wall of his barn and the other three sides created with fencing. He has 45 m of fencing. What is the largest area he can enclose?

10. An open box is made out of a rectangular piece of cardboard measuring 25 cm by 20 cm. A square of side x cm is cut out from each corner and the remaining card is folded to form the box.

    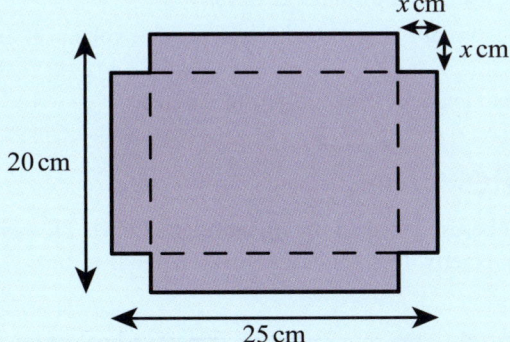

    a Show that the volume of the box is given by $V = 4x^3 - 90x^2 + 500x$.
    b Find the maximum possible volume of the box.

11. A solid cuboid has a square base of length x cm and height h cm. The volume of the cuboid is 460 cm^3.
    a Write an expression for h in terms of x, and hence find an expression for the total surface area of the cuboid in terms of x only.
    b Find the value of x for which the cuboid has the smallest possible surface area.

12. Small sugar boxes are made in the shape of a cuboid with the base measuring x cm by $2x$ cm and height h cm. The volume of each box is $225\,\text{cm}^3$.
    a. Find an expression for the surface are of the box in terms of x only.
    b. Hence find the dimensions for the box which uses the smallest amount of packaging.
13. A plastic bottle is modelled as a closed cylinder with radius r cm and height h cm. The capacity of the bottle needs to be $500\,\text{cm}^3$.
    a. Show that the surface area of the bottle is given by $S = 2\pi r^2 + \dfrac{1000}{r}$.
    b. Find the value of r and h so that the bottle uses the minimum possible amount of plastic.
    c. State one limitation of this model.
14. The surface area of a solid cylinder is $200\pi\,\text{cm}^2$.
    a. Show that the volume of the cylinder is given by $V = 100\pi r - \pi r^3$, where r cm is the radius of the cylinder.
    b. Hence find the largest possible volume of the cylinder.
15. An independent fashion designer believes that if they produce x hats in a week they will make a profit of $£\left(23 - 0.5x - \dfrac{2}{x^2}\right)$ per hat. How many hats should they produce per week in order to maximize their profit?
16. When x key rings are produced, the production cost, $\$C$, is modelled by $C = 9 - \dfrac{x}{6}$ per key ring. The key rings are then sold for $\$\left(22 - \dfrac{x}{3}\right)$ per key ring.
    a. Assuming all the key rings are sold, show that the profit, $\$P$, is given by $13x - \dfrac{x^2}{6}$.
    b. How many key rings should be produced in order to maximize the profit?
17. a. A manufacturer of microchips believes that if he sets the price of a microchip at $\$x$ he will make a profit of $\$x - 4$ per microchip and sell $1000 - 100x$ microchips each day. Based on this model, what price should he charge to maximize his total profit?
    b. An alternative model suggests that the number of microchips sold will be $\dfrac{1000}{(x+1)^2}$. Explain one advantage of this model compared to the one described in **a**.
    c. Based on the alternative model, find the price the manufacturer should charge to maximize his profit.
18. A drone of mass m kg (which is larger than 0.25) can travel at a constant speed of $4 - \dfrac{1}{m}$ metres per second for a time of $\dfrac{200}{m}$ seconds.
    a. Find an expression for the distance travelled by a drone of mass m.
    b. What mass should the manufacturer select to maximize the distance the drone can travel?
19. An open cylindrical container has total outside surface area $3850\pi\,\text{cm}^2$. Find the maximum possible volume of the container.
20. An ice cream cone has outside surface area $23\,\text{cm}^2$. Find the maximum possible volume of the cone.

16C Trapezoidal rule

You know that a definite integral can be used to find the area under a curve. However, many functions cannot be integrated exactly. In such cases the area needs to be estimated.

All estimation methods rely on splitting the area into lots of small parts whose areas can be found approximately, for example squares, rectangles or sectors of circles. In the **trapezoidal rule** the area is split into several trapezoids by using lines parallel to the y-axis.

16C Trapezoidal rule

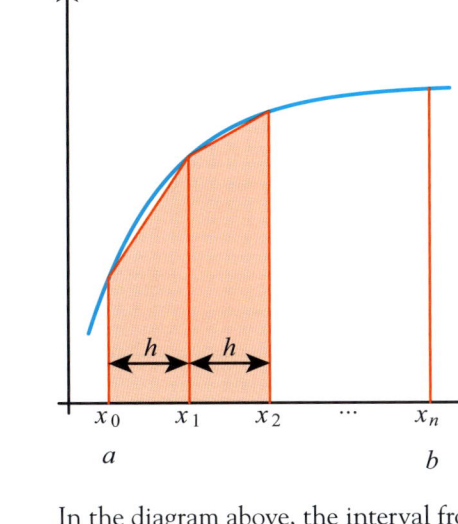

In the diagram above, the interval from $x = a$ to $x = b$ is divided into n strips. The width of each strip is $h = \dfrac{b-a}{n}$. This is the height of each of the trapezoids. The base lengths of the trapezoids are the vertical lines, which correspond to the y-coordinates.

For example, the trapezoid between x_2 and x_3 has base lengths y_2 and y_3, so its area is $\dfrac{1}{2}h(y_2 + y_3)$.

Adding the areas of all the trapezoids gives the following approximate value for the area.

> **Tip**
> The formula may look a bit complicated. If you create a table of values first, it becomes much clearer.

KEY POINT 16.3

Trapezoidal rule with n strips:
$$\int_a^b f(x)\,dx \approx \frac{1}{2}h\left[y_0 + y_n + 2(y_1 + y_2 + \ldots + y_{n-1})\right]$$
where $x_0 = a$, $x_n = b$, $h = \dfrac{b-a}{n}$ and $y_i = f(x_i)$.

> **Tip**
> You can generate the table of values on your calculator.

🌐 British English uses the term *trapezium* instead of *trapezoid* for a shape with two parallel sides. In US and Canadian English, *trapezium* is used for an irregular quadrilateral.

WORKED EXAMPLE 16.6

Use the trapezoidal rule with four strips to estimate: $\int_1^3 \dfrac{1}{x}\,dx$.

Split the interval from 1 to 3 into four equal parts and create a table of values

The width of each strip should be $h = \dfrac{3-1}{4} = 0.5$

x	1	1.5	2	2.5	3
$y = \dfrac{1}{x}$	1	$\dfrac{2}{3}$	$\dfrac{1}{2}$	$\dfrac{2}{5}$	$\dfrac{1}{3}$

Use the trapezoidal rule formula

$$\int_1^3 \frac{1}{x}\,dx \approx \frac{1}{2}(0.5)\left[\left(1 + \frac{1}{3}\right) + 2\left(\frac{2}{3} + \frac{1}{2} + \frac{2}{5}\right)\right]$$
$$= 1.12$$

CONCEPTS – APPROXIMATION

The actual value of $\int_1^3 \frac{1}{x} dx$ is $\ln 3 \approx 1.099$. Can you think of some situations where the trapezoidal rule with four strips gives a sufficiently accurate **approximation** and of some situations where you would need greater accuracy? A better approximation could be obtained by using more strips, but this would take longer to calculate. Whenever we use an approximation, there is a balance to be struck between accuracy and efficiency.

The trapezoidal rule may also be required to estimate an area when you do not actually know the equation of the boundary curve. You may only have the values of some of the y-coordinates, but that is enough to estimate the area.

WORKED EXAMPLE 16.7

A geographer measures the depth of a 5-metre-wide river at one metre intervals:

Distance from starting bank (m)	Depth (m)
0	0
1	1.3
2	1.5
3	1.9
4	0.8
5	0

Use this information to estimate the cross-sectional area of the river at this point.

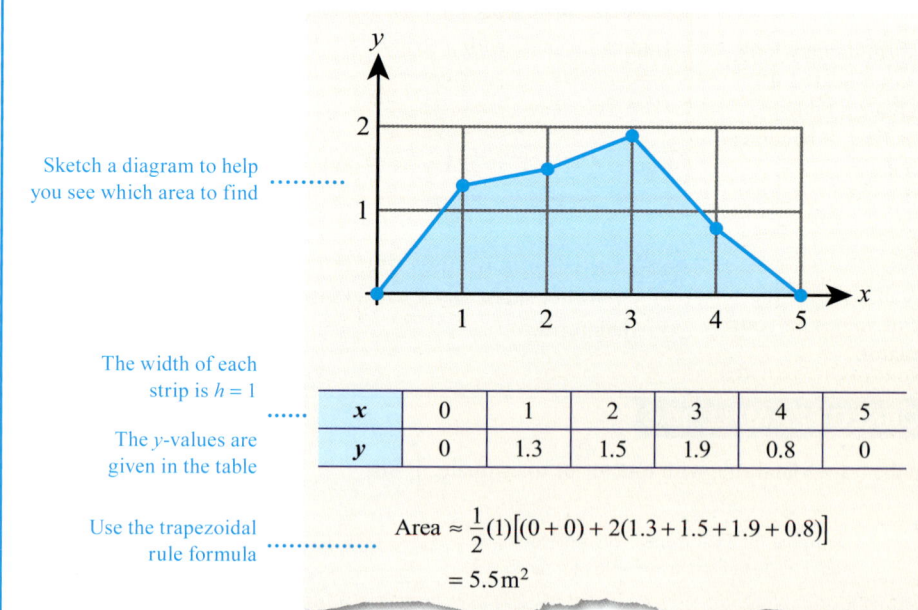

Sketch a diagram to help you see which area to find

The width of each strip is $h = 1$

The y-values are given in the table

x	0	1	2	3	4	5
y	0	1.3	1.5	1.9	0.8	0

Use the trapezoidal rule formula

$$\text{Area} \approx \frac{1}{2}(1)[(0+0) + 2(1.3+1.5+1.9+0.8)]$$
$$= 5.5 \, m^2$$

You are the Researcher

Your calculator estimates areas using more advanced versions of the trapezoidal rule. One common algorithm is called Simpson's rule. How does this work and why is it considered an improvement on the trapezoidal rule?

16C Trapezoidal rule

> Some of the earliest recorded mathematics came from ancient Mesopotamia (similar to modern Iraq), where irregular areas were calculated for the purpose of deciding how much tax to charge farmers.

Exercise 16C

For questions 1 to 4, use the method demonstrated in Worked Example 16.6 to find an approximate value of each integral.

1. a $\int_0^4 5e^x \, dx$, 4 strips   b $\int_0^5 3e^x \, dx$, 5 strips

2. a $\int_0^3 \dfrac{1}{x+2} \, dx$, 6 strips   b $\int_0^2 \dfrac{1}{x+5} \, dx$, 4 strips

3. a $\int_2^5 \ln x \, dx$, 4 strips   b $\int_2^4 \ln x \, dx$, 5 strips

4. a $\int_1^2 \ln(3x) \, dx$, 3 strips   b $\int_1^2 \ln(2x) \, dx$, 4 strips

In questions 5 to 7, you are given some coordinates on the curve $y = f(x)$. Use the trapezoidal rule, as demonstrated in Worked Example 16.7, to estimate the area between the curve and the x-axis.

5. a

x	0	1	2	3	4
y	3	3.5	4.2	3.8	3.2

  b

x	0	1	2	3	4
y	5.3	4.2	3.5	3.6	4.8

6. a

x	4	4.3	4.6	4.9	5.2
y	0	1.6	2.9	1.2	0

  b

x	2	2.4	2.8	3.2	3.6
y	0	4.1	5.7	3.8	0

7. a

x	1.2	1.4	1.6	1.8
y	7.5	6.3	5.7	4.2

  b

x	0.9	1.2	1.5	1.8
y	3	3.5	4.2	4.7

8. Use the trapezoidal rule with four strips to find an approximate value of $\int_4^8 \sqrt{x-4} \, dx$.

9. Use the trapezoidal rule with five strips to estimate the value of $\int_0^3 \sqrt{2x+1} \, dx$.

10 The diagram shows part of the graph of $y = f(x)$ and the table shows some of the values of $f(x)$.

x	0.5	1	1.5	2	2.5	3	3.5
$f(x)$	4.1	5.4	5.0	4.6	4.1	3.6	3.4

Use this information to estimate the shaded area.

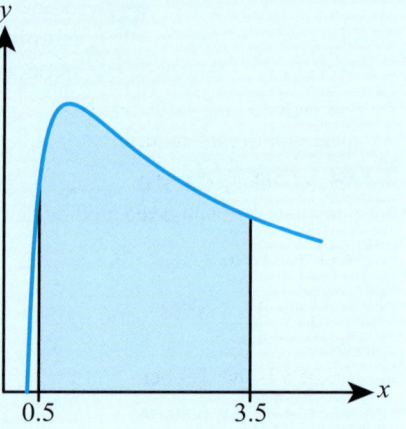

11 The entrance to a tunnel is 6 m wide. The height of the tunnel is measured at 1.5 m intervals, starting from one end, giving the following results:

Distance from one end (m)	0	1.5	3	4.5	6
Height (m)	0.3	1.8	3.2	2.1	0.3

a Estimate the cross-sectional area of the tunnel.

b Is the real cross-sectional area larger or smaller than your estimate? Explain your answer.

12 Find the percentage error when the value of $\int_0^3 (3x^2 - x^3)\, dx$ is approximated using the trapezoidal rule with five strips.

13 The width of a lake (in metres) is measured at two-metre intervals, and the results are shown in the diagram. Estimate the area of the lake.

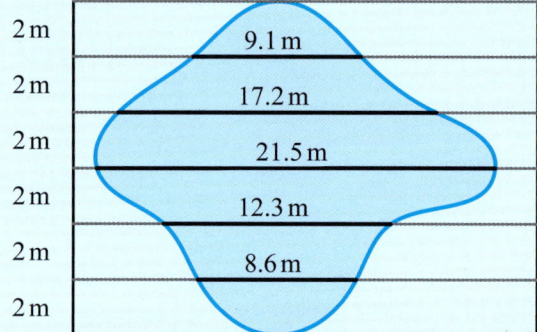

14 A forest is shown on a map, with some of the coordinates marked around the edge. The distances are given in metres.

Estimate the area of the forest.

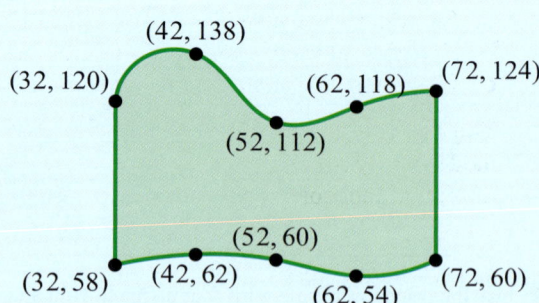

15 a Use the trapezoidal rule with five intervals to estimate the value of $\int_2^4 \sqrt{x - 2}\, dx$.

b Is your answer an underestimate or an overestimate of the actual value of the integral?

Mixed Practice

Checklist

- You should know that, if the graph of $y = f(x)$ has a local maximum or local minimum point at $x = a$, then $f'(a) = 0$.
- You should be able to use technology to sketch the graph of $f'(x)$ and find its zeros. You should also be able to find local maximum and minimum points from the graph of $f(x)$.
- You should know that the largest (or smallest) value of a function occurs either at one of the local maximum (or minimum) points, or at one of the endpoints of the domain.
- You should be able to use the trapezoidal rule to approximate areas:

  $$\int_a^b f(x)\,dx \approx \frac{1}{2}h\left[y_0 + y_n + 2(y_1 + y_2 + \ldots + y_{n-1})\right]$$

  where n is the number of strips, $x_0 = a$, $x_n = b$, $h = \frac{b-a}{n}$ and $y_i = f(x_i)$.

Mixed Practice

1 a Find the coordinates of the point A on the curve with equation $y = 3x^2 - 4x$ where the tangent is horizontal.
  b Sketch the curve and show the position of point A.

2 The graph of $y = 4x^2 - bx$ has a horizontal tangent when $x = -2$. Find the value of b.

3 A function is given by $f(x) = 4x^3 - 5x$.
  a Sketch the graph of $y = f'(x)$.
  b Hence solve the equation $f'(x) = 0$.

4 Given that $y = 4x^2 - \frac{5}{x}$, use your graphical calculator to solve the equation $\frac{dy}{dx} = 0$.

5 Find the coordinates of the local minimum point on the graph with equation $y = 4x^3 - 3x + 8$.

6 a Sketch the graph of $y = 9x^2 - x^4$ for $0 \leq x \leq 3$, showing the coordinates of the maximum and minimum points.
  b Use the trapezoidal rule with six strips to estimate the area between the graph and the x-axis.

7 Use the trapezoidal rule with five strips to find an approximate value of $\int_2^{12} \ln\left(\frac{x}{2}\right) dx$.

8 A rectangle with sides w cm and h cm has perimeter 88 cm.
  a Express h in terms of w and hence show that the area of the rectangle is given by $A = 44w - w^2$.
  b Find the values of w and h for which the rectangle has the maximum possible area.
  c Find the maximum possible area of the rectangle.

9 The fuel consumption of a car, L litres per 100 km, varies with speed, v km h^{-1} according to the model model $L = 17.3 + 0.03(x - 70) + 0.0001(x - 70)^4$. At what speed should the car be driven in order to minimize fuel consumption?

10 A shipping container is to be made with six rectangular faces, as shown in the diagram.

The dimensions of the container are
length $2x$
width x
height y.

All of the measurements are in metres.

The total length of all 12 edges is 48 metres.

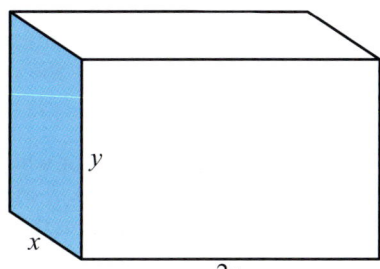

   a Show that $y = 12 - 3x$.
   b Show that the volume $V\,\text{m}^3$ of the container is given by $V = 24x^2 - 6x^3$.
   c Find $\dfrac{dV}{dx}$.
   d Find the value of x for which V is a maximum.
   e Find the maximum volume of the container.
   f Find the length and height of the container for which the volume is a maximum.

The shipping container is to be painted. One litre of paint covers an area of $15\,\text{m}^2$. Paint comes in tins containing four litres.

   g Calculate the number of tins required to paint the shipping container.

Mathematical Studies SL May 2012 TZ2 Paper 2 Q5

11 Find the coordinates of the point on the graph of $y = x^2 - \dfrac{3}{x}$ where the gradient is zero.

12 The curve with equation $y = ax^2 - 48x + b$ has a horizontal tangent at $(8, 21)$. Find the values of a and b.

13 The graph of $y = 3x^2 - \dfrac{a}{x}$ has a horizontal tangent when $x = -1$. Find the coordinates of the local minimum point on the graph.

14 a Use the trapezoidal rule with four intervals to estimate the value of $\displaystyle\int_0^4 e^{-x}\,dx$.

   b Use your calculator to evaluate the integral and hence find the percentage error in your estimate.

15 A function is defined on the domain $0 \leqslant x \leqslant 3$ by $f(x) = x^3 - 3x^2 + 2x$.
   a Find the coordinates of the local maximum point of the function.
   b Find the greatest value of the function on this domain.

16 Find the minimum value of $0.2x^4 - 3x^3 + 7.5x^2 + 1.3x + 1$ for $0 \leqslant x \leqslant 10$.

17 Metal bars can be produced either in the shape of a square based cuboid or a cylinder. The volume of the bars is fixed at $300\,\text{cm}^3$. Find the shape and the dimensions of the bar with the smallest possible surface area.

18 A parcel is in the shape of a rectangular prism, as shown in the diagram. It has a length l cm, width w cm and height of 20 cm. The total volume of the parcel is $3000\,\text{cm}^3$.
   a Express the volume of the parcel in terms of l and w.
   b Show that $l = \dfrac{150}{w}$.

The parcel is tied up using a length of string that fits **exactly** around the parcel, as shown in the diagram.

   c Show that the length of string, S cm, required to tie up the parcel can be written as

$$S = 40 + 4w + \dfrac{300}{w},\ 0 < w \leqslant 20.$$

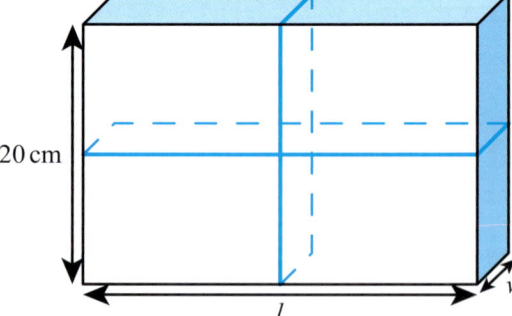

   d Draw the graph of S for $0 < w \leqslant 20$ and $0 < S \leqslant 500$, clearly showing the local minimum point. Use a scale of 2 cm to represent 5 units on the horizontal axis w (cm), and a scale of 2 cm to represent 100 units on the vertical axis S (cm).
   e Find $\dfrac{dS}{dw}$.
   f Find the value of w for which S is a minimum.
   g Write down the value, l, of the parcel for which the length of string is a minimum.
   h Find the minimum length of string required to tie up the parcel.

Mathematical Studies SL May 2014 TZ2 Paper 2 Q5

Mixed Practice

19 A dog food manufacturer has to cut production costs. They wish to use as little aluminium as possible in the construction of cylindrical cans. In the following diagram, h represents the height of the can in cm and x, the radius of the base of the can in cm. The volume of the dog food cans is $600\,\text{cm}^3$.

  a Show that $h = \dfrac{600}{\pi x^2}$.

  b i Find an expression for the curved surface area of the can, in terms of x. Simplify your answer.

  ii Hence write down an expression for A, the total surface area of the can, in terms of x.

  c Differentiate A in terms of x.

  d Find the value of x that makes A a minimum.

  e Calculate the minimum total surface area of the dog food can.

  Mathematical Studies SL May 2010 TZ2 Paper 2 Q5

20 The graph of $y = ax^3 + 2x^2 + bx + 45$ has a horizontal tangent at $(2, 5)$.
Find the second point on the graph where the tangent is horizontal.

21 The temperature in an office is monitored over a period of 24 hours. It is found that the temperature, $T\,°C$, can be modelled as a cubic function of t, the time in hours after midnight. The minimum temperature of $4\,°C$ was recorded at 2 am. The maximum temperature of $28\,°C$ was recorded at 3 pm.

  a Find the cubic model for how the temperature changes with time. Give the coefficients to two significant figures.

  b What temperature does this model predict for 11.50 pm? State one limitation of this model.

22 a is 20 to one significant figure. b is 22 to two significant figures.

The function f is defined by $f(x) = (x - 22)(26 - x)$.

Find the largest value of $f(a) - f(b)$.

23 A field is represented on a map. Several points around the edge of the field are marked and their coordinates noted:

$(7, 8), (8, 6), (9, 4), (10, 3), (11, 3), (12, 3),$
$(12, 13), (11, 12), (10, 12), (9, 11), (8, 9)$

  a Use these coordinates to estimate the area of the field on the map.

  b The units on the map are centimetres, and the map scale is 1 unit : 25 m. Find an estimate of the actual area of the field.

Applications and interpretation SL: Practice Paper 1

1 hour 30 minutes, maximum mark for the paper [80 marks]

1. A biology student measured the lengths of 200 leaves collected from the local park. The results are displayed in the cumulative frequency diagram.

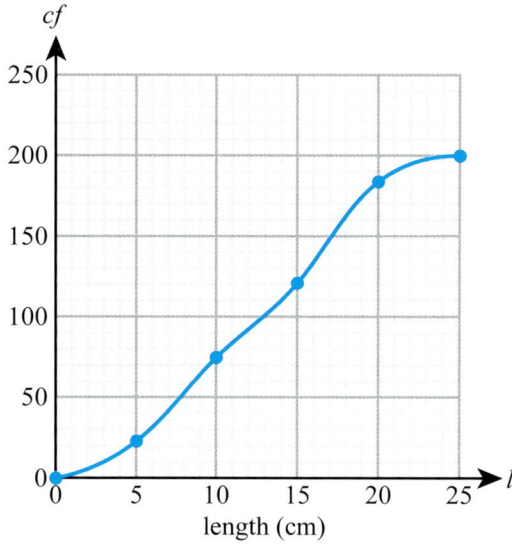

   a Estimate the median leaf length.
   b Estimate the number of leaves which were longer than 26 cm.
   c What is the probability that a randomly selected leaf is longer than 25 cm?

   [6]

2. Given $f(x) = \ln\left(\dfrac{x^2 + 1}{3}\right)$, $x \in \mathbb{R}$,

   a find the range of f
   b state the value of f(1), giving your answer to 2 decimal places
   c solve the equation $f(x) = 5$.

   [5]

3. A theatre is designed so that each row of seats has d more seats than the row in front.
   The front row has a seats.
   There are 25 rows of seats in all.
   The tenth row has 285 seats and the entire theatre contains 9000 seats.

   a Write down two equations in a and d.
   b Hence find the values of a and d.

   [5]

4 The table shows the number of students from two schools studying Maths SL and Maths HL.

	School 1	School 2
Maths SL	20	30
Maths HL	10	20

   a How many students are there in the two schools combined?
   b What is the probability that a randomly selected student is from School 1 and studies Maths HL?
   c Given that a student is from School 2, what is the probability that they study Maths SL?
   [5]

5 Sami wants to take out a loan for AUS 20 000.
   She is offered the loan on the following terms:
   - 5 year loan
   - 12% interest rate compounded monthly
   - monthly repayments.

   If she accepts this loan, find:
   a how much her monthly repayments would be
   b the total amount of interest she would pay over the lifetime of the loan.
   [6]

6 Pierre wants to investigate whether there is any association between preference for type of wine and gender.
   He samples 50 wine drinkers with the following results:

	Red	White	Rosé
Male	13	11	4
Female	6	7	9

   Pierre used these results to conduct a χ^2 test for independence at the 5% significance level.
   a State the null and alternative hypotheses.
   b Calculate the p-value.
   c State, giving a reason, what Pierre should conclude from this test.
   [6]

7 A closed box is made in the shape of a prism, height h, that has a cross section that is a sector of a circle.
   The sector has radius r and angle 60°.
   a Show that the volume is given by
   $$V = \frac{\pi r^2 h}{k}$$
   where k is a whole number to be found.

   The radius and height are measured to two significant figures and found to be
   $r = 14$ cm
   $h = 8.5$ cm
   b Find the upper and lower bounds of the volume.
   [5]

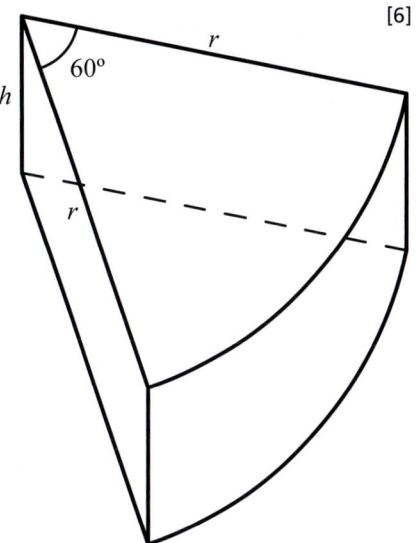

8 A particular species of tree is prone to a disease. A model of the form $D = 10 - 9.5(0.78^t)$, where D is the number of diseased trees in thousands and t is the time in years, is proposed to model the spread of the disease.

a Find the number of infected trees

  i initially

  ii after 2 years.

b Calculate the time for the number of infected trees to reach 6000.

c What does the model predict is the long-term behaviour of the disease?

[6]

9 A square-based pyramid has height 50 cm. The sloping edges make an angle of 75° with the base. Find:

a the length of each sloping edge

b the length of the diagonal of the square base

c the volume of the pyramid.

[7]

10 A scientist wants to investigate whether the mean length of mature perch in river A, μ_A, is different from the mean length of mature perch in river B, μ_B. He gathers the following data:

Length of perch in river A (cm)	Length of perch in river B (cm)
17.1	16.8
16.5	17.8
17.8	17.5
18.1	19.6
16.3	18.8
19.0	20.1
17.4	18.7
18.7	19.5
16.9	
17.0	

He carries out a t-test at the 10% level to compare the means.

a State the null and alternative hypotheses.

b Calculate the p-value.

c State, giving a reason, what the biologist should conclude from this.

d State two assumptions he needed to make to carry out this test.

[8]

11 Points $A(2, 8)$, $B(3, 2)$, $C(6, 6)$, $D(7, 3)$, $E(10, 8)$ represent the location of mobile phone masts.

Mobile phone users will automatically receive a signal from the mast they are closest to at any given time.

The following lines are drawn to start forming a Voronoi diagram:

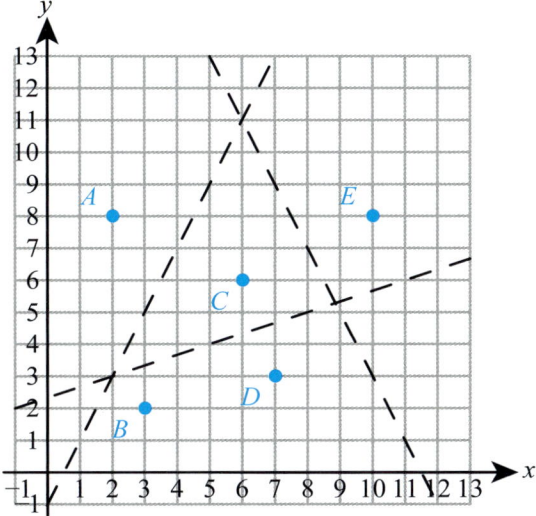

a Find the equation of the missing edge of the Voronoi cell containing C, giving your answer in the form $ax + by + c = 0$ where $a, b, c \in \mathbb{Z}$.

b Explain from which mast someone at point (4, 5) will receive a signal.

c State at which point someone must be to potentially be receiving a signal from mast A, C and D.

[6]

12 A small business owner thinks that her weekly profits, P (thousands of pounds), can be modelled by a function of the form $P = ax^2 + bx + c$ where x is the number of hundred units produced.

She knows that her maximum profit occurs when output is 450 units and that she breaks even when output is 700 units. She also knows that the business will lose £3500 in any week when there is no production.

a Find the values of a, b and c.

b Calculate her maximum weekly profit.

[7]

13 The rate of growth of a great white shark, $\frac{dl}{dt}$, is modelled as being inversely proportional to the square root of its age, t years. When a shark is four years old its growth rate is 0.14 m year^{-1}. A newborn shark is 1.2 m long.

a Find an equation relating a shark's growth rate to its age.

b Find the length of a 25-year-old shark predicted by this model.

c Explain what this model says about the growth of sharks as they get older.

[8]

Applications and interpretation: Practice Paper 2

1 hour 30 minutes, maximum mark for the paper [80 marks]

1 *[11 marks]*

The line l_1 has the equation $3x + 2y = 12$. It crosses the y-axis at the point N.

a Find:
   i   the gradient of l_1
   ii the coordinates of N.

[3]

The line l_2 passes through the point $P(3, -5)$ and is perpendicular to l_1.

b Find the equation of l_2 in the form $ax + by + c = 0$.

[3]

The point of intersection of l_1 and l_2 is Q.

c Find the coordinates of Q.

[1]

d Calculate the distances
   i   NQ
   ii PQ.

[2]

e Find the area of triangle NPQ.

[2]

2 *[13 marks]*

In an archery competition, each arrow scores either 0, 2, 4, 6, 8 or 10 points. The probability distribution of Michele's score from each arrow in the competition is shown in the table.

Score	0	2	4	6	8	10
Probability	0	0.04	0.07	p	q	0.62

a Write down an equation in p and q.

[1]

b Given that the expected value of Michele's score is 8.56, write down a second equation in p and q.

[3]

c Hence find the value of p and q.

[1]

When an archer scores 10 points they are said to have 'hit the bullseye'. Michele shoots 15 arrows in the first round.

d Find the probability that she hits the bullseye
   i   8 times
   ii at least 10 times.

[3]

e For Michele's 15 shots in the first round, find the
  i mean number of bullseyes
  ii standard deviation of the number of bullseyes.
[2]

There are 4 further such rounds in the competition.
f Find the probability that Michele hits the bullseye at least 10 times in more than 3 of the 5 rounds.
[3]

3 *[11 marks]*
The surface area of a solid cuboid, S, can be represented by the function $S(V) = kV^c$, where V is the volume of the cuboid and k and c are constants.
a Find the value of k and c.
[3]

b The domain of the function S is $0 \leqslant V \leqslant 125$.
  i Sketch the graph of S.
  ii On the same axes sketch the graph of the inverse function S^{-1}.
[4]

c State the
  i range of S^{-1}
  ii domain of S^{-1}.
[2]

d i Calculate the value of $S^{-1}(54)$.
  ii In the context of this question, explain the meaning of the value of $S^{-1}(54)$.
[2]

4 *[16 marks]*
A geographer proposes that the depth, d metres, of a river can be modelled by a function of the form $d = p \sin qx$, where x is the distance from the left bank of the river.
The river is 8 m wide and the depth halfway across the river is 2.3 m.
a Find the value of p and q.
[3]

b Calculate the depth the model predicts 2 m from the left bank.
[1]

The geographer checks the prediction by measuring the depth of the river 2 m from the left bank. He finds the depth is 2.5 m.
c Calculate the percentage error in the model's prediction.
[2]

In view of the extra measurement, the geographer decides that a model of the form $d = ax^3 + bx^2 + cx$ would be more appropriate.
d Find, to three significant figures, the values of a, b and c.
[4]

e Sketch the function found in part **d** for $0 \leqslant x \leqslant 8$.
[2]

f Use this model to find
  i the maximum depth of the river
  ii the cross-sectional area of the river.
[4]

5 *[15 marks]*

The shipping costs a business pays vary according to the size of the shipment. The following data is available:

Size of shipment, x (thousands of items)	Cost of shipment, y (thousands of euros)
76.4	6.0
102	6.9
125	8.2
130	8.5
154	9.2
180	10.6
226	11.8
253	12.3
288	12.7
315	13.2

The owner of the business believes the cost can be modelled by a linear function.

a Calculate the value of the Pearson's product-moment correlation coefficient, r.

[2]

b Explain whether this value of r supports the use of a linear model.

[2]

The owner thinks the model can be improved by making it a piecewise linear model of the form

$$y = \begin{cases} ax + b & 0 \leq x \leq 200 \\ cx + d & x > 200 \end{cases}$$

c Using linear regression of y on x, determine the values of a, b, c and d.

[5]

d Use this model to predict the cost of a shipment of

  i 90 000 items **ii** 200 000 items **iii** 330 000 items.

[3]

e Comment on the reliability of the predictions in part **d**.

[3]

6 *[14 marks]*

The point P with x-coordinate $p > 0$ lies on the curve $y = \frac{1}{x}$.

a i Find, in terms of p, the gradient of the tangent to the curve at P.

  ii Show that the equation of the tangent to the curve at P is $x + p^2 y - 2p = 0$.

[5]

The tangent to the curve at P intersects the x-axis at Q and the y-axis at R.

b i Find the coordinates of Q and R.

  ii Find the area of the triangle OQR, where O is the origin.

[4]

c Show that the distance QR is given by $2\sqrt{p^2 + \frac{1}{p^2}}$.

[3]

d Find the minimum distance QR.

[2]

Answers

Toolkit: Using technology

Computer algebra systems (CAS)
a $\sqrt{1-x^2}$
b 1

Spreadsheets
b $370\,\text{m}, 20\,\text{m s}^{-1}$

Dynamic geometry packages
a $AB = 2\,MC$
b 1:1

Programming
c Welcome to Mathematics for the International Baccalaureate. We hope that you enjoy this book and learn lots of fun maths! (This is an example of a ROT13 Caesar Shift Cipher)

Toolkit: Algebra Practice

1 a $11x + 15y$
   b $10x^2 + 7xy$
   c $2x + 3$
   d $x + 4$
   e xy^2z
   f $63x^2y$
   g $8x^3$
   h $\dfrac{x}{6}$
   i $\dfrac{1+2x}{y}$
   j $\dfrac{y}{3z}$
   k $\dfrac{x}{2}$
   l -1
   m $\dfrac{x^2}{15}$
   n $\dfrac{2(x+1)}{5}$
   o $\dfrac{2x}{y}$

2 a $2x^2 - 6x$
   b $x^2 + 6x + 9$
   c $x^2 + x - 20$
   d $x^3 + 3x^2 + 2x$
   e $x^2 + y^2 + 2xy - 1$
   f $x^3 + 9x^2 + 23x + 15$

3 a $4(3 - 2y)$
   b $3(x - 2y)$
   c $7x(x - 2)$
   d $5xy(x + 2y)$
   e $(2z + 5)(x + y)$
   f $(3x - 1)(x - 2)$

4 a $x = -\dfrac{7}{8}$
   b $x = 4$
   c $x = -8$
   d $x = 5$
   e $x < -\dfrac{13}{2}$
   f $x \geqslant \dfrac{5}{3}$

5 a $x = 6, y = 4$
   b $x = 3, y = 1$
   c $x = -1, y = 2$

6 a -2
   b 9
   c 12
   d 2
   e 12.5
   f 5

7 a $x = \dfrac{y-4}{2}$
   b $x = \dfrac{3y}{2-y}$
   c $x = \dfrac{1-2y}{y-1}$
   d $x = \dfrac{by}{y-a}$
   e $x = \pm\sqrt{a^2 + 4}$
   f $x = -\dfrac{1}{2y}$

8 a $2 - \sqrt{2}$
   b $2\sqrt{15}$
   c 4
   d $3\sqrt{3}$
   e $3 - 2\sqrt{2}$
   f -1

9 a $\sqrt{2}$
   b $\sqrt{2} + 1$
   c $\dfrac{1 + 3\sqrt{3}}{13}$

10 a $x = -3$ or -2
   b $x = \pm\dfrac{3}{2}$
   c $x = -1 \pm \sqrt{6}$

11 a $\dfrac{b-a}{ab}$
   b $\dfrac{2x+5}{x^2}$
   c $\dfrac{x^2 - x + 13}{x - 1}$
   d $\dfrac{2}{(1-x)(1+x)}$
   e $\dfrac{25 - 2a^2}{a(5-a)}$
   f $\dfrac{2(x^2 - 2)}{(x-2)(x-1)}$

Chapter 1 Prior Knowledge

1. **a** 81   **b** 40
2. **a** 3.4271×10^2 **b** 8.56×10^{-3}
3. 2^6

Exercise 1A

1. **a** x^6 **b** x^{12}
2. **a** y^6 **b** z^{10}
3. **a** a^7 **b** a^{11}
4. **a** 5^{17} **b** 2^{24}
5. **a** x **b** x^3
6. **a** y^4 **b** z^6
7. **a** b^6 **b** b^8
8. **a** 11^8 **b** 7^5
9. **a** x^{15} **b** x^{32}
10. **a** y^{16} **b** z^{25}
11. **a** c^{14} **b** c^{14}
12. **a** 3^{50} **b** 13^{28}
13. **a** $48x^7$ **b** $15x^7$
14. **a** $3a^3$ **b** $5b^4$
15. **a** $20x^5y^3z$ **b** $12x^8yz^5$
16. **a** $2x^5$ **b** $3x^6$
17. **a** $\dfrac{1}{2x^3}$ **b** $\dfrac{1}{4x}$
18. **a** $\dfrac{5}{3x^2}$ **b** $\dfrac{3x^2}{4}$
19. **a** $7x^2y^3$ **b** $\dfrac{2x^3z}{y}$
20. **a** $\dfrac{1}{10}$ **b** $\dfrac{1}{7}$
21. **a** $\dfrac{1}{27}$ **b** $\dfrac{1}{25}$
22. **a** $\dfrac{4}{3}$ **b** $\dfrac{7}{5}$
23. **a** $\dfrac{9}{4}$ **b** $\dfrac{125}{8}$
24. **a** $\dfrac{7}{9}$ **b** $\dfrac{5}{16}$
25. **a** $\dfrac{6}{x}$ **b** $\dfrac{10}{x^4}$
26. **a** $\dfrac{8}{9}$ **b** $\dfrac{81}{64}$
27. **a** $27u^{-6}$ **b** $32v^{-15}$
28. **a** $\dfrac{1}{4}a^{10}$ **b** $\dfrac{1}{27}b^{21}$
29. **a** $25x^4y^6$ **b** $81a^8b^{-8}$
30. **a** $\dfrac{x^3}{27}$ **b** $\dfrac{25}{x^4}$
31. **a** $\dfrac{27x^6}{8y^9}$ **b** $\dfrac{25u^2v^6}{49b^8}$
32. **a** $\dfrac{16v^4}{9u^2}$ **b** $\dfrac{27b^6}{8a^9}$
33. **a** $2x - 7x^6$ **b** $3y + 5y^3$
34. **a** $5u^2 + 6uv^2$ **b** $5a^2b^5c - 4ac^2$
35. **a** $5p - 3p^{-1}q^2$ **b** $2s^2t^{-2} + 3s^3t^2$
36. **a** 3^8 **b** 3^{24}
37. **a** 2^9 **b** 5^{10}
38. **a** 2^5 **b** 3^{11}
39. **a** 2^{17} **b** 2^7
40. **a** $x = 4$ **b** $x = 3$
41. **a** $x = -1$ **b** $x = 6$
42. **a** $x = \dfrac{7}{3}$ **b** $x = \dfrac{9}{2}$
43. **a** $x = -4$ **b** $x = -3$
44. $2x + 4x^2$
45. x^5y^2
46. $\dfrac{b^6}{8a^3}$
47. **a** $n = \dfrac{1000}{\sqrt{D}}$ **b** $250\,000$
    c $10 million
48. **a** $k_A = 8$, $k_B = 40$ **b** $\dfrac{n}{5}$
    c method B
49. $x = 2$
50. $x = 5$
51. $x = -1$
52. $x = -1$
53. $x = -2$
54. **a** $R = 3.2T^2$ **b** $250\,K$
55. **a** $\dfrac{3}{v}$ **b** $1.5v$
    c 40 km per hour
56. $x = 1$, $y = -3$
57. $x = 4$
58. $x = 6$
59. $x = 1$, 3 or -5
60. 2^{7000}
61. 7

Answers

Exercise 1B

1. a 32 000 b 6 920 000
2. a 0.048 b 0.000 985
3. a 6.1207×10^2 b 3.07691×10^3
4. a 3.0617×10^{-3} b 2.219×10^{-2}
5. a 6.8×10^7 b 9.6×10^{11}
6. a 1×10^0 b 1.2×10^{-5}
7. a 2.5×10^{21} b 3.6×10^{13}
8. a 2×10^{-2} b 4×10^{-4}
9. a 5×10^0 b 2.5×10^2
10. a 2.1×10^{11} b 3.1×10^9
11. a 2.1×10^5 b 3.02×10^8
12. a 7.6×10^5 b 8.91×10^{14}
13. a 4.01×10^4 b 6.13×10^{13}
20. a 1.22×10^8 b 400
    c 4×10^2
21. 6×10^{-57}
22. 1.99×10^{-23} g
23. a 1.5×10^{-14} m b 1.77×10^{-42} m^3
24. a 2.98 b 7.41×10^8
    c Europe
25. a 1.5 b $a + b + 1$
26. a 4 b $a - b - 1$
27. $r = p + q + 1$

Exercise 1C

1. a 1 b 2
2. a 5 b 6
3. a 0 b −1
4. a −2 b −4
5. a 1 b 2
6. a 3 b 4
7. a 0 b −1
8. a −2 b −3
9. a 100 b 1000
10. a 50 002 b 332
11. a 1.55 b −1.99
12. a e^2 b e^5
13. a e^{y+1} b e^{y^2}
14. a $\frac{1}{2}e^{y-3} - 2$ b $2e^{2y+1} + 12$
15. a $5\log x$ b $5\log x$
16. a $4\log 3x$ b $\frac{1}{\log 2x}$
17. a 1 b $\frac{1}{\ln x}$
18. a −15 b 17
19. a 4.5 b −1.5
20. a 13 b 7
21. a 3 b 7
22. a 8 b 9
23. a 25 b 81
24. a $\frac{1}{3}$ b $\frac{1}{6}$
25. a $\frac{1}{32}$ b $\frac{1}{64}$
26. a 2.10 b 3.15
27. a 5.50 b 2.88
28. a −0.301 b −1.40
29. a log 5 b log 7
30. a log 0.2 b log 0.06
31. a ln 3 b ln 7
32. a $2 + \log 7$ b $\log 13 - 4$
33. a $1 + \log 7$ b $\log 13 - 2$
34. a $\ln(k - 2) - 2$ b $\ln(2k + 1) + 5$
35. a 0.349 b −0.398
36. a 1.26 b 0.847
37. 10 000
38. 332
39. $a + b$
40. 0.699
41. 0.824
42. 0.845
43. log 1.6
44. a 7.60 b 0.0126
45. a 10 b 6.93 days
46. a i 1000 b $10\ln 3 = 11.0$ hours
      ii 1220
47. 4.62 years
48. a 57.0 decibels b 67.0 decibels
    c Increases the noise level by 10 decibels.
    d 10^{-3} W m^{-2}
49. a $e^{\ln 20}$ b $\frac{\ln 7}{\ln 20}$
50. $x = 100, y = 10$ or $x = -100, y = -10$
51. 1

Chapter 1 Mixed Practice

1. **a** 9.3×10^3 cm **b** $5\,280\,000\,\text{cm}^2$
2. $\dfrac{9y}{x}$
3. $\dfrac{y^6}{9x^4}$
4. **a** 4 000 000 **b** 1 000 000 **c** 16
5. $x = 3$
9. 500 s
10. $x = 99$
11. $x = 0.5e^3$
12. $x = 0.693$
13. $x = 0.531$
14. $x = \ln\left(\dfrac{y+1}{5}\right)$
15. **a** $m = 3, n = 4$ **b** $x = 7.5$
16. **a** 2.8 **b** $a + b + 1$
17. **a** 1.2 **b** $a - b$
18. 0.0259 to 0.0326
19. **a** $9.42\,\text{m s}^{-1}$ **b** Yes
20. **a** 3
    b Strength increases by one.
    c 3160 m (It would have had to be measured using a special damped seismograph.)
21. $x = \log\left(\dfrac{2}{3}\right) \approx -0.176$
22. $x = 1, y = -2$
23. $x = 10, y = 0.1$

Chapter 2 Prior Knowledge

1. $x = 4, y = -1$
2. **a** 80 **b** 36
3. $x = \pm 2$

Exercise 2A

1. **a** 49 **b** 68
2. **a** −22 **b** −40
3. **a** 2 **b** 7
4. **a** −0.5 **b** −2.5
5. **a** $u_1 = -37, d = 7$ **b** $u_1 = -43, d = 9$
6. **a** $8 - 3(n-1)$ **b** $10 + 4(n-1)$
7. **a** $3 + 2(n-1)$ **b** $10 + 4(n-1)$
8. **a** $1 + 5(n-1)$ **b** $20 - 3(n-1)$
9. **a** 34 **b** 24
10. **a** 13 **b** 29
11. **a** 372 **b** 910
12. **a** −11 **b** −236
13. **a** 352 **b** 636
14. **a** 184 **b** 390
15. **a** −45 **b** 0
16. **a** 32 **b** 99
17. **a** 105 **b** 205
18. **a** 806 **b** 354
19. **a** 13 **b** 10
20. **a** 16 **b** 16
21. **a** 216 **b** 2230
22. **a** 4 **b** 31 **c** 465
23. **a** 8 **b** 305
24. **a** 5 **b** 112
25. **a** £312 **b** £420
26. £300
27. 14
28. **a** 387 **b** 731
29. 140
30. **a** 21 **b** 798
31. **a** $u_1 = -7, d = 7$ **b** 945
32. **a** 296 **b** 390
33. 632
34. $x = 2$
35. 42
36. 42
37. 960
38. **a** 876 **b** 1679 **c** 480
39. 408.5
40. 53
41. −20
42. **a** Day 41 **b** 7995 minutes

Answers

43 a 19 days b 45 days
44 a 3, 7 b 199
45 $2n + 3$
46 0
47 70 336
48 735
49 0.2 m
50 −1
52 b 150

Exercise 2B

1 a 20 971 520 b 1458
2 a −15 625 b −1 441 792
3 a $\frac{1}{32}$ b $-\frac{2}{729} \approx 0.00274$
4 a $u_1 = 7, r = 2$ b $u_1 = \frac{4}{3}, r = 3$
5 a $u_1 = 3, r = \pm 2$ b $u_1 = \frac{5}{9}, r = \pm 3$
6 a $u_1 = -3, r = 2$ or $u_1 = 3, r = -2$
   b $u_1 = 7168, r = \frac{1}{2}$ or $u_1 = -7168, r = -\frac{1}{2}$
7 a 6 b 8
8 a 13 b 8
9 a 5465 b 4095
10 a 190.5 b 242
11 a 153.75 b $\frac{1456}{9}$
12 a 363 b 2800
13 a 76 560 b 324 753
14 a 68 796 b 488 280 000
15 a $\frac{9}{6}$ b $\frac{24827}{999}$
16 a 1 b 255
17 a 2 b 96 c 3069
18 a 1.5 b 182.25
19 1920 cm²
20 0.0375 mg ml⁻¹
21 15.5
22 0.671 > 5
23 a 3580 m³ b 11 days
24 255
25 a 9.22×10^{18} b 1.84×10^{19}
   c 2450 years
26 $x^{2-n} y^{n+1}$
27 3069
28 −1
29 88 572
30 a 0.246 m b 3.9 m
   c The height is so small that measurement error and other inaccuracies would be overpowering.

Exercise 2C

1 a $2382.03 b $6077.45
2 a $580.65 b $141.48
3 a $6416.79 b $1115.87
4 a 48 years b 5 years
5 a 173 months b 77 months
6 a 7.18% b 4.81%
7 a 14.9% b 7.18%
8 a $418.41 b $128.85
9 a £13 311.16 b £7119.14
10 a $97 b $294
11 a £737.42 b £2993.68
12 a $103 b $5130
13 a £94 b £2450
14 a €1051.14 b €579.64
15 a $598.74 b $4253.82
16 £900.41
17 £14 071.00
18 $8839.90
19 €16 360.02
20 £8874.11
21 a monthly b £28.55
22 £6960
23

Year	Start-year value ($)	Depreciation expense ($)	End-year value ($)
1	20 000	6 000	14 000
2	14 000	4 200	9 800
3	9 800	2 940	6 860
4	6 860	2 058	4 802
5	4 802	1 441	3 361
6	3 361	1 008	2 353
7	2 353	706	1 647
8	1 647	147	1 500

24 15.4%
25 12.7%

26 a 2.44% b 12.8%
27 a 250 billion marks
   b 0.0024 marks c 0.0288 marks

Chapter 2 Mixed Practice

1 a €6847.26 b 7.18%
2 a i d_n b i $\frac{1}{2}$
    ii b_n ii $-\frac{3}{256}$
3 a 2 b −4 c 300
4 a 4 b 512 c 43 690
5 8
6 10
7 a 48.8 cm b 9
8 a €563.50 b 6.25 years
9 a 8.04 billion b 2033
10 a i 20 b 7290
    ii 10
11 a i 7 ii 16
    c 2 d 25
    e $n = 498$ f $n = 36$
12 $x = 4$
13 b 74
14 8190
15 16 months
16 $a^{2n} b^{3-n}$
17 $11.95 million
18 $10 450
19 8.63%
20 $1212.27
21 22 years
22 a $a_1 = 6$
    b i $a_2 = 8$ ii $a_3 = 10$
    c 2
    d i 22 ii 594 m
    e 28
    f 16 m
23 b 32
24 a B − 12th day b A − 6th day
25 a £68 500 b £1 402 500 c £43 800
26 3

Chapter 3 Prior Knowledge

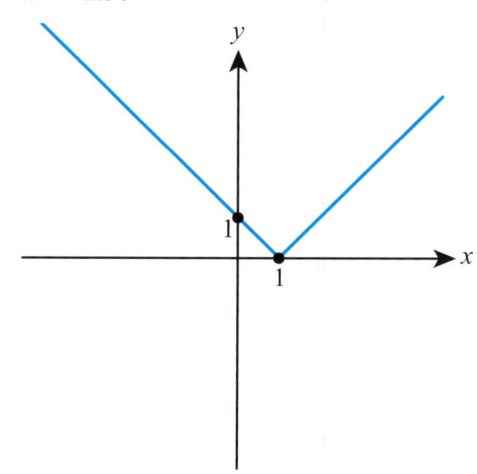

2 $x \geq \frac{3}{2}$
3 0
4 $x = -2.30$
5

Exercise 3A

1 a 9 b 14
2 a 7 b 47
3 a −17 b 19
4 a −15 b −44
5 a Yes b Yes
6 a No b No
7 a Yes b No
8 a No b Yes
9 a Yes b Yes

Answers

10 a ℝ b ℝ
11 a $n \neq 0$ b $x \neq 0$
12 a $x \geq -5$ b $x \geq 3.5$
13 a $x > -0.6$ b $x > 4$
14 a $x \neq 2.5$ b $x \neq -3$
15 a $f(x) \geq -2$ b $g(x) \geq 7$
16 a $f(x) \leq 18$ b $g(x) < 4$
17 a $f(x) \geq 0$ b $g(x) > 0$
18 a $f(x) \geq 2$ b $g(x) < 3$
19 a 40 b -7
20 a 1 b 2
21 a -3 b -1

22 a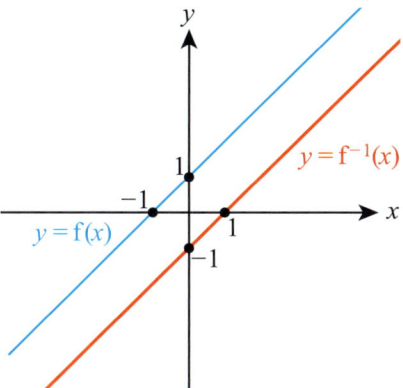

b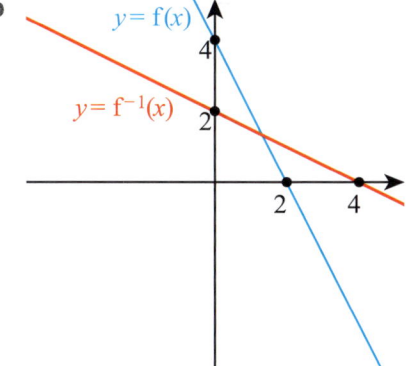

23 a No inverse b No inverse

24 a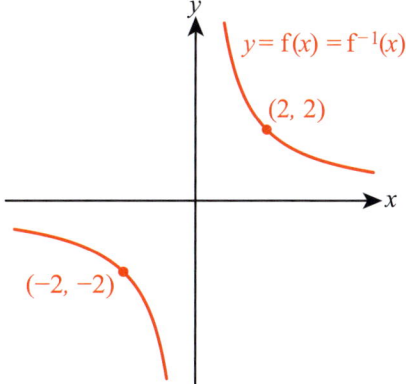

b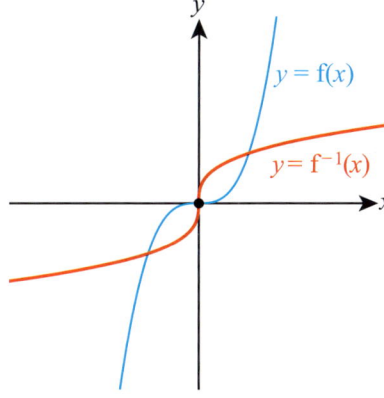

25 a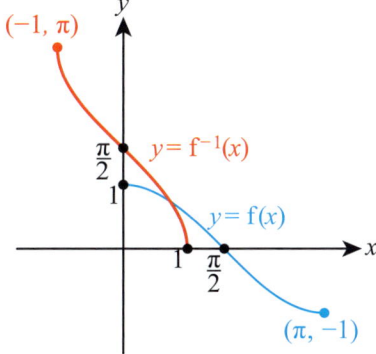

b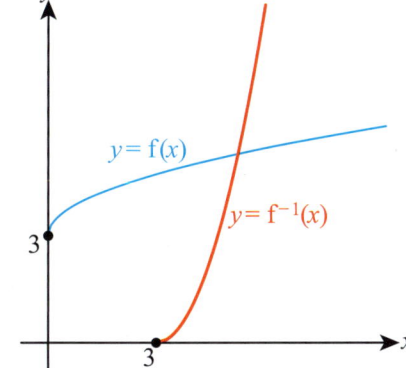

26 a −13 **b** $x = 3$
27 $x = 41$
28 a 5.7 m s^{-1}
   b No, the car cannot accelerate uniformly for that long / The model predicts an unreasonable speed of 114 m s^{-1}.
29 a $x \neq 5$ **b** $\frac{1}{3}$
30 a $-\frac{5}{4}$ **b** $q(x) \geq -2$
   c $x = 4$ or -4
31 a 4.95 billion
   b E.g. smartphones are likely to get replaced by another technology
32 a $f(x) = 1.3x$
   b Amount in pounds if x is the amount in dollars.
33 a 1 **b** $x = 2$
34 a $x \geq 2.5$ **b** $f(x) \geq 0$ **c** $x = 7$
35 a 60 **b** 129.9; 130
   c It predicts a non-integer number of fish after 15 months.
36 a 14 **b** 4

37
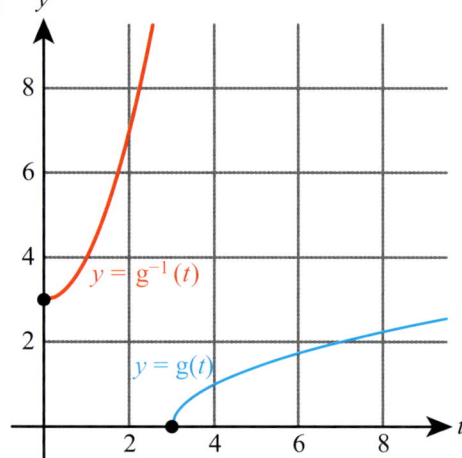

38 a $x > 0$ **b** 4 **c** $\frac{1}{9}$
39 $x > 5$
40 a $x < \frac{7}{3}$ **b** $x = -31$
41 a 19 **b** $f(x) \geq 4$
   c 1 is not in the range

42 a 9 **b** 2.5
   c
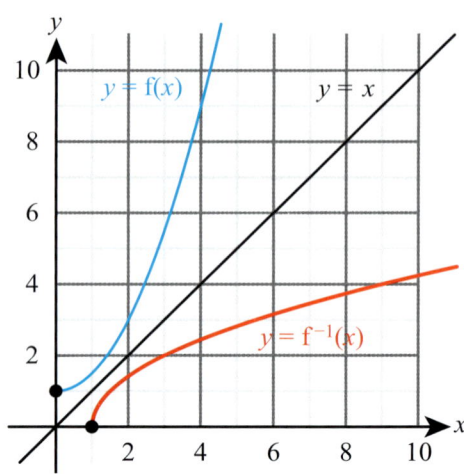

43 a 0.182 **b** 0.892
44 a $x \geq 1, x \neq 17$ **b** $f(x) \geq \frac{3}{4}$ or $f(x) < 0$
45 a and b
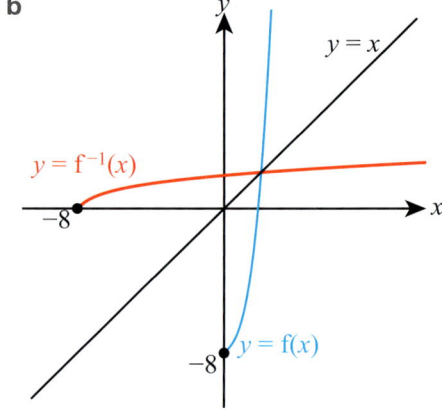

   c $x = 2$

Exercise 3B

1 a
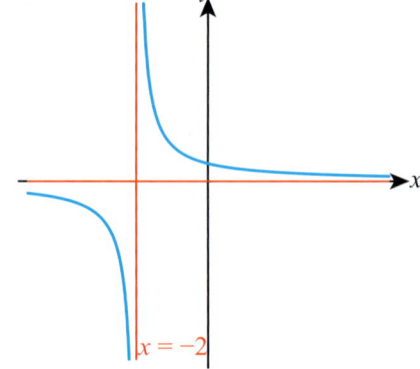

Answers

b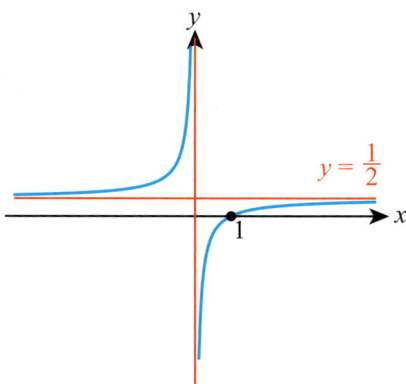

2 a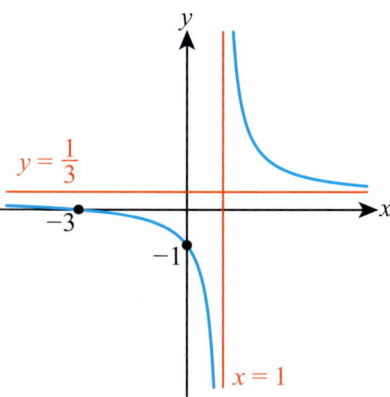

b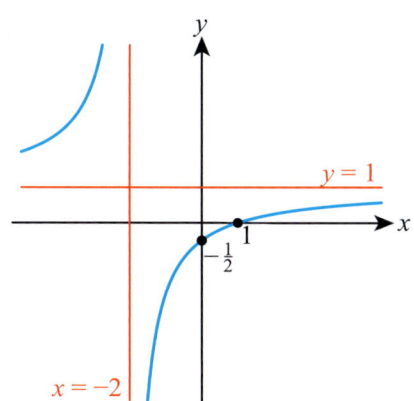

3 a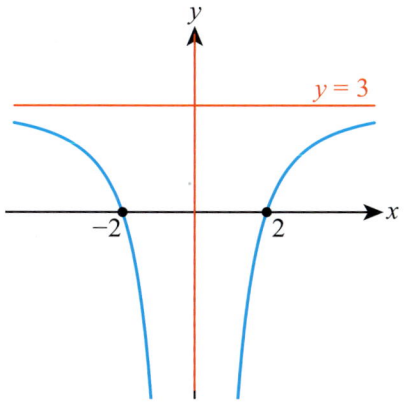

b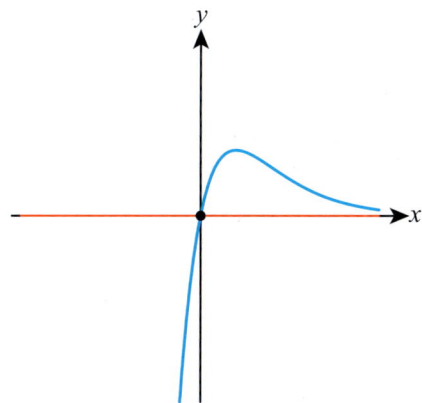

4 a $f(x) \geq \dfrac{11}{3}$ **b** $g(x) \leq \dfrac{81}{8}$

5 a $f(x) \geq \ln(0.75)$ **b** $g(x) \geq \ln(4.5)$

6 a $-1.17 \leq f(x) \leq 1.17$ **b** $-1.57 \leq g(x) < e$

7 a $(0.7, 0.55)$; $x = 0.7$

b $\left(-\dfrac{2}{7}, \dfrac{39}{7}\right)$; $x = -\dfrac{2}{7}$

8 a $\left(\dfrac{-2 \pm \sqrt{2}}{2}, 1.75\right)$, $(-1, 2)$; $x = -1$

b $(0, -1)$, $(1, -1)$, $\left(\dfrac{1}{2}, -\dfrac{3}{4}\right)$; $x = \dfrac{1}{2}$

9 a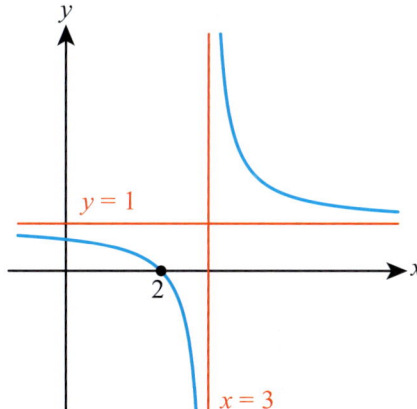

b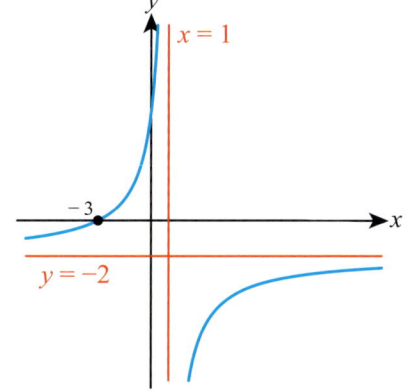

10 a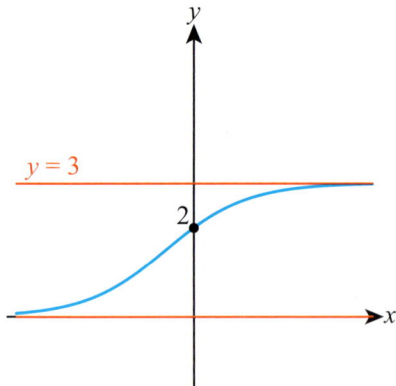

b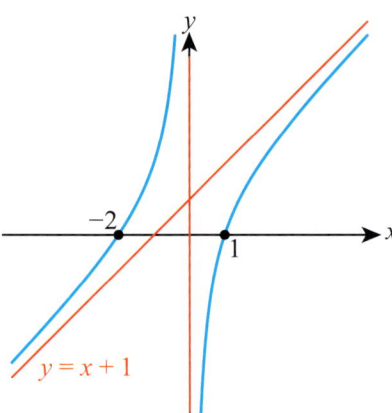

11 a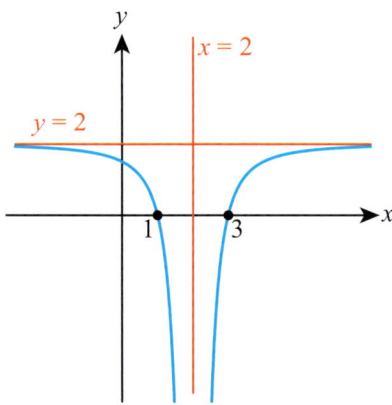

b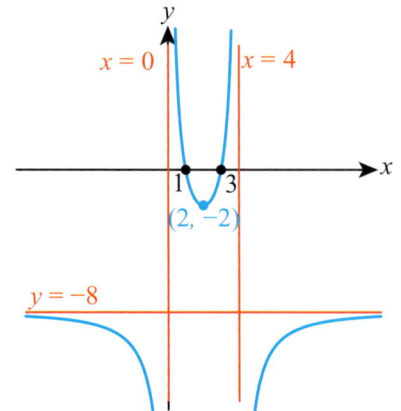

12 a $(-1.49, -4.78)$ **b** $(-0.661, 0.421)$

13 a $(-2, 7), (-1, 5), (1, 7)$
b $(-2, 23), (1, 5), (2, 3), (3, 3)$

14 a $\left(-\sqrt{2}, -2\sqrt{2} - 2\right), \left(\sqrt{2}, 2\sqrt{2} - 2\right)$
b $\left(1 - \sqrt{2}, 7 - \sqrt{2}\right), \left(1 + \sqrt{2}, 7 + \sqrt{2}\right)$

15 a $x = 0.040, 1.78$ **b** $x = 0.213, 1.632$
16 a $x = 1, 6.71$ **b** $x = -3.48, 2.48$
17 a $x = 0.063, 1.59$ **b** $x = 0.288, 49.0$
18 a $x = -2.18, 0.580$ **b** $x = -1.67, 0.977$

19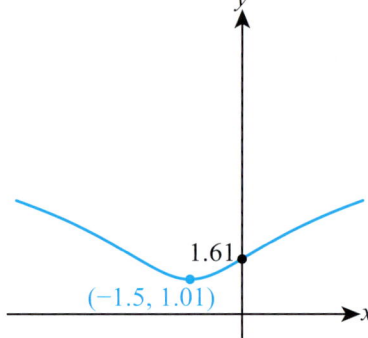

20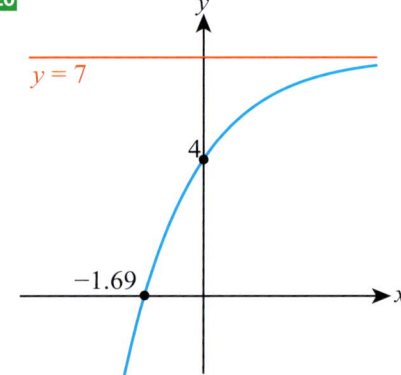

$(-1.69, 0), (0, 4), y = 7$

21 a $x \neq -2$

b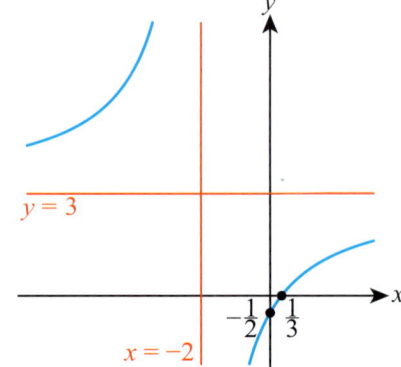

$x = -2, y = 3$

22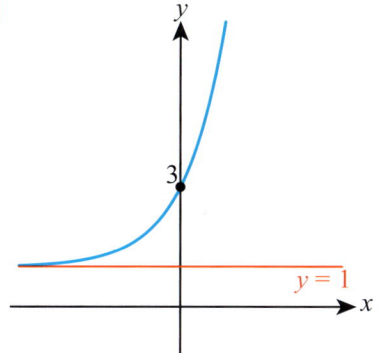

23 (1.84, 3.16)
24 (−1, 2.5)
25 235
26 $x = -1$
27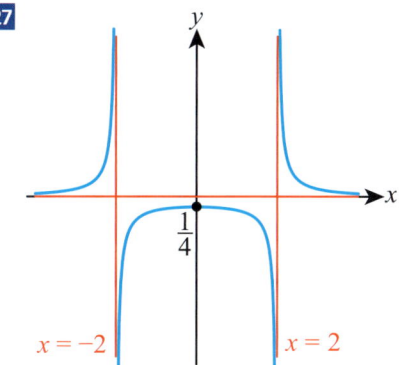

$x = -2$, $x = 2$, $y = 0$

28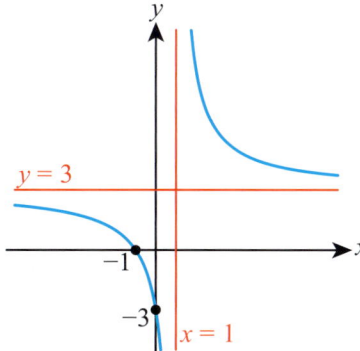

29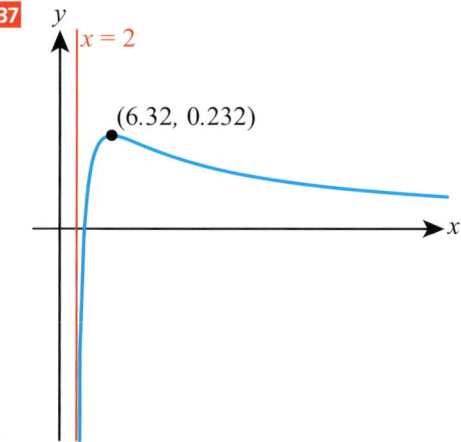

30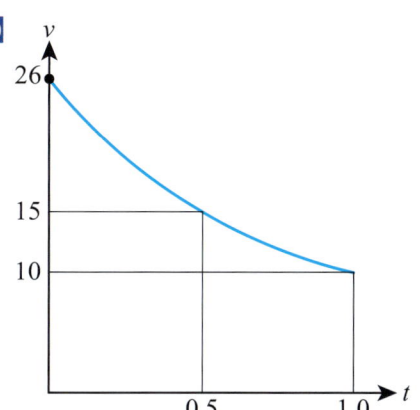

31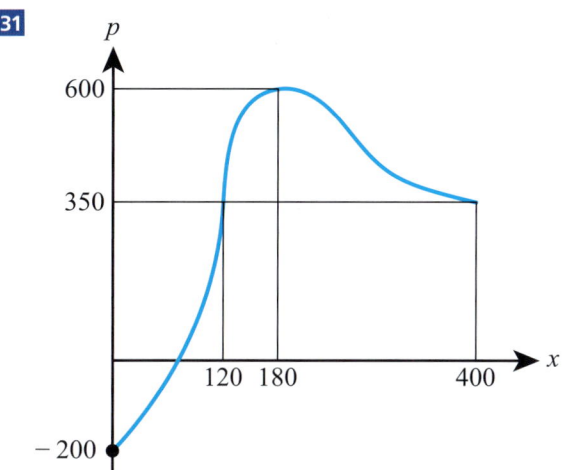

32 $x = 0.755$
33 $x = -2.20, -0.714, 1.91$
34 −2.41, 0.414, 2
35 −1.41, 1.41
36 0.920
37

38 $x \geq 2$

Chapter 3 Mixed Practice

1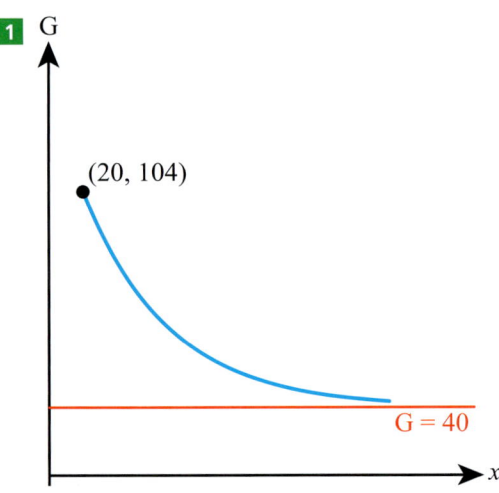

   b $3538.09

2 $x = 1.86, 4.54$

3 a $x \geqslant -5$ **b** $x = -2.38$

4 $(-1.61, 0.399)$ and $(0.361, 2.87)$

5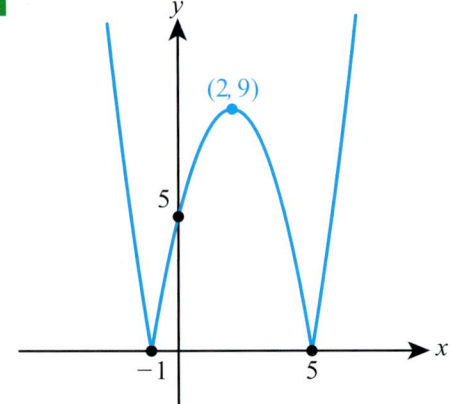

6 a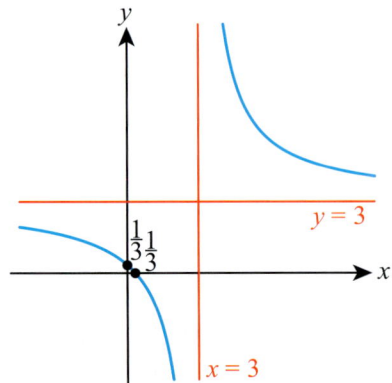

   b $x \neq 3$, $f(x) \neq 3$

7 a $20.0\,\text{m s}^{-1}$ **b** $0.630, 17.1\,\text{s}$ **c** $5\,\text{s}$

8 b $1.30\,\text{s}$

9 a $100°C$ **b** $95.3°C$ **c** $8.82\,\text{km}$

10 a $f(x) = 1.8x + 32$

   b Temperature in Celsius if x is temperature in Fahrenheit.

11 a $p = -2, q = 4$

   b i $x \neq 2$

      ii $g(x) > 0$

      iii $x = 2$

12 a $f(x) \geqslant 20$ **b** 12

13 $x = -5.24, 3.24$

14 $x = 1, 2.41$

15 $x \neq \pm 3$, $h(x) \leqslant -2$ or $h(x) > 0$

16 $-9.25 \leqslant f(x) < 3.$

17 a 4.89 **b** $x = 28.9$

18 a $-19 < g(x) \leqslant 8$ **b** $x = -1, x = 1$

   c -20 is not in the range of g

19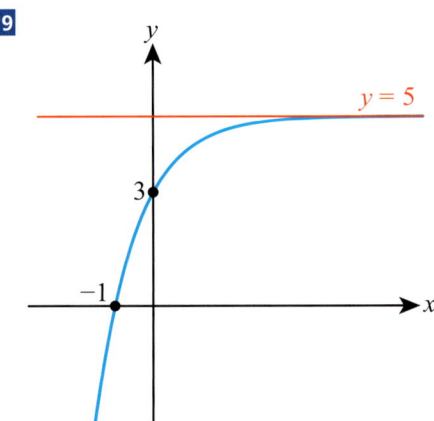

20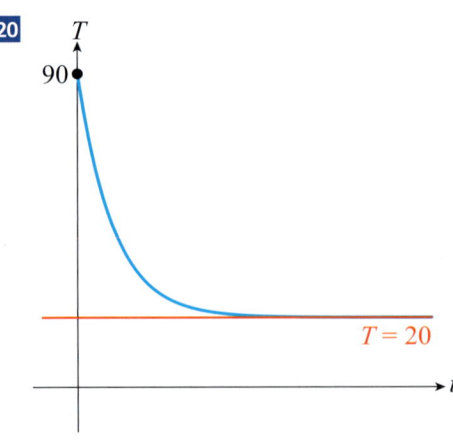

Answers

21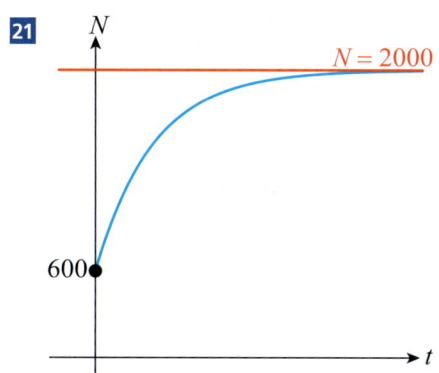

22 a 1.10 m s^{-1}
   b e.g. The car will stop by then
23 $x = -2.50, -1.51, 0.440$
24 a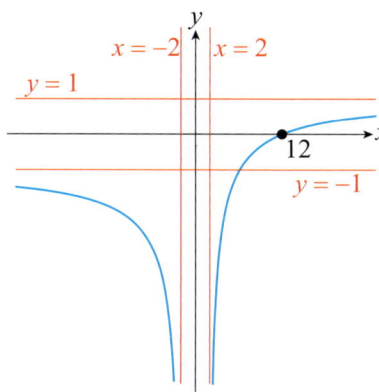

   b i $(12, 0)$
   ii $x = \pm 2, y = \pm 1$
25 $x = 2.27, 4.47$ **26** -1.34
27 a $x \neq 0, 4$
   b $f(x) \leqslant -5.15 \text{ or } f(x) \geqslant 3.40$
28 a $x > 0, x \neq e^{-3}$
   b $g(x) < 0 \text{ or } g(x) \geqslant 0.271$
29 a $x > 2, g(x) \in \mathbb{R}$
   b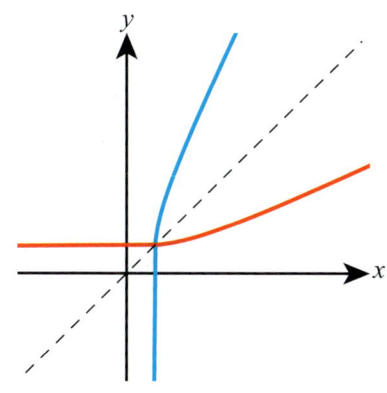
   c $x = 2.12$

Chapter 4 Prior Knowledge

1 24 cm
2 a 10   **b** $(-1, 1)$

Exercise 4A

1 a 1 **b** 2
2 a -3 **b** -1
3 a $\dfrac{1}{2}$ **b** $-\dfrac{1}{2}$
4 a $(2, 0), (0, -6)$ **b** $(4, 0), (0, 8)$
5 a $(5, 0), (0, 2.5)$ **b** $(-3, 0), (0, 2)$
6 a $(3, 0), (0, 4)$ **b** $(4.5, 0), (0, -3)$
7 a $2x - y + 3 = 0$ **b** $5x - y + 1 = 0$
8 a $2x + y - 4 = 0$ **b** $3x + y + 7 = 0$
9 a $x - 2y + 7 = 0$ **b** $x + 3y - 9 = 0$
10 a $y = 2x + 5; 2, 5$ **b** $y = 3x + 4; 3, 4$
11 a $y = -\dfrac{2}{3}x - 2; -\dfrac{2}{3}, -2$ **b** $y = -\dfrac{5}{2}x + 5; -\dfrac{5}{2}, 5$
12 a $y = -0.6x + 1.4; -0.6, 1.4$
   b $y = -5.5x - 2.5; -5.5, -2.5$
13 a $y - 4 = 2(x - 1)$ **b** $y - 2 = 3(x - 5)$
14 a $y - 3 = -5(x + 1)$ **b** $y + 1 = -2(x - 2)$
15 a $y + 1 = \dfrac{2}{3}(x - 1)$ **b** $y - 1 = -\dfrac{3}{4}(x - 3)$
16 a $2x - y - 5 = 0$ **b** $2x - y + 7 = 0$
17 a $x + y - 10 = 0$ **b** $3x + y + 2 = 0$
18 a $x + 2y - 7 = 0$ **b** $x + 2y - 4 = 0$
19 a $3x + 4y - 13 = 0$ **b** $8x + 5y - 19 = 0$
20 a $y = 3x + 4$ **b** $y = -x + 9$
21 a $y - 5 = 1.5(x - 5)$ **b** $y + 3 = 0.5(x + 1)$
22 a $x - y - 6 = 0$ **b** $2x - y - 7 = 0$
23 a $x + 3y - 4 = 0$ **b** $x + 5y + 8 = 0$
24 a $2x - 5y = 0$ **b** $3x + 4y + 2 = 0$
25 a $y = -x + 5$ **b** $y = -\dfrac{1}{3}x + 8$
26 a $y - 5 = -4(x - 1)$ **b** $y - 4 = 3(x - 2)$
27 a $5x + y + 3 = 0$ **b** $2x + y - 7 = 0$
28 a $x - 2y + 5 = 0$ **b** $x - 3y - 10 = 0$
29 a $2x - 5y + 37 = 0$ **b** $2x + 3y + 3 = 0$
30 a $(1, 2)$ **b** $(5, 1)$
31 a $(1, 3)$ **b** $(2, 5)$
32 a $(2, 5)$ **b** $(-3, -4)$

Answers

33 a $\left(\frac{2}{5}, \frac{11}{5}\right)$ b $\left(\frac{11}{5}, \frac{2}{5}\right)$

34 a $\frac{7}{4}$ b $y = -\frac{4}{7}x + \frac{53}{7}$

35 a $-\frac{4}{3}$ b $4x + 3y = 36$

36 a $-\frac{5}{7}$ b $\frac{17}{5}$

37 a $\frac{1}{5}$ b $x - 5y = -8$

38 a $-\frac{7}{2}$ b No
   c $y = \frac{2}{7}x - \frac{51}{7}$

39 0.733

40 a (8, 11)   b 8
   c $x = 8$ d 56

41 a $y = -\frac{3}{2}x - 6$ b $x + 4y = -1$
   c (−4.6, 0.9)

42 6.10 m

43 a $0.5t + 0.1$ b 9.8 s

44 a N/m   b 0.018 N
   c larger   d 0.467 m

45 a $C = 0.01m + 5$ b $6.80
   c 500 minutes

46 a $P = 10n - 2000$ b 350
   c $P = 8n - 1200$ d 400, fewer

48 1390 m

Exercise 4B

1 a 5   b 13
2 a 13   b 10
3 a $\sqrt{29}$ b $\sqrt{85}$
4 a 3   b 11
5 a 6   b 9
6 a $\sqrt{98}$ b $\sqrt{65}$
7 a (4, −1)   b (4, 6)
8 a (3, −4, 3)   b (−1, 2, −1)
9 a (5.5, 2, −3)   b (−5.5, 3.5, 7)
10 a (1.5, 0.5, 5.5)   b $\sqrt{171}$
11 (6, 3, −8)
12 $a = 3, b = 20, c = 4.5$
13 $\sqrt{196.5}$
14 $4.52 \, \text{m s}^{-1}$
15 $\sqrt{30}$
16 $k = \pm 2$

17 $a = 2.97$

18 a $a = -2.2, b = -8.6$ b $y + 5 = -\frac{1}{16}(x - 4)$

19 a $\frac{4}{7}$ b $7x + 4y = -20$
   c (0, −5)   d $\sqrt{65}$

20 a 1.8 m   b 3.33 m

21 a (3, 7)   b (6, 5) and (0, 9)

22 a $y + x = 9$ b $B(4, 5) \, D(5, 4)$
   c 5

23 a $\sqrt{50} \approx 7.07 \, \text{m}$ b $\sqrt{74} \approx 8.60 \, \text{m}$

Chapter 4 Mixed Practice

1 a   i (2, −2)   b $k = -\frac{2}{3}$
   ii $\frac{3}{2}$
   iii $-\frac{2}{3}$

2 a $s = 6, t = -2$ b $4x + 5y = 23$

3 b $\frac{3}{2}$ c $-\frac{2}{3}$ d 4

4 a −1   b $y = x + 3$ c 4.5

5 a (2.5, −1)   b 9.22
   c $y = \frac{7}{6}x - \frac{47}{12}$

6 a (2.5, −1, 4)   b 9.43

7 a $P(0, 3) \, Q(6, 0)$ b $\sqrt{45}$
   c (2, 2)

8 a $-\frac{7}{4}$ b $k = 2$ c $d = 1$

11 ±20

12 a $p = 1, q = -18$ b 27.5

13 1.5

14 4.37

15 b $y = -2x + 5$ c $S(1, 3)$

16 10 m

17 a

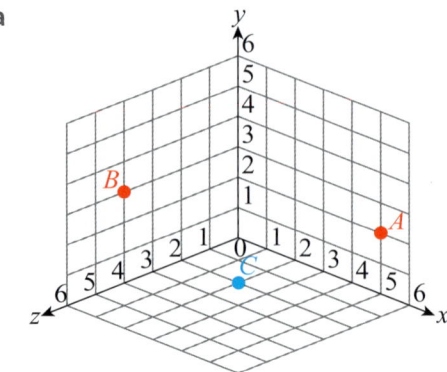

   b (2.5, 2.5, 2)   c 6.48

Answers

18 a $\dfrac{1}{2}$ b (−1, 5)
c $2x + y = 3$ e 4.47
19 a 42 b 15.2 c 5.51
20 4.32
21 a $y = 3x - 13$ b (6, 5) and (4, −1)
23 135 m
24 4 m

Chapter 5 Prior Knowledge

1 95°
2 12.4 cm
3 60
4 $\alpha = 140°, \beta = 85°$
5 050°
6 Volume = 1570 cm^3, Surface area = 785 cm^2

Exercise 5A

1 a 40.7 cm^2, 24.4 cm^3 b 1580 m^2, 5880 m^3
2 a 8.04 m^2, 2.14 m^3 b 0.126 m^2, 0.00419 m^3
3 a 170 m^2, 294 m^3 b 101 km^2, 134 km^3
4 a 47.9 cm^2, 38.6 cm^3
   b 4.40 mm^2, 0.293 mm^3
5 a 16π cm^2, $\dfrac{32\pi}{3}$ cm^3 b 36π m^2, 36π m^3
6 a 64π m^2, $\dfrac{256\pi}{3}$ m^3 b 288π m^2, 432π m^3
7 a $\dfrac{3\pi}{4}$ m^2, $\dfrac{\pi}{6}$ m^3 b $\dfrac{20\pi}{9}$ mm^2, $\dfrac{4\pi}{3}$ mm^3
8 a 90π cm^2, 100π cm^3 b 36π cm^2, 16π cm^3
9 a 6 cm^3 b 16 cm^3
10 a 20 cm^3 b 18 cm^3
11 a 26.7 cm^3 b 2.67 cm^3
12 a 64 mm^3, 144 mm^2 b 336 mm^3, 365 mm^2
13 a 2 mm^3, 13.8 mm^2 b 320 mm^3, 319 mm^2
14 a 5.33 cm^3, 33.9 cm^2 b 816 cm^3, 577 cm^2
15 a 75.4 cm^3, 109 cm^2 b 20.9 cm^3, 46.4 cm^2
16 707 cm^2
17 96.5 cm^2
18 8.31 m^3
19 134 cm^3
20 3210 cm^3
21 3400 cm^2, 8120 cm^3
22 4.21×10^4 cm^3
23 a 12.0 cm b 1100 cm^2
24 a 3.61 cm b 36.9 cm
25 a 192 cm^3 b 3.58 cm
26 a 15 cm b 1330 cm^2
27 a 82.5 cm^2 b 474 cm^2
   c 594 cm^3
28 55.4 cm^2, 23.0 cm^3
29 1140 mm^2, 3170 mm^3
30 1880 mm^3, 889 mm^2
31 a 13 cm b 815 cm^2
32 242 m^2
33 a 151 mm^2, 134 mm^3 b 12.9 mm
34 a 9.35 m^3 b 0.194 m^3

Exercise 5B

1 a 29.7° b 58.0°
2 a 53.1° b 39.5°
3 a 40.1° b 53.1°
4 a 2.5 b 3.86
5 a 7.83 b 11.5
6 a 9.40 b 14.1
7 a 17.4 b 26.3
8 a 2.68 b 25.7
9 a 344 b 4.23
10 a 71.6° b 78.7°
11 a 18.4° b 15.9°
12 a 21.8° b 12.5°
13 a 7.13° b 10.6°
14 a 29.7° b 45°
15 a 10.5° b 14.5°
16 a 8.49 cm b 6.53 mm
17 a 3.42 cm b 3.29 mm
18 a 3.31 cm b 4.10 mm
19 a 37.9° b 50.4°
20 a 39.9° b 32.9°
21 a 20.7° b 51.2°
22 a $\sqrt{37}$ cm b 7 mm
23 a 5.06 cm b 6.40 mm
24 a 5.23 cm b 6.43 mm
25 a 39.4° b 41.6°
26 a 42.6° b 62.8°

Answers

27 a 53.8°   b 97.3°
28 a 28.5 cm²   b 10.5 mm²
29 a 127 cm²   b 1.73 mm²
30 a 631 cm²   b 1710 mm²
31 a 52.2°   b 61.8°
32 a 8.30 cm   b 5.23 mm
33 a 7.04 cm   b 8.36 mm
34 4.16 cm
35 32.5°
36 a (0, 3), (−6, 0)   b 26.6°
37 a

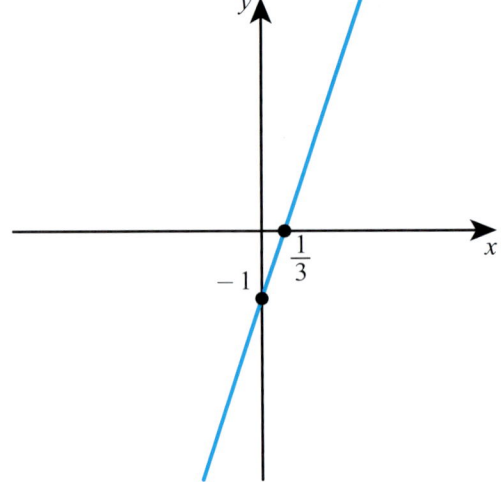

   b 71.6°
38 6.98 cm
39 58.5°, 78.5°
40 87.4°
41 a 18.6°   b 32.0
42 19.0
43 $x = 11.8$ cm, $\theta = 15.5°$
44 38.7°
45 b 49.4°
46 82.2°
47 16.2 cm
48 5.69 cm
49 29.0°
50 47.0°
51 21.0°
52 5.63 cm
53 15.2
54 $\theta = 52.0°$, $AB = 8.70$
55 9.98
56 $h = \dfrac{d \tan 40}{\tan 50 - \tan 40}$
57 a $x = \dfrac{h}{\tan 30}$, $y = \dfrac{4-h}{\tan 10}$
   b 3.73

Exercise 5C

1 a 35.5°   b 59.0°
2 a 16.7°   b 51.3°
3 a 55.6°   b 18.9°
4 a 48.0°   b 82.6°
5 a 68.9°   b 59.8°
6 a 67.4°   b 41.6°
7 a 49.3°   b 48.2°
8 a 69.3°   b 46.6°
9 a 2.73 m   b 8.22 m
10 a 36.9°   b 42.8°
11 a 36.6°   b 46.5°
12 a 31.2°   b 15.0°
13 a 45.3°   b 45.3°
14 a 50.6°   b 61.5°
15 a 17.8°   b 67.4°
16 a 54.2°   b 43.4°
17 a 1.16 km, 243°   b 2.71 km, 320°
18 a 4.00 km, 267°   b 7.62 km, 168°
19 a 3.97 km, 317°   b 16.5 km, 331°
20 57.1 m
21 a 15.8   b 18.4°
22 a

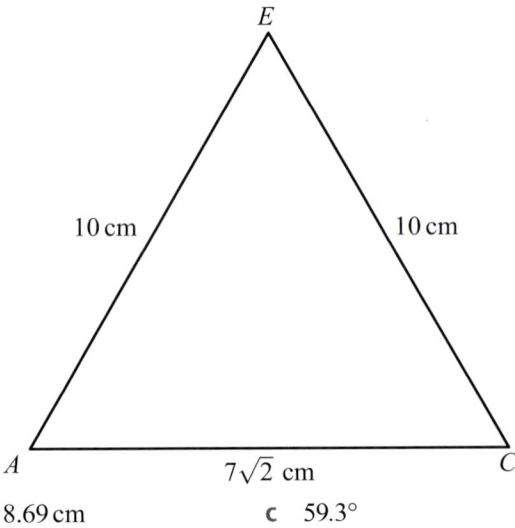

   b 8.69 cm   c 59.3°
23 56.3°

Answers

24 a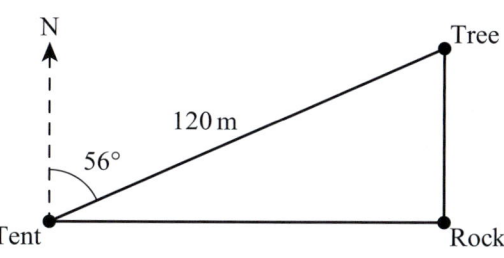

  b 67.1 m

25 a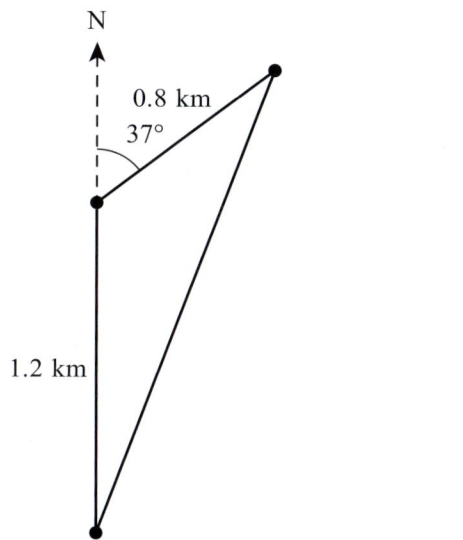

  b 1.90 km

26 13.8 m

27 42.3 m

28 3.18 m

29 11.4 m

30 a 12.1 cm **b** 35.4° **c** 41.9°

31 a 13.02 m **b** 3.11°

32 a 7.81 cm **b** 45.2°

33 191 m, 273°

34 2.92 km, 008.6°

35 146 m

36 a 9.51 m **b** 10.9 m **c** 48.3°

37 3.96 m

38 a 9 – assumes lengths given are internal or thin glass or no large objects in the space

  b 359 000 W

  c yes – maximum angle is 51.8°

39 a 2.45 m **b** 42.9 m³

  c 75.5% **d** 94.0%

40 38.9 cm²

41 39.9°

42 92.3 m

Chapter 5 Mixed Practice

1 35.0 m

2 a $AC = 22.6$ cm, $AG = 27.7$ cm

  b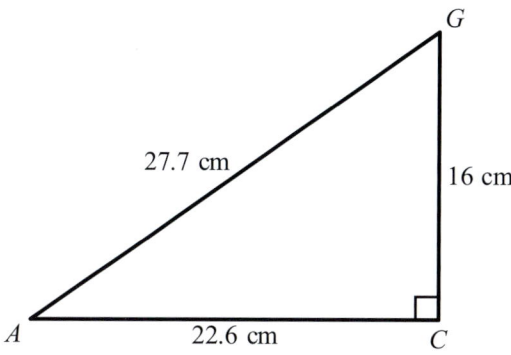

  c 35.3°

3 a 32.5 cm **b** 24.1 cm **c** 29.1 cm

4 45.2°

5 a 6 **b** $-\frac{1}{2}$ **c** 26.6°

6 a 12.9 cm

  b 80.5°

7 12.3 cm

8 3.07 cm

9 21.8°

10 32.4°

11 37.4°

12 18.3 cm

13 55.4 m

14 17.9 m

15 a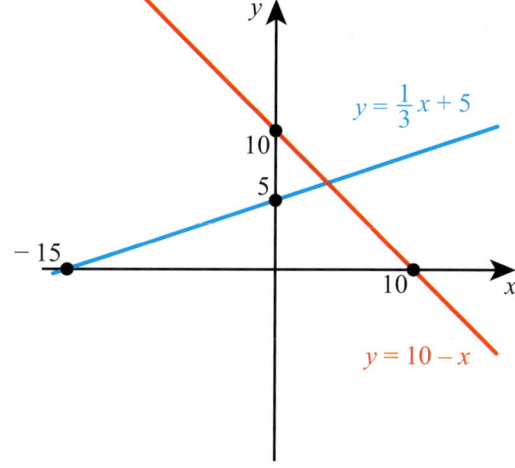

  b (3.75, 6.25)

  c 63.4°

16 23 900 cm³
17 a 65.9° b 21.0°
18 5.46
19 21
20 17.6 m
21 a 131 cm³ b 313 cm³
22 a 5.12 cm b 9.84 cm c 31.4°
23 a 5 cm
   b 9.43 cm
   c 58.0°
   a 720 m
   b 72 300 m²
   c 88.3°
24 a i 22.5 m
     ii

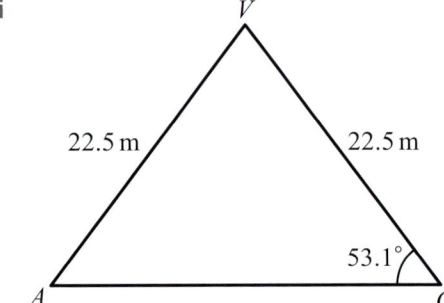

   c 27.0 m
   e 41 600 m³
   f 44 900 kg
25 22.5

Chapter 6 Prior Knowledge

1 a 4.75 b 5.5 c 6 d 9
2 a 1.5 b (0, 7)

Exercise 6A

1 a 9 lions, 21 tigers
   b 9 strawberry, 11 chocolate
2 a 24 boys, 16 girls
   b 18 HL, 27 SL
3 a 10 football, 14 hockey, 16 basketball
   b 6 cod, 9 haddock, 5 mackerel
4 a 24 chairs, 9 tables, 4 beds
   b 8 oak, 7 willow, 4 chestnut

5 a all households in Germany
   b convenience sampling
   c e.g. Households in the city may have fewer pets than in the countryside.
6 a the number/proportion of pupils in each year group
   b quota sampling
   c i keep
     ii discard
7 a convenience sampling
   b all residents of the village
   c e.g. People using public transport may have different views from those who drive.
   d would need access to all the residents
8 Not necessarily correct, it could just be an extreme sample.
9 a continuous
   b They would all be destroyed.
   c list in serial number order, select every 20th
10 a quota
   b more representative of the scarves sold
   c 12 red, 12 green, 10 blue, 6 white
11 a quota
   b more representative of the population
   c difficult to compile a list of all the animals
12 a continuous
   b Basketball team are likely to be taller than average.
   c systematic sampling
   d Some samples not possible, e.g. it is not possible to select two students adjacent on the list.
13 5 cats, 9 dogs and 6 fish
14 a

Gender/age	12	13	14
Boys	4	5	5
Girls	0	4	2

   b no
15 a discrete
   b no
   c convenience
   d e.g. Different species may live in different parts of the field.
16 a i possible
     ii not necessarily true
     iii possible/quite likely

Answers

  b iii would be unlikely
  c discard –32, keep 155
17 a 65%
  b i 84
     ii 65%
18 a 100 million
  b e.g. No fish die or are born; captured cod mix thoroughly with all other cod.

Exercise 6B

1 a $16.5 \leq t < 18.5$; $20.5 \leq t < 22.5$; $22.5 < t$ 24.5; 3, 12, 3, 1, 1
       20 observations
  b $14.5 \leq t < 17.5$; $20.5 \leq t < 23.5$; 10, 8, 2
       20 observations
2 a $300 \leq n \leq 499$; $500 \leq n \leq 699$; $700 \leq n \leq 899$; 11
       40 observations
  b $200 \leq n \leq 399$; $400 \leq n \leq 599$; $600 \leq n \leq 899$; 20
       40 observations
3 a mean = 3.86, med = 3, mode = 1
  b mean = 5.57, med = 6, mode = 6
4 a mean = 4.7, med = 4.5, mode = 3
  b mean = 31.2, med = 24.5, mode = 24
5 a $x = 10$ b $x = 7$
6 a $x = \frac{31}{9}$ b $x = 7.5$
7 a $y = 10$ b $y = 52$
8 a $y = 4.6$ b $y = 2.2$
9 a i 13   b i 20.5
     ii $0 \leq x < 10$ ii $25 \leq x < 35$
10 a i 13.5   b i 10.2
      ii $12.5 \leq x < 14.5$ ii $11.5 \leq x < 15.5$
11 a 10.4   b 4.96
12 a 5.24   b 32.8
13 a $x < 5$ or $x > 125$ b $x < 10$ or $x > 26$
14 a $x < 4.5$, $Q_3 > 12.5$ b $x < 15.5$ or $x > 63.5$
15 a mean = 46, sd = 8
  b mean = 62, sd = 18
16 a median = 25, IQR = 13
  b median = 0.5, IQR = 2.1
17 a med = 360, IQR = 180
  b med = 5, IQR = 2.6

18 a mean = 8, sd = 3
  b mean = 41, sd = 3
19 a 48   b 19   c 6.17, 1.31
20 a 31   b $1.4 \leq m < 1.6$ c 1.34
21 a 28
  b $15.5 \leq t < 17.5$
  c 17.3 s; used midpoints, actual times not available
22 a 4.25
  b 1.5
  c The second artist's songs are longer on average and their lengths are more consistent.
23 a 4   b 3   c yes (11)
24 a 67
  b 23 cm
  c 24.7 cm
  d
$20.5 \leq l < 23.5$	$23.5 \leq l < 26.5$	$26.5 \leq l < 29.5$
23	30	14

  e 44
  f no
25 a $a = 12.5$ b 13.4
26 a $x = 5.5$ b 30.9
27 63.8
28 62
29 a $\frac{161 + 24a}{50}$ b $a = 10$
30 a $x = 9.5$
  b $Q_1 = 5$, $Q_2 = 5$, $Q_3 = 6.5$
  c yes
31 a £440, £268   b median
  c $576, $351
32 a mean = 4, sd = 2.94
  b mean = 2012, sd = 8.83
33 10.6°C, 2°C
34 mean = 9.18 km, var = 11.9 km^2
35 med = –75, IQR = 42
36 a med = 42, IQR = 8   b $m = 58$
37 74%
38 a $p = 12$ b 0.968
39 $p = 7$, $q = 4$
40 $c = 30$
41 10
42 (4, 8) or (5, 7)

Exercise 6C

You might find that different GDCs or programs give slightly different values for the quartiles, resulting in slightly different answers from those given here. In an exam, all feasible answers would be allowed.

1 a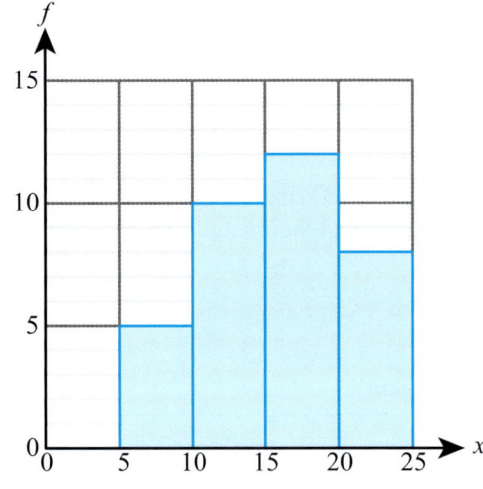

b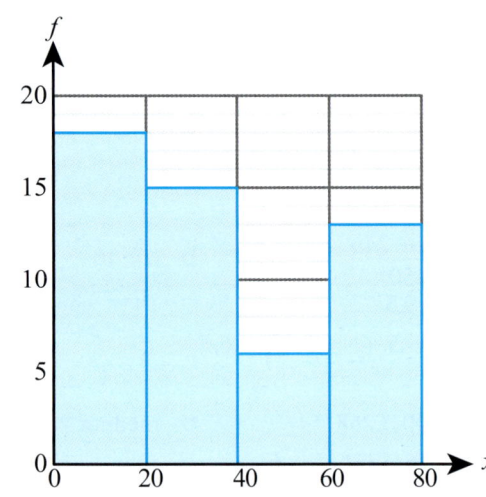

2 a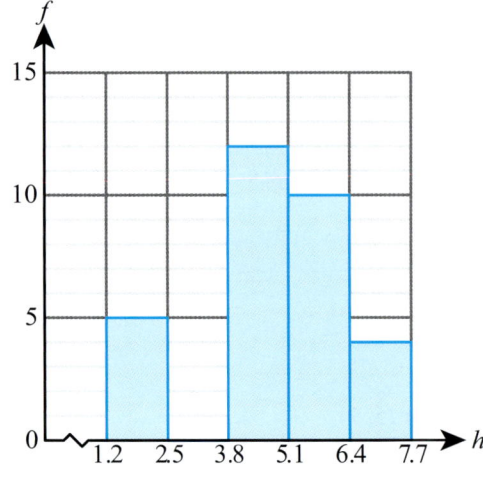

b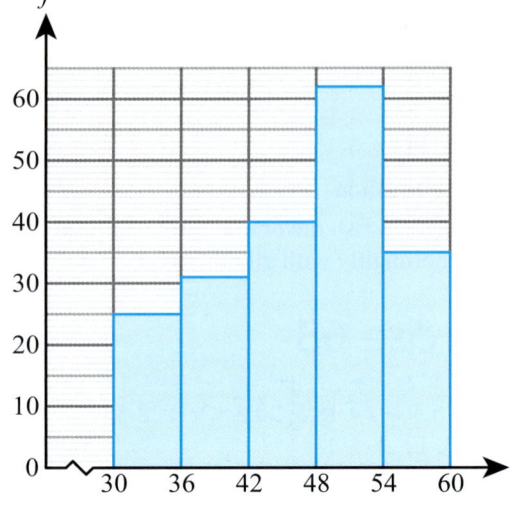

3 a 16   **b** 30
4 a 23   **b** 4
5 a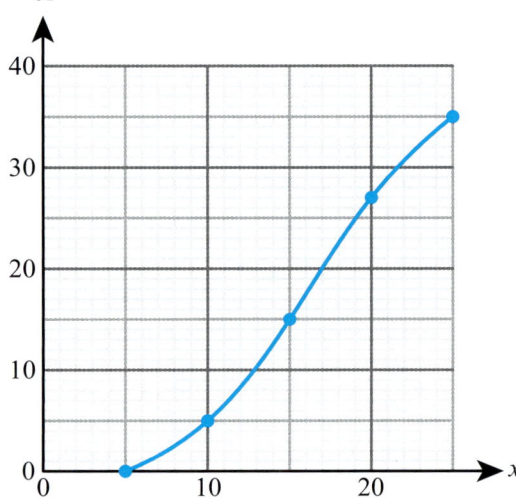

b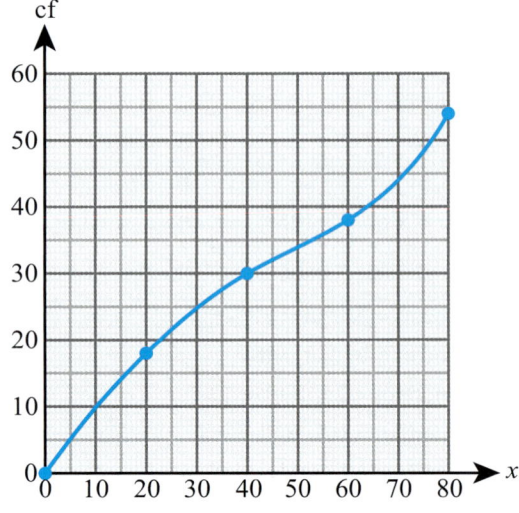

Answers

6 a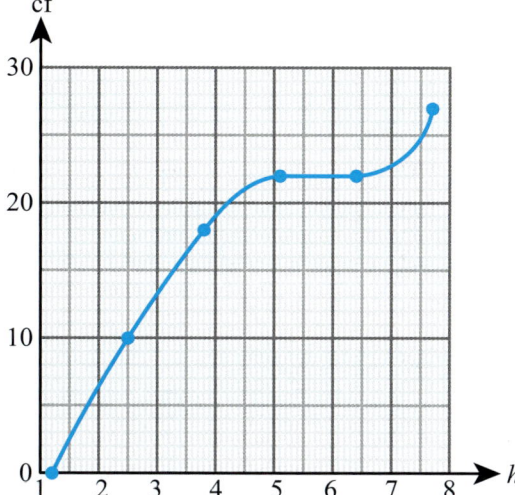

b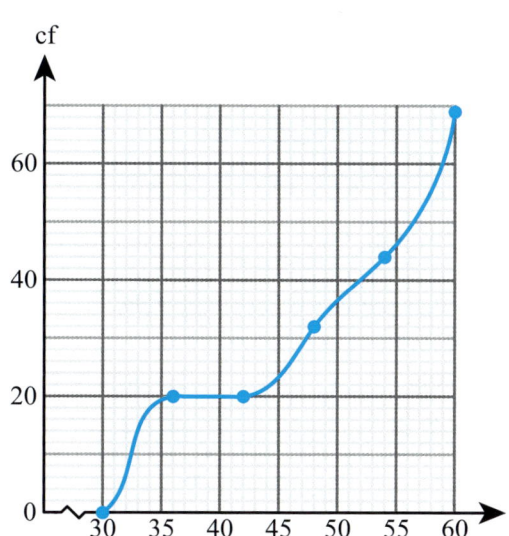

7 a i 13
   ii 20
   iii 6
   iv 21
 b i 4.6
   ii 6.5
   iii 2.3
   iv 6.6

8 a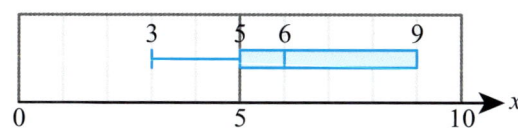

b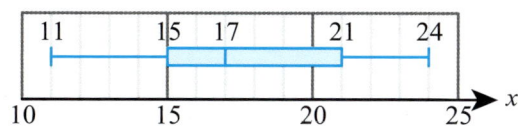

9 a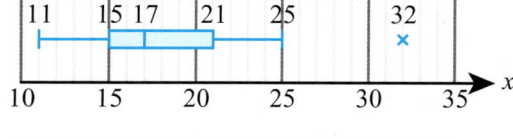

b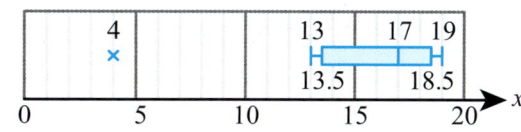

10 a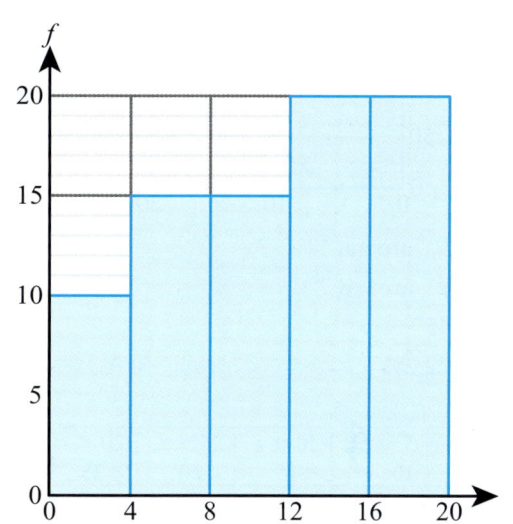

b 25

11 a 105

b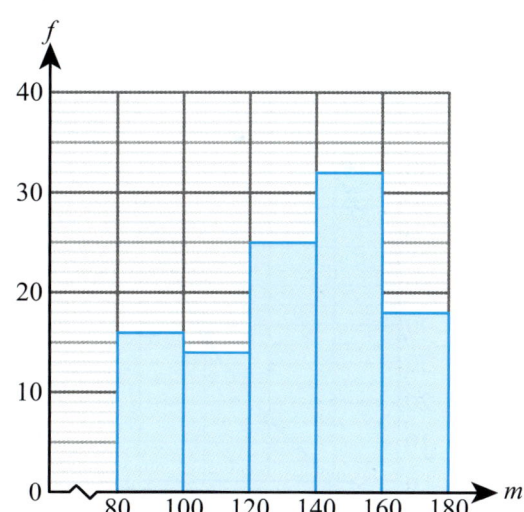

c 46%

12 a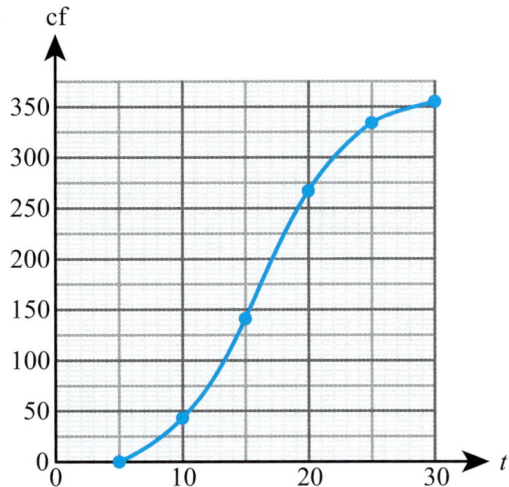

b i around 17°C

ii around 7°C

13 a 45

b 22.2%

c

Time (min)	$5 \leq t < 10$	$10 \leq t < 15$	$15 \leq t < 20$	$20 \leq t < 25$	$25 \leq t < 30$
Freq	7	9	15	10	4

d 16.9 minutes

14 a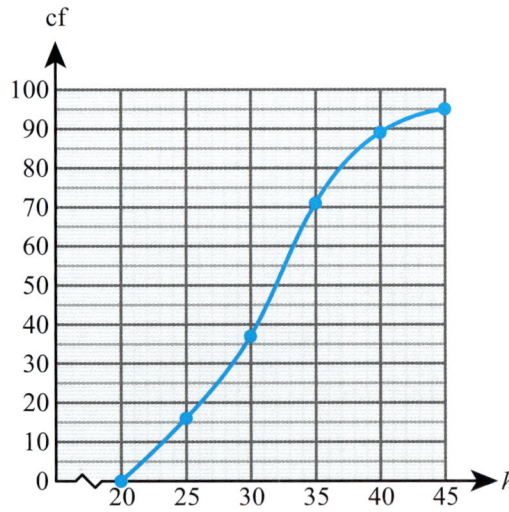

b median ≈ 31 cm, IQR ≈ 9 cm

c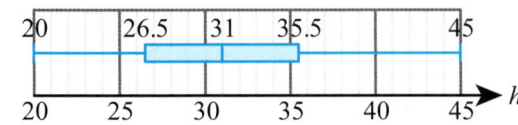

15 a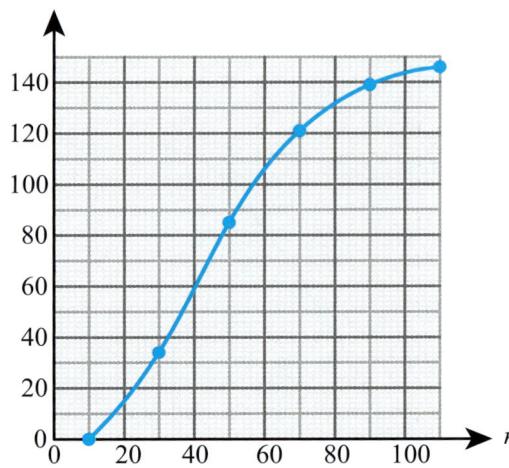

b 40

c 45, 32

d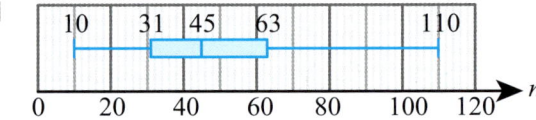

e Overall, fewer candidates take History SL. History has a larger spread of numbers than maths (based on the IQR).

16 159 cm

17 a 160   **b** 90%

c 75   **d** 72, 18

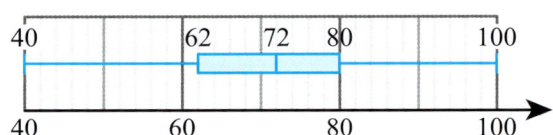

e The second school has a higher median score, but more variation in the scores.

18 a $Q_1 = 1, Q_2 = 2, Q_3 = 3$

b 2

c 7 is an outlier

d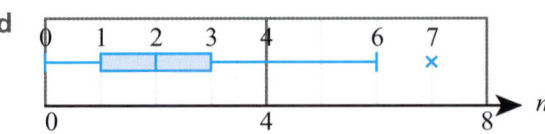

19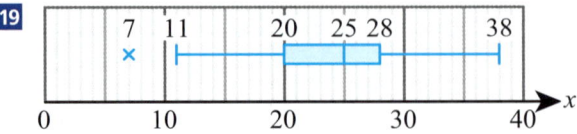

Answers

20 a

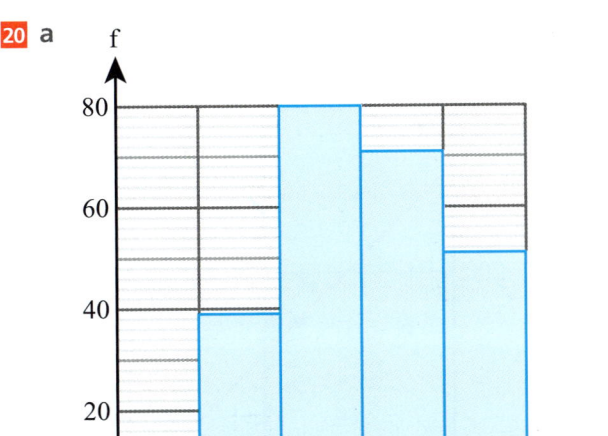

b 59.9

20 A2, B3, C1

Exercise 6D

1 a

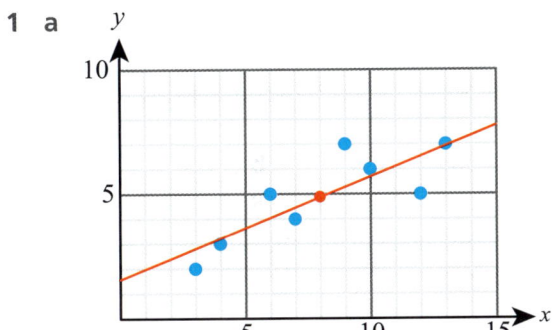

b

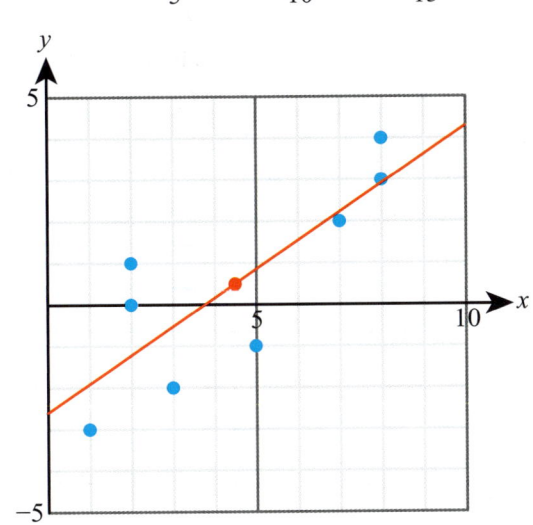

2 a

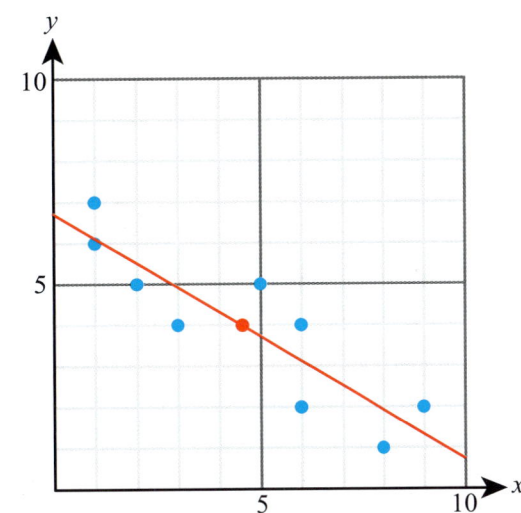

b

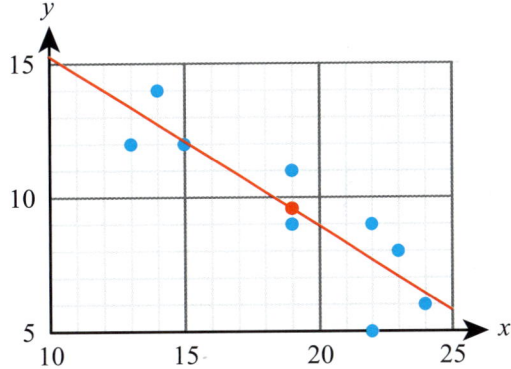

3 a

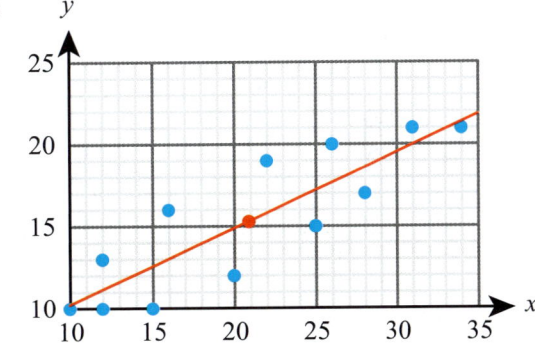

b

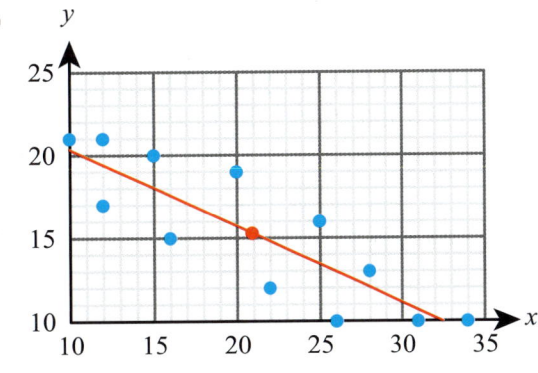

4 a weak positive b strong positive
5 a strong negative b weak negative
6 a no correlation (circle)
  b no correlation (V-shape)
7 a no correlation (slightly scattered around a vertical line)
  b no correlation (slightly scattered around a horizontal line)
8 a i $r = 0.828$ b i $r = -0.886$
    ii strong positive   ii strong negative
    iii yes   iii yes
9 a i $r = 0.542$
    ii weak/moderate positive
    iii no
  b i $r = -0.595$
    ii weak/moderate negative
    iii yes
10 a $y = 0.413x + 1.57$ b $y = 0.690x - 2.60$
11 a $y = -0.589x + 6.72$ b $y = -0.632x + 21.6$
12 a i $y = 21.9$ ii reliable
   b i $y = -4.26$
     ii not reliable (extrapolation)
13 a i $y = 73.5$
     ii not reliable (no correlation)
   b i $y = 87.9$
     ii not reliable (no correlation)
14 a, c, d

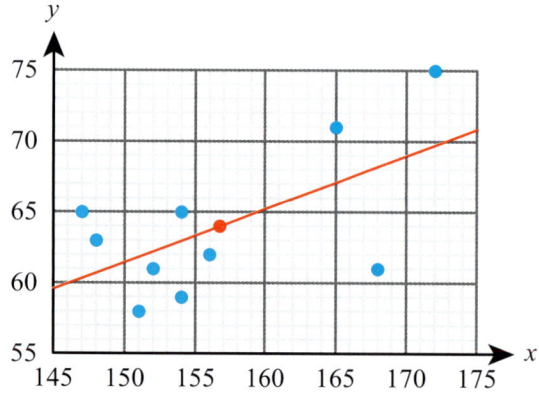

  b weak positive
  c 156.7 cm, 64 cm
  e arm length ≈ 61.5 cm
  f i appropriate
    ii not appropriate (extrapolation)
    iii not appropriate (different age from sample)

15 a, d

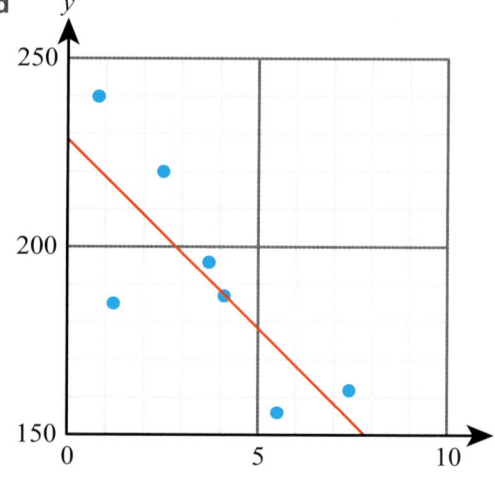

  b weak negative
  c 3.6 km, $192 000
  e average price ≈ $161 000
16 A1, B2, C3, D4
17 a 0.688   b $y = 0.418x + 18.1$
   c 46.5 (or 47)
   d no (correlation does not imply causation)
18 a 0.745
   b The larger the spend on advertising, the larger the profit.
   c $y = 10.8x + 188$
   d i $1270   ii $2350
   e the first one (no extrapolation required)
19 a −0.0619
   b $y = -0.370x + 51.6$
   c no (no correlation)
20 a

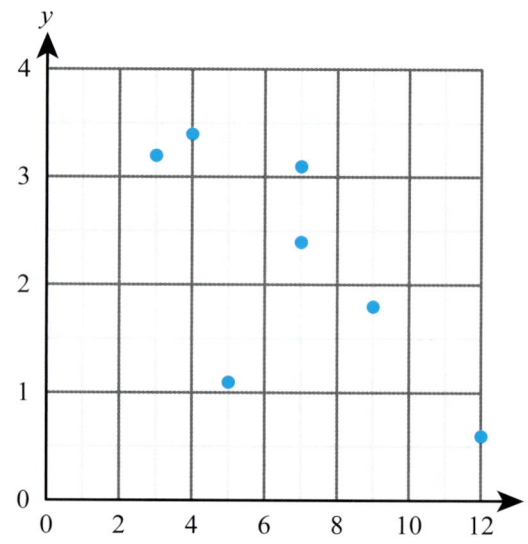

   b −0.695

Answers

c   It shows statistically significant negative correlation.

21 a

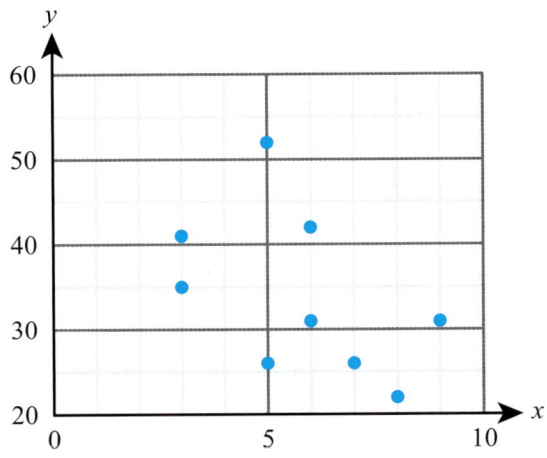

b   weak negative   c   −0.480
d   not significant

22 a   moderate positive correlation
b   yes (there is correlation and value is within range of data)
c   40.8 cm

23 a   0.820   b $m = 0.631t + 5.30$
c   For every extra minute practice he can expect 0.631 extra marks. With no practice he can expect around 5 marks.

24 a   Positive correlation: the larger the advertising, budget, the larger the profit.
b   no (correlation does not imply cause)
c   i   For each €1000 euros spent on advertising, the profit increases by €3250.
     ii   With no advertising the profit would be €138 000.

25 a

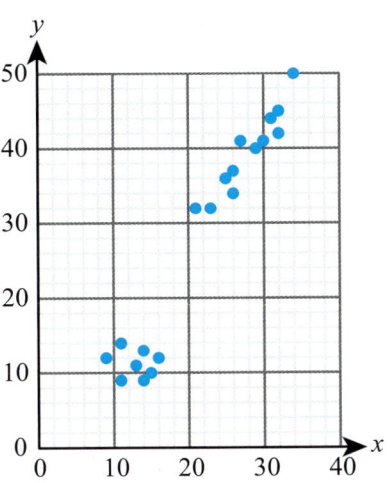

b   summer and winter

c   positive correlation in the summer, no correlation in winter
d   39.5

26 a

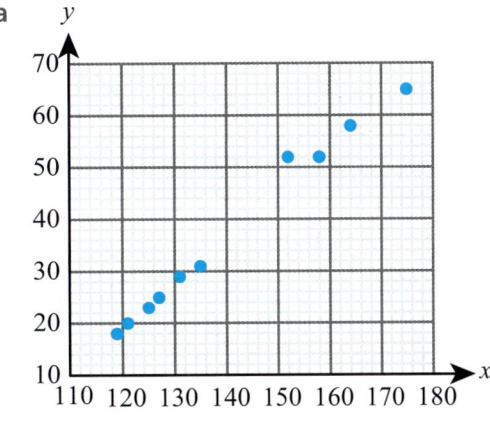

b   children and adults
c   children (126, 24.3), adults (162, 56.8)

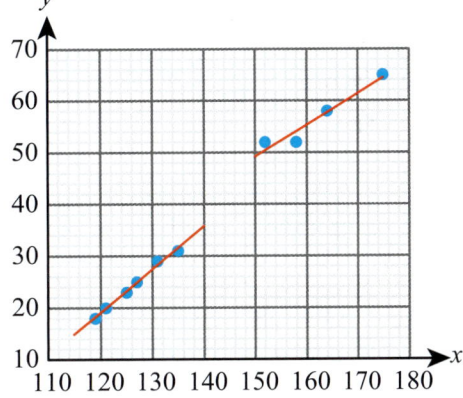

d   19.0 kg

27 a $LQ(x) = 12$, $UQ(x) = 24$, $LQ(y) = 11$, $UQ(y) = 19$

c, d, e

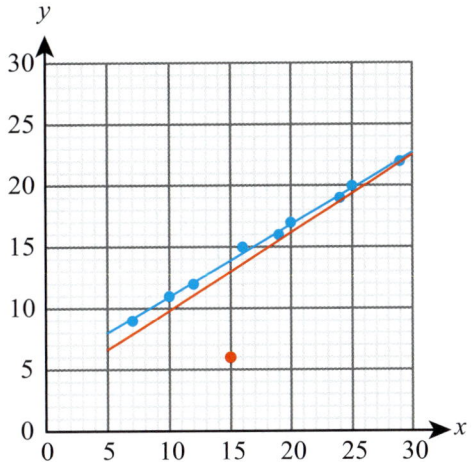

Chapter 6 Mixed Practice

1 a i systematic sampling
  ii e.g. Students may take books out on the same day each week.
 b i Each possible sample of 10 days has an equal chance of being selected.
  ii Representative of the population of all days.
 c i 11
  ii 17.4
  iii 3.17

2 a

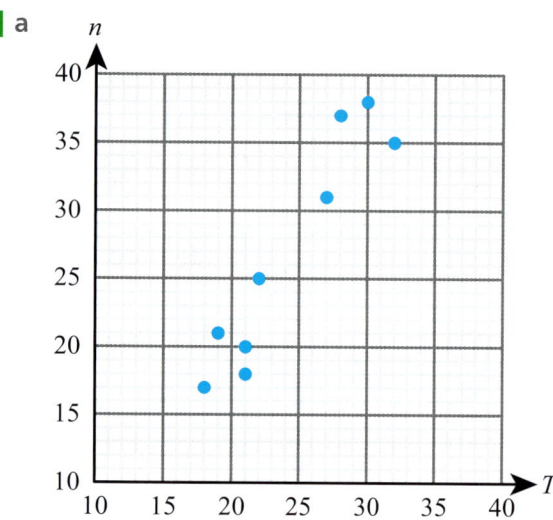

 b strong positive correlation
 c $n = 1.56T - 10.9$
 d 30.0

3 a 2.38 kg
 b

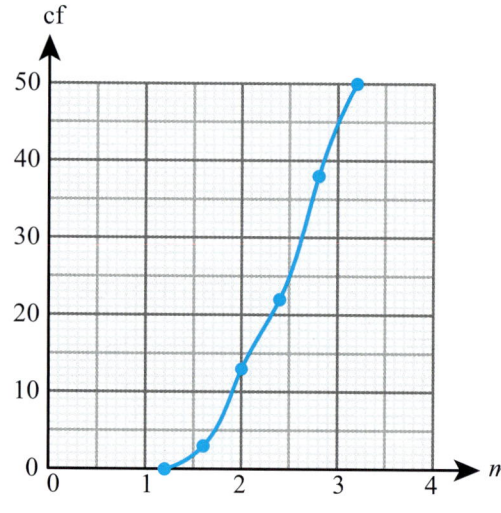

 c 2.5 kg, 0.9 kg
 d

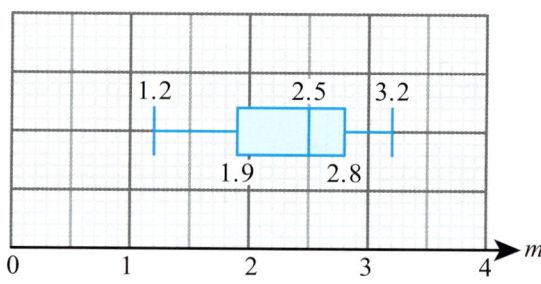

4 a discrete
 b 0
 c i 1.47
  ii 1.5
  iii 1.25

5 a 4
 b

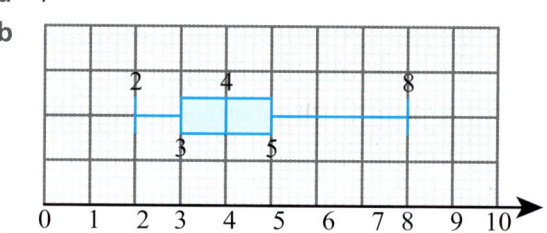

 c 10

6 a 0.996
 b $a = 3.15$, $b = -15.4$
 c 66.5

7 a cf

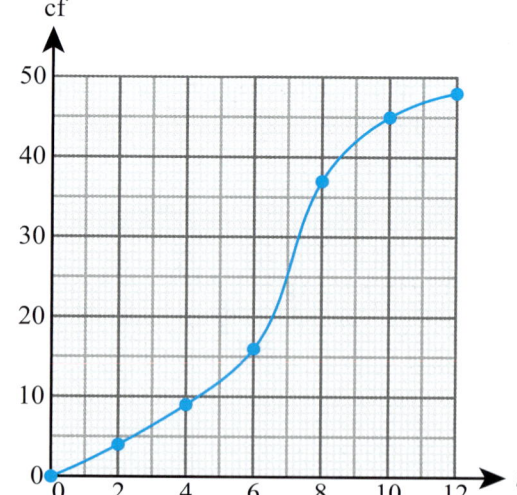

 b i 6.9 minutes
  ii 2.9 minutes
  iii 9.3 minutes

Answers

c

Time	Freq
$0 \leq t < 2$	4
$2 \leq t < 4$	5
$4 \leq t < 6$	7
$6 \leq t < 8$	11
$8 \leq t < 10$	8
$10 \leq t < 12$	3

d 6.21 minutes

8 a med = 46, $Q_1 = 33.5$, $Q_3 = 56$
b yes (93)
c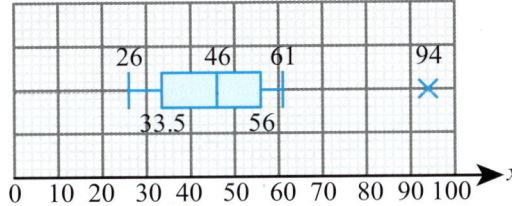

9 a 121 cm, 22.9 cm^2
b 156 cm, 22.9 cm^2
10 mean = \$55.02, sd = \$43.13
11 $x = 6$
12 a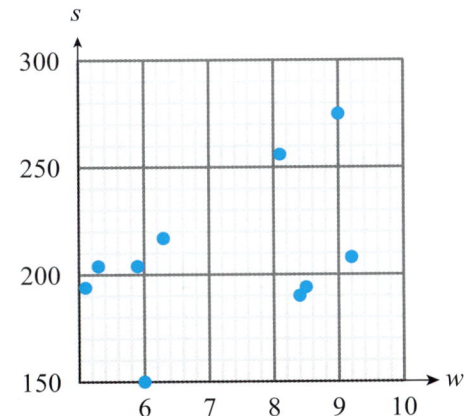

b $w = 19.1s + 99.0$
c 0.994 – strong positive correlation
d 252 g
e 137 g to 175 g; extrapolating from the data so not reliable.
13 a i Athletes generally do better after the programme.
  ii Better athletes improve more.
b 11.6 miles
c i 0.84
  ii $Y = 1.2X + 3.2$

Chapter 7 Prior Knowledge

1 a {2, 3, 4, 5, 6, 7, 8, 9}
  b {5, 8}
2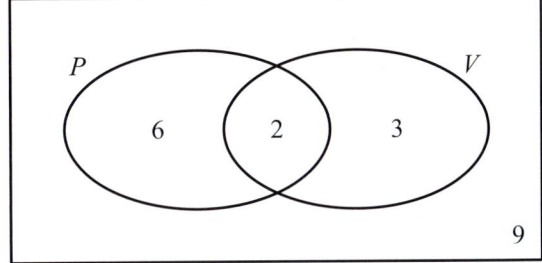

3

Draw 1	Draw 2	
$\frac{6}{10}$ R	$\frac{5}{9}$ R	$\frac{6}{10} \times \frac{5}{9} = \frac{30}{90} = \frac{1}{3}$
	$\frac{4}{9}$ B	$\frac{6}{10} \times \frac{4}{9} = \frac{24}{90} = \frac{4}{15}$
$\frac{4}{10}$ B	$\frac{6}{9}$ R	$\frac{4}{10} \times \frac{6}{9} = \frac{24}{90} = \frac{4}{15}$
	$\frac{3}{9}$ B	$\frac{4}{10} \times \frac{3}{9} = \frac{12}{90} = \frac{2}{15}$

Exercise 7A

1 a $\frac{3}{10}$ b $\frac{2}{15}$
2 a $\frac{1}{5}$ b $\frac{3}{5}$
3 a $\frac{1}{2}$ b $\frac{1}{3}$
4 a $\frac{1}{4}$ b $\frac{1}{13}$
5 a $\frac{1}{26}$ b $\frac{3}{26}$
6 a 0.94 b 0.55
7 a $\frac{11}{20}$ b $\frac{23}{40}$
8 a 0.85 b 0.13
9 a $\frac{47}{120}$ b $\frac{41}{48}$
10 a 0.73 b 0.66
11 a 0.44 b 0.11
12 a 4 b 27
13 a 12 b 6
14 a 4.8 b 7.5

15 a 1.6 **b** 1.5
16 a 0.0743 **b** 66.9
17 7.5
18 15
19 8
20 0.75
21 $1162.50
22 a 1.5 **b** 0.8
23 $\frac{1}{16}$
24 $\frac{78}{25}$

Exercise 7B

1
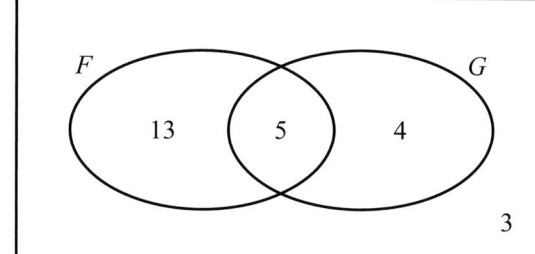
  a $\frac{3}{25}$ **b** $\frac{4}{25}$

2
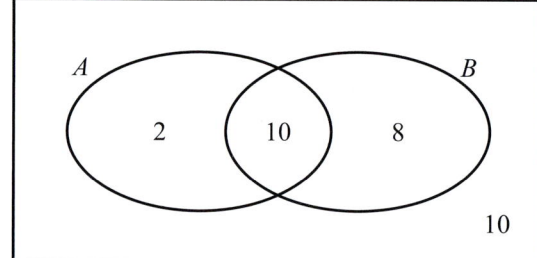
  a $\frac{4}{15}$ **b** $\frac{1}{3}$

3
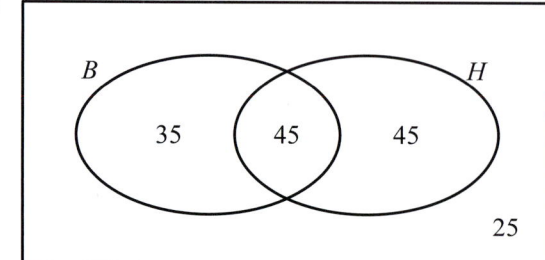
  a $\frac{7}{30}$ **b** $\frac{9}{30}$

4
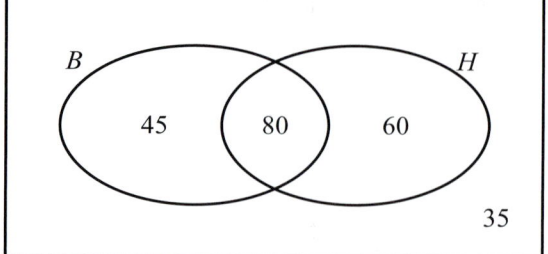
  a $\frac{37}{44}$ **b** $\frac{7}{11}$

5
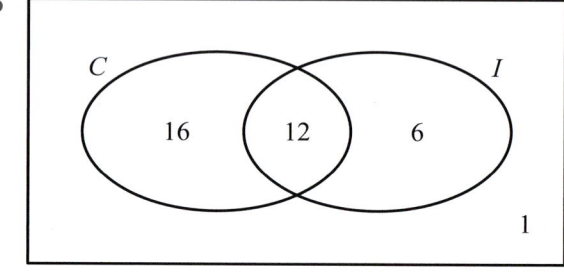
  a $\frac{6}{35}$ **b** $\frac{16}{35}$

6
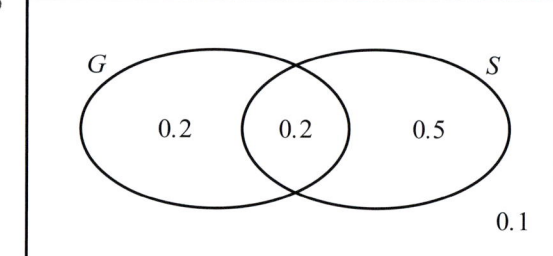
  a 0.1   **b** 0.5

7
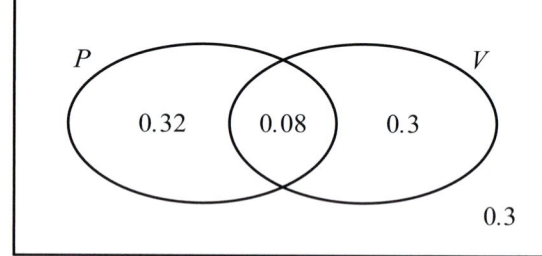
  a 0.3   **b** 0.32

Answers

8 Transport / Late?

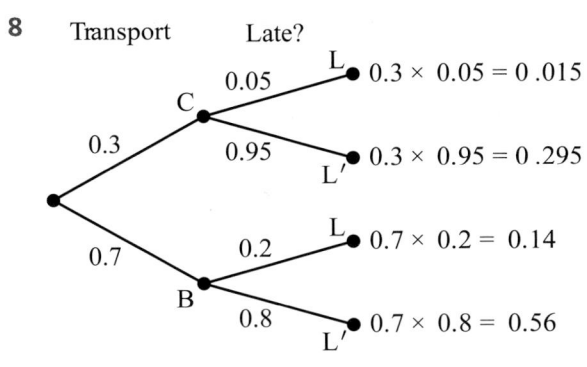

a 0.155   b 0.845

9 Pizza? / Fries?

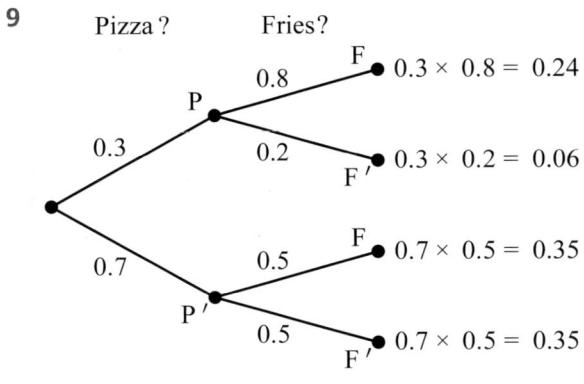

a 0.41   b 0.59

10 Dice / Coin

Dice branch: $\frac{1}{6}$ → 6, $\frac{5}{6}$ → 6′; Coin 0.5 H, 0.5 T
- $\frac{1}{6} \times 0.5 = \frac{1}{12}$
- $\frac{1}{6} \times 0.5 = \frac{1}{12}$
- $\frac{5}{6} \times 0.4 = \frac{1}{3}$
- $\frac{5}{6} \times 0.6 = \frac{1}{2}$

a $\frac{5}{12}$ b $\frac{7}{12}$

11 First / Second

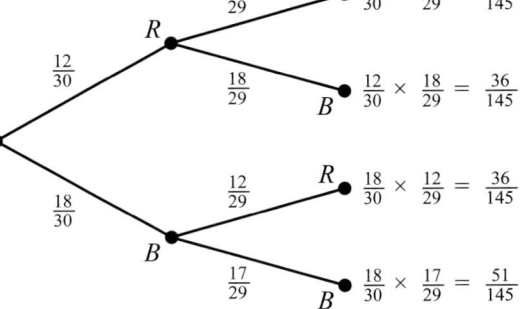

a $\frac{73}{145}$ b $\frac{3}{5}$

12 Revise? / Pass?

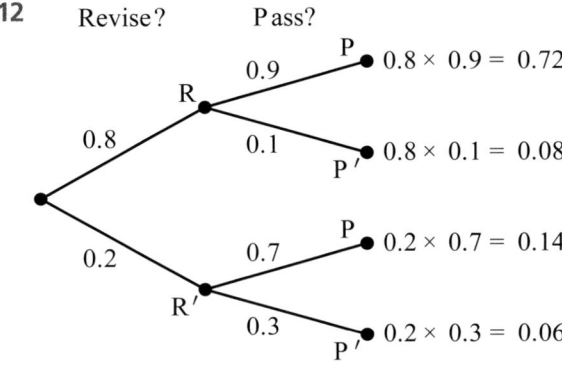

a 0.72   b 0.14

13

	First Dice					
Second Dice	**1**	**2**	**3**	**4**	**5**	**6**
1	2	3	4	5	6	7
2	3	4	5	6	7	8
3	4	5	6	7	8	9
4	5	6	7	8	9	10
5	6	7	8	9	10	11
6	7	8	9	10	11	12

a $\frac{5}{36}$ b $\frac{1}{6}$

14

Second Dice \ First Dice	1	2	3	4	5	6
1	=	>	>	>	>	>
2	<	=	>	>	>	>
3	<	<	=	>	>	>
4	<	<	<	=	>	>
5	<	<	<	<	=	>
6	<	<	<	<	<	=

a $\frac{1}{6}$ b $\frac{5}{12}$

15

	H	T
H	H, H	T, H
T	H, T	T, T

a $\frac{1}{4}$ b $\frac{1}{2}$

16

	B	G
B	B, B	G, B
G	B, G	G, G

a $\frac{1}{2}$ b $\frac{1}{4}$

17

AILT	IALT	LAIT	TAIL
AITL	IATL	LATI	TALI
ALIT	ILAT	LIAT	TIAL
ALTI	ILTA	LITA	TILA
ATIL	ITAL	LTAI	TLAI
ATLI	ITLA	LTIA	TLIA

a $\frac{1}{6}$ b $\frac{1}{2}$

18 a $\frac{38}{109}$ b $\frac{55}{109}$

19 a $\frac{3}{19}$ b $\frac{25}{57}$

20 a $\frac{18}{77}$ b $\frac{16}{77}$

21 a $\frac{23}{65}$ b $\frac{13}{23}$

22 a 0.2 b 0.3

23 a 0.8 b 0.9

24 a 0.5 b 0.7

25 a 0.2 b 0.1

26 a $\frac{11}{15}$ b $\frac{17}{20}$

27 a $\frac{19}{49}$ b $\frac{30}{49}$

28 a $\frac{25}{51}$ b $\frac{1}{17}$

29 a $\frac{3}{31}$ b $\frac{18}{53}$

30 a $\frac{16}{39}$ b $\frac{4}{9}$

31 a 0.14 b 0.24

32 a 0.3 b 0.4

33 a

1	2	3	4	5	6
2	4	6	8	10	12
3	6	9	12	15	18
4	8	12	16	20	24
5	10	15	20	25	30
6	12	18	24	30	36

b 40

34 $\frac{13}{16}$

35 a

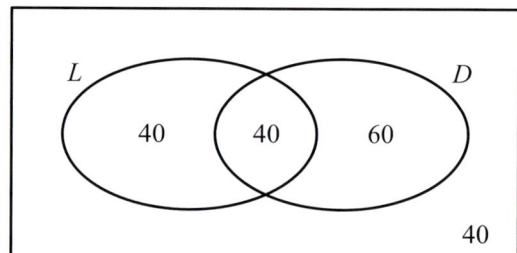

b $\frac{2}{9}$ c $\frac{1}{2}$

36 0.496

37 a i $\frac{1}{8}$ ii $\frac{1}{8}$

b yes

38 a

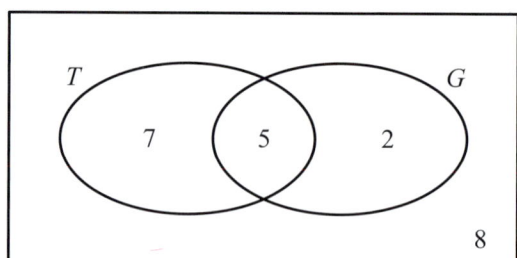

b $\frac{7}{22}$ c $\frac{5}{7}$

39 a 47 b $\frac{83}{130}$ c $\frac{33}{71}$

Answers

40 0.582
41 0.75
42 a $\frac{1}{15}$ b 0.88
43 $\frac{1}{36}$
44 a $\frac{7}{17}$ b $\frac{79}{187}$ c $\frac{26}{187}$
45 0.5
46 0.5
47 a $\frac{1}{5}$ b $\frac{11}{20}$
48 a $\frac{13}{23}$ b $\frac{52}{161}$ c same
49 a 0.226   b 0.001 73
50 a

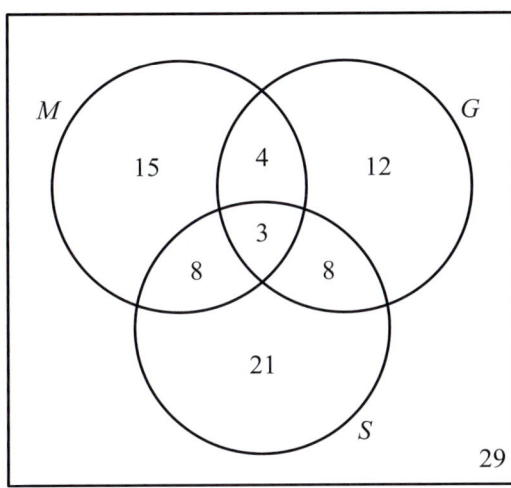

b 0.21   c $\frac{11}{40}$

51 a and b

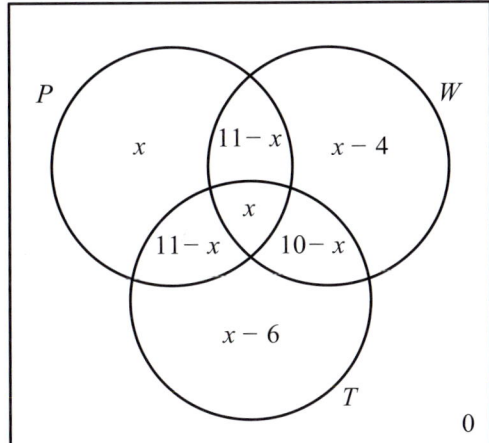

c $\frac{4}{15}$ d $\frac{11}{15}$ e $\frac{1}{4}$

52 a $\frac{1}{3}$ b $\frac{1}{2}$

Chapter 7 Mixed Practice

1 416
2 a $\frac{56}{115}$ b $\frac{21}{115}$ c $\frac{32}{59}$ d $\frac{21}{32}$
3 a $\frac{2}{15}$ b $\frac{37}{60}$
4 a

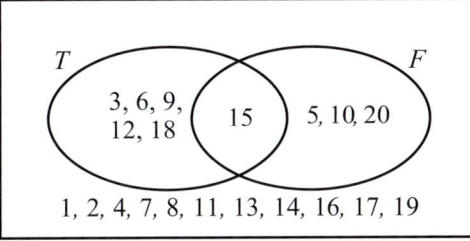

b i $\frac{1}{5}$ ii $\frac{3}{14}$

5 a 0.2   b $\frac{2}{3}$
6 a 0.18   b yes
7 0.92
8 a

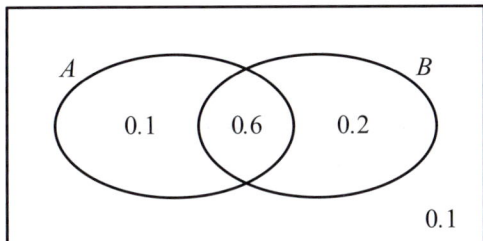

b 0.75
c 0.667

9 a i $\frac{2}{9}$
b $\frac{5}{18}$
   ii $\frac{1}{18}$
10 a 0.143
b 0.111
c 0.238

11 a Toss 1 Toss 2 Toss 3

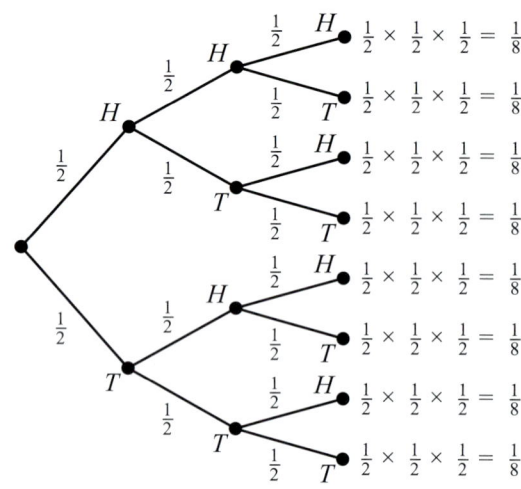

  b $\frac{1}{8}$ **c** $\frac{7}{8}$ **d** $\frac{3}{8}$

12 Asher $\left(\frac{48}{91} > \frac{24}{49}\right)$

13 0.310

14 a $\frac{8}{23}$ **b** $\frac{13}{23}$

15 a $\frac{1}{4}$ **b** $\frac{7}{66}$ **c** $\frac{25}{66}$

16 a

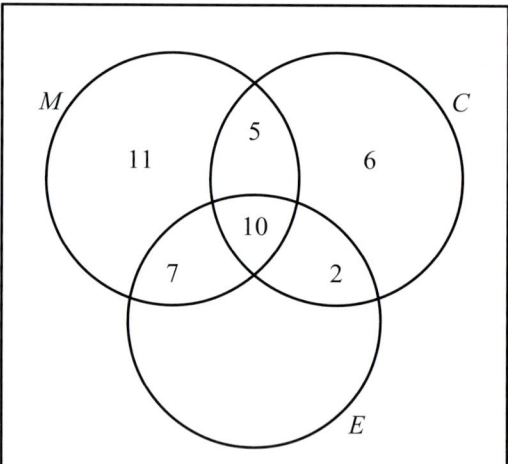

  b 16
  c i 3 **ii** 56
  d i 0.22
    ii 0.05
    iii 0.62
    iv $\frac{31}{39}$

17 b 6 or 15

18 a 15% **b** 60%
  c i 0.442 **d** 0.642

19 a, b

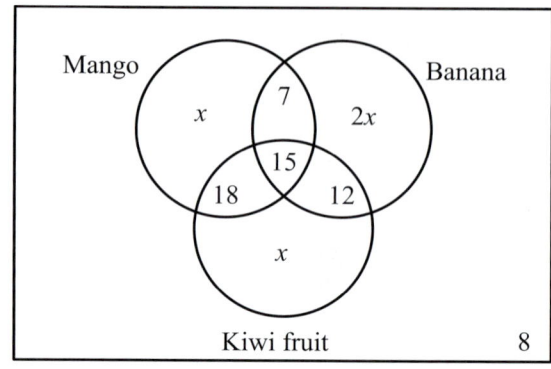

  c $x = 10$
  d i 50 **ii** 82
  e i 0.08 **ii** 0.37 **iii** $\frac{15}{82}$
  f $\frac{14}{2475}$

Chapter 8 Prior Knowledge

1 a $\frac{1}{3}$ **b** $\frac{8}{15}$

2

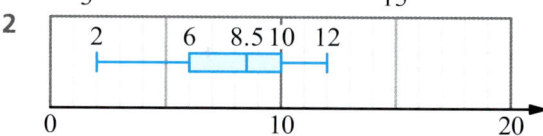

Exercise 8A

1 a

x	0	1	2
$P(X=x)$	$\frac{2}{7}$	$\frac{4}{7}$	$\frac{1}{7}$

  b

x	0	1	2
$P(X=x)$	$\frac{4}{15}$	$\frac{8}{15}$	$\frac{1}{5}$

2 a

x	0	1	2
$P(X=x)$	$\frac{1}{4}$	$\frac{1}{2}$	$\frac{1}{4}$

  b

x	0	1	2
$P(X=x)$	$\frac{1}{4}$	$\frac{1}{2}$	$\frac{1}{4}$

Answers

3 a

x	0	1	2	3
P($X=x$)	$\frac{1}{8}$	$\frac{3}{8}$	$\frac{3}{8}$	$\frac{1}{8}$

b

x	0	1	2	3
P($X=x$)	$\frac{1}{8}$	$\frac{3}{8}$	$\frac{3}{8}$	$\frac{1}{8}$

4 a P($X=x$) = $\frac{1}{6}$ for $x = 1, 2, ..., 6$

b P($X=x$) = $\frac{1}{8}$ for $x = 1, 2, ..., 8$

5 a

x	0	1	2	3
P($X=x$)	$\frac{125}{216}$	$\frac{25}{72}$	$\frac{3}{72}$	$\frac{1}{216}$

b

x	0	1	2	3
P($X=x$)	$\frac{343}{512}$	$\frac{147}{512}$	$\frac{21}{512}$	$\frac{1}{512}$

6 a

x	0	1	2
P($X=x$)	0.64	0.32	0.04

b

x	0	1	2
P($X=x$)	0.09	0.42	0.49

7 a $k = 0.17$
   i 0.77   **ii** 0.416
b $k = 0.26$
   i 0.49   **ii** 0.245

8 a $k = 0.2$
   i 0.4   **ii** 0.75
b $k = 0.36$
   i 0.43   **ii** 0.814

9 a $k = 0.125$
   i 0.6   **ii** 0.833
b $k = 0.2$
   i 0.9   **ii** 0.667

10 a 2.1   **b** 1.9
11 a 2.1   **b** 2.24
12 a 5   **b** 9.2
13 a $k = 0.4$ **b** 0.7   **c** 2.9
14 a $k = 0.2$ **b** 0.4   **c** 6.2
15 a

x	0	1	2
p	$\frac{4}{13}$	$\frac{48}{91}$	$\frac{15}{91}$

b 0.857

16 a $\frac{1}{17}$

b

h	0	1	2
P($H=h$)	$\frac{19}{34}$	$\frac{13}{34}$	$\frac{1}{17}$

c 0.5

17 no
18 $n = 1$
19 $1.50

20 a

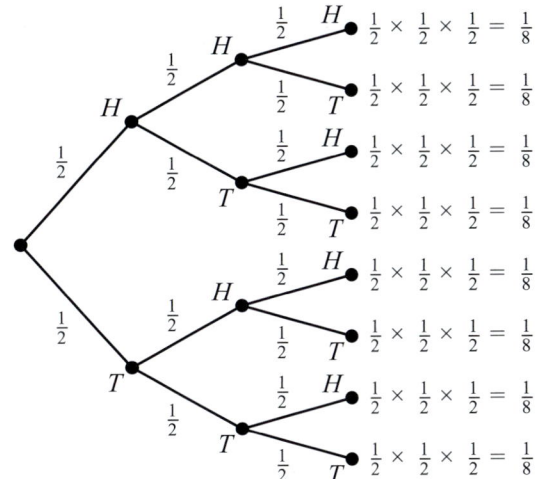

b $\frac{3}{8}$

c

x	0	1	2	3
P($X=x$)	$\frac{1}{8}$	$\frac{3}{8}$	$\frac{3}{8}$	$\frac{1}{8}$

d 1.5

21 a

x	2	3	4	5	6	7	8
P($X=x$)	$\frac{1}{16}$	$\frac{1}{8}$	$\frac{3}{16}$	$\frac{1}{4}$	$\frac{3}{16}$	$\frac{1}{8}$	$\frac{1}{16}$

b 5
22 a 0.4   **b** 2.13
23 a $k = \frac{1}{12}$ **b** 0.75   **c** 5.17
24 a $c = 0.48$ **b** $\frac{3}{11}$ **c** 1.92
25 $a = 0.3$, $b = 0.2$

Exercise 8B

1. a yes, $X \sim B(30, \frac{1}{2})$
   b yes, $X \sim B(45, \frac{1}{6})$
2. a no; number of trials not constant
   b no; number of trials not constant
3. a no; probability not constant
   b no; probability not constant
4. a yes, $X \sim B(50, 0.12)$
   b yes, $X \sim B(40, 0.23)$
5. a no; trials not independent
   b no; trials not independent
6. a 0.160   b 0.180
7. a 0.584   b 0.874
8. a 0.596   b 0.250
9. a 0.173   b 0.136
10. a 0.661   b 0.127
11. a 0.371   b 0.280
12. a 0.571   b 0.280
13. a 0.0792   b 0.231
14. a 0.882   b 0.961
15. a 0.276   b 0.001 69
16. a 6.25, 2.17 b 10, 2.58
17. a 5, 1.58   b 10, 2.24
18. a 6, 2.24   b 3.33, 1.67
19. a $B(10, \frac{1}{6})$ b 0.291 c 0.225
20. a 0.292   b 0.736   c 10.2
21. a 0.160   b 0.872
    c 0.121   d 4.8
22. a 0.983   b 5   c 0.383
23. a All have same probability; employees independent of each other.
    b e.g. May not be independent, as could infect each other.
    c 0.0560
    d 0.0159
24. 0.433
25. a 0.310   b 0.976   c 0.643
26. a 0.650   b 0.765
27. 0.104
28. 0.132
29. a 0.0296   b 0.008 76
    c e.g. A source of faulty components, such as a defective machine, would affect several components.

Exercise 8C

1. a 0.159   b 0.345
2. a 0.726   b 0.274
3. a 0.260   b 0.525
4. a 0.523   b 0.244
5. a 0.389   b 0.552
6. a 0.246   b 0.252
7. a 0.189   b 0.792
8. a 0.133   b 0.132
9. a 13.4   b 6.78
10. a 8.08   b 15.9
11. a 31.0   b 46.0
12. a 28.1   b 46.7
13. no; not symmetrical
14. a

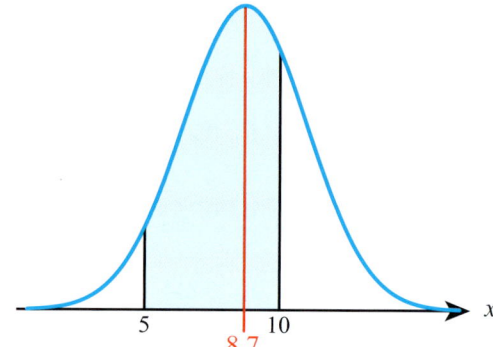

    b 0.660
15. a 0.465   b 0.0228
16. a 0.308   b 0.328
17. a 0.274   b 10.4   c 0.282
18. a 0.0478   b 6.42 minutes
19. a 0.0831   b 6.34 hours   c 62.3
20. a 4.52   b 6.47 m
21. 12.8 s
22. a 15.4   b 6.74
23. 4.61
24. 20.9
25. 15.2
26. predicts 4% get a negative score
27. a symmetrical

Answers

b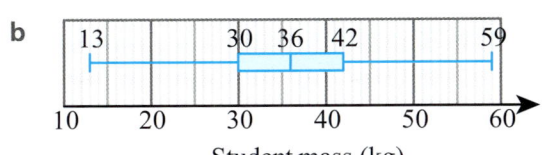
Student mass (kg)

28 a 0.0228 e.g. anticipating the start gun
   b 0.159   c 0.488
   d The same distribution is true in all races. Unlikely to be true.

29 $728

Chapter 8 Mixed Practice

1 a $k = 0.5$ b 0.6   c 2.9
2 a 0.296   b 0.323
3 a 0.347   b 0.227
4 215
5 0.346
6 a $\frac{1}{2}, \frac{1}{3}, \frac{1}{6}$
   b no; expected outcome is not 0
7 $N = 7$
8 a 0.933
   b i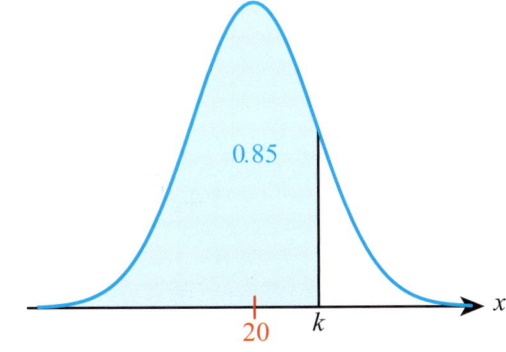
      ii $k = 23.1$
9 a $P = \frac{1}{3}$ b 0.153   c 0.0435
10 a 0.206   b 0.360
11 a 0.356
   b a box of six containing exactly one VL egg
12 a 0.354   b 0.740
13 0.227
14 0.232
15 a 3.52   b 0.06   c 0.456
16 $a = \frac{1}{6}$
17 a 10.8 cm   b 0.698%
18 22.8% of times would be negative.
19 320 ml
20 a $\frac{1}{9}$ b 0.0426
21 a 0.925   b $k = 20.4$
22 a 0.835   b $k = 1006.58$ c $a = 6.58$
23 a i 1   ii 0.0579
   b ii $a = 0.05, b = 0.02$
   c Bill (0.19)
24 a i 0.845   ii 1.69
   b That Josie's second throw has the same distribution as the first and is independent. These seem unlikely.
25 $n = 9$
26 10
27 a 0.919   b 0.0561
28 277
29 b 6

Chapter 9 Prior Knowledge

1 $\frac{1}{4}x^{-1} - \frac{1}{2}x^{-2}$
2 $y = 3x + 11$
3 −2

Exercise 9A

1 a 0.6   b 0
2 a 1.5   b 3
3 a 2   b 3
4 a 1   b 1
5 a 0.693
   In further study you might learn how to show that this is actually ln 2
   b 1.10
6 a 0.2   b 0
7 a 0.5   b 0.5
8 a 1   b 10
9 a −0.5   b 0
10 a 0.805   b 0.347
11 a 0   b 4
12 a 6   b 0
13 a 0.5   b 0.0833

14 a −1 **b** −2
15 a 1 **b** 0.693
16 a $\dfrac{dz}{dv}$ **b** $\dfrac{da}{db}$
17 a $\dfrac{dp}{dt}$ **b** $\dfrac{db}{dx}$
18 a $\dfrac{dy}{dn}$ **b** $\dfrac{dt}{df}$
19 a $\dfrac{dh}{dt}$ **b** $\dfrac{dw}{dv}$
20 a $\dfrac{dw}{du}$ **b** $\dfrac{dR}{dT}$
21 a $\dfrac{dy}{dx} = y$ **b** $\dfrac{dy}{dx} = \dfrac{x}{2}$
22 a $\dfrac{ds}{dt} = kt^2$ **b** $\dfrac{dq}{dp} = k\sqrt{q}$
23 a $\dfrac{dP}{dt} = kL(P)$ **b** $\dfrac{dP}{dt} = kA(P)$
24 a $\dfrac{dh}{dx} = \dfrac{1}{h}$ **b** $\dfrac{dr}{d\theta} = k$
25 a $s'(t) = 7$ **b** $q'(x) = 7x$
26 a $\dfrac{d(pH)}{dT} = k$

 b $\dfrac{dC}{dn} = k(1000 - n)$
Technically, the number of items produced is not a continuous quantity so we should not formally use derivatives. However, as long as n is large, as is usually the case in economic models, then approximating it as continuous is reasonable.

27 a $\dfrac{dV}{dt} = V + 1$ **b** 5
28 57.3
29 1
30 a $\dfrac{x^2 - 1}{x - 1}$
 b 2, the gradient of the curve at $x = 1$
31 3
32 $\dfrac{1}{6}$

Exercise 9B

1 a $x > 1$ **b** $x < 0$
2 a $-1 < x < 1$ **b** $x < -1$ or $x > 6$
3 a $0 < x < 90$ or $270 < x < 360$
 b $180 < x < 360$
4 a

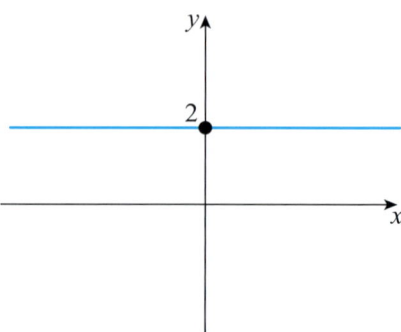

 b

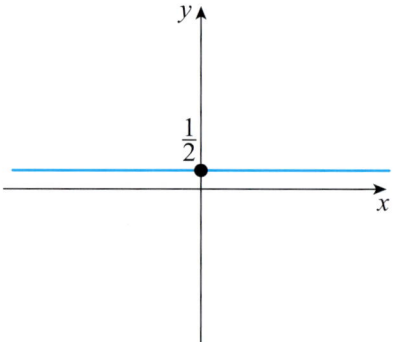

5 a

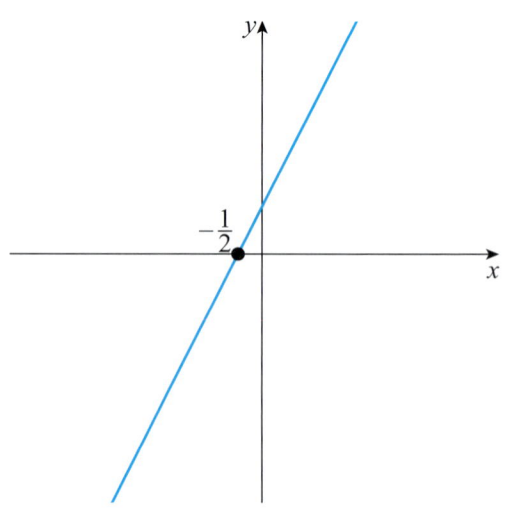

b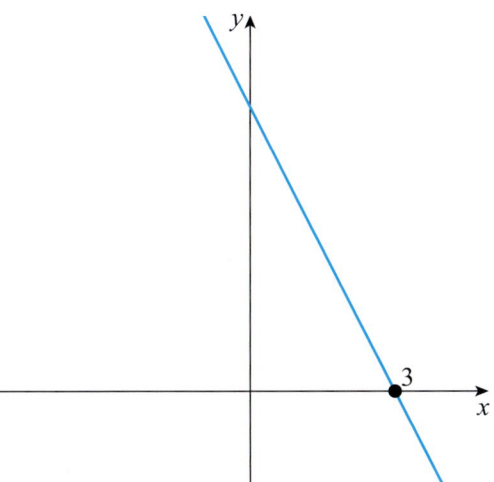

6 a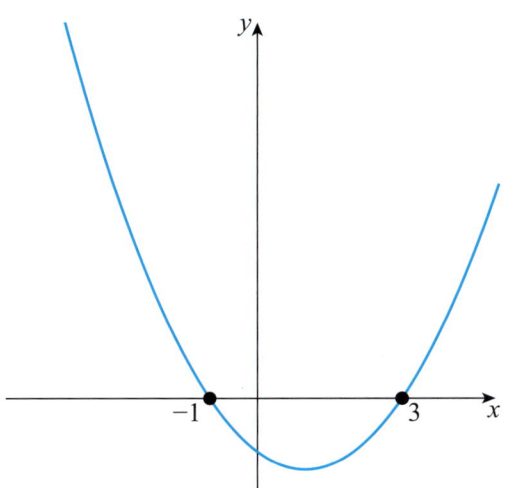

b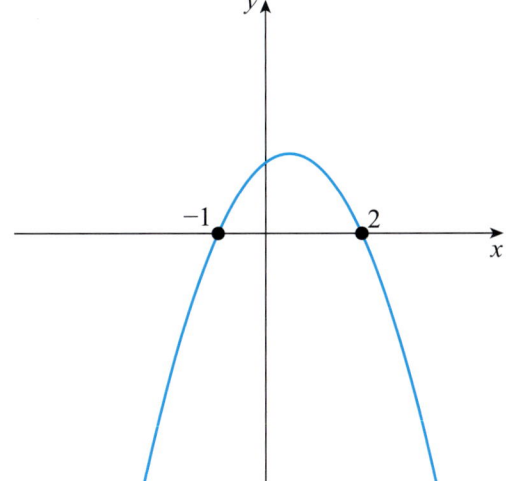

7 a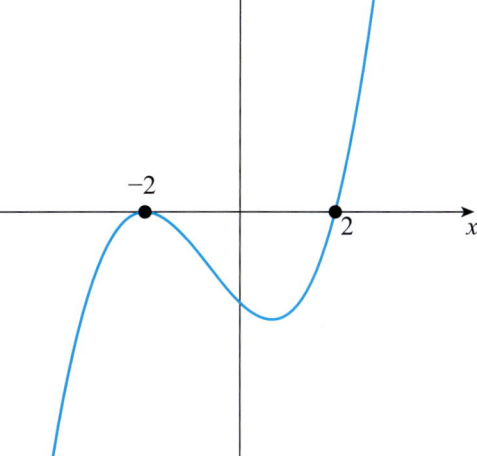

b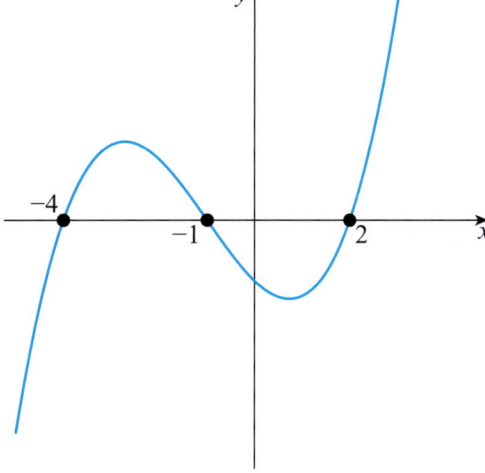

8 a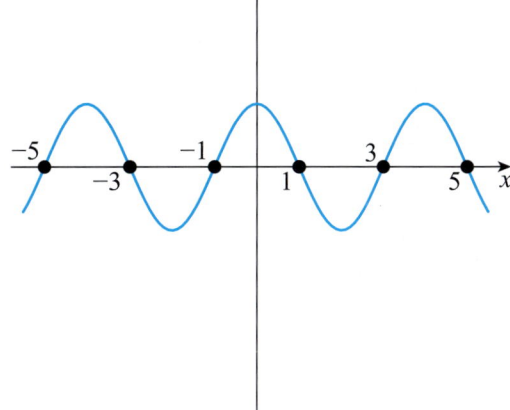

b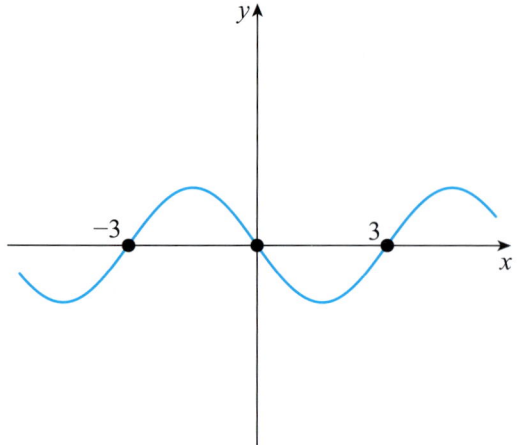

9 a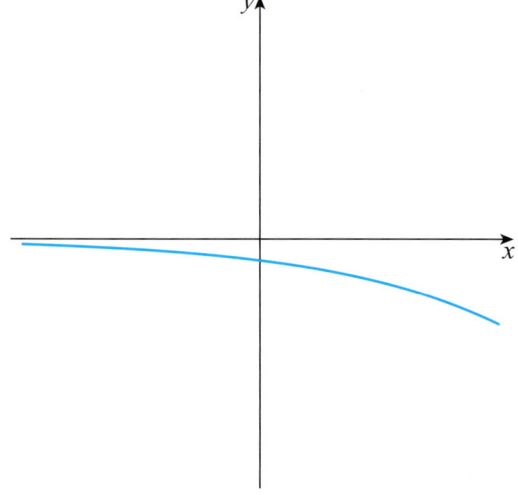

b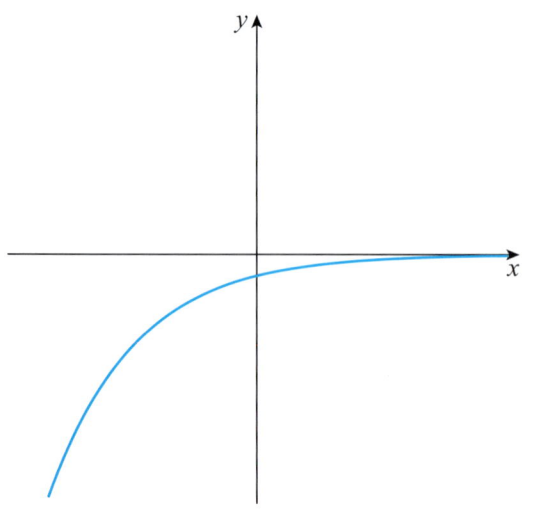

10 a $x > -1$ **b** $x < 3$
11 a $x < -3$ or $x > 3$ **b** $-4 < x < 0$

12 a $x < -1$ or $x > 0$ **b** $x < -\frac{1}{2}$ or $x > \frac{1}{2}$

13 a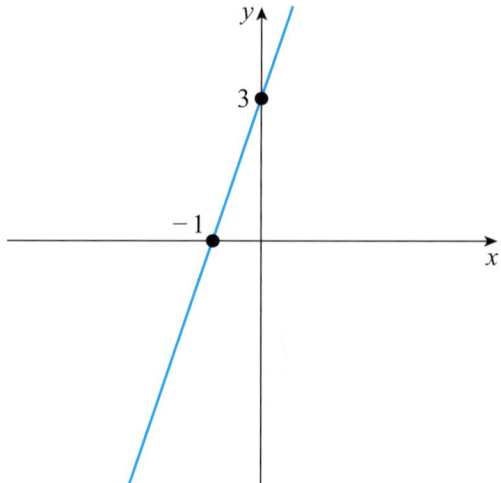

b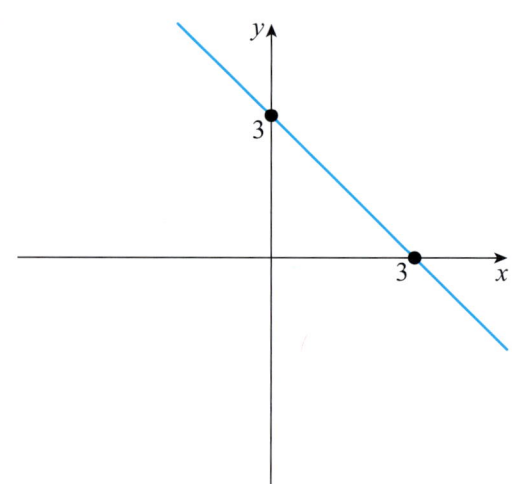

14 a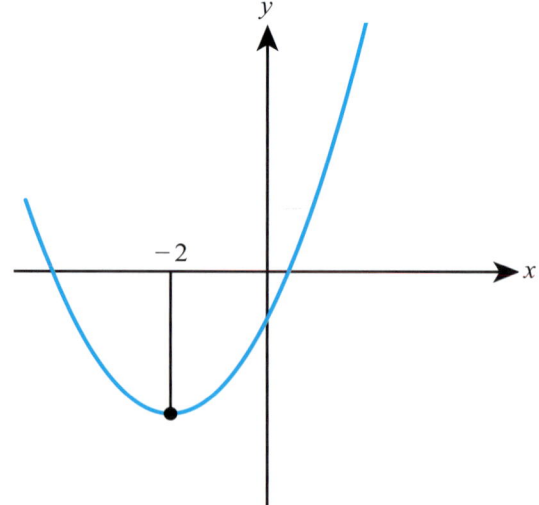

Answers

b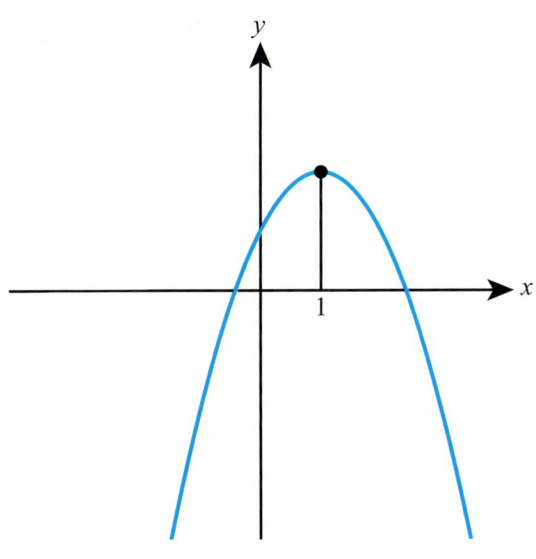

15 a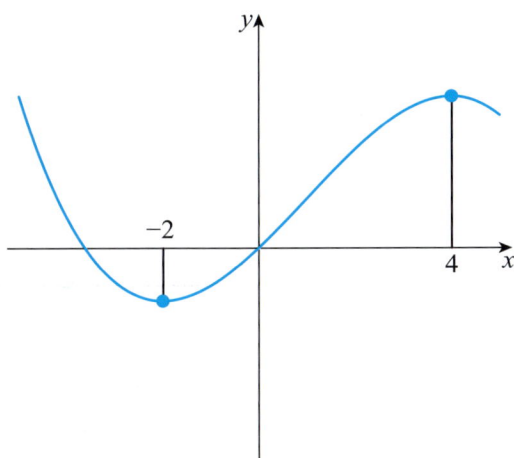

b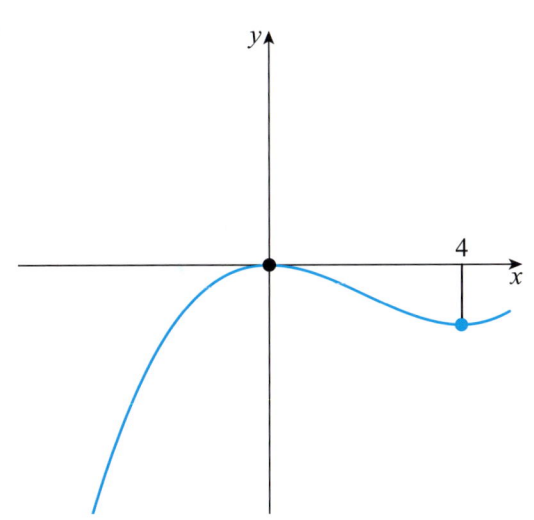

16 a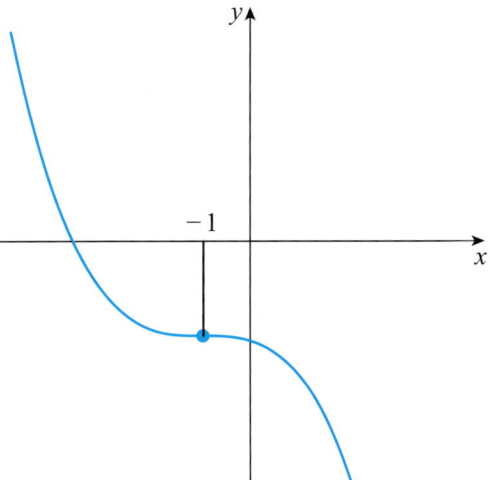

b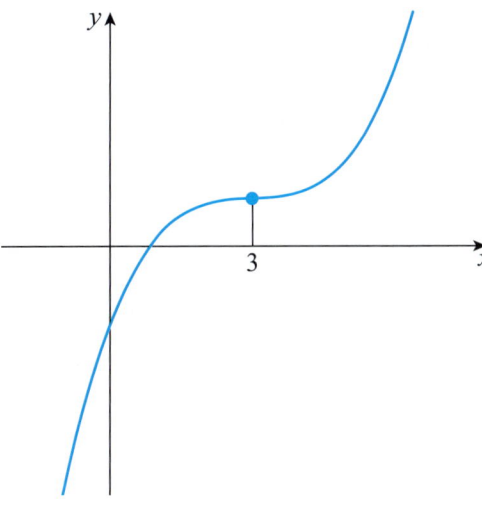

17 a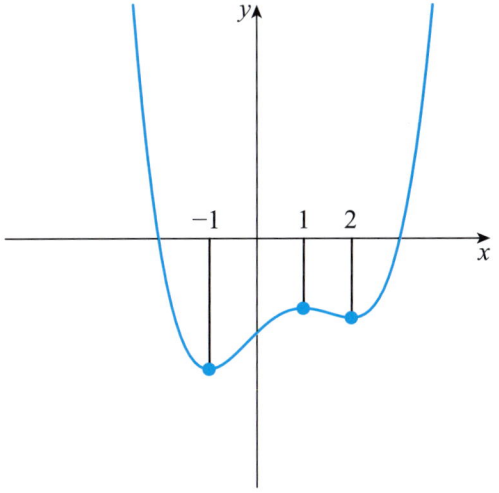

b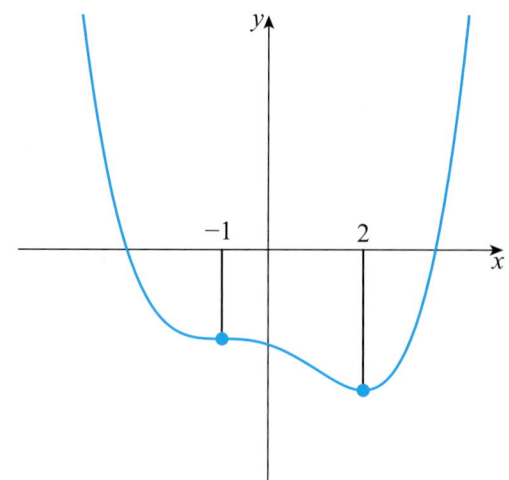

18 a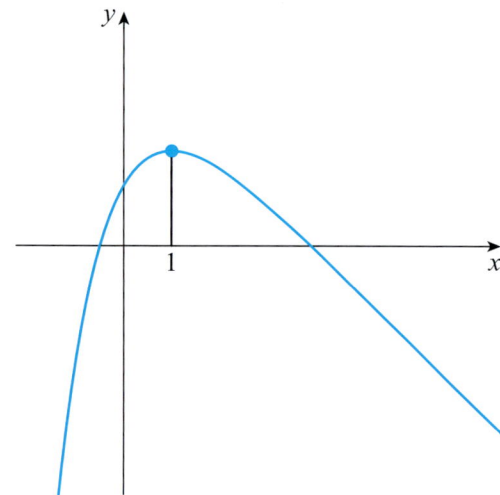

b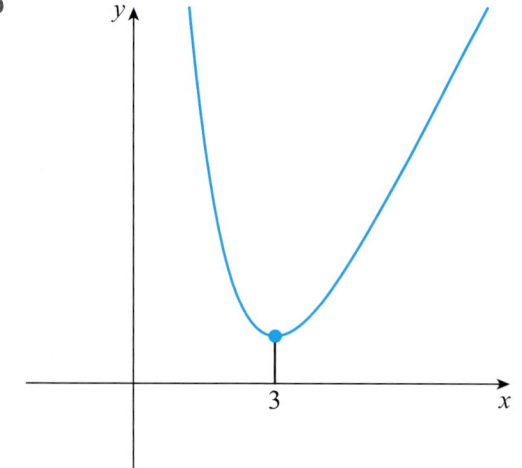

19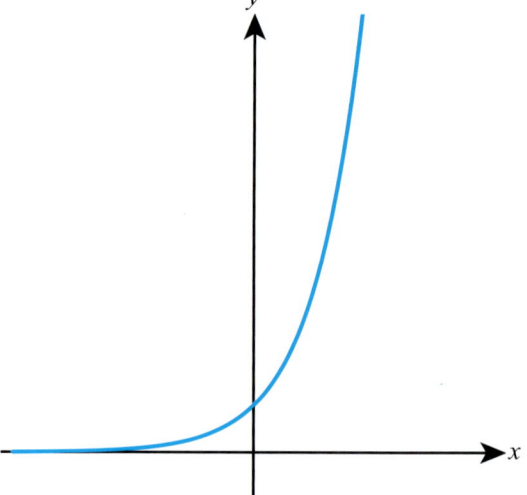

20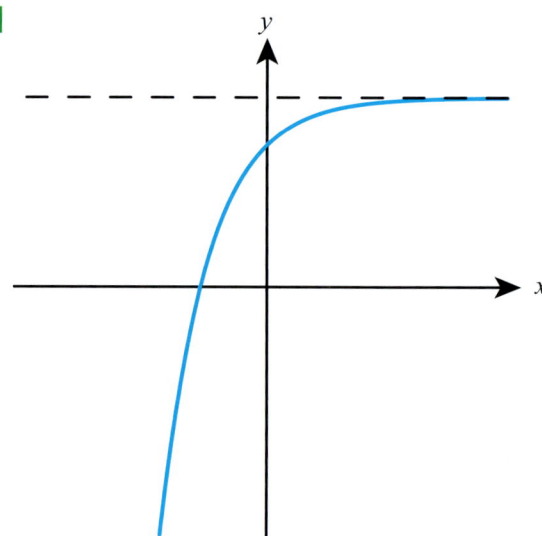

21 a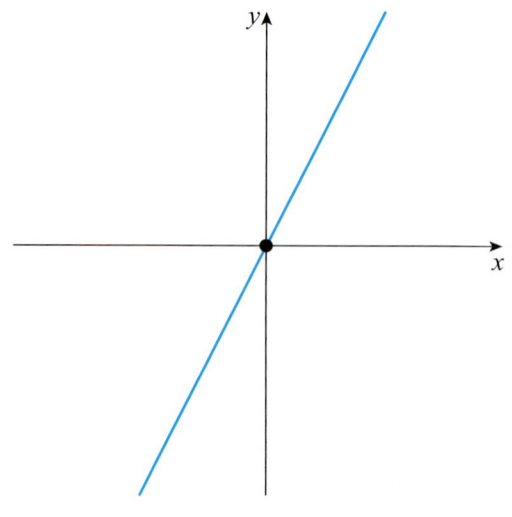

Answers

b

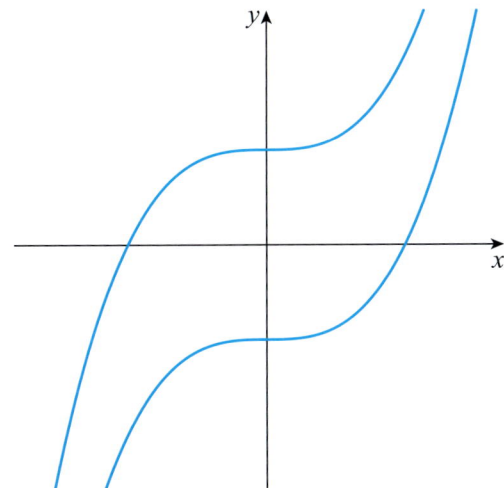

22 a

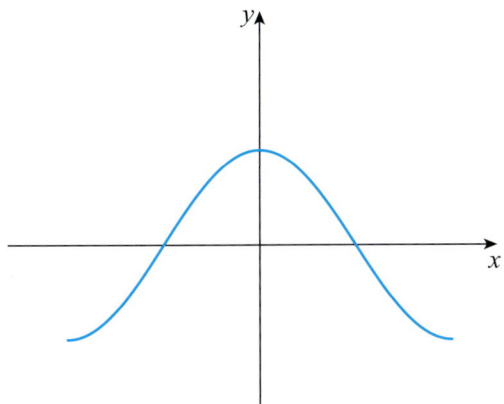

b

[graph]

c Gradient specifies shape but not vertical position.

23 a

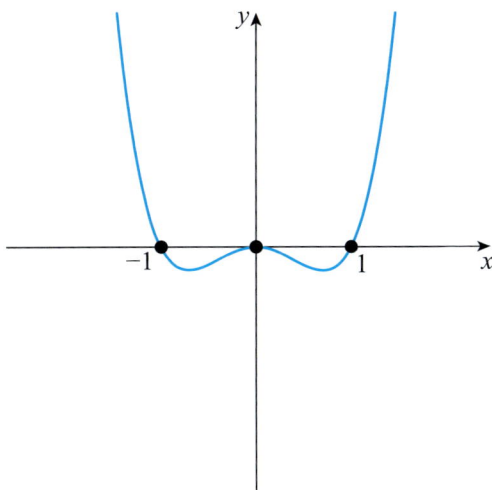

b $x < -1$ or $x > 1$

c $x < -0.707$ or $0 < x < 0.707$

24 **a** and **e**, **b** and **d**, **f** and **c**

25 **a** and **c**, **f** and **b**, **d** and **e**

Exercise 9C

1 a $f'(x) = 4x^3$ **b** $g'(x) = 6x^5$

2 a $h'(u) = -u^{-2}$ **b** $z'(t) = -4t^{-5}$

3 a $\dfrac{dy}{dx} = 8x^7$ **b** $\dfrac{dp}{dq} = 1$

4 a $\dfrac{dz}{dt} = -5t^{-6}$ **b** $\dfrac{ds}{dr} = -10r^{-11}$

5 a $f'(x) = -4$ **b** $g'(x) = 14x$

6 a $\dfrac{dy}{dx} = 6$ **b** $\dfrac{dy}{dx} = 15x^4$

7 a $\dfrac{dy}{dx} = 0$ **b** $\dfrac{dy}{dx} = 0$

8 a $g'(x) = -x^{-2}$ **b** $h'(x) = -3x^{-4}$

9 a $\dfrac{dz}{dx} = -6x^{-3}$ **b** $\dfrac{dy}{dt} = -50t^{-6}$

10 a $\dfrac{dy}{dx} = -x^{-5}$ **b** $\dfrac{dy}{dx} = -x^{-6}$

11 a $f'(x) = \dfrac{3}{2x^2}$ **b** $f'(x) = \dfrac{5}{x^4}$

12 a $f'(x) = 2x - 4$ **b** $g'(x) = 4x - 5$

13 a $\dfrac{dy}{dx} = 9x^2 - 10x + 7$ **b** $\dfrac{dy}{dx} = -4x^3 + 12x - 2$

14 a $\dfrac{dy}{dx} = 2x^3$ **b** $\dfrac{dy}{dx} = -\dfrac{9}{2}x^5$

15 a $\dfrac{dy}{dx} = 3 - \dfrac{3}{4}x^2$ **b** $\dfrac{dy}{dx} = -6x^2 + x^3$

16 a $f'(x) = 8x^3 - 15x^2$
  b $g'(x) = 3x^2 + 6x - 9$

17 a $g'(x) = 2x + 2$ **b** $f'(x) = 2x - 1$

18 a $h'(x) = -\dfrac{2}{x^2} - \dfrac{2}{x^3}$
  b $g'(x) = 8x - \dfrac{18}{x^3}$

19 a a **b** $2ax + b$
20 a $2ax + 3 - a$ **b** $3x^2 + b^2$
21 a $2x$ **b** $3a^2 x^2$
22 a $2ax^{2a-1}$ **b** $-ax^{-a-1} + bx^{-b-1}$
23 a $7a + \dfrac{6b}{x^3}$ **b** $10b^2 x - \dfrac{3a}{cx^2}$
24 a $2a^2 x$ **b** $18a^2 x$
25 a $2x + a + b$ **b** $2abx + b^2 + a^2$
26 a $-\dfrac{10}{x^2}$ **b** $\dfrac{5}{x^2}$
27 a $\dfrac{4}{x^3} - \dfrac{1}{x^2}$ **b** $-\dfrac{3}{x^4} - \dfrac{4}{x^3}$
28 a $0.5 + 3x$ **b** $7x^3 + 4x^7$
29 a $x < 1$ **b** $x < -4$
30 a $x > 6$ **b** $x > 0$
31 $6 + \dfrac{4}{t^2}$
32 $1 - 2m^{-2}$
33 $\dfrac{3}{2}k$
34 $x > \dfrac{1}{2}$
35 $x > -\dfrac{b}{2}$
36 -1
37 $1 - 3x^2$
38 a $-\dfrac{k}{r^2}$ **b** $-\dfrac{V^2}{k}$ **c** 4
39 a $\dfrac{dA}{dL} = q + 2qL$
  b $q > 0$

Exercise 9D

1 a 16   **b** 20
2 a $\dfrac{1}{8}$ **b** 16.75
3 a 32   **b** -7
4 a $\dfrac{23}{9}$ **b** 100.4
5 a 108   **b** 6
6 a 24   **b** $-\dfrac{3}{16}$
7 a 1   **b** ± 2
8 a ± 2 **b** $\pm \dfrac{1}{2}$
9 a $-1, -3$ **b** $\pm \dfrac{1}{\sqrt{2}}$
10 a $4x - y = 1$ **b** $y - x = 0$
11 a $2x + y = 7$ **b** $7x + y = 2$
12 a $x + 3y = 6$ **b** $2x - y = 3$
13 a $y = -x + 15$ **b** $y = 2x - 12$
14 a $y = -x$ **b** $y = -x + 3$
15 a $y = \dfrac{1}{3}x - \dfrac{1}{3}$ **b** $y = \dfrac{1}{16}x + \dfrac{81}{16}$
16 a $y = 2x - 3$ **b** $y = x + 1$
17 a $y = \dfrac{x}{4} + 1$ **b** $y = -\dfrac{x}{16} + \dfrac{3}{4}$
18 a $y = -\dfrac{x}{4} + \dfrac{3}{4}$ **b** $y = -\dfrac{x}{32} + \dfrac{7}{32}$
19 a $y = x - 1$ **b** $y = x - 1$
20 a $y \approx 2.77x - 1.55$ **b** $y \approx 1.10x + 1$
21 a $y = 14 - 4x$ **b** $y = 1 - x$
22 a $8x + 5y = 47$ **b** $4x + 11y = 126$
23 a $x + y \approx 2.10$ **b** $x + 2y \approx 13.4$
24 a $y = -x$ **b** $y = x - 2$
25 a -0.984 **b** -0.894
26 a 2   **b** 0.693
27 a $-3, -1$ **b** $0, -2$
28 a $4x^3 - 1$ **b** -1 **c** $y = x$
29 a $3x^2 - x^{-2}$ **b** 2   **c** $y = 2x$
30 $y = \dfrac{11}{4}x - \dfrac{9}{4}$
31 $y = 4x + \dfrac{17}{2}$
32 $x = 0$
33 $\left(-\dfrac{3}{2}, \dfrac{9}{4}\right)$
34 $\left(-\dfrac{1}{8}, -8\right)$

Answers

35 $y + 2x = -1$

36 $y = \frac{1}{4}x + \frac{15}{4}$

37 $y = \frac{4}{3} - \frac{x}{3}$ or $y = -\frac{4}{3} - \frac{x}{3}$

38 $(2, 6), (-2, -6)$

39 $(-3, -2), (1, 6)$

40 $(2, 0.5)$

41 $\frac{2}{3}$ or 2

42 $(1, 1), (3, 9)$

44 $a = 20$

Chapter 9 Mixed Practice

1 a $8x - 1$ b $(2, 15)$

2 a

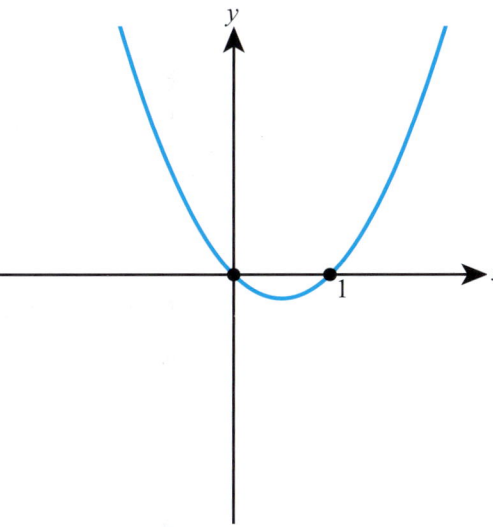

b $x > \frac{1}{2}$

c

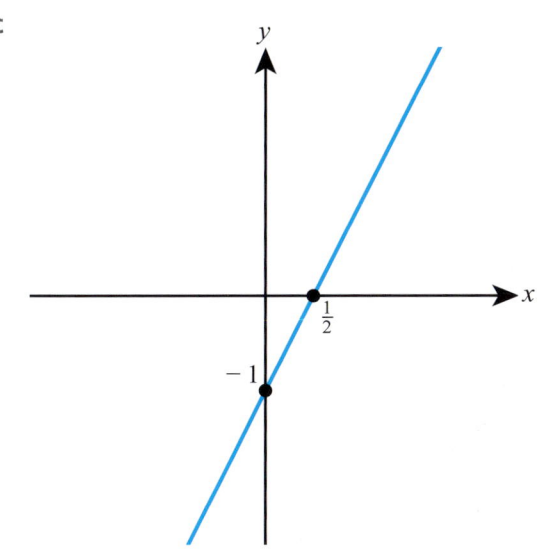

3 a $f'(x) = 6x^2 + 10x + 4$
 b $f'(-1) = 0$ c $y = 2$

4 -0.5

5 $y = 27x - 58$

6 a $12 + 10t$
 b $V(6) = 302$, $\frac{dV}{dt}(6) = 72$

After 6 minutes, there is $302\,m^3$ of water in the tank, and the volume of water in the tank is increasing at a rate of $72\,m^3$ per minute.

 c $\frac{dV}{dt}(10) = 112$, so the volume is increasing faster after 10 minutes than after 6 minutes.

7 a $2 - 2t$ b i 0.75   ii 1
 c $0 < t < 1$

8 a 6   b $x + 8x^{-2}$
 c 0   d $y = 6$
 e, f

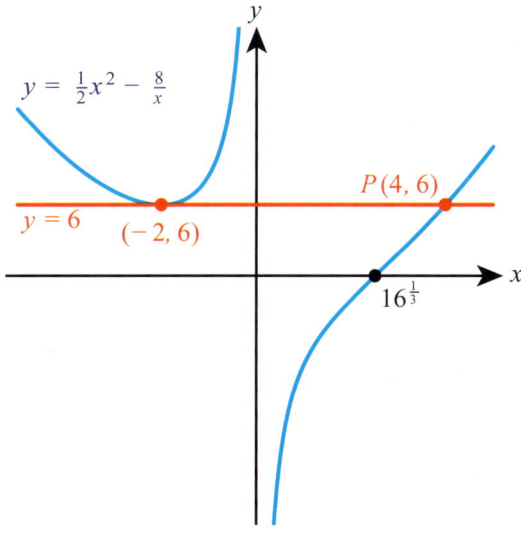

 g $(4, 6)$

9 $(1, -3), (-1, 9)$

10 a $(0, 0), (4, -32)$ b $y = -8x$

11 a $y = 6x$, $y = -6x - 12$
 b $(-1, -6)$

12 $p = -2, c = -3$

13 a $30t - 3t^2$ b $72, -72$
 c The profit is increasing after 6 months, but decreasing after 12 months.

14 a i 2   ii $2x + 3$ b $-\frac{1}{2}$

c

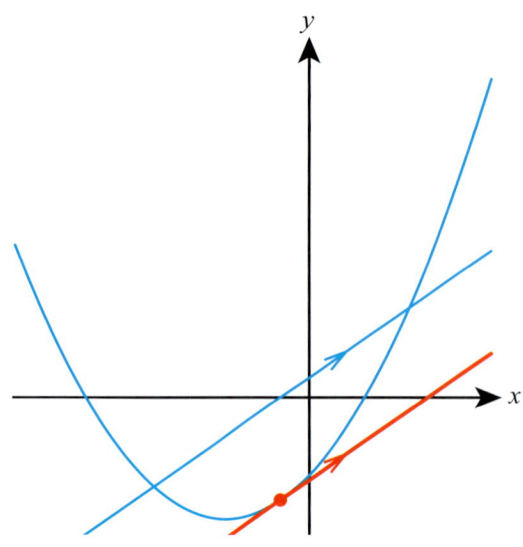

15 a i $0 < x < \frac{4}{3}$

ii

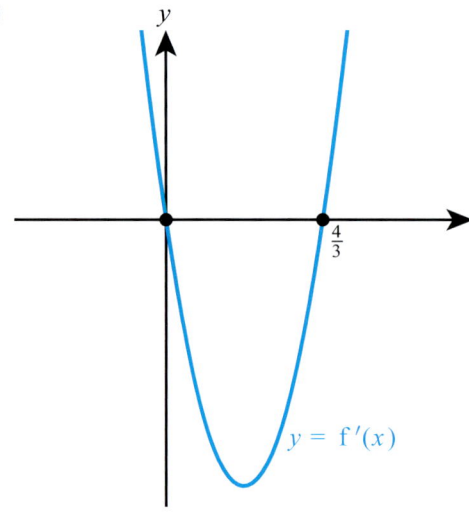

$y = f'(x)$

b i $x < 0$ and $0 < x < 2$

ii

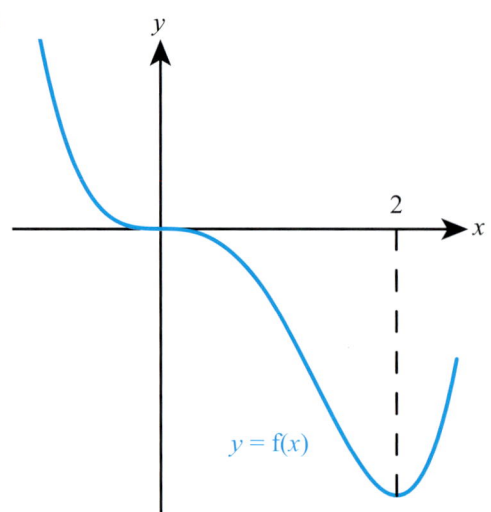

$y = f(x)$

16 a 3   b $2x - 1$

c (2, 4)   d $\left(\frac{1}{3}, \frac{16}{9}\right)$

e $y = 5x - 7$ f $\left(\frac{1}{2}, \frac{7}{4}\right)$, gradient = 0

17 a $2x + 1$ b 3 and −2

18 $a = 2, b = -5$

19 $a = -8, b = 13$

20 (1.5, 6)

21 (0.701, 1.47)

You need to make good use of technology in this question!

Chapter 10
Prior Knowledge

1 $x^{-2} - 4x^{-3} + 4x^{-4}$ 2 $f'(x) = -12x^{-4} + 2$

Exercise 10A

1 a $f(x) = \frac{1}{4}x^4 + c$ b $f(x) = \frac{1}{6}x^6 + c$

2 a $f(x) = -x^{-1} + c$

b $f(x) = -\frac{1}{2}x^{-2} + c$

3 a $f(x) = 1$ b $f(x) = x + c$

4 a $y = x^3 + c$ b $y = -x^5 + c$

5 a $y = -\frac{7}{4}x^4 + c$ b $y = \frac{3}{8}x^8 + c$

6 a $y = \frac{1}{4}x^6 + c$ b $y = -\frac{1}{6}x^{10} + c$

7 a $y = x^{-3} + c$ b $y = -x^{-5} + c$

8 a $y = -\frac{3}{2}x^{-4} + c$ b $y = 2x^{-2} + c$

9 a $y = \frac{2}{5}x^{-1} + c$ b $y = -\frac{1}{4}x^{-7} + c$

10 a $y = x^3 - 2x^2 + 5x + c$

b $y = \frac{7}{5}x^5 + 2x^3 - 2x + c$

11 a $y = \frac{1}{2}x^2 - \frac{2}{3}x^6 + c$

b $y = \frac{3}{2}x^4 - \frac{5}{8}x^8 + c$

Answers

12 a $y = \frac{1}{4}x^3 + \frac{7}{6}x^2 + c$

b $y = \frac{4}{5}x - \frac{2}{15}x^5 + c$

13 a $y = -\frac{1}{2}x^{-4} + \frac{3}{2}x^2 + c$

b $y = 5x + 3x^{-3} + c$

14 a $y = -\frac{5}{2}x^{-1} + \frac{2}{7}x^{-5} + c$

b $y = -\frac{2}{3}x^{-2} + \frac{1}{15}x^{-6} + c$

15 a $\frac{1}{4}x^4 + \frac{5}{3}x^3 + c$ **b** $3x^2 - x^3 + c$

16 a $\frac{1}{3}x^3 - \frac{1}{2}x^2 - 6x + c$ **b** $\frac{5}{2}x^2 - \frac{1}{3}x^3 - 4x + c$

17 a $3x^3 - 6x - x^{-1} + c$

b $\frac{4}{3}x^3 - 4x^{-1} - \frac{1}{5}x^{-5} + c$

18 a $\frac{3}{2}x^2 - 2x + c$ **b** $\frac{5}{2}x^2 - 3x + c$

19 a $-x^{-2} + \frac{7}{6}x^{-3} + c$ **b** $-\frac{1}{2}x^{-1} + \frac{3}{8}x^{-2} + c$

20 a $y = x^3 + 6$ **b** $y = x^5 + 5$

21 a $y = \frac{1}{4}x^4 - 6x^2 + 10$ **b** $y = 3x - 2x^5 + 4$

22 a $y = -3x^{-2} + 2x^2 - \frac{5}{3}$

b $y = 3x^3 + 2x^{-1} - 4$

23 $-\frac{4}{3t} + \frac{1}{2t^4} + c$ **24** $y = x^3 - 4x + 7$

25 $y = -\frac{4}{x} - x^3 + 10$

26 $\frac{3x^4}{4} - \frac{2x^3}{3} + \frac{3x^2}{2} - 2x + c$

27 $\frac{z^4}{4} + \frac{z^2}{2} + c$ **28** $\frac{x^3}{9} + \frac{2}{3x} + c$

29 a $k = 0.2$ **b** 3 kg

30 a 40 litres   **b** 4   **c** no

Exercise 10B

1 a 97.6   **b** 2.25
2 a −0.625   **b** −145.5
3 a 4   **b** 36
4 a 7.5   **b** 4.5
5 21.3
6 0.25
7 30
8 4.5
9 $\frac{32}{3}$
10 18
11 a (2.5, 6.25)   **b** 18.2
12 12.7 litres
13 3600 g
14 a 58 g   **b** 60 g
   c It suggests it takes forever for all the sand to fall through / sand is infinitely divisible.
15 18.75
16 $12x^2 - \frac{8}{x^3}$
17 $\frac{16}{3}$
18 $af(a) - A$

Chapter 10 Mixed Practice

1 $\frac{x^4}{4} + \frac{3}{x} + c$

2 $\frac{4}{3}x^3 - \frac{3}{2}x^2 + 5x + c$

3 0.661
4 $y = x^3 - 4x^3 + 6$
5 $a = 8$
6 5.7
7 a (2, 0), (5, 0)   **b** 6.75
8 a $y = 1.6x - 3.2$ **c** 1.07
9 b 1.5
10 1.5
11 1000 cm³
12 21.5 kg
13 47
14 a e.g. no energy lost to surroundings.
   b 160 calories
15 $y = \frac{5x^3}{3} + 3$
16 0
17 $6x + \frac{2}{x^2}$
18 a (3, 0)   **b** 18

Core SL content: Review Exercise

1 a 4   b 83   c 900

2 a $\frac{2}{3}$ b 2   c $3x + 2y = 31$

3 a

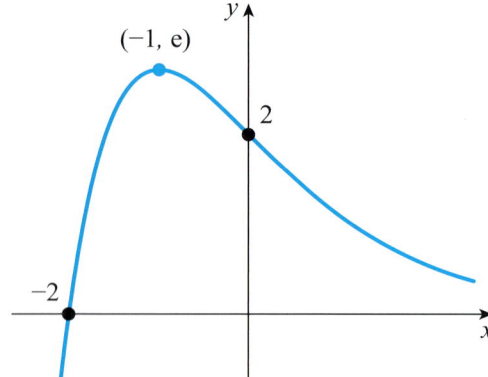

   b $y = 0$ c $y \leqslant 2.72$

4 a $a = 2.5; x \geqslant 2.5$ b $f(x) \geqslant 0$
   c 18.5

5 a $38.9°$ b $95.5\,\text{cm}^2$

6 a 4.5   b 0.305
   c 28.2

7 a $(-2, 0), (2, 0)$ b 15.6

8 a 4.5   b 2
   c $\frac{7}{10}$

9 a $9.56\,\text{cm}^3$ b $88.9\,\text{g}$
   d $52.8°$ e $30.9\,\text{cm}^2$

10 a and c

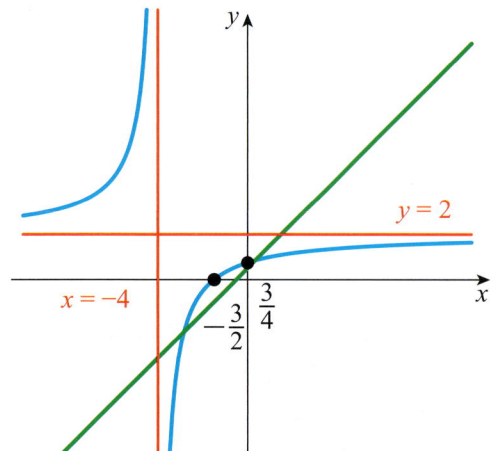

   b $x = -4$
   d $(-2.85078, -2.35078)$
     OR $(0.35078, 0.85078)$
   e 1
   f $y = -x - 5$

11 $x - 11y + 112 = 0$

12 a 1000   b -2

13 a $8x^7 - 24x^{-2}$ b $56x^6 + 48x^{-3}$

14 a $18r^3$ b $\frac{18(1-r^{15})}{1-r}$
   c 0.315

15 3.6%

16 $\frac{60}{41}$

17 a 12   b 0.0733
   c 0.764

18 a -2
   c $y = -2x - 3$ d $(-3, 3)$
   e $\sqrt{20}$ f 10

19 a i $1295.12\,\text{cm}^3$ ii 6
     iii I $431\,\text{cm}^3$
        II $4.31 \times 10^{-4}\,\text{m}^3$
   b i I $73.5°$
     II $55.8\,\text{m}$
     ii $55.0\,\text{m}$ iii $217\,\text{m}$

20 a i 0.985
     ii strong positive
   b $y = 260x + 699$
   c 4077 USD
   d e.g. $3952 < 4077$
   e i $304x$
     ii $304x - (260x + 699)$
     iii 16

21 a 0.159
   b i 0.119
     ii 0.394

22 a 3
   b $\frac{7}{26}$
   c $\frac{8}{15}$
   d $\frac{172}{325}$

Answers

Chapter 11 Prior Knowledge

1. **a** 1269 **b** 1360
2. **a** 6420 **b** 16

Exercise 11A

1. **a** 32.8 **b** 654.0
2. **a** 0.2 **b** 0.1
3. **a** 32.76 **b** 654.04
4. **a** 0.25 **b** 0.05
5. **a** 32.762 **b** 654.038
6. **a** 0.249 **b** 0.052
7. **a** 30 **b** 700
8. **a** 0.2 **b** 0.05
9. **a** 33 **b** 650
10. **a** 0.25 **b** 0.52
11. **a** 32.8 **b** 654
12. **a** 0.249 **b** 0.0517
13. **a** 110 (2 s.f.) **b** 12.3 (3 s.f.)
14. **a** 0.16 (2 s.f.) **b** 3 (1 s.f.)
15. **a** $12.25 \leq x < 12.35$ **b** $0.75 \leq x < 0.85$
16. **a** $0.485 \leq x < 0.495$ **b** $5.725 \leq x < 5.735$
17. **a** $55 \leq x < 65$ **b** $0.25 \leq x < 0.35$
18. **a** $465 \leq x < 475$ **b** $23.5 \leq x < 24.5$
19. **a** $11.675 \leq a + b < 12.685$
    b $3.575 \leq c + d < 3.685$
20. **a** $0.75 \leq pq < 3.75$
    b $2.7225 \leq rs < 3.2825$
21. **a** $3.315 \leq a - b < 4.325$
    b $3.115 \leq c - d < 3.225$
22. **a** $100 \leq \dfrac{p}{q} < 500$
    b $0.1089 \leq \dfrac{r}{s} < 0.1313$
23. **a** 0.298% **b** 0.118%
24. **a** 0.147% **b** 0.121%
25. **a** 6.25% **b** 3.66%
26. **a** 5.14% **b** 3.34%
27. **a** 46.512 cm³
    b **i** 46.51 cm³ **ii** 47 cm³
28. **a** $r = 2.5$ (2 s.f.) **b** $r = 3$ (1 s.f.)
29. **a** **i** 128.4 **ii** 130
    b 1.27%
30. **a** 329.54 ml **b** 0.140%
31. $84.5 \text{ cm} \leq l < 85.5 \text{ cm}$
32. 777 g
33. **a** $11.292 < AC < 11.441$
    b 1.325%
34. $12.7027 < I < 14$
35. $1.875 < a < 4.25$
36. **a** 37.5% **b** 150%
37. **a** **i** 0.9
       ii 1.2
       iii −0.675
    b **ii** and **iii** are wrong. Proportions must be between 0 and 1 inclusive
38. **a** $1.445 \leq a < 1.455$
    b $27.86 \leq 10^a < 28.51$
    c 30 to one significant figure (maximum significance that includes all possible values)
39. Assuming limiting factor is weight rather than e.g. volume or number of trucks available. 9 trucks
40. $237.5 \text{ cm}^2 \leq A < 525 \text{ cm}^2$

Exercise 11B

1. **a** $11 037.26 **b** $831.90
2. **a** $9648.31 **b** $10 598.34
3. **a** $94 841.43 **b** $87 410.24
4. **a** $89 654.86 **b** $98 150.01
5. **a** €5742.79 **b** €1347.86
6. **a** €5133.14 **b** €423.69
7. $4414.91
8. **a** 12 **b** $1013.20
9. $1203.38
10. **a** A: €42.03 B: €31.87
    b €105.24
    c It depends on the utility of the money used in the payment holiday
11. **a** 27 months **b** $186.65
12. **a** B (€557.48 cf. €592.76)
    b A (€177 828 cf. €200 693)
13. Yes: repayments of $300 in first two years leave repayments of $270.29 in remaining three years

Chapter 11 Mixed Practice

1. a 12.5   b 12
2. 19.09%
3. 1.01%
4. £3719.95
5. a 1380 m   b 743 m   c 3.49%
6. a 1.75
   b i $x = 2, y = 1, z = 50$
      ii 1.98
   c 13.1%
7. a 0.00125
   b i 0.0013   ii 0.001
   c 60%
8. Age and height in metres
9. a $3.1031 < \pi < 3.1895$
   b 0.135%
   c 1.52%
10. $1.360 < t < 1.474$
11. 9
12. a 27.2%   b 300%
13. a i −0.5
       ii 0.149
       iii 1.22
    b i and iii are wrong. Probabilities must be between 0 and 1 inclusive
14. a 22.7%   b $7.39
15. a 99 months   b ¥202.81
16. Option B: £343 739 vs £351 486
17. a $648.96   b $12 500
18. a €66.61   b 33.2%   c 72.4%
19. 604.75
20. a $2.55 \leq a < 2.65$
    b $354.8 \leq 10^a < 446.7$
    c 400 to one significant figure (maximum significance that includes all possible values)

Chapter 12 Prior Knowledge

1. $x = \frac{39}{7}, y = -\frac{1}{7}$
2. a $3x^2 + 11x - 11 = 0$ b $3x^2 - x - 4 = 0$

Exercise 12A

1. a $x = 5, y = -2$ b $x = \frac{34}{23}, y = -\frac{26}{23}$
2. a $x = -\frac{5}{19}, y = -\frac{65}{38}$ b $x = -3, y = 4$
3. a $x = \frac{17}{37}, y = \frac{16}{37}, z = \frac{39}{37}$
   b $x = -11, y = 64, z = 88$
4. a $x = -2, y = 1, z = 0$
   b $x = \frac{2}{5}, y = -\frac{1}{7}, z = \frac{52}{35}$
5. a $a = -3, b = 2, c = -5, d = 1$
   b $a = \frac{61}{29}, b = -1, c = \frac{239}{29}, d = \frac{17}{29}$
6. a $a = \frac{121}{8}, b = \frac{61}{8}, c = \frac{13}{8}, d = -\frac{19}{8}$
   b $a = 2, b = -1, c = 3, d = 4$
7. a $b + g = 60; g = 2b$ b 40
8. a $3w + 7g = 14.10; 5w + 4g = 12$
   b $2.70
9. $a = 25, d = -1.8$
10. $a = 3.4, d = 0.2$
11. $t = \$10.50, d = \$5.25, p = \$6$
12. $A = \$3900, B = \$3600, C = \$2500$
13. $c = 4, f = 5, s = 2$
14. £301.75
15. 0.2
16. 754
17. $x = \pm 1, y = \pm 3, z = \pm 4$
18. 5

Exercise 12B

1. a $x = 0.438, 4.56$ b $x = -2.14, 1.64$
2. a $x = 1.5$ b $x = \frac{1}{3}$
3. a $x = -2.09, 0.797, 1.80$
   b $x = -2.21, 0.539, 1.68$
4. a $x = -1, \frac{4}{3}$ b $x = -2, 2.5$
5. a $x = -4.72, -1.16, 0.364, 1.51$
   b $x = -0.798, 0.174, 0.872, 2.75$
6. a $x = -1.30, 2.30, 3$ b $x = -2, \frac{3}{2}$
7. a No real solutions   b No real solutions

Answers

8 **a** $x = -1.49$ **b** $x = 1.84$
9 **a** $x = -4.89, -1.10$ **b** No real solutions
10 8 cm
11 $r = 2.5, -3.5$
12 $r = 2$
13 **b** $x = -3.61, 1.11$
14 **a** $x^3 - 3x - 1 = 0$
    b $x = -1.53, -0.347, 1.88$
15 **b** $x = -1.42, 8.42$
    c $x = 8.42$ since length must be positive
16 **a** $5x^2 + 2x - 19.8 = 0$
    b $x = 1.8$ m
17 $x = 5$ or 10
18 **c** $r = 6.50$ cm or $r = 3.07$ cm
19 4
20 20
21 $(4, 5)$ or $(-5, -4)$
22 **a** $\frac{1}{2}n(n-3)$ **b** 10

Chapter 12 Mixed Practice

1 4 cm
2 **b** $x = -3.19, 4.69$
3 **b** $x = -3.70, -3, 2.70$
4 **a** $x^2 + x - 10 = 0$ **b** $2.70, -3.70$
5 1.65 cm
6 **a** $8s + 6f = 142; s + f = 20$
    b 11
7 $2
8 39
9 **a** $x + y = 10000$ **b** 39 AUD
    c $12x + 5y = 108800$ **d** $x = 8400, y = 1600$
10 $u_1 = -11, d = 3$
11 **a** $A = x^2 + 2x$
    b $x = 9.5$
    c 42 m
12 8
13 0.5 or -1.5
14 -3
15 **b** $x = 2.73$
16 40 or 61
17 **a** $V = 2x^3 + 7x^2 + 6x$ **b** $1.5 \times 3.5 \times 6$

18 **b** 80
    c **ii** Solutions to $n^2 + n - 4200$ are not whole numbers.
19 0.3
20 24
21 3.28
22 0.264 or -1.26
23 $x = 1, y = -1, z = 2$
24 20

Chapter 13 Prior Knowledge

1 $y = \frac{7}{5}x + \frac{8}{5}$
2 $1, 0.264, -1.26$
3 $x = -\frac{5}{9}, y = \frac{22}{9}, z = \frac{65}{9}$
4 1.46

Exercise 13A

1 **a** $y = 2x - 6$ **b** $y = x + 3$
2 **a** $y = -3x + 15$ **b** $y = -1.5x + 18$
3 **a** $y = 0.5x + 0.5$ **b** $y = 4x - 16$
4 **a** $y = -2.4x + 9.4$ **b** $y = -2x + 4$
5 **a** $y = 5x - 11.5$ **b** $y = \frac{7}{3}x + 1$
6 **a** $y = -4x + 27.7$ **b** $y = -0.4x - 4.82$
7 **a** **i** Each year the car will travel 8200 miles.
        ii The car had travelled 29 400 miles when it was bought.
    b 70,400
    c Ninth
8 **a** $C = 1.25d + 2.50$
    b $6.50
    c No, the cost would be $10.63.
9 **a** **i** $k = 100$ **ii** $m = -8$
    b $0 \leq t \leq 12.5$
10 **a** $C = -50d + 670$ **b** £430 000
11 **a** $a = 0.225, b = 3.63$
    b **i** The change in weight per week
        ii The weight of the baby at birth

12 a $C = \begin{cases} 0.12e + 10 & \text{for } 0 \leq e \leq 250 \\ 0.15e + 2.5 & \text{for } e > 250 \end{cases}$

  b £47.50

13 a i $C = 7.5m + 80$ **ii** $C = 5m + 125$

  b $m > 18$ miles

14 a $D(p) = 1125 - 15p$

  b $S(p) = 12p - 360$

  c $p = \$28.33$

15 a 1.5 m   **b i** $a = 0.18$

  c 2.1 m   **ii** $b = -1.3$

  d 21.1 m

16 a $T = \begin{cases} 0 & 0 \leq I \leq 12.5 \\ 0.2I - 2.5 & 12.5 < I \leq 50 \\ 0.4I - 12.5 & 50 < I \leq 150 \\ 0.45I - 20 & I > 150 \end{cases}$

  b i £4 500   **ii** £15 500

  c £155 556

Exercise 13B

1 a $y = x^2 + 5x - 3$ **b** $y = 2x^2 - 3x + 5$
2 a $y = 3x^2 - 2x - 4$ **b** $y = -x^2 + 4x + 2$
3 a $y = -2x^2 + 6x + 5$ **b** $y = 0.5x^2 - 3x + 2$
4 a $y = x^2 + 3.2x - 5.4$ **b** $y = -0.3x^2 - 2x + 4.1$
5 a $y = 4x^2 - 8x - 12$ **b** $y = -x^2 + 3x + 10$
6 a $y = -2x^2 + 12x - 8$ **b** $y = 3x^2 + 12x - 5$
7 a $(-5, 0)$
  b $b = 3, c = -10$
  c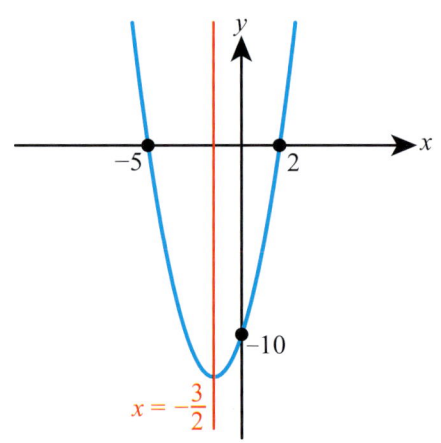

8 a $x = -3$
  b $0 = 30.25a - 5.5b + c$
    $0 = 0.25a - 0.5b + c$
    $12.5 = 9a - 3b + c$
  c $a = -2, b = -12, c = -5.5$
  d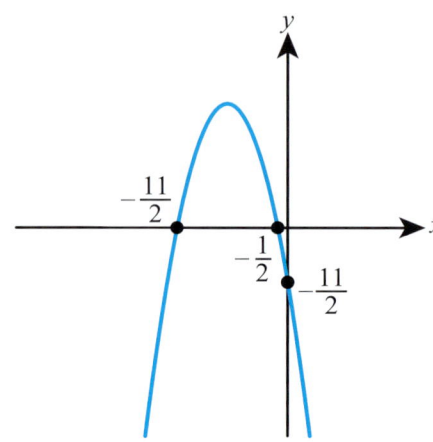

9 a $1350 = 2500a + 50b + c$
    $5100 = 40000a + 200b + c$
    $3100 = 160000a + 400b + c$
  b $a = -0.1, b = 50, c = -900$
  c If the business produces no items it will make a loss of £900.
  d i 250   **ii** £5350

10 10.8 cm

11 a $c = 1.45$
  b i $1.4 = -\dfrac{b}{2a}$
    $8.41a + 2.9b + 1.45 = 0$
    ii $a = -5, b = 14$
  c $t = 0.281, 2.52$ s

12 a 3.88 m^2
  b Assumes that R is always growing at the same rate, e.g. ignoring temperature.
  c The algae will eventually fill the pond, so its surface area will no longer increase, but in the model it continues to grow indefinitely.

Answers

Exercise 13C

1 a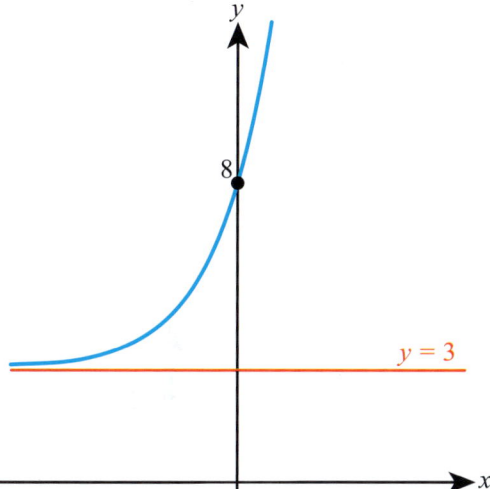

b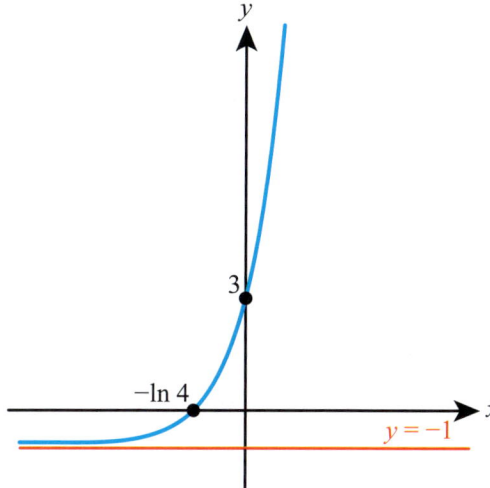

2 a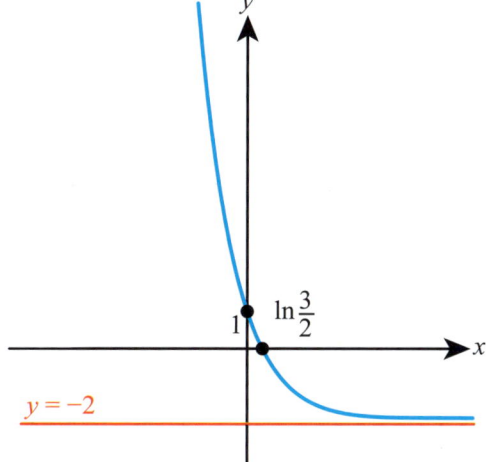

b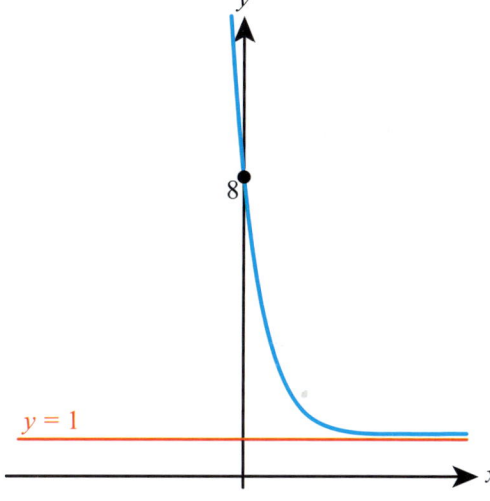

3 a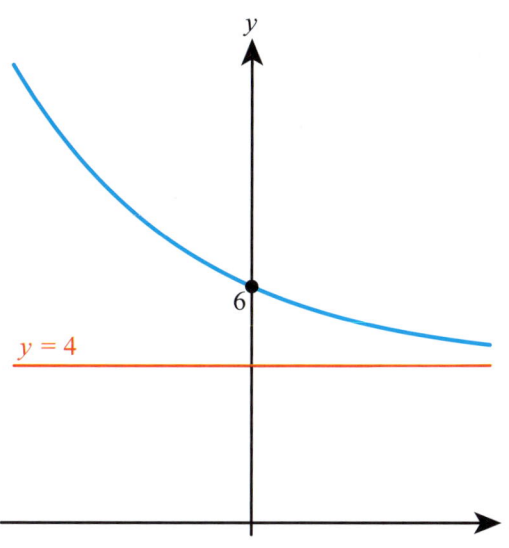

b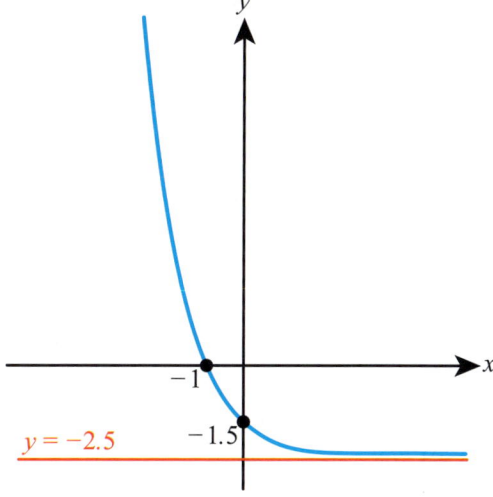

4 a

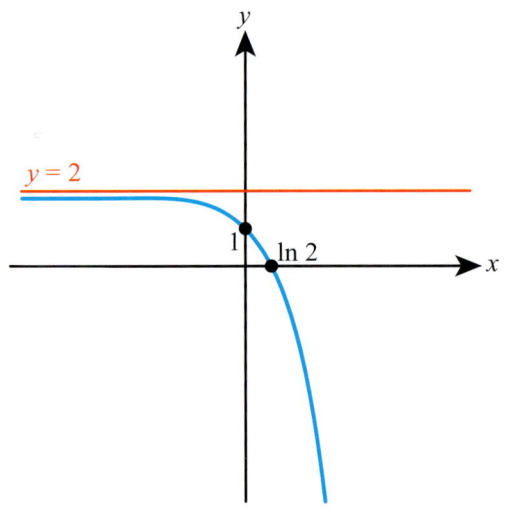

b

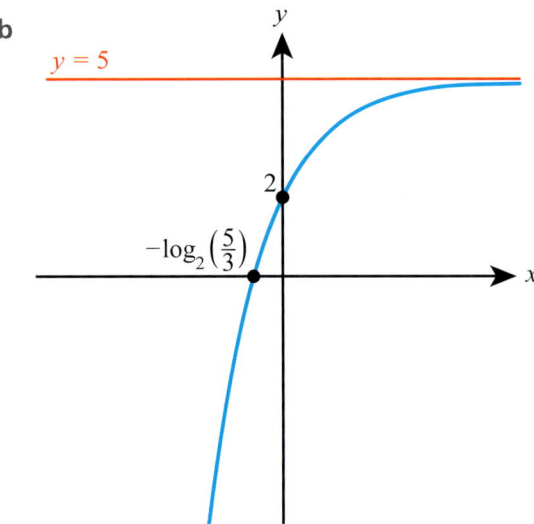

5 a $y = 2 \times 3^x - 1$ **b** $y = 4^x + 1.5$
6 a $y = 10 \times 5^{-x} + 6$ **b** $y = 6 \times 2^{-x} - 5$
7 a $y = 1.2^x + 3$ **b** $y = 4 \times 0.5^x + 2$
8 a $y = 5e^{-x} - 2$ **b** $y = e^x + 4$
9 a 4.68 billion
  b 2%
  c 5.27 billion
10 a $1.5 \,\text{mg}\, l^{-1}$
   b $0.556 \,\text{mg}\, l^{-1}$
   c 9 hours and 2 minutes
11 a $k = 2.5, r = 0.206$ **b** 1 205 000
12 a $k = 45.1, a = 1.15$ **b** 27.2 s
13 a $V = 8500 \times 0.8^t$
   b $912.68
   c $c = 8000, d = 500$

14 a $c = 0.0462$ **b** 99.7 hours
15 $50 \,\text{m s}^{-1}$
16 a 66.6 °C
   b i No change
     ii Increase but stay less than 1

Exercise 13D

1 a $y = 4x^2$ **b** $y = 0.5x^4$
2 a $y = 3\sqrt[3]{x}$ **b** $y = 1.5\sqrt{x}$
3 a $y = \dfrac{5}{x}$ **b** $y = \dfrac{2}{x^3}$
4 a $y = \dfrac{15}{\sqrt{x}}$ **b** $y = \dfrac{12}{\sqrt[4]{x}}$
5 a $y = x^3 + 2x^2 - 5x + 4$
  b $y = 3x^3 - 4x^2 - x + 2$
6 a $y = 2x^3 + 4x^2 - 3x - 6$
  b $y = -x^3 - 3x^2 + 2$
7 a $y = -2x^3 + 6x^2 + 1$
  b $y = 0.5x^3 - 3x^2 + 0.3x + 10$
8 a $y = -1.2x^3 + 4x - 2.5$
  b $y = 3x^3 + 2x^2 - 4.7x - 3.8$
9 a $F = 2.4x$ **b i** 9.6 N
        ii 8.33 cm
10 a $T = 0.2\sqrt{l}$ **b i** 1.55 s
         ii 16 cm
11 a 5.63×10^6 Watts **b** $17.0 \,\text{m s}^{-1}$
12 a $P = \dfrac{900\,000}{V}$ **b i** 7200 Pa
         ii 11.25 cm³
13 29
14 a $-1 = -8a + 4b - 2c + d$
     $-8 = -a + b - c + d$
     $2 = a + b + c + d$
     $-5 = 8a + 4b + 2c + d$
   b $a = -2, b = 0, c = 7, d = -3$
15 a $T = -0.00342d^3 + 0.251d^2 + 1.04d - 0.265$
   b i 11.4 **ii** 3.88%
   c 252 years
16 $x = kz^{\frac{3}{2}}$
17 17.4%
18 $f(5) = 38$
19 7 days
20 $f(x) = kx(x-1)(x-2) + x$

Answers

Exercise 13E

1. **a** i 3 ii 90 iii $y = 2$
   iv 5 v 1 vi $(0, 2)$
   b i 2 ii 720 iii $y = -5$
   iv -3 v -7 vi $(0, -5)$
2. **a** i 0.5 ii 1080 iii $y = -2$
   iv -1.5 v -2.5 vi $(0, -1.5)$
   b i 5 ii 4 iii $y = 3$
   iv 8 v -2 vi $(0, 8)$
3. **a** 3 **b** 8 **c** 2
4. **a** 4 **b** $\frac{1}{2}$ **c** -1
5. **a** 50 m **b** 1.8 m **c** 5 m
6. **a** i 1.68 m ii 0.52 m
   b $300° \, s^{-1}$
   c 0.81 m
7. **a** 152 cm **b** 0.4 s **c** 140 cm
8. **a** $p = 40$, $q = 18$, $r = 0$
   b 10.8 s
9. 13.7 °C
10. 3.30 m

Exercise 13F

1. **a** F $y = 2 \times 3^{-x} + 1$
   b F $y = 4 \times 3^x - 2$
2. **a** C $y = x^3 - 3x^2 + x + 1$
   b C $y = -x^3 - x^2 + 4x + 2$
3. **a** E $y = 3\cos(180x) - 2$
   b E $y = -2\cos(18x) + 1$
4. **a** D $y = -\sin(0.5x) + 3$
   b D $y = 10\sin(90x) + 4$
5. **a** $P = 67 \times 1.01^n$
   b Population growth rate may change over the longer term due to, e.g. changes in immigration, birth rate, death rate, etc.
   There might be large errors by the time we get to 2100.
6. **a** $0 \leqslant t \leqslant 55$
   b An exponential model so that the rate of decrease slows down and the pressure tends to zero.
7. **a** 0 °C
   b $T = 80 \times 0.87^t + 20$

Chapter 13 Mixed Practice

1. **a** 16% **b** 52%
   c 70 minutes
2. **a** $c = 3$ **b** $6 = -\dfrac{b}{2a}$
   $12 = 36a + 6b$
   c $a = -\dfrac{1}{3}$, $b = 4$
3. **a** $p = 1.5$, $q = 3.5$ **b** $y = 3.5$
4. **a** $V = 0.4T$ **b** $V = 117 \, cm^3$
5. **a** $m = 90$
   b Change per year in the number of apartments
   c 60
   d Number of apartments initially
6. **a** $-8p + 4q - 2r = -8$
   $p + q + r = -2$
   $8p + 4q + 2r = 0$
   b $p = 1$, $q = -1$, $r = -2$
7. **a** i 3 ii 180°
   b i $a = 2$ ii $b = 1$ iii $c = -1$
   c 5
8. **a** $1.00 **b** $2.00
   c $350 \leqslant w < 500$
9. **a** $a = 4$, $b = 5$ **b** 2
   c $c = -\dfrac{1}{2}$
10. 36%
11. **a** i 40 °C ii 20 °C iii 10 °C
    c

    d i $94 = p + q$ ii $54 = \dfrac{p}{2} + q$
    e $p = 80$, $q = 14$ **f** $y = 14$
    g i $r = -0.878$ ii $y = 71.6 - 11.7t$
    h 36.7 **i** 52.8%

12 a 1.5   **b** $C = 2.5$
  c 3 hours 19 minutes
13 a i 100 m   **ii** 50 m
  b ii 2.4
  c

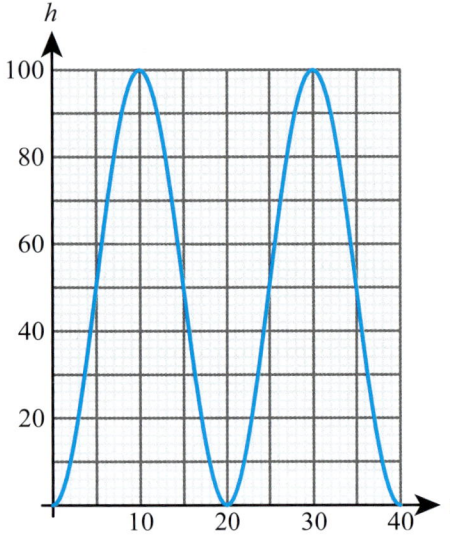

  d $a = -50$, $b = 18$, $c = 50$

Chapter 14 Prior Knowledge

1 a 10.4   **b** $(-0.5, -1)$
2 $2x + 5y = 27$

Exercise 14A

1 a 3.49 cm, 8.73 cm²   **b** 9.07 cm, 36.3 cm²
2 a 6.98 cm, 14.0 cm²   **b** 14.7 cm, 51.3 cm²
3 a 33.5 cm, 134 cm²   **b** 25.3 cm, 63.3 cm²
4 a 20.3 cm², 28.4 cm   **b** 4.32 cm², 15.5 cm
5 a 43.3 cm², 34.2 cm   **b** 15.4 cm², 19.1 cm
6 a 176 cm², 57.1 cm   **b** 8.81 cm², 13.3 cm
7 per = 20.9 cm, area = 19.5 cm²
8 per = 27.5 cm, area = 47.0 cm²
9 10.1 cm
10 a 80.2°   **b** 17.5 cm²
11 40.1°
12 15.5 cm
13 a 11.6 cm   **b** 38.2 cm
14 area = 57.1 cm², per = 30.3 cm
15 6.85 cm
16 1.32 cm, 13.7 cm
17 0.219 cm
18 per = 16.6 cm, area = 3.98 cm²
19 $x = 3$
20 $r = 23.4$, $\theta = 123$
21 per = 33.5 cm, area = 78.6 cm

Exercise 14B

1 a $x + 2y = 2$ **b** $x - 3y = 39$
2 a $3x - 4y = -60$ **b** $5x + 2y = 60$
3 a $5x - 4y = 6$ **b** $4x + 3y = -13$
4 a $12x + 3y = 11$ **b** $12x - 4y = -41$
5 a $y = -2x + 22$ **b** $y = -3x + 40$
6 a $y = \frac{3}{2}x + \frac{5}{4}$ **b** $y = -\frac{2}{5}x + 2$
7 a $y = \frac{2}{3}x + \frac{1}{12}$ **b** $y = -\frac{1}{2}x + \frac{53}{10}$
8 a $y = -x - \frac{26}{3}$ **b** $y = 2x - \frac{16}{3}$
9 a i A, B, C, D, E **ii** F, G, H
  iii $[FG], [GH]$ **iv** No finite cells
  b i A, B, C, D, E, F **ii** G, H, I, J, K, L
  iii $[GH], [HI], [IJ], [JG], [JL], [LK], [KI]$
  iv $GHIJ, IJLK$
10 a i A **ii** F
  b i B **ii** C
11 a i B and C **ii** $x = 4$
  b i A and B **ii** $y = 3 - x$
  c i D and C **ii** $6y + 4x = 25$
12 a

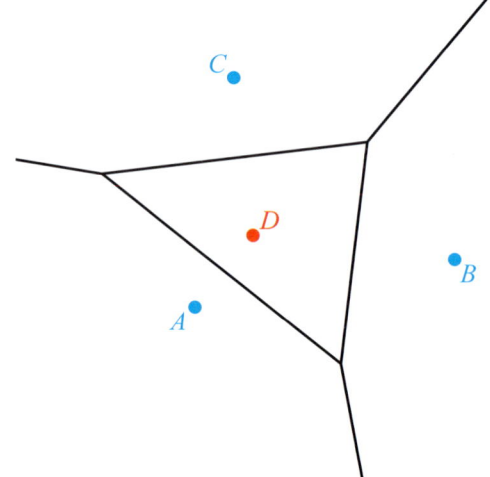

Answers

b
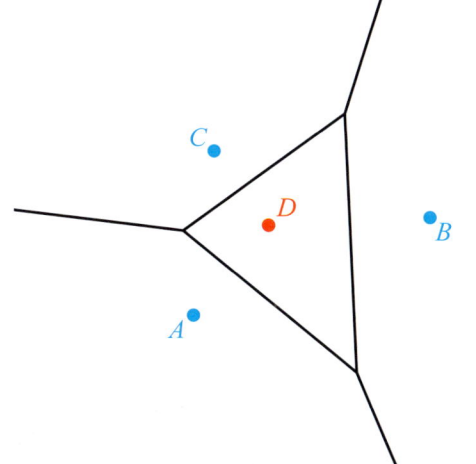

13 a
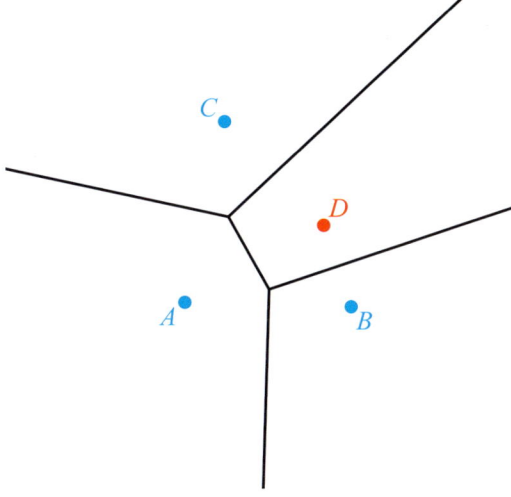

b
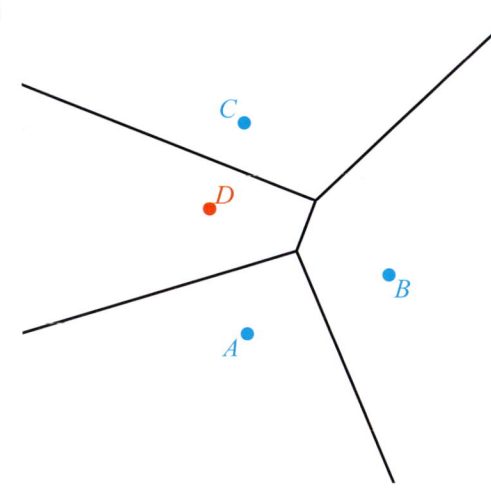

14 a
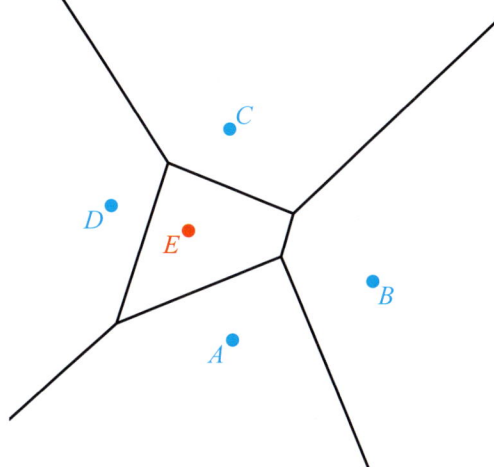

b
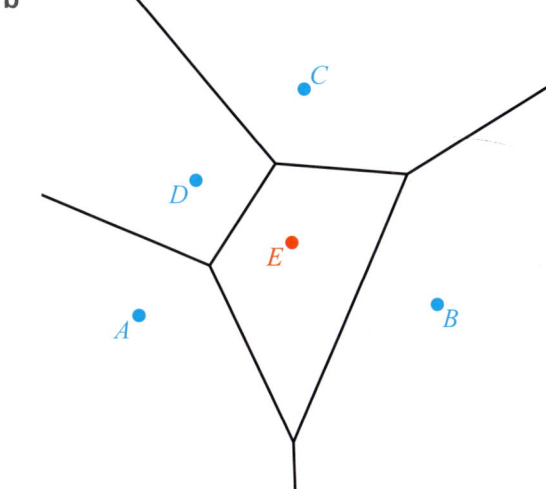

15 $y = 2x - 2.5$

16 a $x = 5.5$ **b** $22\,°C$
    c $1025\,mbar$ **d** $31\,mm$

17 a   **i** F, G
       ii A, B, G, I
   b B, C, D, H, I

18 C; highest mean contamination at sites in that region

19 a   **i** D **ii** C **iii** D
   b   Contains all stores served by distribution centre A

c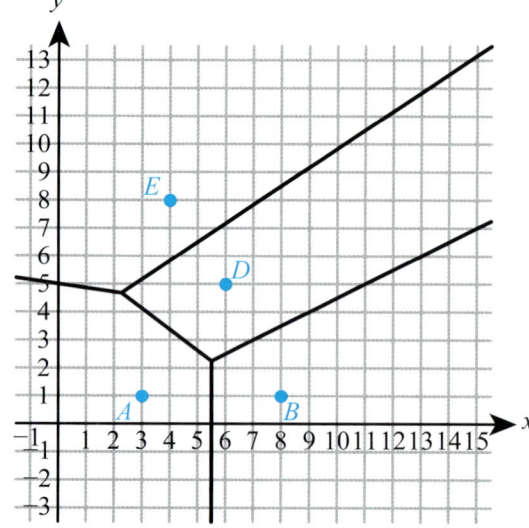

d (14, 1) to B

20 a i $y = 2.5 - 0.5x$
  ii $x = 3$
  iii $y = 4 - x$
b (3, 1)
d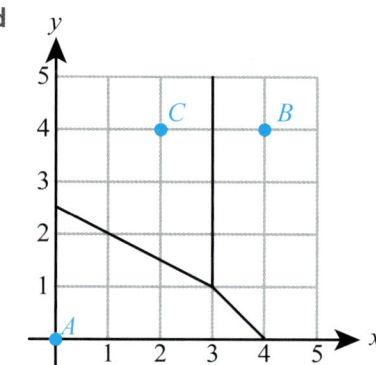
e B

21 b (7.83, 5.67)
  c (7, 5)

d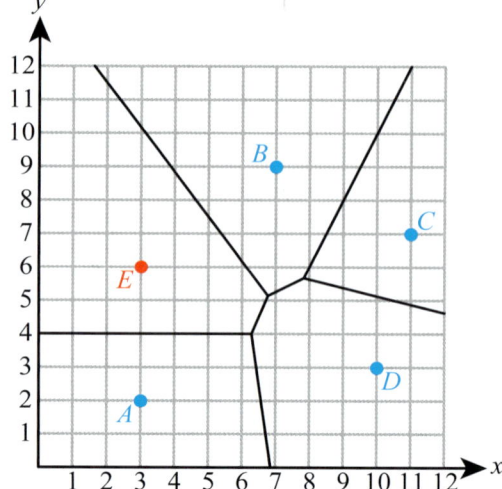

e 412 m

22 a $3x - 2y = 1$ **b** D
  c (9, 13)

23 a D **b** (3, 6)
  c 2400 m²   **d** $y = -x + 11$
  e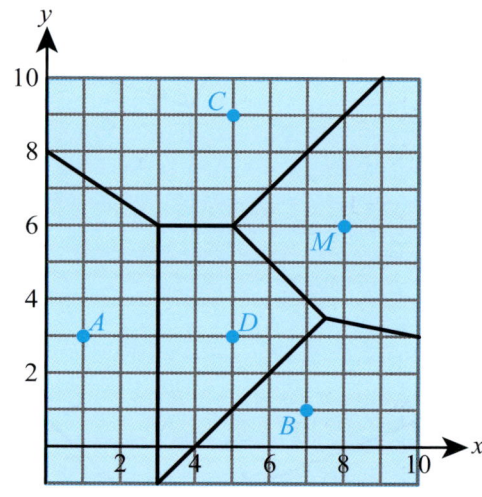

24 a (25, 5)   **b** 25.5 km
  c e.g. the towns have no size – they are just a point

25 a (31, 29) to be at the vertex furthest from all towns on a Voronoi diagram
  b Moves further away from B (i.e. larger x and y coordinates)

Chapter 14 Mixed Practice

1 a 27.9 cm²   **b** 25.6 cm
2 8
3 1.18 m³

Answers

4 a $y = -7.42x + 1440$ b 142 km

5 a 5390 m   b $\frac{5}{2}$

c $4x + 10y = 55$

d It is the same distance from A, C and D.

6 a 7 km², 316 people per km²

b 0.194

c B – cell has the highest probability of cholera

7 a (15, 25)   b 15.8 km

8 a $6x + 4y = 29$

b
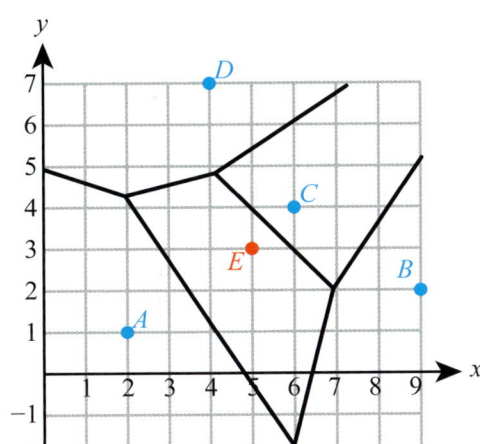

9 25 600

10 a i $y = 1$

ii $x = 3$

iii $y = 10 - 3x$

b (3, 1)

c
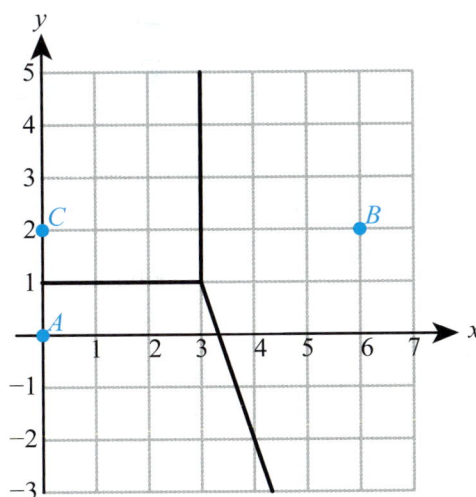

d (3, 1)

e
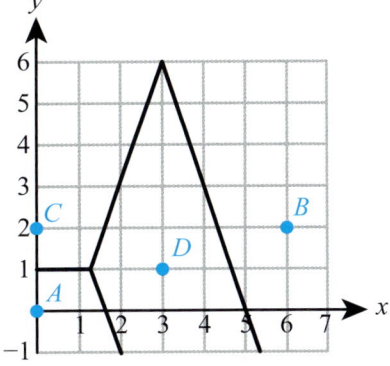

f 16.7%

11 a i 23 °C   ii 28 °C   iii 25 °C

b
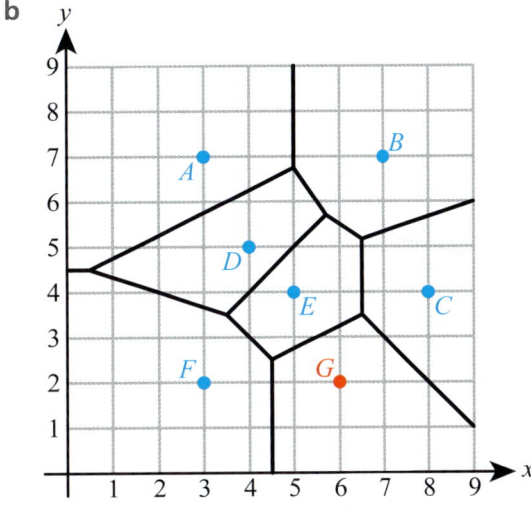

c (6, 3), to 24 °C

12 6 cm

13 4.0 cm and 3.4 cm

14 a 44.0°   b 1.18 cm²

15 b 58.3°   c 23.9 cm

16 a 82.3°   b 111 cm

17 a 130°   b 36.1%

c 18.2 °C

18 (3.1, 1.9) (although (1.25, 3.75) is also locally stable)

19 a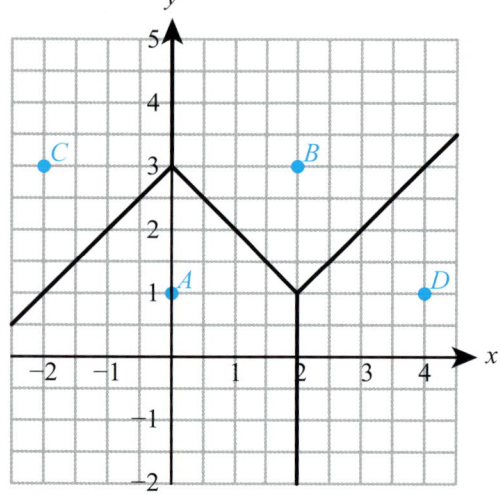

b 8 km

20 a $a^2 + b^2 = 1$ **b** $(1+a, b)$

c $y - \dfrac{b}{2} = -\dfrac{1+a}{b}\left(x - \dfrac{1+a}{2}\right)$

Chapter 15 Prior Knowledge

1. 18.4
2. 0.0280
3. 0.388
4. **a** 0.662
   b Significant positive (linear) correlation

Exercise 15A

1. **a** $H_0: \mu = 23.4$, $H_1: \mu > 23.4$, one-tail
   b $H_0: \mu = 8.3$, $H_1: \mu > 8.3$, one-tail
2. **a** $H_0: \mu = 300$, $H_1: \mu \neq 300$, two-tail
   b $H_0: \mu = 163$, $H_1: \mu \neq 163$, two-tail
3. **a** H_0: There is no correlation between temperature and the number of people
   H_1: There is a negative correlation between the two; one-tail
   b H_0: There is no correlation between rainfall and temperature
   H_1: There is correlation between the two; two-tail
4. **a** Insufficient evidence that the mean is greater than 13.4
   b Insufficient evidence that the mean is smaller than 26
5. **a** Sufficient evidence that the mean is not 2.6
   b Insufficient evidence that the mean is not 8.5
6. **a** Sufficient evidence that the data does not come from the distribution $N(12, 3.4^2)$
   b Insufficient evidence that the data does not come from the distribution $N(53, 8^2)$
7. **a** **i** Expected: 25, 50, 25, $\chi^2 = 7.26$
       ii 2
       iii $p = 0.0265$, reject H_0
   b **i** Expected: 50, 30, 20, $\chi^2 = 1.00$
       ii 2
       iii $p = 0.607$, do not reject H_0
8. **a** **i** Expected: 10, 10, 10, 10, 10, $\chi^2 = 8.00$
       ii 4
       iii $p = 0.0916$, do not reject H_0
   b **i** Expected: 15, 15, 15, 15, $\chi^2 = 10.5$
       ii 3
       iii $p = 0.0145$, reject H_0
9. **a** **i** Expected: 66.7, 50, 40, 43.4, $\chi^2 = 18.7$
       ii 3
       iii $p = 0.000320$, reject H_0
   b **i** Expected: 53.3, 26.7, 32, 48, $\chi^2 = 7.38$
       ii 3
       iii $p = 0.0606$, do not reject H_0
10. **a** $\chi^2 = 6.81$, do not reject H_0
    b $\chi^2 = 6.43$, do not reject H_0
11. **a** $\chi^2 = 10.1$, reject H_0
    b $\chi^2 = 1.69$, do not reject H_0
12. **a** $\chi^2 = 9.13$, reject H_0
    b $\chi^2 = 5.12$, do not reject H_0
13. **a** $\chi^2 = 5.51$, insufficient evidence that they are not independent
    b $\chi^2 = 33.2$, sufficient evidence that they are not independent
14. **a** $\chi^2 = 0.695$, insufficient evidence that they are not independent
    b $\chi^2 = 11.2$, sufficient evidence that they are not independent
15. **a** $\chi^2 = 42.8$, sufficient evidence that they are not independent
    b $\chi^2 = 5.19$, insufficient evidence that they are not independent
16. $\chi^2 = 12.14$, $p = 0.00231$, sufficient evidence that the age and food choices are not independent

Answers

17 $\chi^2 = 4.41$, $p = 0.110$, insufficient evidence that type of insect depends on location

18 a 10, 20, 30; 2   **b** 1.93

    c Insufficient evidence that the ratio is different from 1:2:3

19 a H_0: Each course is equally likely, H_1: Each course is not equally likely

    b 20, 20, 20, 20; 3

    c $p = 0.00119$, sufficient evidence that each course is not equally likely

20 $p = 8.55 \times 10^{-23}$, sufficient evidence that city and mode of transport are dependent

21 $\chi^2 = 11.6$, $p = 0.0407$, insufficient evidence that the dice is not fair

22 a H_0: The data comes from the distribution $B(3, 0.7)$
H_1: The data does not come from the distribution $B(3, 0.7)$

    b 5.40, 37.8, 88.2, 68.6; 3

    c 3.57

    d Insufficient evidence that the data does not come from the distribution $B(3, 0.7)$

23 a Binomial, $n = 6$, $p = 0.5$

    b 9.38, 56.3, 141, 188, 141, 56.3, 9.38

    c $\chi^2 = 7.82$; insufficient evidence that the coins are biased

24 a $X \sim B(4, 0.5)$

    b H_0: The data comes from the distribution $B(4, 0.5)$
H_1: The data does not come from this distribution

    c 4

    d $\chi^2 = 13.4$, $p = 0.00948$, sufficient evidence that the data does not come from $B(4, 0.5)$

25 a 0.440, 0.334, 0.0668

    b 3

    c H_0: The data comes from the distribution $N(5.8, 0.8^2)$
H_1: The data does not come from the distribution $N(5.8, 0.8^2)$

    d $\chi^2 = 2.82$, $p = 0.420$, insufficient evidence to reject H_0

26 a 14.1, 7.09, 7.62, 7.09, 14.1

    b 4

    c 30.7

    d Sufficient evidence that the times do not follow the distribution $N(23, 2.6^2)$

27 $\chi^2 = 5.46$, $p = 0.243$, insufficient evidence that the model is not appropriate

28 a $\chi^2 = 14.0$, $p = 0.00733$, sufficient evidence that diet and age are dependent

    b $\chi^2 = 15.9$, $p = 0.0141$, insufficient evidence that diet and age are dependent

29 a $\chi^2 = 3.37$, $p = 0.185$, insufficient evidence that favourite science depends on gender

    b $\chi^2 = 10.1$, $p = 0.00636$, sufficient evidence that favourite science depends on gender

Exercise 15B

1 a $t = 0.278$, $p = 0.395$
Do not reject H_0

    b $t = 1.95$, $p = 0.0492$
Reject H_0

2 a $t = -1.63$, $p = 0.0823$
Reject H_0

    b $t = -1.19$, $p = 0.143$
Do not reject H_0

3 a $t = -2.78$, $p = 0.0239$
Do not reject H_0

    b $t = 3.03$, $p = 0.0163$
Reject H_0

4 a $t = 10$, $p = 2.46 \times 10^{-10}$
Reject H_0

    b $t = 2$, $p = 0.0262$, reject H_0

5 a $t = -0.232$, $p = 0.409$, do not reject H_0.

    b $t = -0.861$, $p = 0.202$, do not reject H_0.

6 a $t = 0.253$, $p = 0.806$, do not reject H_0.

    b $t = -3.35$, $p = 0.0285$, reject H_0

7 a $t = 1.55$, $p = 0.0781$, do not reject H_0

    b $t = 15.49$, $p < 10^{-10}$, reject H_0

8 a $t = -0.885$, $p = 0.189$, do not reject H_0

    b $t = -2.30$, $p = 0.0118$, reject H_0

9 a $t = -2.97$, $p = 0.00331$, reject H_0

    b $t = 1.34$, $p = 0.196$, do not reject H_0

10 $t = 0.579$, $p = 0.290$, do not reject H_0

11 $t = -1.49$, $p = 0.152$, do not reject H_0

12 $t = -1.229$, $p = 0.1251$, do not reject H_0

13 $H_0: \mu = 14.3$. $H_1: \mu > 14.3$
$t = 1.91$, $p = 0.0330$, reject H_0

14 a $t = 0.314$, $p = 0.381$, do not reject H_0

    b The times are normally distributed.

15 a The amounts are normally distributed.
  b $t = -3.37$, $p = 0.00213$, reject H_0
16 $t = 0.673$, $p = 0.514$, do not reject H_0
17 a The times of two groups are normally distributed and have a common variance.
  b $H_0: \mu_1 = \mu_2$, $H_1: \mu_1 < \mu_2$ where μ_1 and μ_2 are population mean times of the two groups
  c $t = -2.20$, $p = 0.0162$
    i Reject H_0
    ii Do not reject H_0
18 a $H_0: \mu = 0$, $H_1: \mu > 0$ where μ is the mean change in share price
  b The changes in price are normally distributed.
  c $t = 0.582$, $p = 0.288$, do not reject H_0
19 a equal variances
  b $t = -0.9$, $p = 0.197$, do not reject H_0
20 a $t = 1.90$, $p = 0.0365$, reject H_0
  b $-8, 4, -12, -11, -11, -15, 0, 2$
  c $H_0: \mu = 0$, $H_1: \mu < 0$, where μ is the mean change in blood pressure
  d $t = 2.68$, $p = 0.0125 < 0.05$, reject H_0
  e The differences are distributed normally.
  f The first test required that the observations are independent, which is not true.

Exercise 15C

1 a 0.857 **b** 0.762
2 a −0.617 **b** −0.983
3 a 0.286 **b** −0.833
4 a −0.0357 **b** −0.357
5 a −0.891 **b** −0.873
6 a −0.593 **b** −0.593
7 a 0.795 **b** −0.807
8 a 0.343 **b** 0
9 a B **b** C
  c A
10 a i 0.879
    ii 0.886
  b Strong positive correlation
11 a 0.929
  b As temperature increases, it appears that the number of people visiting increases.

12 a

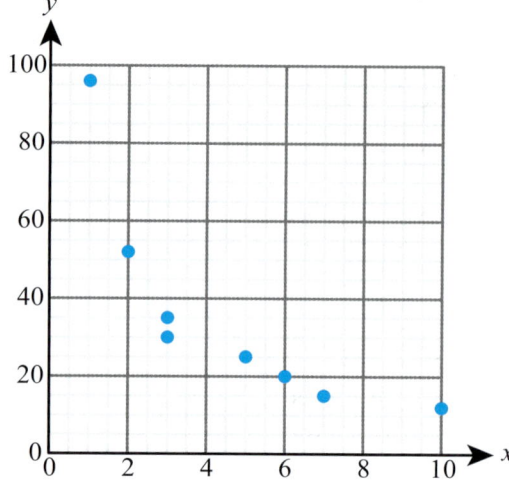

  b Appears non-linear
  c −0.994, strong negative association
13 H_0: There is no correlation, H_1: There is positive correlation, $r_s = 0.303$. Do not reject H_0
14 0.332; no significant evidence of correlation between time playing video games and sleep
15 a −0.952 **b** No
16 a 0.8857
  b Insufficient evidence of correlation, $r_s < 0.886$
17 Insufficient evidence of correlation, $r_s = -0.505$
18 a 0.806 **b** No **c** Yes
19 a 0.738
  b Sufficient evidence of positive correlation
20 a Increases
  b No change
  c No change

Chapter 15 Mixed Practice

1 a 0.946 **b** 1
  c As c increases V increases, but not in a perfectly linear pattern
2 a 0.0668
  b Sufficient evidence to support claim ($p = 0.00186$)
3 Insufficient evidence to reject H_0 ($p = 0.681$)
4 a H_0: The dice is unbiased (all outcomes have equal probability), H_1: The dice is biased
  b all = 50 **c** 5
  d Insufficient evidence that the dice is biased, $p = 0.123 > 0.10$

Answers

5 a 0.857

b H_0: There is no correlation between hours of sunshine and temperature, H_1: There is a positive correlation

c There is sufficient evidence of positive correlation.

6 Sufficient evidence that the mean time is greater than 17.5 minutes ($p = 0.0852 < 0.10$)

7 a $H_0: \mu = 16$, $H_1: \mu < 16$

b Insufficient evidence of decrease ($p = 0.0538 > 0.05$)

8 a H_0: The grades are independent of the school
H_1: The grades depend on the school

b $p = 0.198$, insufficient evidence that grades depend on school

9 H_0: The times are the same after training,
H_1: The times have decreased after training.
Insufficient evidence that times have improved ($p = 0.277$)

10 H_0: The data comes from the given distribution
H_1: The data does not come from this distribution
Insufficient evidence to reject claim ($p = 0.465$)

11 a 791

b H_0: The average weight of a loaf is 800 g,
H_1: The average weight of a loaf is less than 800 g. $t = -2.63$, $p = 0.0137$. Sufficient evidence to reject H_0 and say the average weight of a loaf is less than 800 g.

12 a not random (observations may not be independent)

b $H_0: \mu_1 = \mu_2$, $H_1: \mu_1 \neq \mu_2$

c The masses of the eggs of both chickens are distributed normally with equal variance.

d $p = 0.666$, insufficient evidence that the means of egg masses for the two chickens are different

13 a -0.511

b H_0: no correlation, H_1: some correlation

c Insufficient evidence of correlation ($0.511 < 0.564$)

14 a $t = 1.689$, $p = 0.103$, do not reject H_0

b The lifetimes follow a normal distribution with equal variances.

15 a $B(5, 0.45)$

b 5.03, 20.6, 33.7, 27.6, 13.1

c 4

d $p = 0.0249$, sufficient evidence that Roy's belief is incorrect

16 a H_0: All the coins are fair (P(tails = 0.5))
H_1: (Some of) the coins are biased

b 6.25, 43.75, 131.25, 218.75, 218.75, 131.75, 43.75, 6.25

c $p = 0.0401$, $\chi^2 = 14.7$

d Insufficient evidence that the coins are not fair

17 a Mean is 1, which is consistent with the claimed $B(5,0.2)$ distribution but does not prove it must be that

b i H_0: data is drawn from a $B(5,0.2)$ distribution
H_1: data is not drawn from a $B(5,0.2)$ distribution

ii $\chi^2 = 0.588$

iii $p = 0.899$

iv There is no significant evidence to doubt the manufacturer's claim.

18 a $t = -1.70$, $p = 0.0754$
No: there is insufficient evidence to reject H_0

b i H_0: There is no correlation
H_1: There is negative correlation

ii -0.857

iii Reject H_0 (as age increases there is a tendency for 100 m time to decrease)

19 a $t = -1.19$, $p = 0.118 > 0.05$

b $\chi^2 = 9.79$, $p = 0.0440 < 0.05$

c They could come from a normal distribution with a different mean (or different variance).

d Scores are discrete, normal distribution is continuous.

20 a $t = 2.51$, $p = 0.0129$. Reject H_0 (significant change from 3.3)

b $r_S = 0.696$, do not reject H_0 (no significant evidence of increasing frequency with score)

c $\chi^2 = 1.93$, $p = 0.858$, do not reject H_0 (no significant evidence of non-random pattern)

Chapter 16 Prior Knowledge

1 $f'(x) = 2x$ **2** 3 **3** 5.75

Exercise 16A

1 a 8 **b** 10

2 a −6   b $-\dfrac{9}{2}$
3 a $\dfrac{1}{18}$ b $\dfrac{1}{4}$
4 a −12   b −16
5 a 0, 2.67   b −1.73, 1.73
6 a −0.779, 0.178, 0.601   b 0.607
7 a 1.10   b −0.632, 0.632
8 a i   max (−0.577, 1.77), min (0.577, 0.23)
      ii 1.77
   b i   max (0, 5), min (2.67, −4.48)
      ii 5
9 a i   max (−0.167, −2.99), min (0, −3)
      ii −2.25
   b i   max (0.667, −0.852), min (0, −1)
      ii 1
10 a i   max (0, 1), min (1.58, −5.25)
      ii 1
   b i   max (−0.707, 1.25), (0.707, 1.25), min (0, 1)
      ii 1.25
11 a i   min (0.630, 1.53)
      ii 4
   b i   local maximum (0,2), local minimum (0.750, 1.90)
      ii 4
12 $b = -2, c = 3$
13 $b = -4, c = 1$
14 a −0.329, 1.54   b 0.909
15 (0, 0), (3, 27)
16 (−1, −5)
17 a (0.180, 7.81)   b −32
18 108.9
19 $13.5 \leq f(x) \leq 30.1$
20 $f(x) \geq -30$
21 $a = 2, b = 5$
22 $a = 3, b = 6$
23 (−1.49, 20.9)

Exercise 16B

1 $2.5 million
2 $1726
3 77.1 km h^{-1}
4 3.15 hours
5 a 40 cm   b 100 cm^2
6 a 42 cm^2 b 25.9 cm
7 108 cm^3
8 a $P = 2x + \dfrac{72}{x}$ b 24 cm
9 253 m^2
10 b 821 cm^3
11 a $h = \dfrac{460}{x^2}, S = 2x^2 + \dfrac{1840}{x}$
   b 9.73
12 a $4x^2 + \dfrac{675}{x}$
   b 4.39 cm × 8.77 cm × 5.85 cm
13 b $r = 4.30, h = 8.60$
   c e.g. doesn't take into account the neck of the bottle
14 b 1210 cm^3
15 23
16 b 39
17 a $7
   b It does not predict negative sales when $x > 10$
   c $9
18 a $\dfrac{800}{m} - \dfrac{200}{m^2}$ b 0.5 kg
19 102 000 cm^3
20 12.9 cm^3

Exercise 16C

1 a 290   b 478
2 a 0.921   b 0.337
3 a 3.65   b 2.16
4 a 1.48   b 1.08
5 a 14.6   b 16.4
6 a 1.71   b 5.44
7 a 3.57   b 3.47
8 5.15
9 5.82
10 13.2
11 a 11.1 m^2 b Larger (the tunnel is curved)
12 4%
13 137 m^2
14 2250 m^2
15 a 1.84   b Underestimate

Chapter 16 Mixed Practice

1 a $\left(\dfrac{2}{3}, -\dfrac{4}{3}\right)$

b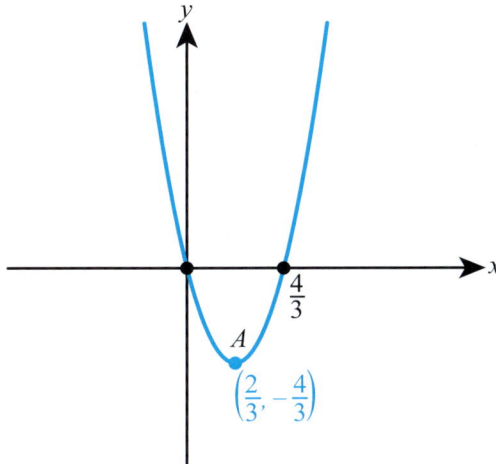

2 −16

3 a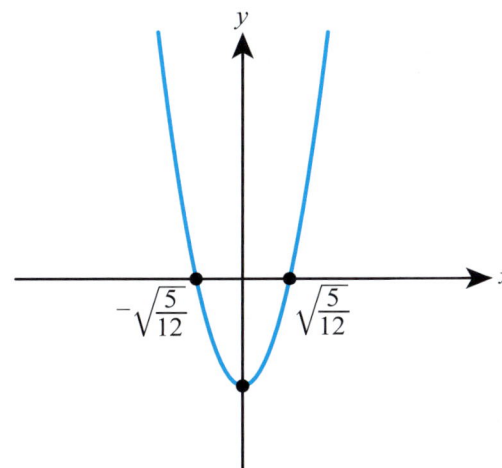

b $\pm\sqrt{\dfrac{5}{12}}$

4 −0.855

5 (0.5, 7)

6 a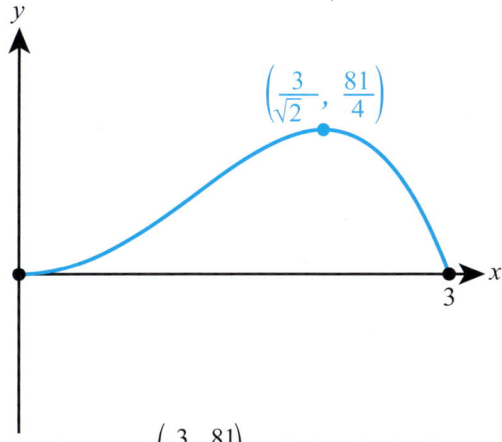

maximum at $\left(\dfrac{3}{\sqrt{2}}, \dfrac{81}{4}\right)$, mimima in the interval at the end points (0, 0) and (3, 0)

b 31.3

7 11.4

8 a $h = 44 - w$

b $w = 22$, $h = 22$

c $484\,\text{cm}^2$

9 $65.8\,\text{km}\,\text{h}^{-1}$

10 c $48x - 18x^2$ d $2.67\,\text{m}$

e $56.9\,\text{m}^3$ f $5.33\,\text{m}, 4\,\text{m}$

g 7

11 (−1.14, 3.93)

12 $a = 3$, $b = 213$

13 (−1, 9)

14 a 1.06   b 0.982, 8.20%

15 a (0.423, 0.385)   b 6

16 −256

17 cylinder, $r = 3.63\,\text{cm}$, $h = 7.26\,\text{cm}$

18 a $V = 20lw$

d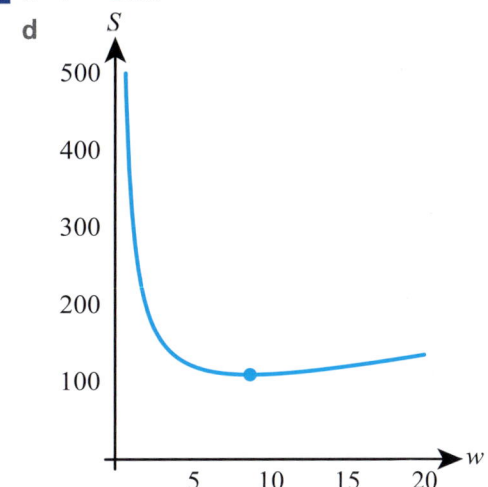

- e $4 - \dfrac{300}{w^2}$
- f 8.66
- g 17.3
- h 109 cm

19 b i $\dfrac{1200}{x}$ ii $A = 2\pi x^2 + \dfrac{1200}{x}$

- c $\dfrac{dA}{dx} = 4\pi x - \dfrac{1200}{x^2}$ d 4.57
- e 394 cm^2

20 (−2.67, 107)

21 a $T = -0.022t^3 + 0.56t^2 - 2.0t + 5.9$
- b −21.5 °C; it predicts a much lower temperature at midnight on the next day than the previous day

22 6.25

23 a 33 units^2 b 20 625 m^2

Applications and interpretation SL: Practice Paper 1

1 a 13 cm   b 38
- c $\dfrac{5}{8}$ [6]

2 a f(x) > −1.10   b −0.41
- c x = 21.1   [5]

3 a $a + 9d = 285$
  $25a + 300d = 9000$
- b $a = 60, d = 25$ [5]

4 a 80   b $\dfrac{1}{8}$ c $\dfrac{3}{5}$ [5]

5 a AUS 444.89
- b AUS 6,693.40   [6]

6 a H_0: gender and favourite type of wine are independent
  H_1: gender and favourite type of wine are not independent
- b $p = 0.0935$
- c $0.0935 > 0.05$ so do not reject H_0. Insufficient evidence at the 5% level that gender and wine preference are not independent.   [6]

7 a $k = 6$
- b $806 \leqslant V < 941$ [5]

8 a i 500   ii 4220
- b 3.48 years
- c Continues to spread, tending to an upper limit of 10 000 (which is possibly the full population of this species)   [6]

9 a 51.8 cm
- b 26.8 cm
- c 5980 cm^3 [7]

10 a $H_0: \mu_A = \mu_B$
  $H_1: \mu_A \neq \mu_B$
- b $p = 0.0338$
- c $0.0338 < 0.05$ so reject H_0. There is sufficient evidence at the 10% level that the mean length of perch in the two rivers is different.
- d Both populations are normally distributed. The two populations have equal variance.   [8]

11 a $6x + 8y = 59$
- b C as that point is in the Voronoi cell containing C.
- c (2, 3)   [6]

12 a $a = -0.25, b = 2.25, c = -3.5$
- b £1560   [7]

13 a $\dfrac{dl}{dt} = \dfrac{0.28}{\sqrt{t}}$
- b 4 m
- c Rate of growth slows but sharks continue to grow throughout their lifetime.   [8]

Applications and interpretation SL: Practice Paper 2

1 a i $-\dfrac{3}{2}$ ii (0, 6)   [3]
- b $2x - 3y = 21$ [3]
- c (6, −3)   [1]
- d i $\sqrt{117} \approx 10.8$ ii $\sqrt{13} \approx 3.61$ [2]
- e 19.5   [2]

2 a $p + q = 0.27$ [1]
- b $3p + 4q = 1$ [3]
- c $p = 0.08, q = 0.19$ [1]
- d i 0.161   ii 0.467   [3]
- e i 9.3   ii 1.88   [2]
- f 0.149   [3]

3 a $k = 6, c = \frac{2}{3}$ [3]
  b y [4]

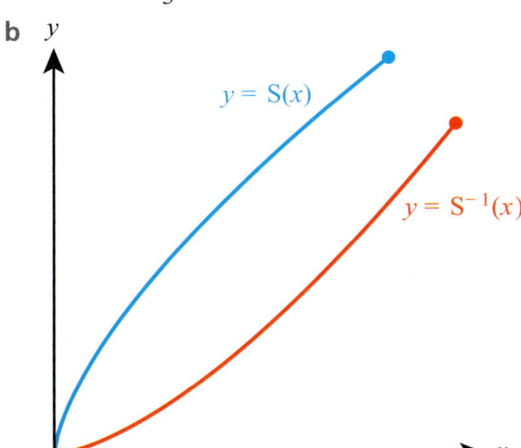

  c i $0 \leq S^{-1} \leq 125$
     ii $0 \leq S \leq 150$ [2]
  d i 27
     ii The volume of the cuboid when the surface area is 54 [2]
4 a $p = 2.3, q = 22.5$ [3]
  b 1.63 m [1]
  c 34.9% [2]
  d $a = 0.0323, b = -0.531, c = 2.18$ [4]
  e y [2]

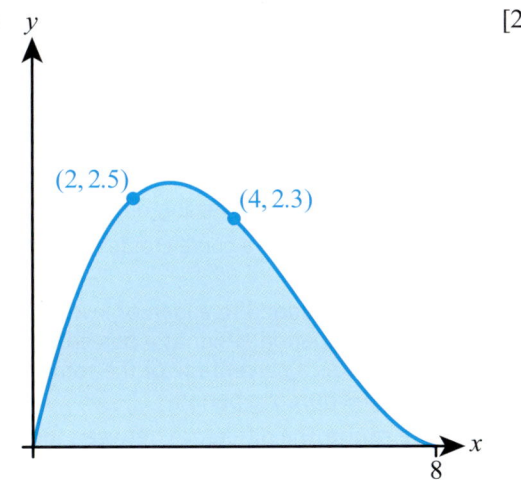

  f i 2.65 m
     ii 12.2 m^2 [4]
5 a 0.978 [2]
  b Yes, since value of r close to 1 so points lie near to straight line. [2]
  c $a = 0.0444, b = 2.56$
     $c = 0.0152, d = 8.40$ [5]
  d i 6560 euros
     ii 11 400 euros
     ii 13 400 euros [3]
  e Part i and ii are reliable as within the range of the given data. Part iii is potentially unreliable as had to extrapolate. [3]
6 a i $-\dfrac{1}{p^2}$ [5]
  b i $Q(2p, 0), R\left(0, \dfrac{2}{p}\right)$
     ii 2 [4]
  d $2\sqrt{2} \approx 2.83$ [2]

Glossary

χ^2 statistic In a χ^2 test the value that measures how far the observed data values are from what would be expected if the null hypothesis were true

Alternative hypothesis In a hypothesis test, the stated difference from the null hypothesis that is to be investigated

Amortization The process of paying off debt through regular payments

Amplitude Half the distance between the maximum and minimum values of a sinusoidal function

Annuity A fixed sum of money paid at regular intervals, typically for the rest of the recipient's life

Arithmetic sequence A sequence with a common difference between each term

Arithmetic series The sum of the terms of an arithmetic sequence

Asymptotes Lines to which a graph tends but that it never reaches

Axis of symmetry The vertical line through the vertex of a parabola

Base The number b in the expression b^x

Biased A description of a sample that is not a good representation of a population

Cells (on a Voronoi diagram) The regions containing the points which are closer to a given site than any other site

Contingency table A table showing the observed frequencies of two variables

Continuous Data that can take any value in a given range

Cubic model A function of the form $y = ax^3 + bx^2 + cx + d$

Decimal places The number of digits after the decimal point

Definite integration Integration with limits – this results in a numerical answer (or an answer dependent on the given limits) and no constant of integration

Degrees of freedom (v) The number of independent values in the hypothesis test

Depreciate A decrease in value of an asset

Derivative A function that gives the gradient at any point of the original function (also called the slope function or gradient function)

Differentiation The process of finding the derivative of a function

Direct proportion Two quantities are directly proportional when one is a constant multiple of the other

Discrete Data that can only take distinct values

Discrete random variable A variable with discrete output that depends on chance

Edges (on a Voronoi diagram) The boundaries of a cell

Event A combination of outcomes

Expected frequencies The frequencies of each group in a frequency table or for each variable in a contingency table assuming the null hypothesis is true

Exponent The number x in the expression b^x

Exponential decay The process of a quantity decreasing at a rate proportional to its current value

Exponential equation An equation with the variable in the power (or exponent)

Exponential growth The process of a quantity increasing at a rate proportional to its current value

Exponential model A function with the variable in the exponent

Geometric sequence A sequence with a common ratio between each term

Geometric series The sum of the terms of a geometric sequence

Gradient function See **derivative**

Hypothesis testing A statistical test that determines whether or not sample data provide sufficient evidence to reject the default assumption (the null hypothesis)

Incremental algorithm An algorithm for building a Voronoi diagram one site at a time

Indefinite integration Integration without limits – this results in an expression in the variable of integration (often x) and a constant of integration

Inflation rate The rate at which prices increase over time

Initial value The value of a quantity when $t = 0$

Integration The process of reversing differentiation

Intercept A point at which a curve crosses one of the coordinate axes

Interest The amount added to a loan or investment, calculated in each period either as a percentage of the initial sum or as a percentage of the total value at the end of the previous period

Interquartile range The difference between the upper and lower quartiles

Inverse proportion Two quantities are inversely proportional when one is a constant multiple of the reciprocal of the other

Limit of a function The value that f(x) approaches as x tends to the given value

Linear model A function of the form $y = mx + c$

Local maximum point A point where a function has a larger value than at any other points nearby

Local minimum point A point where a function has a smaller value than at any other points nearby

Glossary

Long term behaviour The value of a function when x gets very large

Lower bound The smallest possible value of a quantity that has been measured to a stated degree of accuracy

Normal to a curve A straight line perpendicular to the tangent at the point of contact with the curve

Null hypothesis The default position in a hypothesis test

Observed frequencies In a χ^2 test the sample values on the variable

One-tailed test A hypothesis test with a critical region on only one end of the distribution

Outcomes The possible results of a trial

p-value Assuming the null hypothesis is correct, the probability of the observed sample value, or more extreme

Parabola The shape of the graph of a quadratic function

Period The smallest value of x after which a sinusoidal function repeats

Perpendicular bisector The perpendicular bisector of the line segment connecting points A and B is the line which is perpendicular to AB and passes through its midpoint

Piecewise linear model A model consisting of different linear functions on different parts of the domain

Polynomial A function that is the sum of terms involving non-negative integer powers of x

Pooled sample t-test A two-sample t-test conducted when the two populations are assumed to have equal variance

Pooled variance In a two-sample t-test, an estimate of each of the common variance of each of the two populations formed by combining both samples

Population The complete set of individuals or items of interest in a particular investigation

Principal The initial value of a loan or investment

Principal axis The horizontal line halfway between the maximum and minimum values of a sinusoidal function

Quadratic model A function of the form $y = ax^2 + bx + c$

Quartiles The points one quarter and three quarters of the way through an ordered data set

Range The difference between the largest and smallest value in a data set

Relative frequency The ratio of the frequency of a particular outcome to the total frequency of all outcomes

Roots of an equation The solutions of an equation

Sample A subset of a population

Sample space The set of all possible outcomes

Significance level In a hypothesis test the value specifying the probability that is sufficiently small to provide evidence against the null hypothesis

Significant figures The number of digits in a number that are needed to express the number to a stated degree of accuracy

Simple interest The amount added to an investment or loan, calculated in each period as a percentage of the initial sum

Sinusoidal model A function of the form $y = a\sin(bx) + d$ or $y = a\cos(bx) + d$

Sites (on a Voronoi diagram) The given points around which cells are formed

Slope function See **derivative**

Spearman's rank correlation coefficient A measure of the agreement between the rank order of two variables

Standard deviation A measure of dispersion, which can be thought of as the mean distance of each point from the mean

Standard index form A number in the form $a \times 10^k$ where $1 \leqslant a < 10$ and $k \in \mathbb{Z}$

Subtended The angle at the centre of a circle subtended by an arc is the angle between the two radii extending from each end of the arc to the centre

t-statistic In a t-test the value that measures how far the sample mean is from what would be expected if the null hypothesis were true

Tangent to a curve A straight line that touches the curve at the given point but does not intersect the curve again (near that point)

Toxic waste dump problem A problem in which the object is to find the point which is as far as possible from any of the sites

Trapezoidal rule A rule for approximating the value of a definite integral using trapezoids of equal width

Trial A repeatable process that produces results

Turning point A local maximum or minimum point

Two-tailed test A hypothesis test with a critical region on either end of the distribution

Upper bound The largest possible value of a quantity that has been measured to a stated degree of accuracy

Value in real terms The value of an asset taking into account the impact of inflation

Variance The square of the standard deviation

Vertex (or vertices) of the graph The point(s) where the graph reaches a maximum or minimum point and changes direction

Vertices (on a Voronoi diagram) The points at which the edges of the cells intersect

Voronoi diagram A diagram that separates an area into regions based on proximity to given initial points (sites)

Zeros of a function/zeros of a polynomial The values of x for which $f(x) = 0$

Index

3D shapes *see* three-dimensional shapes
abstract reasoning xv
accumulations 256–7
accuracy
    levels of 278–80, 283–4
    of predictions 309, 334
    see also approximation; estimation
algebra
    basic skills xxiv–xxv
    computer algebra systems (CAS) xx
    definition 2
    exponential equations 7
algorithms 279, 297, 353, 414
alternative hypothesis 372–3
amortization 287–90
amplitude 328–9
angles
    of depression 119
    of elevation 119–20
    see also triangles; trigonometry
annuities 289–90
approximation 40, 136, 278, 285
    decimal places 280
    percentage errors 282–3
    rounded numbers 280–3
    significant figures 280
    trapezoidal rule 414
    see also estimation
arcs 344
area
    of a sector 345–6
    spatial relationships 343
    surface 94–5
    of a triangle 107–8, 345
    upper and lower bounds 281
arithmetic sequences 24–9
    applications of 28–9
    common difference 25–7
    sigma notation 27
    sum of integers 26–8
arithmetic series 25
assumptions xvii, 23, 29, 189, 209
    see also hypothesis testing
asymptotes 62, 318–19
average *see* central tendency
axiom xii
axis
    principal 328–9
    of symmetry 315–16

base 4
bearings 118–19
bias, in sampling 132–3
binomial distribution 207–10
bivariate data 160
box-and-whisker diagrams 153

calculations
    levels of accuracy 278–80
    number line 281
    rounded numbers 280–3
    significant figures 12

calculus 282, 404
    trapezoidal rule 412–14
    see also differentiation; integration
Cartesian coordinates 78
cell sequences 22–3
central tendency 140–2
chi-squared (χ^2) tests 372, 375–9
circles, arcs and sectors 344–6
coefficients 325
common difference, arithmetic
    sequences 25–7
common ratio, geometric sequences 34–5
compound interest 39, 288–90
computer algebra systems (CAS) xx
computer algorithms 279
cone
    surface area 94
    volume 95
conjecture x–xi
contingency tables 378–9
continuous data 132
contrapositive statements xvi
convenience sampling 134–5
converse xvi
coordinate geometry
    Cartesian coordinates 78
    midpoint 87
    Pythagoras' theorem 86–7
    three-dimensional 86–7
    two-dimensional 74–82
correlation 159–61
cosine 101, 328
    rule 105–6, 345
cubic functions 325–6
cumulative frequency graphs 151–2
curves *see* parabola
cycle of mathematical inquiry x

data
    bivariate 160
    contingency tables 378–9
    continuous 132
    correlation 159–61
    degrees of freedom 375
    discrete 132, 142–3, 202–3
    effect of constant changes 145
    frequency distributions 139, 141, 375–9
    frequency tables 375–7
    goodness of fit 375
    grouped 141–2
    measure of dispersion 143–4
    measures of central tendency 140–2
    modal class 142
    outliers 133–4, 144
    patterns in 164
    presenting 150–3, 159–66
    quartiles 142–3
    regression 162–4
    reliability of 132–3
    summarizing 139–40
    validity 133, 144
    see also statistics

data points, plotting 332
decimal places 280
definite integrals 262–4, 412
depreciation 40
derivative function 225, 406
differential equations 320
differentiation 222–3, 406
    anti-differentiation 258–60
    derivatives 225, 230–3, 240–3, 245–50
    gradient of a curve 225–7, 231–3,
        245–50, 406–8
    limits 224–5
    maximum and minimum points 407
    rate of change 227, 257
    see also integration
direct proportion 323–4
discrete data 132, 142–3, 202–3
discrete random variable 202–3
dispersion 143–4
distribution
    binomial 207–10
    normal 212–15, 372–3, 377, 385–7
dynamic geometry packages xxi

e (number) 15
elevation 119–20
encryption algorithms 297
enlargement symmetry 63
equations
    defining quantities 299
    differential 320
    linear 298–9
    normal to a curve 248–9
    polynomial 297, 301–3
    quadratic 316
    roots of 67–8, 301–3
    simultaneous 298–9
    solving using graphs 67–8
    solving using technology 297
    of straight lines 76–82, 310
    see also exponential equations
error analysis 282–3
estimation 283–5
exponential
    decay 318–19
    equations 7, 17
    growth 318–19
    models 318–21
exponents
    applied to fractions 5–6
    laws of 4–8, 20, 320–1
    negative 5
    simplifying expressions 4–7

Fermi, Enrico 284
financial mathematics
    amortization 287–90
    annuities 289–90
    compound interest 39, 288–90
    depreciation 40
    inflation 40
    interest 28–9, 39, 287–90

Index

levels of accuracy 278–80
modelling 278–85, 290
simple interest 28–9
forces 257
fractal 63
fractions, with exponents 5–6
frequency distributions 139, 141, 375–9
frequency tables 375–7
function notation 50
functions 48
derivative 406
domain 52
graphical representation 50–1, 53, 55–6
inputs 49–52
inverse 54–5
and mathematical modelling 53–4
outputs 49, 52–3
range of 52
zeros of 67

Gauss, Carl Friedrich 26
generalizations 22, 24
geometric sequences 33–6
applications of 36, 39–40
common ratio 34–5
financial applications of 39–40
geometric series 34–5
geometry 92–3
arcs and sectors 344–6
origins of 118
perpendicular bisectors 349–54
spatial relationships 343, 349–58
Voronoi diagram 342–3, 349–58
see also coordinate geometry; geometric sequences; trigonometry
Godfrey, Charles 67
gradient of a curve 225–7, 231–3, 245–50, 406–8
gradient of a straight line 76–80, 310
Graham's Number 11
graphical interpretation, of derivatives 230–3
graphical representation
and data 150–4
functions 50–1, 53, 55–6
solving equations 68
graphs 61–8
amplitude 328–9
asymptotes 62
axis of symmetry 315–16
box-and-whisker diagrams 153
cubic functions 325–6
cumulative frequency 151–2
direct proportion 323–4
equations of straight lines 76–82
exponential models 318–19
finding points of intersection 66–7
gradient of a curve 225–7, 231–3, 245–50, 406–8
gradient of a straight line 76–80, 310
histograms 150–1
intercepts 62, 77–80
intersection of two lines 82
inverse proportion 324–5
parabola 315–16

parallel lines 80
period of 328–9
perpendicular lines 80–2
plotting data points 332
principal axis 328–9
proportion 323
quadratic models 314–16
scatter diagrams 159–61, 391
simultaneous equations 82
sine 328
sinusoidal models 328–9, 332
sketching 65–7
solving equations using 67–8
symmetries 63
tangents 225, 246–7
turning points 314
vertices 62, 315
x-intercept 77–9, 315–16, 319, 328
y-intercept 62, 77–9, 315–16, 319, 325, 328

histograms 150–1
horizontal asymptote 318–19
hypotenuse 101
hypothesis testing 370–9, 385–8, 391–4
alternative hypothesis 372–3
chi-squared (χ^2) tests 372, 375–9
null hypothesis 372–3, 375, 378, 385
one-tailed test 373
Pearson's correlation coefficient 161–2, 391–3
p-value 374–5, 385
significance level 373–7, 385–8
Spearman's rank correlation 391–4
t-tests 385–8
two-tailed test 373

implication xvi
incremental algorithm 353
indices see **exponents**
infinity, reasoning with xiv
inflation 40
integration
accumulations 256–7
anti-differentiation 258–60
definite integrals 262–4, 412
trapezoidal rule 412–14
see also differentiation
interest
amortization 287–90
compound 39, 288–90
simple 28–9
see also financial mathematics
interquartile range 143
inverse
functions 54–5
proportion 324–5
irrational numbers 15

Lagrange's notation 226
large numbers 11
laws of exponents 4
Leibniz, Gottfried 225, 259
level of significance 373–7, 385–8
limits 224–5
linear equations 298–9
linear models 310–11

lines of symmetry 63
local maximum points 407–8
local minimum points 407–8
logarithms 14, 20, 52
evaluating 16
natural logarithm 15–16
logic xv–xvi
lower bounds, of rounded numbers 281–2

mathematical
induction 24
inquiry, cycle of x
mean 140–2
median 140
modal class 142
mode 140, 142
modelling xvii–xix, 22
accuracy of predictions 309, 334
arithmetic sequences 28–9
assumptions xvii, 23, 29, 189, 209
cubic models 325–6
differential equations 320
exponential models 318–21
Fermi estimate 284
geometric sequences 36
linear 310
mathematical model of a process 53–4
normal distribution as 214
patterns 23–4
piecewise linear models 310–11, 332, 334
quadratic models 314–16
qualitative and quantitative results 230
representations 308–10
sinusoidal models 328–9, 332
skills 332–4
straight-line models 75
using equations 297
variables in 309, 314
see also financial mathematics
modulus sign 282–3

Napier, John 14
natural logarithm 15–16
nearest neighbour interpolation 355
negation xvi
negative exponents 5
negative gradient 76
negative numbers, and the domain of a function 52
neural networks 297
normal distribution 212–15, 372–3, 377, 385–7
normals 248–50
notation xv
***n* terms**
arithmetic sequences 24–5
geometric sequences 33–5, 321
null hypothesis 372–3, 375, 378, 385
number line 281

Occam's razor xviii, 326
one-tailed test 373
operations, large numbers 11
oscillating patterns 332
outliers 133–4, 144

Index

parabola 315–16
paradox xv
parallel lines 80
patterns 23–4, 30
Pearson's correlation coefficient 161–2, 391–3
percentage errors 282–3
perpendicular
    bisectors 349–54
    lines 80–2
perspective 119
piecewise linear models 310–11, 332, 334
point-gradient form 79
Polya, George ix–x
polynomial equations 297, 301–3
pooled variance 387
positive gradient 76
powers *see* exponents
predictions
    accuracy of 309, 334
    see also modelling
principal axis 328–9
probability
    binomial distribution 207–10
    combined events 186–7
    complementary events 181
    concepts in 179–82
    conditional 187–8
    discrete random variable 202–3
    distributions 201–15
    events 180
    independent events 188–9
    mutually exclusive events 187
    normal distribution 212–15
    outcomes 180
    relative frequency 180
    sample space 180–1
    techniques 184–5
    trials 180
problem solving ix–xii
programming xxi
proportion 323–5
p-value 374–5, 385
pyramid
    finding angles 115–16
    surface area 96
    volume 96–7
Pythagoras' theorem xiii, 86–7, 96, 116

Q-Q plots 214
quadratic
    equations 316
    models 314–16
quadrilateral 413
quantities, defining 299
quota sampling 135

Ramanujan xiv
random sampling 134, 375
rate of change 227, 257
reasoning xv
representations 12, 203, 308–10
representative sampling 375
right-angled triangles 101, 113
    see also trigonometry

roots 62
    of an equation 67–8, 301–3
rotational symmetry 63
rounded numbers 280–3
    upper and lower bounds of 281–2
Russell, Bertrand xii–xiii

sample space diagrams 185
sampling 132–3
    bias in 132–3
    pooled variance 387
    random 375
    representative 375
    techniques 134
    see also normal distribution
scatter diagrams 159–61, 391
sector, area of 345–6
sequences 22–3
    standard notation 24
    see also arithmetic sequences; geometric sequences
sets xv
shapes, three-dimensional solids 94–5
Shapiro-Wilk test 214
sigma notation
    arithmetic sequences 27
    geometric sequences 35–6
significance level 373–7, 385–8
significant figures 12, 280
simple interest 28–9
simplifying exponent expressions 4–7
simultaneous equations 82, 298–9
sine 101, 328
    rule 103–4, 345
sinusoidal models 328–9, 332
spatial relationships 343, 349–58
Spearman's rank correlation 391–4
spreadsheets xx
square root 14
    see also logarithms
standard deviation 143
standard index form 11–12
standard notation, sequences 24
statistics 130
    approximation 136
    misleading 130, 133
    outliers 133–4, 144
    sampling 132–4, 375
    validity 133, 144
    see also data; hypothesis testing
straight-line models 75, 310
stratified sampling 135–6
surface area
    three-dimensional solids 94–5
    see also area
syllogism xvi
symmetries 63
systematic measurement error 282
systematic sampling 135

tangents 101, 225, 246–7, 249–50, 406
tan rule 106, 345
technology xix–xxi
    computer algebra systems (CAS) xx
    dynamic geometry packages xxi

    Graphical Display Calculator (GDC) xix–xx
    programming xxi
    terminology xvi
tetration 11
three-dimensional shapes
    finding angles 113–18
    surface area 94–5
    volume 94–5
toxic waste dump problem 356
trajectory 75
translational symmetry 63
trapezium 413
trapezoidal rule 412–14
tree diagrams 184–5, 202
triangles
    area of 107–8, 345
    in the real world 92–3
    right-angled 101, 113
    see also coordinate geometry
trigonometry 92, 101–8, 308
    applications of 113–20
    area of a triangle 107–8
    bearings 118–19
    cosine 101
    cosine rule 105–6
    finding angles 101–2, 113–18
    measuring angles 118–19
    sine 101
    sine rule 103–4
    tan rule 106
    three-dimensional shapes 113–14
true value 282–3
t-tests 385–8
two-tailed test 373

upper bounds, of rounded numbers 281–2

validity 133, 144
value in real terms 40
variables
    correlation between 391–4
    modelling xvii–xviii, 309
variance 143
velocity 257
Venn diagrams 184, 186, 188
    and sets xv
vertices 62, 315
volume, three-dimensional solids 94–5
Voronoi diagrams 342–3, 350–8

Watson selection test xv

x-axis 62
x-intercept 77–9, 315–16, 319, 328

y-axis 62
y-intercept 62, 77–9, 315–16, 319, 325, 328

Zeno's paradoxes 227
zero xv, 4, 52, 224
    of a function 67
    gradient of a curve as 406
    of a polynomial 301
Zipf's Law 53